CARBON NANOMATERIAL-BASED ADSORBENTS FOR WATER PURIFICATION

CARBON NANOMATERIAL-BASED ADSORBENTS FOR WATER PURIFICATION

Fundamentals and Applications

SUPRAKAS SINHA RAY
Centre for Nanostructures and Advanced Materials, DSI-CSIR Nanotechnology Innovation Centre, Council for Scientific and Industrial Research, Pretoria, South Africa; Department of Chemical Sciences, University of Johannesburg, Johannesburg, South Africa

RASHI GUSAIN
Department of Chemical Sciences, University of Johannesburg, Johannesburg, South Africa

NEERAJ KUMAR
Centre for Nanostructures and Advanced Materials, DSI-CSIR Nanotechnology Innovation Centre, Council for Scientific and Industrial Research, Pretoria, South Africa

ELSEVIER

Elsevier
Radarweg 29, PO Box 211, 1000 AE Amsterdam, Netherlands
The Boulevard, Langford Lane, Kidlington, Oxford OX5 1GB, United Kingdom
50 Hampshire Street, 5th Floor, Cambridge, MA 02139, United States

Notices
Knowledge and best practice in this field are constantly changing. As new research and experience broaden our
understanding, changes in research methods, professional practices, or medical treatment may become necessary.

Practitioners and researchers must always rely on their own experience and knowledge in evaluating and using any
information, methods, compounds, or experiments described herein. In using such information or methods they
should be mindful of their own safety and the safety of others, including parties for whom they have a professional
responsibility.

To the fullest extent of the law, neither the Publisher nor the authors, contributors, or editors, assume any liability
for any injury and/or damage to persons or property as a matter of products liability, negligence or otherwise, or
from any use or operation of any methods, products, instructions, or ideas contained in the material herein.

British Library Cataloguing-in-Publication Data
A catalogue record for this book is available from the British Library

Library of Congress Cataloging-in-Publication Data
A catalog record for this book is available from the Library of Congress

ISBN: 978-0-12-821959-1

For Information on all Elsevier publications
visit our website at https://www.elsevier.com/books-and-journals

Publisher: Matthew Deans
Acquisitions Editor: Simon Holt
Editorial Project Manager: Mariana C. Henriques
Production Project Manager: Poulouse Joseph
Cover Designer: Greg Harris

Typeset by MPS Limited, Chennai, India

Dedicated to our parents

CONTENTS

About the authors

Prof. Suprakas Sinha Ray is a chief researcher in polymer nanocomposites at the Council for Scientific and Industrial Research (CSIR) with a PhD in physical chemistry from the University of Calcutta in 2001 and manager of the Centre for Nanostructures and Advanced Materials, DSI-CSIR Nanotechnology Innovation Centre. His current research focuses on polymer-based advanced nanostructured materials and their applications. He is one of the most active and highly cited authors in the field of polymer nanocomposite materials and he has recently been rated by Thomson Reuters as being in the top 1% most impactful and influential scientists and top 50 high-impact chemists (out of 2 million chemists worldwide).

He is the author of 5 books, coauthor of 3 books, author of 35 book chapters on various aspects of polymer-based nanostructured materials and their applications, and the author or coauthor of 325 articles in high-impact international journals and 30 articles in national and international conference proceedings. He also has six patents (including applications) and seven new demonstrated (commercialized) technologies shared with colleagues, collaborators, and industrial partners. So far his team has commercialized 19 different products. His honors and awards include South Africa's most prestigious 2016 National Science and Technology Award (NSTF); prestigious 2014 CSIR-wide Leadership award; prestigious 2014 CSIR Human Capital development award; and prestigious 2013 Morand Lambla Awardee (top award in the field of polymer processing worldwide), International Polymer Processing Society, United States. He has been appointed as Extraordinary Professor, University of Pretoria and Distinguished Visiting Professor, University of Johannesburg. Currently, he is serving as an associate editor/editorial board member of the *RSC Advances (Associate Editor)*, *Journal of Nanoscience and Nanotechnology (Associate Editor)*, *Advanced Science Letters*, *International Journal of Plastic Films and Sheeting*, *Applied Nanoscience (Associate Editor)*, and *Macromolecular Materials and Engineering*.

Dr. Rashi Gusain received her PhD in chemical science from CSIR-Indian Institute of Petroleum, India. In 2016 she started her first postdoc at the Indian Institute of Technology Madras, India and later she joined the group of Prof. Oleg Antzutkin at Luleå University of Technology, Sweden. Currently, she is working as a postdoctoral fellow with Prof. Suprakas Sinha Ray at the University of Johannesburg and also as a visiting researcher at CSIR-Pretoria. Her expertise is in organic and inorganic synthesis, ionic liquids, nanomaterials, tribology, photocatalysis, and water purification. Her current research interest is the development of inorganic materials for the photodegradation of organic water contaminants and CO_2 capture and conversion.

Dr. Neeraj Kumar is currently a postdoctoral researcher with Prof. Suprakas Sinha Ray at CSIR, Pretoria, South Africa. He obtained his PhD in chemistry from the University of Johannesburg in 2017. His research interests focus on controllable synthesis of nanostructural layered materials, especially carbon nanomaterials and metal chalcogenides, and their modifications for environmental applications including water treatments, adsorption, photocatalysis, and CO_2 capture and conversion.

Preface

The deterioration of water quality and unavailability of drinkable water are pressing challenges worldwide. The removal of toxic organic and inorganic pollutants from water is indispensable for a clean environment, as a response to water scarcity, and for human society. Adsorption-based water technologies are among the most favored and widely used because of their high efficiency at low cost, without relying on a complex infrastructure. In recent years, carbon nanomaterials (CNMs), such as graphene and derivatives, carbon nanotubes, carbon nanofibers, nanoporous carbon, fullerenes, graphitic carbon nitride, and nanodiamonds, have been extensively exploited as adsorbents due to their extraordinary surface properties, easy modification, large specific surface area, controlled structural varieties, high chemical stability, porosity, low density, ease of regeneration, and reusability. Graphene oxides and other oxidized carbons provide strong acidity and abundant functional groups, and demonstrate excellent adsorption of cationic and basic compounds via electrostatic and hydrogen bonding interactions, while their pristine counterparts exhibit hydrophobic surfaces and offer high adsorption via strong $\pi-\pi$ interactions.

The published review articles, reports, and books either concentrate on specific adsorbents or deliver only short descriptions with limited examples. Thus there is still a need for a comprehensive authored book addressing the adsorptive performance of all CNMs used for water treatment, along with their pros and cons, and mechanisms, and summarizing the details of recent trends and developments. This book focuses on the utilization of carbon-based nanomaterials as nanoadsorbents for wastewater treatment. For efficient adsorption, carbon-based nanoadsorbents should have high specific areas, eco-friendly synthetic approaches, low energy demands, and fast adsorption kinetics. The synthesis of a variety of carbon-based nanoadsorbents, as well as their adsorption kinetics, isotherms, and adsorption mechanisms with respect to different water contaminants (e.g., dyes, heavy metal ions, pharmaceutical compounds, and pesticides), is examined, and a perspective on future developments for these nanoadsorbents is presented.

This book provides a thorough overview of the state-of-the-art in CNMs, including significant past and recent advances, as well as future strategies for the use of carbon-based nanoadsorbents in water treatment. This book primarily emphasizes the fundamentals of adsorption, its mechanical aspects, synthesis and properties of CNMs, and adsorption performances of CNMs and their nanocomposites with inorganic materials and organics, that is, polymers. Structural engineering and activation processes produce materials with enhanced adsorptive properties and separation efficiencies. Furthermore, the formation of CNMs with 2D and 3D macro-/microstructures and high porosities is a potential approach to improve adsorption performances and extend CNM use at the industrial level. This book also addresses some vital issues that persist with regard to these adsorbents, which could shape future research and industrial application of carbon-based nanoadsorbents in water security.

This book is ideal for water scientists, material scientists, researchers, and engineers (chemical and civil) including under- and postgraduate students who are interested in this exciting field of research. This book will also help industrial researchers and R&D managers who want to bring advanced CNM-based adsorbents into the market.

The authors would like to thank the Department of Science and Innovation, the Council for Scientific and Industrial Research, and University of Johannesburg, South Africa, for financial support. We express our sincerest appreciation to all colleagues, postdoctoral fellows, and students for their valuable contributions as well as the reviewers for their critical evaluation of the proposal and chapters. We also thank the authors and publishers for their permission to reproduce their published works. Our special thanks go to Simon Holt, Mariana Pizzolatto Henriques, Poulouse Joseph and Swapna Praveen at Elsevier for their encouragement, cooperation, suggestions, and advice during various phases of preparation, organization, and production of this book. Finally, we would like to thank our CEO, Dr. T. Dlamini, for his continued support and encouragement.

Prof. Suprakas Sinha Ray★
Centre for Nanostructures and Advanced Materials, DSI-CSIR
Nanotechnology Innovation Centre, Council for Scientific and
Industrial Research, Pretoria, South Africa
Department of Chemical Sciences, University of Johannesburg,
Johannesburg, South Africa
ORCID: 0000-0002-0007-2595

Dr. Rashi Gusain★
Department of Chemical Sciences, University of Johannesburg,
Johannesburg, South Africa
ORCID: 0000-0002-7340-7237

Dr. Neeraj Kumar★
Centre for Nanostructures and Advanced Materials,
DSI-CSIR Nanotechnology Innovation Centre, Council for Scientific
and Industrial Research, Pretoria, South Africa
ORCID: 0000-0001-5019-6329

★All the authors equally contributed.

CHAPTER ONE

Introduction

1.1 Introduction

Water and oxygen are the most vital constituents needed to survive on Earth. Almost 70% of the planet is covered by water, by oceans, icecaps, glaciers, rivers, and lakes [1]. Despite this, less than 1% of the total water on Earth is accessible or available for consumption as freshwater from rivers, lakes, wells, or ponds. Rapid urbanization, civilization, population growth, and industrialization significantly pollute natural resources directly and indirectly and are becoming a threat to future generations. This gives birth to the global problem of clean water scarcity [2]. Untreated industrial effluents from textiles, paper, pharmaceuticals, batteries, rubber, printing, leather tanning, food processing, oils, etc., containing organic [dyes, aromatic phenols, pesticides, pharmaceuticals, polycyclic aromatic hydrocarbons, and personal care products (PPCPs)] and inorganic (heavy metal ions) contaminants, are introduced into the water bodies and jeopardize natural resources [3–13].

The introduction of foreign harmful substances, for example, unwanted chemicals and microorganisms, into water bodies, such as river, sea, ocean, lakes, and streams, and thus contaminating the water, degrading the quality of water, and imposing toxicity on the ecosystem is termed as water pollution. It is a dreadful threat, which can lead the world towards destruction. Marine species are directly affected by the water pollution, whereas animals and humans can be affected directly as well as indirectly, due to contaminated water. Consumption of contaminated water for drinking and other purposes directly affects the human health, whereas dependence on the vegetation and other food sources (e.g., fish, seafood) which have grown in contaminated water, indirectly affects the health. Water pollution is of increasing interest to researchers, scientists, and governments worldwide. There is also a need to increase awareness regarding wastewater management and to learn about the water pollution sources and their effects, as well as effective water treatment approaches.

Carbon Nanomaterial-Based Adsorbents for Water Purification.
DOI: https://doi.org/10.1016/B978-0-12-821959-1.00001-5

1.2 Water pollution sources

1.2.1 Industrial waste

Industry is primarily responsible for environmental pollution. The production and processing units of industry consume a vast amount of water and release lots of waste into the water sources. Many industries, such as steel, paper, textile, mining, leather, etc., produce lots of toxic chemicals and pollutants as waste, which is later discharged directly into the environment without proper treatment [14−19]. Dyes, acids, alkali, metal ions (Hg(II), Pb(II), Cd(II), etc.), sulfates, nitrates, phosphates, halogens, and other chemicals from these industries are directly dumped into the freshwater bodies and damage the water quality. These toxic chemicals can change the water quality, pH, color, and mineral composition, which is known as eutrophication and is a hazard to the aquatic organisms.

1.2.2 Sewage and domestic waste

Sewage and household or domestic waste carries harmful microorganisms and chemicals that can give rise to several serious health issues. Domestic wastewater includes the water used in daily household chores such as bathing, washing, cleaning, etc., and primarily consists of soap, detergent, PPCPs, pesticides, etc., as toxic chemical substances. Sewage wastewater generally includes excreted urine and faeces. Sewage and domestic waste should be chemically treated before discharge into the freshwater bodies. The drains containing such wastewater produce pathogens and thereby several diseases. Microorganisms present in sewage and domestic waste can cause severe deadly diseases and become the breeding grounds for other microorganisms and dangerous creatures, which act as carriers. Malaria is one such examples, developing from water and the female anopheles mosquito act as a carrier for the *Plasmodium* parasite.

1.2.3 Agricultural waste

In agriculture, the excess use of pesticides, manure, fertilizers, and ploughing of the land can contribute to water pollution. Pesticides and fertilizers are used to protect the crop from infection or damage due to insects and bacteria and are very helpful for the growth of plants. Excess use of pesticides can be harmful to the plants and animals. Inappropriate disposal of these pesticides contributes to water and soil pollution. During rain and irrigation, these chemicals mix with the water and flow down into the water resources causing severe damage to the water quality and aquatic species. Dichlorodiphenyltrichloroethane is one of the most common examples of pesticides. Pesticides are generally nonbiodegradable and are persistent organic pollutants, which stay in the environment for a longer period. These chemicals enter into the food chain and trigger damage to the various organs such as liver, reproductive system, and respiratory system [20,21]. Many pesticides are carcinogenic, mutagenic, endocrine disruptors, teratogenic, and are lethal for the ecosystem [21−24].

1.2.4 Oil spillage

A large amount of oil enters into the seas, oceans, rivers, and other water bodies by accidental oil spillage, which initiates many problems for the local aquatic and marine organisms. Oil is not miscible in water and makes it unsafe to drink. Oil creates a layer on the water as a water-in-oil emulsion, which reduces the oxygen supply and the amount of sunlight penetrating the water, and can destroy marine wildlife. It is primarily released from storage facilities, natural release from the ocean floor through fractures, accidents or leakage during transportation, and untreated disposal of oil waste.

1.2.5 Radioactive waste

Radioactive substances are generally used in the medical, nuclear, mining, and defense industries as well as other scientific practices. They can be found in watches, X-ray machines, television sets, etc. Radium is commonly used in watches and medicine, whereas uranium is the most widely used material in nuclear energy power plants. The waste of nuclear power plants contains several radioactive materials and needs to be disposed of immediately to prevent further any nuclear incidents. If the disposal of the nuclear power plant or other radioactive waste is not done correctly, it can be hazardous for the environment. Radioactive substances are highly toxic and their exposure can cause severe adverse effects to the human body such as cancer. The disposal of radioactive waste into the water streams provides a pathway for the radioactive substances to enter the food chain of humans, animals, and other marine wildlife, which is hazardous. Therefore accidental or intentional release of radioactive substance into any water body is a threat to life.

1.2.6 Urbanization

The increasing population and urbanization are major factors contributing to every kind of pollution. With the increase in population, the demand for food, housing, and clothing are also increasing. This is giving rise to the development of more cities and towns with a price of deforestation, the use of more fertilizer to produce more food, more garbage production causing landfill, inadequate sewer management, and more chemical waste from the industries to meet the demands of more materials. All of these wastes are disposed into the water streams. This is making a visual and negative impact on water quality.

1.2.7 Effect of water pollution

The significant effect of water pollution is the scarcity of fresh and clean water. This affects the quality of potable water, food, the health of all living beings in the ecosystem, and cultivation. Direct or indirect disposal of any kind of wastewater into the water bodies will introduce different types of chemicals, such as textile dyes, heavy metal ion, petrochemicals, pesticides, phosphates, nitrates, halogenated substances and pharmaceuticals. All these contaminants can be permissible in the water within a specific limit, standardized by the World Health Organization (WHO) and the United States Environmental Protection Agency (US EPA). Beyond such limits, it is hazardous for the flora, fauna, and biota of the ecosystem. In humans, the use of contaminated water for drinking or anything else can invite several diseases. For example, fluoride is commonly used for dental care and bones, whereas a fluoride concentration of more than 0.5 mg/L in water can lead to adverse effects to human health and cause fluorosis [25].

Similarly, many heavy metal ions such as arsenic (As), lead (Pb), mercury (Hg), and cadmium (Cd), etc., are widely used in industries, insecticides, and fertilizers and are introduced into the water bodies via the disposal of wastewater. Several heavy metal ions, especially As, Pb, Hg, and Cd, are acute carcinogens and teratogens, which can have adverse effects on the central nervous system, kidney, respiratory system, reproduction system, as well as cause congenital disabilities [26]. Minamata disease at Minamata Bay is the best-known example of Hg pollution, which was spread through eating the Hg-contaminated fish and took around the life of 2000 people [27]. Itai-Itai disease, nephrosis, and nephritis originate from cadmium contamination [28,29]. Also, the effluents of textile, paper, and pulp industries are majorly contributing to the dye pollution in water. Organic dyes are hazardous and have a negative impact on human health if contaminated water is consumed [30]. Dyes also change the color of the water and form a layer on the top of the water, which reduces the photosynthesis of water plants. The intake of dye molecules from contaminated water can also cause adverse effects on the body, such as damage of the reproductive system, respiratory system, kidneys, and neurological system.

Similarly, many microorganisms can grow in low-quality water, mostly in sewage wastewater and give birth to waterborne diseases. *Shigella* species, Escherichia coli, *Salmonella* species, and *Vibrio cholera* are some of the most concerning microorganisms that grow in the polluted water [31]. On drinking such water, it causes typhoid, gastroenteritis, diarrhoea, dysentery, and cholera. Groundwater contamination is causing deaths and diseases worldwide (more than 14,000 people every day) and the majority of them are kids and elderly people [32]. As per one report, about 1.9—5.6 million children died globally due to diarrhoeal and malnutrition disease [33].

Plants that are growing in the contaminated water can also be severely affected by the pollutants. Plants can absorb the toxic substances from the contaminated water through their roots. Later those pollutants can easily enter the food web and animals or human can also get harmed on eating those plants. Additionally, those pollutants can deteriorate the plant health and function. The acids are another form of contaminants, which are disposed of by industries or in domestic sources to the water bodies and lower the pH of the water. Ammonia from the fertilizers is also acidic and contaminates the groundwater. This acidification of water reduces the growth of decomposing microorganisms like fungi and bacteria in water, which are responsible for the decomposition of organic substances in the nutrients. Therefore acid contamination in water declines the nutrient concentration required for plant growth. Detergent is rich in phosphates and is the most used chemical in households and industries for cleaning purposes. It is one of the most common pollutants in domestic effluents. Plants can efficiently uptake the phosphates from contaminated groundwater and show adverse effects, such as growth retardation, CO_2 fixation, destruction of chlorophyll, photosynthesis, destruction of cell membranes, and so on.

Also, the alkaline contamination in water reduces the uptake of several essential bases, which result in the death of the plant. Therefore the changes in pH on acidification or alkalization reduce the plant life or damage the function of plant cells. Several contaminants increase water turbidity or make a layer on the water (e.g., oil spillage), which reduce the photosynthesis activity of plants. Agricultural runoff can cause an excess of inorganic nutrients in the water leading to eutrophication, which encourages the growth of algae. This, in turn, causes oxygen depletion for marine species and can lead to the death of fish due to suffocation.

1.3 Statement of the problem and objective of the book

Availability of fresh and clean water is a basic necessity to survive a healthy life. However, the consumption of polluted water can cause death. Regular increases in the population, industrialization, and urbanization are the primary contributing factors to water pollution. The use of polluted water has become a threat to the survival of humans, animals, and aquatic ecosystem. Water pollution has now become a global challenge and the efficient, sustainable and economical remediation techniques are of interest to various researchers. The treatment of wastewater before discharge into the water bodies is the best way to prevent water pollution. Several techniques, such as photocatalysis, membrane filtration, reverse osmosis, adsorption, electrochemical, and so on, have been used to purify the wastewater [12,34–39]. Adsorption is an efficient, economic, sustainable, and straightforward approach to clean the wastewater and also to perform many water remediation processes. The objective of this book is to provide a thorough knowledge of the different types of water contaminants and their hazardous effects, and the removal of water contaminants by adsorption processes using various carbon-based nanoadsorbents.

In recent years, carbon nanomaterials (CNMs), such as graphene and derivatives, carbon nanotubes, carbon nanofibers, nanoporous carbon, fullerenes, graphitic carbon nitride, and nanodiamonds, have been extensively exploited as adsorbents due to their extraordinary surface properties, easy modification, large specific surface area, abundant availability, economical feasibility, controlled structural varieties, high chemical stability, porosity, low density, ease of regeneration, and reusability. Graphene oxides and other oxidized carbons provide strong acidity and abundant functional groups, and demonstrate excellent adsorption of cationic and basic compounds via electrostatic and hydrogen bonding interactions, while their pristine counterparts exhibit hydrophobic surfaces and offer high adsorption via strong $\pi-\pi$ interactions.

This book provides a thorough overview of the state-of-the-art in CNMs, including significant past and recent advances, as well as future strategies for the use of carbon-based nanoadsorbents in water treatment. This book primarily emphasizes the fundamentals of adsorption, its mechanical aspects, isotherm, kinetics, synthesis, properties of CNMs, and the adsorption performances of CNMs and their nanocomposites with inorganic and organic materials. Structural engineering and activation processes produce materials with enhanced adsorptive properties and separation efficiencies. Furthermore, the formation of CNMs with two-dimensional and three-dimensional macro-/microstructures with high porosities is a potential approach to improve adsorption performances and extend CNMs' use at the industrial level.

Many reviews have been published (the majority on graphene-based nanomaterials) concerning the adsorption potential of carbon-based materials [40–47]. However, reports that focus on the synthesis, adsorption kinetics, and adsorption potential of carbon-based nanomaterials with varying size and morphology are still scarce. The published reports either concentrate on specific adsorbents or deliver only short descriptions with limited examples. Thus there is still a need for comprehensive data addressing the adsorptive performance of all CNMs used for water treatment, along with their pros and cons, mechanisms, and summarising the details of recent trends and developments. From this perspective this book will be focused on the utilization of carbon-based nanomaterials as nanoadsorbents for wastewater treatment. For efficient adsorption, carbon-based nanoadsorbents should have high specific areas, eco-friendly synthetic approaches, low energy demands, and fast adsorption kinetics. The synthesis of a variety of carbon-based nanoadsorbents, as well as their adsorption kinetics, isotherms, and adsorption mechanisms concerning different water contaminants (e.g., dyes, heavy metal ions, pharmaceutical compounds, and pesticides) is examined and a perspective on future developments for these nanoadsorbents is presented. This book also addresses some vital issues that persist about these adsorbents, which could shape the future research and industrial application of carbon-based nanoadsorbents in water security.

1.4 Conclusion

Water is the essence of the existence of life on earth. However, the availability of clean freshwater resources is declining globally, due to an excessive increase in population, urbanization, and industrialization. In many parts of the world, the poor quality of water is making life worse. Water pollution is a threat to the environment and affects the ecosystem adversely. There is a need to enhance the awareness regarding the preservation of water, wastewater management, and wastewater treatment. Effluents from many industries and households are still directly discharged into the water bodies, which are causing serious waterborne diseases and other negative impacts on human health. The various water pollution sources such as industry, sewage, domestic, radioactive, oil spillage, urbanization, and agriculture have been explained along with the negative impacts on the environment. The prevention of hazardous effects of water pollutants is difficult but water pollution can be controlled. This book provides the enormous awareness regarding the different wastewater treatment procedures and briefly and elaborately explains the application of adsorption using carbon-based nanoadsorbents in water remediation. In the scientific community, carbon-based nanomaterials are gaining popularity as nanoadsorbents for water treatment due to their size and shape-dependent properties, environmentally benign nature, abundance, and ease of handling. Research on the adsorption of water contaminants using carbon-based nanomaterials (>2500 published articles) is growing rapidly. In this context, it is vital to study the current state-of-the-art concerning the use of carbon-based nanomaterials for water treatments through adsorption methods, and to explore new dimensions and emerging trends, which could help to provide new directions in this active research area.

References

[1] Kundzewicz ZW, Mata LJ, Arnell N, Doll P, Kabat P, Jimenez B, et al. Freshwater resources and their management. In: Parry ML, Canziani OF, Palutikof JP, van der Linden PJ, Hanson CE, editors. Climate change 2007: impacts, adaptation and vulnerability. Contribution of Working Group II to the Fourth Assessment Report of the Intergovernmental Panel on Climate Change. Cambridge University Press; 2007. p. 173–210.

[2] Shannon MA, Bohn PW, Elimelech M, Georgiadis JG, Marinas BJ, Mayes AM. Science and technology for water purification in the coming decades. Nature 2008;452:301.

[3] Forgacs E, Cserhati T, Oros G. Removal of synthetic dyes from wastewaters: a review. Environ Int 2004;30:953–71.

[4] Sheikh B, Cooper R, Israel K. Hygienic evaluation of reclaimed water used to irrigate food crops-a case study. Water Sci Technol 1999;40:261–7.

[5] Fick J, Söderström H, Lindberg RH, Phan C, Tysklind M, Larsson D. Contamination of surface, ground, and drinking water from pharmaceutical production. Environ Toxicol Chem 2009;28:2522–7.

[6] Gavrilescu M. Fate of pesticides in the environment and its bioremediation. Eng Life Sci 2005;5:497−526.

[7] Rodriguez-Mozaz S, de Alda MJL, Barceló D. Monitoring of estrogens, pesticides and bisphenol A in natural waters and drinking water treatment plants by solid-phase extraction-liquid chromatography-mass spectrometry. J Chromatogr A 2004;1045:85−92.

[8] Fosso-Kankeu E, Mittal H, Waanders F, Ray SS. Thermodynamic properties and adsorption behaviour of hydrogel nanocomposites for cadmium removal from mine effluents. J Ind Eng Chem 2017;48:151−61.

[9] Song JY, Jhung SH. Adsorption of pharmaceuticals and personal care products over metal-organic frameworks functionalized with hydroxyl groups: quantitative analyses of H-bonding in adsorption. Chem Eng J 2017;322:366−74.

[10] M. Ntakadzeni, W. W. Anku, N. Kumar, P. P. Govender, L. Reddy, PEGylated MoS_2 nanosheets: a dual functional photocatalyst for photodegradation of organic dyes and photoreduction of chromium from aqueous solution, Bull. Chem. React. Eng. Catal. 14, 2019, 142−152.

[11] Mukwevho N, Kumar N, Fosso-Kankeu E, Waanders F, Bunt J, Ray S. Visible light-excitable ZnO/2D graphitic-C3 N4 heterostructure for the photodegradation of naphthalene. Desalination Water Treat 2019;163:286−96.

[12] Gusain R, Kumar N, Ray SS. Recent advances in carbon nanomaterial-based adsorbents for water purification. Coordinat Chem Rev 2020;403:213111.

[13] Kumar N, Mittal H, Parashar V, Ray SS, Ngila JC. Efficient removal of rhodamine 6G dye from aqueous solution using nickel sulphide incorporated polyacrylamide grafted gum karaya bionanocomposite hydrogel. RSC Adv 2016;6:21929−39.

[14] Senthil Kumar P, Saravanan A. 11 - Sustainable wastewater treatments in textile sector. In: Muthu SS, editor. Sustain Fibres Text. Woodhead Publishing; 2017. p. 323−46.

[15] Ketchum E. Air and water pollution in the iron and steel industry: a case study: the calumet industrial region. J Geogr 1967;66:500−6.

[16] Ochieng G, Seanego E, Nkwonta O. Impacts of mining on water resources in South Africa: A review. Sci Res Essays 2010;5:3351−7.

[17] Dixit S, Yadav A, Dwivedi P, Das M. Toxic hazards of leather industry and technologies to combat threat: A review. J Clean Prod 2015;87:39−49.

[18] Kumar N, Reddy L, Parashar V, Ngila JC. Controlled synthesis of microsheets of ZnAl layered double hydroxides hexagonal nanoplates for efficient removal of Cr(VI) ions and anionic dye from water. J Environ Chem Eng 2017;5:1718−31.

[19] Gusain R, Kumar N, Fosso-Kankeu E, Ray SS. Efficient removal of Pb(II) and Cd(II) from industrial mine water by a hierarchical MoS_2/SH-MWCNT nanocomposite. ACS Omega 2019;4:13922−35.

[20] Mokarizadeh A, Faryabi MR, Rezvanfar MA, Abdollahi M. A comprehensive review of pesticides and the immune dysregulation: mechanisms, evidence and consequences. Toxicol Mecha Methods 2015;25:258−78.

[21] Al-Saleh IA. Pesticides: a review article. J Environ Pathol Toxicol Oncol Off Organ Int Soc Environ Toxicol Cancer 1994;13:151−61.

[22] Mnif W, Hassine AIH, Bouaziz A, Bartegi A, Thomas O, Roig B. Effect of endocrine disruptor pesticides: a review. Int J Environ Res Public Health 2011;8:2265−303.

[23] Waters MD, Simmon VF, Valencia R, Mitchell AD, Jorgenson TA. An overview of short-term tests for the mutagenic and carcinogenic potential of pesticides. J Environ Sci Health Part B 1980;15:867−906.

[24] Yang X, Li S, Wang Z, Lee SM, Wang LH, Wang R. Constraining the teratogenicity of pesticide pollution by a synthetic nanoreceptor. Chem Asian J 2018;13:41−5.

[25] Meenakshi, Maheshwari RC. Fluoride in drinking water and its removal. J Hazard Mater 2006;137:456−63.

[26] Mohod CV, Dhote J. Review of heavy metals in drinking water and their effect on human health. Int J Innov Res Sci Eng Technol 2013;2:2992−6.

[27] Almeida P, Stearns LB. Political opportunities and local grassroots environmental movements: the case of minamata. Soc Probl 1998;45:37−60.

[28] Inaba T, Kobayashi E, Suwazono Y, Uetani M, Oishi M, Nakagawa H, et al. Estimation of cumulative cadmium intake causing Itai-itai disease. Toxicol Lett 2005;159:192−201.

[29] Sorahan T, Adams RG, Waterhouse JA. Analysis of mortality from nephritis and nephrosis among nickel-cadmium battery workers. J Occupat Environ Med 1983;25:609−12.

[30] Correia VM, Stephenson T, Judd SJ. Characterisation of textile wastewaters - a review. Environ Technol 1994;15:917−29.

[31] Adetunde L, Glover R. ILC. Navrongo Campus Upper-East Reg Ghana Curr Res J Biol Sci 2010;2:361−4.

[32] Gupta, A. Water pollution assessment, modeling and management, first ed., Pointer Publishers, Jaipur, 2016.

[33] Steiner TS, Samie A, Guerrant RL. infectious diarrhea: new pathogens and new challenges in developed and developing areas. Clin. Infect. Dis. 2006;43:408−10.

[34] Kumar N, Mittal H, Reddy L, Nair P, Ngila JC, Parashar V. Morphogenesis of ZnO nanostructures: role of acetate ($COOH^-$) and nitrate (NO_3^-) ligand donors from zinc salt precursors in synthesis and morphology dependent photocatalytic properties. RSC Adv 2015;5:38801−9.

[35] Kumar N, Sinha Ray S, Ngila JC. Ionic liquid-assisted synthesis of Ag/Ag_2Te nanocrystals via a hydrothermal route for enhanced photocatalytic performance. N J Chem 2017;41:14618−26.

[36] Umukoro EH, Kumar N, Ngila JC, Arotiba OA. Expanded graphite supported p-n MoS_2-SnO_2 heterojunction nanocomposite electrode for enhanced photo-electrocatalytic degradation of a pharmaceutical pollutant. J Electroanal Chem 2018;827:193−203.

[37] Kumar N, Fosso-Kankeu E, Ray SS. Controllable disorder engineering in oxygen-incorporated MoS_2 ultrathin nanosheets for efficient hydrogen evolution. ACS Appl Mater Interfaces 2019;11:19141−55.

[38] Ama OM, Kumar N, Adams FV, Ray SS. Efficient and cost-effective photoelectrochemical degradation of dyes in wastewater over an exfoliated graphite-MoO_3 nanocomposite electrode. Electrocatalysis 2018;9:623−31.

[39] Mukwevho N, Fosso-Kankeu E, Waanders F, Kumar N, Ray SS, Yangkou Mbianda X. Photocatalytic activity of $Gd_2O_2CO_3 \cdot ZnO \cdot CuO$ nanocomposite used for the degradation of phenanthrene. SN Appl Sci 2018;1:10.

[40] Upadhyay RK, Soin N, Roy SS. Role of graphene/metal oxide composites as photocatalysts, adsorbents and disinfectants in water treatment: a review. RSC Adv 2014;4:3823−51.

[41] Perreault F, De Faria AF, Elimelech M. Environmental applications of graphene-based nanomaterials. Chem Soc Rev 2015;44:5861−96.

[42] Upadhyayula VK, Deng S, Mitchell MC, Smith GB. Application of carbon nanotube technology for removal of contaminants in drinking water: a review. Sci Total Environ 2009;408:1−13.

[43] Chen L, Han Q, Li W, Zhou Z, Fang Z, Xu Z, et al. Three-dimensional graphene-based adsorbents in sewage disposal: a review. Environ Sci Pollut Res 2018;25:25840−61.

[44] Yang K, Wang J, Chen X, Zhao Q, Ghaffar A, Chen B. Application of graphene-based materials in water purification: from the nanoscale to specific devices. Environ Sci Nano 2018;5:1264−97.

[45] Kyzas GZ, Deliyanni EA, Bikiaris DN, Mitropoulos AC. Graphene composites as dye adsorbents: review. Chem Eng Res Des 2017;129:75−88.

[46] Sherlala A, Raman A, Bello M, Asghar A. Chemosphere 2018;193:1004−17.

[47] Sarkar B, Mandal S, Tsang YF, Kumar P, Kim K-H, Ok YS. Designer carbon nanotubes for contaminant removal in water and wastewater: a critical review. Sci Total Environ 2018;612:561−81.

CHAPTER TWO

Classification of water contaminants

2.1 Water contaminants

Regular water contamination is a severe issue as it is directly or indirectly affecting the ecosystem in terms of human health and aquatic/marine life. Any foreign substance (either chemical or biological) which enters into the water system and makes an adverse impact on the quality of water, affecting the health of living beings is termed as a water contaminant. Untreated effluents from domestic households, industry, agriculture, pharmacy, etc., contain ecotoxicologically hazardous substances that when released into the water bodies degrade the water quality. Later, this contaminated water affects the flora, fauna, and biota of the aquatic life and these contaminants may also enter into the human food chain on drinking the affected water or the consumption of food from the affected water. Some contaminants, termed as persistent organic pollutants (POP), are resistant to biodegradation, remain in the water bodies for long periods, and are hazardous to the ecosystem. Diarrhea and parasitic intestinal infections are common health problems caused by the consumption of contaminated water and lead to poor digestion and malnutrition [1]. Industry discharges millions of gallons of effluent, full of toxic substances, such as heavy metal ions (HMIs), dyes, highly stable aromatic compounds, drugs, etc., directly into the sea, damaging the ecosystem. In this regard, it is necessary to understand the sources and types of water pollutants along with their harmful effects on the ecosystem. Table 2.1 lists some water pollutants with their permissible limits, sources, and adverse effects on human health [2,3].

2.2 Classification of water contaminants

Water contaminants are various natural and synthetic chemical compounds or ions which enter into the water through various activities and exhibit potential risk to the ecosystem. Water pollutants are broadly divided into two categories: (1) organic pollutants and (2) inorganic pollutants, which further can be subdivided into other categories. Fig. 2.1 gives an overview of the classification of water pollutants. The following sections will discuss the different types of water contaminants in detail.

Carbon Nanomaterial-Based Adsorbents for Water Purification.
DOI: https://doi.org/10.1016/B978-0-12-821959-1.00002-7

Table 2.1 Types of pollutant, their permissible limits, adverse effects, and industrial sources.

Pollutants	Permissible limit (mg/L)	Effects	Source
Cr	(USEPA) 0.5 (EC) 0.05 (WHO)	Necrosis nephritis, carcinogenic effects, imitability, fatigue Irritation of gastrointestinal Mucosa	Mines, electroplating, tannery, textile, industries, mineral sources
Pb	0.015 (USEPA) 0.5 (EC) 0.01 (WHO)	Toxic to humans and aquatic fauna, phytotoxic, tiredness, irritability, anemia, high doses lead to metabolic poison, neurotoxicity, brain damage and hypertension	Batteries, automobile emission, pesticides, mining, mineral weathering, old pipelines, paint, burning of coal
As	0.01 (USEPA) 0.01 (EC) 0.01 (WHO)	Hepatotoxicity, immunotoxic, causes melanosis, keratosis, hyperpigmentation, liver damage, vomiting, and diarrhea in humans, toxicological and carcinogenic effects, modulation of coreceptor expression, genotoxicity	Fungicides, pesticides, insecticide, metal smelters, toys, paints, electric waste, geological origin
Hg	0.002 (USEPA) 0.001 (EC) 0.001 (WHO)	Poisonous, gingivitis, tremors, disturbs the cholesterol, causes mutagenic changes, neurotoxic spontaneous abortion, protoplasm poisoning	Pesticides, paper industries, paints, mining, batteries, volcanoes, dental filling, fishes
Cd	0.005 (USEPA) 0.005 (EC) 0.003 (WHO)	Acute effects in children, causes severe problem to kidneys and bones, emphysema bronchitis, anemia, nephrotoxicity, pulmonotoxicity	Electroplating, welding, batteries Nuclear fission plant, fertilizer, pesticides
Zn	5 (USEPA) NA (EC) 3 (WHO)	Lack of muscular coordination, anemia, abdominal pain, corrosive effect on skin and nervous membrane, hemotoxicity, phytotoxic	Refineries, metal plating, plumbing, brass manufacturing industries
Cu	1.3 (USEPA) 2 (EC) 2 (WHO)	Mucosal irritation and corrosion, central nervous system irritation, depression, phytotoxic, toxic to aquatic fauna	Mining, chemical industry, pesticide production, metal piping, electroplating

(continued)

Table 2.1 (Continued)

Pollutants	Permissible limit (mg/L)	Effects	Source
Ni	0.1 (USEPA) 0.1 (EC) 0.07 (WHO)	Eczema of hands, phytotoxic, causes DNA damage, damages fauna	Electroplating, battery industries, zinc base casting
Mn	0.05 (USEPA) 0.001 (EC) 0.4 (WHO)	Causes damage to central nervous system	Welding, ferromanganese production, fuel addition
Fluorides	2−4 (USEPA) 1.0−1.5 (WHO)	Cause dental and skeletal fluorosis, neurological disorders	Steel manufacturing, coal combustion, phosphate fertilizer production, geological origin
Pesticides	DDT: 0.001 Carbofuran: 0.004 Simazine: 0.04	Cancer, liver and kidney problems, cardiovascular/ neurological/reproductive/ disorders	Industrial pollution, contamination in soft drinks, farming effluents
Halogenated organics	CCl_4: 0.005 TCE: 0.005	CCl_4: carcinogenic, high toxicity to liver and kidney TCE: lung/liver tumor	Chlorination, manufacturing plants, home and agricultural effluents

CCL, Carbon tetrachloride; *DDT*, dichlorodiphenyltrichloroethane; *EC*, European Community; *TCE*, trichloroethylene; *USEPA*, United States Environment Protection Agency; *WHO*, World Health Organization.

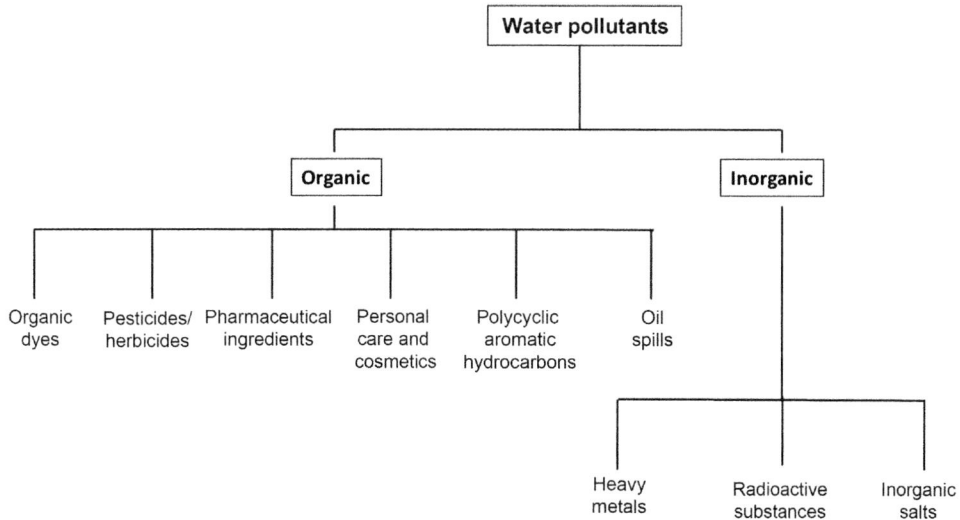

Figure 2.1 Classification of water contaminants.

2.3 Organic dyes

Dyes are complex organic molecules that consist of chromophores and auxochromes. The chromophore is responsible for the production of color while the auxochrome is a functional group attached to the chromophore that helps the dye to bind to the object. Carboxylic acid, amino, sulfonic acid, and hydroxyl groups are typical examples of auxochromes. Auxochromes do not impart or participate in color production, but their presence shifts color and also influences dye solubility. For many decades, dyes have been used on fabrics to impart colors which are not easy to remove and are resistant to heat, detergent, etc. [4]. More than 100,000 types of organic dyes are commercially available in the market with annual production of 7×10^5 tonnes per year [5]. The use of dyes is ubiquitous in the leather tanning, paper, and textile industries for coloring paper and fabric [6–8]. However, the majority of the dye does not fix to the material during the dyeing process and remains in the solution, which is later disposed of into the water system, without proper treatment. The presence of coloring materials and an oily scum on the surface of the water causes a foul smell and disrupts the aquatic ecosystem by preventing the penetration of the sunlight necessary for photosynthesis [9]. Thus the toxic nature and adverse effects of the dye constituents are becoming a threat to natural resources by changing the pH level and increasing the total organic carbon, chemical oxygen demand, and biological oxygen demand values of the affected water [10]. Intake of dye-contaminated water by human being and animals, even at only ppm level, can be mutagenic, carcinogenic, and cause severe dysfunction in the central nervous system, kidneys, lungs, and reproductive system [11,12]. Therefore regular discharge of untreated dye effluents in water resources has a negative impact on water quality and the life of those living beings who are consuming it for living purposes.

Dyes can be classified based on several parameters, such as source, chemical structure, chemical nature, and particle charge on dissolving in an aqueous medium (Fig. 2.2) [4]. Based on their sources, dyes can be classified as natural dyes or synthetic dyes. In ancient time, natural dyes were extracted from plants, animals, insects, and mineral sources. Turmeric, barberry, henna, onion, jackfruit, teak, etc., are popular examples of natural dyes. But with the passing of time and rapid increase in population, natural dyes could not meet the industrial requirements and hence industries have become dependent on synthetic dyes. Furthermore, dyes can also be broadly classified based on the chemical structure and chemical nature. Based on chemical structure, dyes are classified by the presence of the typical group, for example, azo, nitroso, nitro, etc., in the dye. However, based on their chemical nature, dyes are classified according to their reaction application mode [4]. On dissolving the dyes in an aqueous medium, particle charges may be generated and these can be divided primarily into

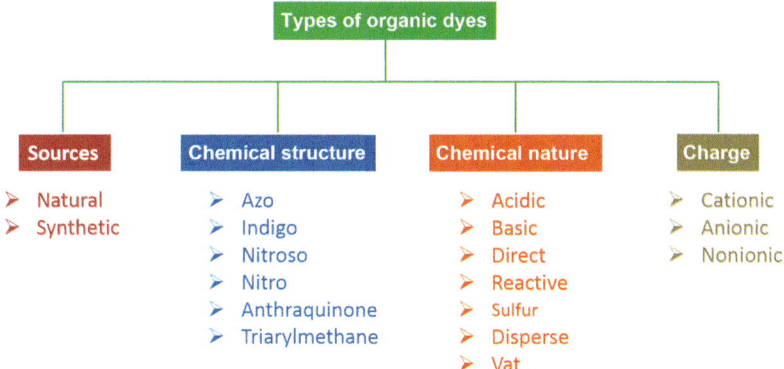

Figure 2.2 Classification of organic dyes.

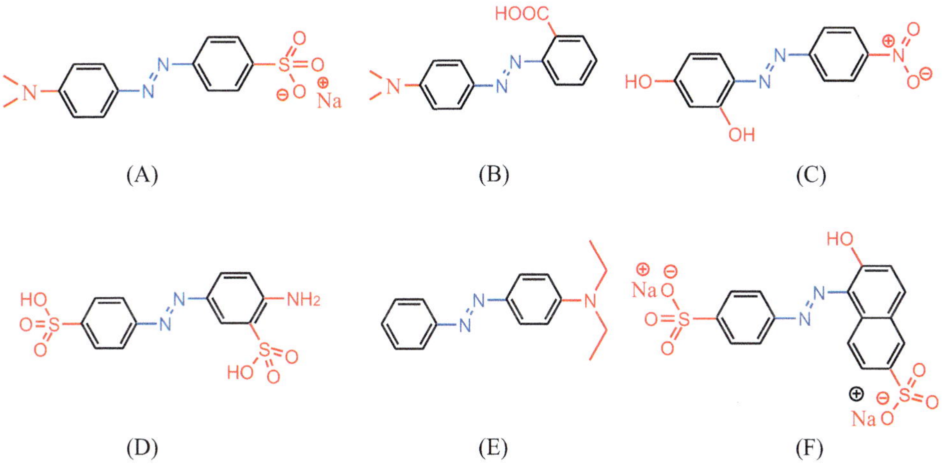

Figure 2.3 Molecular structure of azo dyes: (A) methyl orange; (B) methyl red; (C) azo violet; (D) fast yellow B; (E) oil yellow DE; and (F) sunset yellow FCF. The blue color represents the chromophore group and the red color represents the auxochrome group.

three categories: cationic dyes, anionic dyes, and nonionic dyes. Acidic, direct, and reactive dyes fall under the category of anionic dyes and basic dyes come under cationic dyes. Here, the classification of dyes is briefly discussed, based on chemical structure.

2.3.1 Azo dyes

Azo dyes exhibit at least one $R-N=N-R'$ functional group, which is also the chromophore group of the dye and is responsible for the coloring [13]. Fig. 2.3 shows the molecular structure of few azo dyes. They generally absorb in the visible range

(350−650 nm) of the solar spectrum and are the most popular dyes in food, textile, cosmetics, inks, and pharmaceutical industries. Generally, all azo dyes are water-soluble and can be synthetic or natural. The largest amount of azo dyes are consumed by the textile industry but not all the dye is adsorbed onto the fabric's surface. Approximately 10%−15% of the dye remains in the solution and generates colored wastewater which may be released into water bodies without proper treatment [14]. Thus regular industrial effluents discharge with azo dye residuals become a threat to human health and aquatic life as a few azo dyes or their metabolites (e.g., arylamines) are highly toxic and carcinogenic [15]. Benzidine and its derivatives are one of the major azo reduction products of the azo dyes, and these are potentially highly carcinogenic and toxic in nature. Also, azo dyes are toxic to fishes, algae, aquatic plants, and other organisms in the water.

2.3.2 Indigo dye

Indigo dye, also known as 2,2′-bis(2,3-dihydro-3-oxoindolyliden) (Fig. 2.4), exhibits a distinctive blue color. In ancient times, all blue dyes belonged to the indigo dye category, which was a natural dye extracted from the plants. However, with industrial development, indigo dyes are mostly synthesized now. It is not soluble in water, ether, or alcohol, but can readily dissolve in nitrobenzene, sulfuric acid, and chloroform. It is most commonly used by the textile industry to color cotton yarn, that is, to produce denim clothes [16]. It is also used to dye wool and silk fabrics. Indigo dyes display an absorbance peak at a wavelength of 600−620 nm in visible light [17]. Indigo carmine dye is also used in the pharmaceutical industry for medical diagnostic purposes [18]. The discharge of indigo dye-contaminated effluents into the water streams changes the water color and creates a foul smell. Additionally, it is toxic to aquatic species as it breaks down into aromatic amines which are hazardous [19,20]. It can damage the cornea and conjunctiva and is also carcinogenic with acute toxicity [21]. Gastrointestinal irritations with diarrhea, nausea, and vomiting are other harmful effects of indigo dye exposure [22].

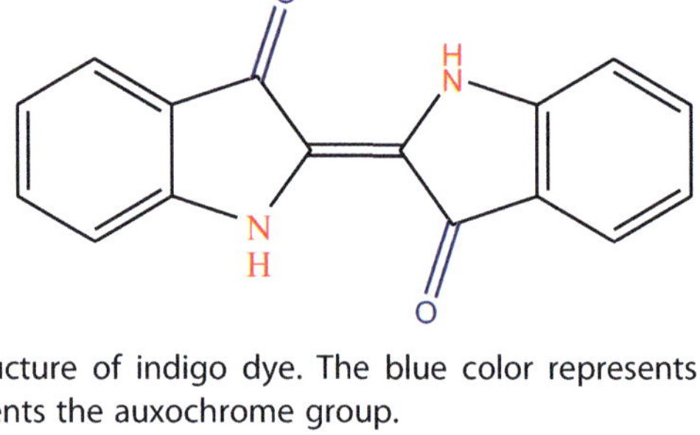

Figure 2.4 Chemical structure of indigo dye. The blue color represents the chromophore group, and the red color represents the auxochrome group.

2.3.3 Nitroso dye

Nitroso dyes are a class of organic dyes in which a nitroso group (NO) is present at the ortho position of the electron–donating groups, such as hydroxyl (OH) or amine (NH$_2$) group. Fuchs first discovered it in 1875 [23]. They are generally blue or green in color. They are used to dye or print cotton, woolen, and silk fabric. The most common example of a nitroso dye is 1-nitroso-2-naphthol (Fig. 2.5A). 1-Nitroso-2-naphthol on reacting with ferrous sulfate in an alkaline medium makes an iron(II) complex derivative (Fig. 2.5B) which is known as Pigment Green B and is widely used for wallpaper and paints of yellowish—green color due to its light fastness properties [24]. Similarly, an iron(II) complex of 2-naphthol-6-sulfonic acid nitroso dye (Fig. 2.5C) is known as Naphthol Green BN which is largely used to dye paper and leather [23].

2.3.4 Nitro dyes

Nitro dyes comprise at least one nitro group which is present at the ortho or para position of electron-donating group (mostly hydroxyl, imino, or amino groups) in the aromatic ring. Nitroaniline and nitrophenol (see Fig. 2.6A and B) are the simplest examples of nitro dyes and are yellow in color. Picric acid (see Fig. 2.6C) is the oldest nitro dye, which was discovered in 1771 by treating indigo with nitric acid [24]. In the early period, nitro dyes were used to color natural animal fibers such as silk and wool [13]. Nitrodiphenyamine (Fig. 2.6D) is another popular example of a nitro dye used to color dense fibers such as polyester. Nitro dyes are also engaged in optical data storage and inkjet printing [25,26]. Nitro dyes are also used as food colorings,

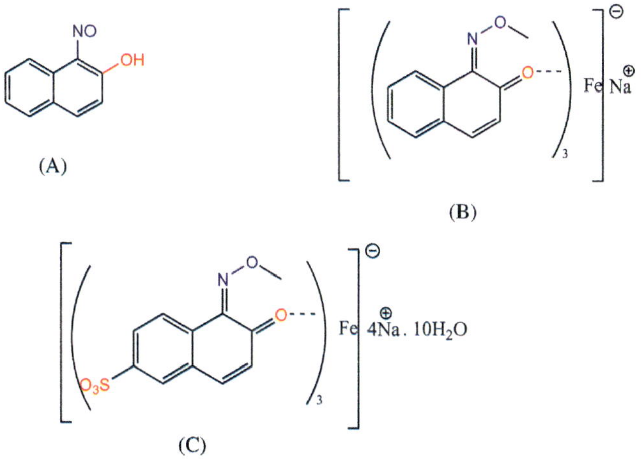

Figure 2.5 (A) 1-Nitroso-2-naphthol, (B) Pigment Green B, and (C) Naphthol Green BN. Blue color represents the chromophore group and the red color represents the auxochrome group.

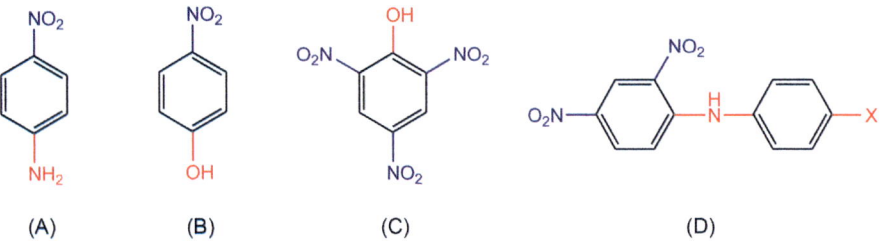

Figure 2.6 (A) *p*-Nitroaniline, (B) *p*-nitrophenol, (C) picric acid, and (D) nitrodiphenyamine (Disperse Yellow 14 (X = H), Disperse Yellow 1 (X = OH), and Disperse Yellow 9 (X = NH$_2$)). The blue color represents the chromophore group and the red color represents the auxochrome group.

fungicides, and herbicides [27,28]. Picric acid and hexanitrodiphenylamine nitro dyes are also used as explosives [29]. Exposure of nitro dyes to a human being can cause mutagenesis and hence toxic [30].

2.3.5 Anthraquinone dye

After azo dyes, anthraquinone dyes are the second most important class of dye in the textile industry with light fastening characteristics. They include a large group of dyes with a wide range of colors [31]. Anthraquinone dyes are natural as well as synthetic dyes and can provide bright blue shades that cannot be obtained using azo dyes. Anthraquinone dyes exhibit an anthraquinone group (Fig. 2.7A) and, depending on the substitution groups and position, they are valued for producing red, violet, bluish-green, greenish–yellow, and blue dyes. Due to the wide application range of anthraquinone dyes in the textile industry, they might be present in high concentrations in textile industry wastewater effluents. Anthraquinone dyes exhibit a very low rate of decolorization and substantial toxicity with volatile fatty acid degradation and methanogenesis [32]. Additionally, anthraquinone dyes are allergic, carcinogenic, and mutagenic [33−35]. Alizarin (Fig. 2.7B) is the most common anthraquinone dye as a red textile dye and is also use in pigments and painting. Emodin, rhein, purpurin, aloe-emodin, etc. (Fig. 2.7) are the most common examples of anthraquinone dyes.

2.3.6 Triarylmethane dyes

Triarylmethane dyes have a triphenylmethane group (Fig. 2.8A) as the backbone of the organic molecule. Apart from dyeing paper, plastics, and leather, these dyes are also commonly used as food colorings, biocides, and pH indicators. Methyl violet, malachite green, fuchsine, phenolphthalein, and Victoria blue are common examples of triarylmethane dyes. Triarylmethane dyes cause carcinogenesis, chromosomal fractures, mutagenesis, respiratory toxicity, and teratogenicity [36,37].

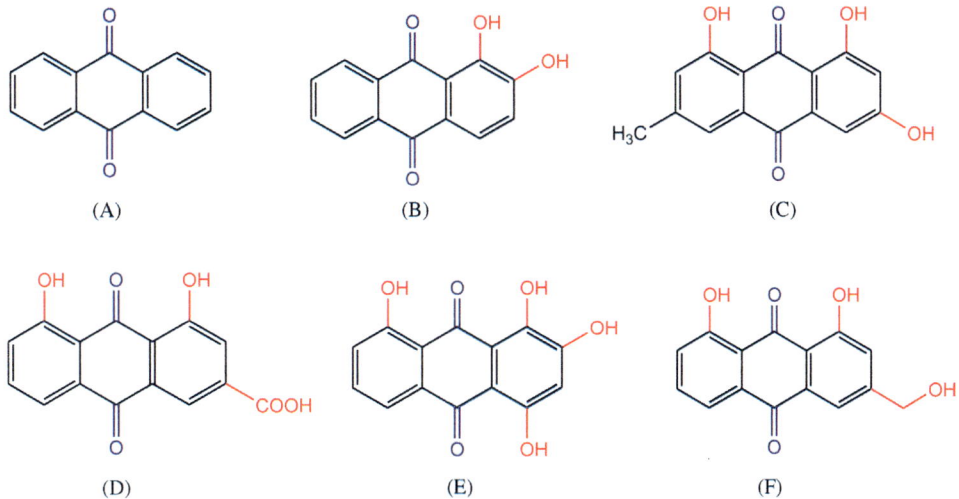

Figure 2.7 (A) Anthraquinone, (B) alizarin, (C) emodin, (D) rhein, (E) purpurin, and (F) aloe-emodin. The blue color represents the chromophore group and the red color represents the auxochrome group.

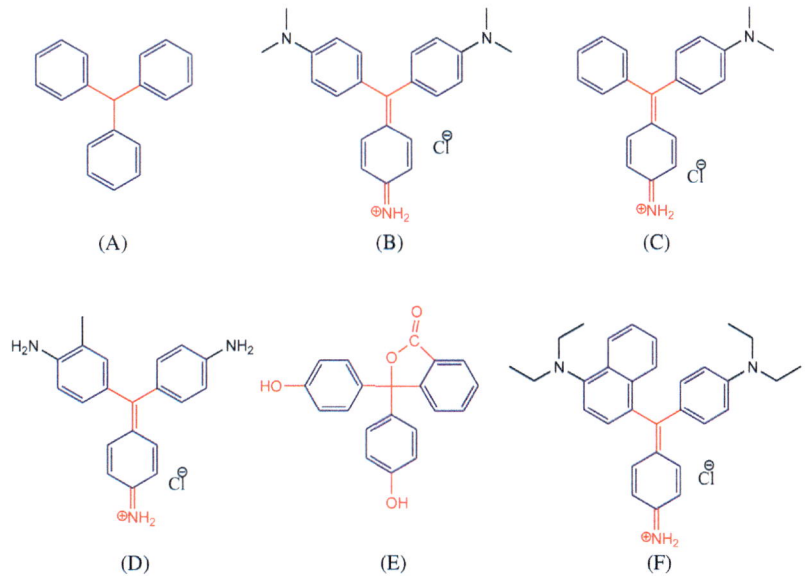

Figure 2.8 (A) Triarylmethane, (B) methyl violet, (C) malachite green, (D) fuchsine, (E) phenolphthalein, and (F) Victoria blue. The blue color represents the chromophore group and the red color represents the auxochrome group.

2.4 Pesticides

Pesticides are organic chemicals which are used to control pests, insects (e.g., mosquitoes, rodents, ticks, etc.), fungus, and herbs and to prevent weeds, disease, and insect infection in crops. This results in an invisible flow of pesticides into the water table [38]. These chemicals are beneficial to humans to one extent as they can control the harmful effects and diseases by killing the pests, but the extensive use of such chemicals contaminates the environment and is threatening to public health. In the United States, the annual usage of conventional pesticides is approximately 1 billion pounds and of this 750−800 million pounds are used in agriculture [39]. Pesticide residue is one of the major sources of organic contamination. Pesticides can easily be dissolved in water and it is then difficult to extract them later. They are considered as one of the major contributing water pollutants. Similar to organic dyes, pesticides can also be classified according to their function, degradability, and chemical structure (Fig. 2.9) [40].

Here, the classification of pesticides is briefly discussed on the basis of chemical structure.

2.4.1 Organophosphate pesticides

Organophosphate pesticide is a well-known insecticide which is generally considered to be esters of phosphoric acids and consists of an $O = P(OR)_3$ functional group. The main objective of organophosphate insecticide is to suppress the activity of

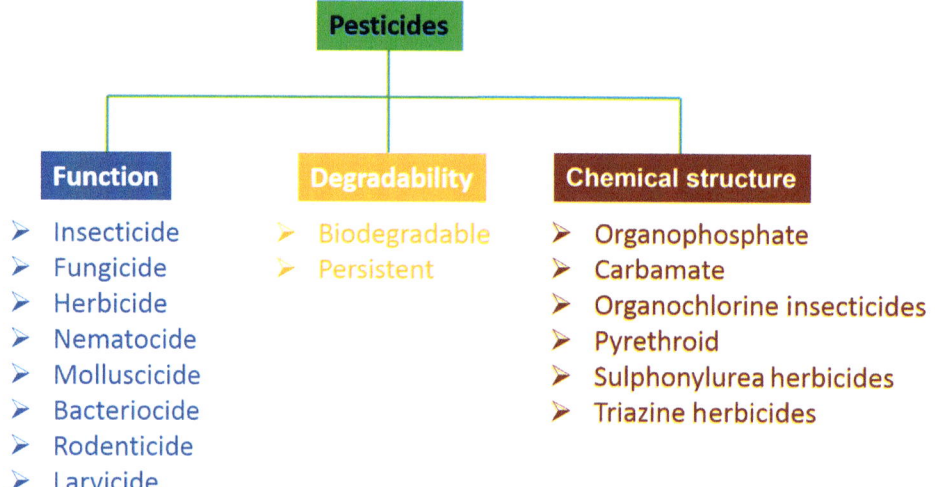

Figure 2.9 Classification of pesticides.

cholinesterase in the nerve tissue of insects, ultimately killing the insect [41]. However, cholinesterase enzyme is also present in animals and acts as a neurotransmitter. Therefore excessive usage of organophosphate insecticide is also toxic to animals and human health. Exposure to organophosphate shows harmful effects on the reproductive and central nervous system and it is also carcinogenic. It is also a potent poison which causes symptoms of muscle cramping, paralysis, fatigue, and fasciculation. Dichlorvos, cyanofenphos, tetrachlorvinphos, trichlorfon, chlorpyrifos, malathion, fenchlorphos, and profenofos are common examples of organophosphate pesticides.

2.4.2 Carbamate pesticides

Carbamate pesticides are ester of carbamic acid (RHNCOOR′) and are readily soluble in water. Similar to organophosphates, Carbamate pesticides also kill insects by inhibiting the activity of cholinesterase enzyme and affect nerve impulse transmission [42]. Carbamate pesticides are not as popular as organophosphate pesticides. They are highly chemically reactive and mainly used in agriculture. Carbamate pesticides poisoning leads to respiratory depression combined with pulmonary edema, which ultimately causes death. The symptoms of carbamate poisoning are similar to organophosphate pesticides poisoning. Carbaryl, aldicarb, isoprocarb, captan, and carbafuran are some examples of carbamate pesticides.

2.4.3 Organochlorine insecticide

The organochlorine insecticides are a class of highly POP with many chlorine substitutions in hydrocarbon compounds. The organochlorine pesticide is popular to control insects in domestic and agricultural purposes and 40% of the total pesticides used belong to this class of pesticides [43]. A few organochlorine pesticides are banned in many countries including in North America, China, Japan, and Europe due to their harmful effects on health and environmental [41]. They also attack the central nervous system and kill insects. However, they have severe adverse effects on animals and human beings. The most commonly used organochloride pesticide for insect control in households and agriculture is 1,1-bis (4-chlorophenyl)-2,2,2-trichloroethane (DDT) which is an extremely hazardous POP. DDT enters the food chain by water contamination and, even at the ppm level, can act as an endocrine disruptor, is carcinogenic, and leads to congenital disabilities upon consumption by pregnant women [44]. Not only affecting humans and other mammals, DDT also disturbs the bird population and environment balance. Short-term exposure of organochloride pesticides may produce vomiting, dizziness, confusion, nausea, headache, salivation, and sweating. However, long-term exposure may severely damage the body organs such as liver and kidney, thyroid, bladder, and central nervous system. It is also carcinogenic. Aldrin, chlordane, dieldrin, heptachlor, toxaphene, lindane are other examples of organochlorine pesticides.

2.4.4 Pyrethroid

A pyrethroid is a synthetic insecticide which was invented in the 1970s, and the chemical structure was adapted from natural pyrethrins, which can be extracted from East African pyrethrum flowers. It has high arthropod toxicity and requires a very small quantity (30 g per 1000 m^2 land) to kill the insects [41]. It exhibits low toxicity toward humans, other mammals, and birds but is highly toxic to aquatic species. Low toxicity toward mammals and low dose requirements mean that it is commonly used in the home, garden, and agriculture to control pests. Pyrethroids are also used by humans to treat lice or flea infestations. Burning, tingling, itching, and stinging are the common symptoms of pyrethroid exposure at the site of application. Phenothrin, resmethrin, permethrin, deltamethrin, tetramethrin, cypermethrin are some examples of pyrethroid pesticides.

2.4.5 Sulfonylurea herbicides

Sulfonylurea herbicides are used to control a wide range of weeds at very low concentration and also have low toxicity toward humans and other mammals. However, they interfere with the acetolactate synthase enzyme present in plants and microorganisms and thus exert toxicity [45] as this directly blocks the biosynthesis of branched-chain amino acids (leucine, valine, and isoleucine) [46]. Sulfonylurea herbicides have a low impact on the environment and are widely acceptable for field applications. However, a few reports confirm that these herbicides are damaging crop rotations [46,47]. These herbicides are developed from triazine herbicides and exhibit a triazine chemical structure. Sulfonylurea herbicides can easily be degraded by means of chemical and biological degradation and are not environment-persistent [48]. Chlorsulfuron, tribenuron-methyl, and metsulfuron-methyl are well-accepted sulfonylurea herbicides.

2.4.6 Triazine pesticides

Symmetrical triazine (s-triazine) pesticides are widely used as herbicides in agriculture fields. The s-triazine exhibits a six-member heterocyclic ring structure in which nitrogen and carbon are at alternative positions. It was found that s-triazine has low toxicity toward humans and animals, but chronic exposure causes carcinogenic effects [49,50]. Exposure of atriazine (2-chloro-4-(ethylamino)-6-(isopropylamino)-s-triazine), a well-known s-triazine herbicide, in agriculture fields inhibits photosynthesis in plants and also increases the chance of arsenic uptake by treated plants [51]. Commonly used s-triazine are atrazine, ametryne, terbutryne, and simazine.

2.5 Pharmaceutical ingredients

Pharmaceutical drugs are essential for the diagnosis and treatment of many diseases in humans and animals. However, pharmaceutical industry effluents contain drug precursors (including pathogenic organisms and radioactive elements) and antibiotics and are often disposed of invisibly into the water bodies. In addition to direct disposal into the water by industry, pharmaceutical active compounds (PhACs) also enter the water system via excretion (urine and feces) after the consumption of medicines, or when disposed of as unused and expired medicines [52]. The most commonly used pharmaceutical drugs are paracetamol, ibuprofen, diclofenac, citalopram, naproxen, acetylsalicylic acid, and carbamazepine [53]. These are nonbiodegradable, contain radioactive elements, and their excess intake via contaminated water has a deleterious effect on health [54]. Although a few pharmaceutical ingredients have a low persistence in the environment, the rate of release of pharmaceutical effluents into water streams is much higher than the rate of transformation [55]. The introduction of active pharmaceutical ingredients into drinking water and the food chain through human and veterinary consumption, excretion, and waste disposal is represented in Fig. 2.10 [56].

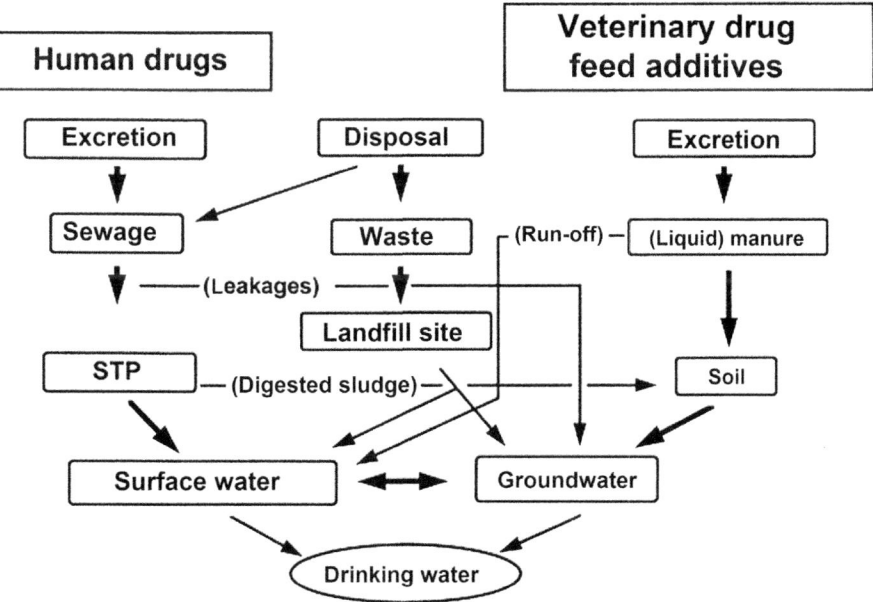

Figure 2.10 Scheme for the main fates of drugs in the environment after application. *STP*, Sewage treatment plant. *Reproduced with permission from Ternes TA. Occurrence of drugs in German sewage treatment plants and rivers. Water Res 1998;32:3245–60. Copyright 2017, Elsevier Science Ltd.*

Although, the pharmaceuticals are necessary to cure the disease but the intake of drugs after a certain limit causes adverse effects. Therefore, a professional practitioner always recommends pharmaceutical drugs in certain amount to avoid the harmful effects of an overdose of drugs. Consuming the drugs-contaminated water may lead to an adverse impact on human health and/or water organisms due to the enormous side effects. Exposure to drugs and drugs ingredients in the water bodies is significantly affecting the fish populations and other aquatic species. The development of antibiotic-resistant pathogens is one of the major threats which is growing due to the regular discharge of antibiotics into the environment through contaminated effluents. Treatment of such antibiotic-resistant pathogens is difficult. The consumption of contaminated pharmaceutical water also affects health via reproduction issues, respiratory disorders, chronic depression, mental retardedness, cancer, and other physical abnormalities [57]. It can also lower the productivity of agricultural land and cause extensive death of livestock. Table 2.2 lists the different types of pharmaceutical drugs, their applications, and side effects.

2.6 Personal care and cosmetics

Personal care products and cosmetics are widely used throughout the world in huge amounts. It results in the continuous invisible release of personal hygiene and cosmetics products or ingredients into the environment. Altogether these impact on the life of living beings as these are bioactive, highly persistent in the environment, and can also bioaccumulate [58,59]. Personal care and cosmetics include any product which is applied on the external part of the body such as skin, nails, hair, lips, and external genital organs, or oral hygiene such as teeth and mucous membrane of the oral cavity, in order to make them clean, protect from germs, prevent bad odor, change in appearance, and keep in good condition [60]. Unlike pharmaceutical drugs, personal care and cosmetic products can be consumed only for external usage. Therefore they are more likely to enter into the environment in large quantities due to human activities, for example, during bathing or washing and give more trouble to the ecological systems [61−63]. Soaps, hair dyes, nail paints, fragrances, emulsifiers, ultraviolet absorbers, acrylates, preservatives, and antioxidants are the common examples of daily use personal care and cosmetic products. Some of these are hazardous for health and heavy exposure of such chemical cosmetics can cause cancer, endocrine disruption, mutation, allergy, and reproductive toxicity [64].

Personal care products generally belong to hygienic practices and rinse off immediately after use, such as shampoos, soaps, toothpastes, and shower gels. However, a few personal care products are leave-on products such as sanitizers, sunscreen lotion, etc. Cosmetics are generally leave-on products, staying on the skin for at least few hours,

Table 2.2 Different types of pharmaceuticals with their application, examples and side effects.

Type of pharmaceuticals	Application	Examples	Side effects
Antiinflammatory and analgesics	Relief from symptoms of pain, high temperature, swelling, stiffness, and fever	Paracetamol, diclofenac, ibuprofen, acetylsalicylic acid	Loss of appetite, nausea, liver and kidney damage, stomach pain, gas, indigestion, diarrhea
Antidepressants	Relief from major depression, anxiety, obsessive-compulsive disorder (OCD) and mood disorders	Benzodiazepines	Nausea, constipation, weight loss, dizziness, urinary retention, irregular heartbeat
Antiepileptic	Reduces all forms of epileptic seizures	Carbamazepine	Depression, dizziness, constipation, clumsiness
Lipid-lowering drugs	Reduces the level of lipids and lipoproteins in the blood	Fibrates Bile acid-binding resins	Nausea, bloating, gastrointestinal disturbance, myositis, acute renal failure
β-Blockers	Inhibit adrenergic response facilitated by β-receptors	Atenolol, metoprolol, propranolol	Bradycardia, worsening bronchial asthma, tiredness, nightmares
Antiulcer	Reduce and neutralize gastric acid secretion, and also heal the ulcers	Ranitidine, famotidine	Headache, bowel upset, dizziness, hepatic failure, gynecomastia
Antihistamines	Relief from allergic rhinitis and other allergies such as sneezing, itching, running nose, skin rashes etc.	Benadryl, chlor-trimeton	Headache, drowsiness, dry mouth, nose, and throat, nausea, loss of appetite, chest congestion, hyperactivity
Antibiotics	Treat bacterial infections	Tetracycline, macrolides, β-lactams, penicillin, quinolones, sulfonamides, fluoro-quinolones, chloramphenicol, imidazole derivatives	Nausea, indigestion, diarrhea, abdominal pain, rashes, loss of appetite

for example, body and face creams, insect repellent, cosmetics, perfumes, and antiperspirants. On removing the cosmetic products and washing or showering, these products enter the sewage treatment plants which are not efficient in removing all the chemicals of personal care and cosmetic ingredients [65,66]. Therefore these chemicals get accumulated with sewage sludge which is further used as fertilizers on crops and provides a pathway for the entry of these chemicals into the food chain [67]. Organic compounds (bisphenol-A, *p*-chloro-*m*-xylenol, triclosan, grease, fat, and surfactants) are the most common examples of personal care and cosmetic pollutants, which not only deteriorate the aquatic environment but also affect the vegetation and human health equally [61,68]. Table 2.3 presents a few examples of personal care and cosmetic contaminants and their toxicity.

2.7 Polyaromatic hydrocarbons

Polycyclic aromatic hydrocarbon (PAH) is a class of organic water pollutants which is receiving a considerable amount of interest worldwide due to its chronic carcinogenic and mutagenic toxicity [80,81]. PAHs are generally lipophilic and exhibit minimum solubility in water. Also, the solubility of PAHs decreases in water with the increase in molecular weight which allows them to settle down and accumulate at the bottom sediments in water resources [82]. PAHs consists of two or more fused aromatic hydrocarbon rings and highly persistent in an environment which decreases with an increase in several aromatic rings. PAHs have majorly industrial manufacturing application as intermediate reactants in photographic products, dyes, pharmaceuticals, thermosetting plastics and other chemical industries [83]. Naphthalene is the simplest example of PAHs. Table 2.4 shows the structure of a few PAHs with their application and toxicity.

Anthropogenic activities which include petrogenic and pyrolytic events and natural emissions are the sources of PAHs in the water sediments. Pyrogenic PAHs sources are the pyrolysis of organic matter with low or no O_2 conditions and incomplete combustion of fossil fuels and biomass [91]. Petrogenic PAHs are generated through crude oil maturation and other petroleum product reactions. However, the natural causes of PAHs emission in the environment include vegetation degradation, synthesis due to bacteria, algae and some, volcanic eruptions, forest fires, and natural petroleum and rare minerals sources. PAHs consist of acute toxicity for all living beings and responsible for inducing tumors production in body, impotence, and reduces immunity [92,93]. Lung cancer is the biggest risk of excess exposure to PAHs [94].

2.8 Oil spills

Oil is a major organic water pollutant which can enter into water bodies from leaks from oil pies, oil spills during transport through ship and refineries, and industrial

Table 2.3 Few examples of personal care and cosmetic contaminants and their application and toxicity.

Personal care and cosmetics ingredients	Application	Toxicity	Reference
Triclosan	Antimicrobial/soaps, fragrances	Endocrine disruptor effects, reproduction, carcinogenic, resistant bacteria, bioaccumulation, aquatic toxicity	[69,70]
Triclocarban	Antimicrobial, soaps, body washes, lotions, detergents, and wipes for its sanitizing properties.	Carcinogenic, endocrine disruptor, teratogenicity, bioaccumulation, weight loss, abortion	[71,72]
Phenol	Soap, cleaning lotions	Mutagenic, corrosive to skin, carcinogenic (tumor promoter), local tissue irritation, irregular pulse, vomiting, darkened urine, liver damage, blood-forming organs damage, collapse	[73]
1,4-Dichlorobenzene	Insect-repellent and deodorizer	Carcinogenic, mutagenic, hematological abnormalities, bioaccumulation, impair kidney and liver function	[74]
Benzylparaben	Preservatives	Estrogenic effects, carcinogenic, inhibited spermatogenesis	[59,75]
Butylparaben	Preservatives	Adverse effect on male reproductive system, carcinogenic	[59,76]
Methylparaben	Antifungal and preservatives	Induce chromosomal damage, carcinogenic, subcutaneous toxicity	[77]
Benzophenone-3	Sunscreen product, UV-filters in cosmetics	Endocrine disruption, hormone-dependent diseases, adverse effect on reproduction and development	[78]
Benzophenone-4	UV-filters in cosmetics, hairspray, shampoo and antiperspirant	Antiestrogenic, antiandrogenic, and estrogenic activity	[79]
4-Methylbenzylidene camphor	UV-filters	Antiestrogenic activity affects the hypothalamus—pituitary—gonadal system in male rats, alter steroid hormone production	[59]

Table 2.4 Few common examples of PAHs and their chemical structure, application, and toxicity.

PAHs	Structure	Application	Toxicity	References
Acenaphthene		Manufacture of pigments, dyes, plastics, pesticides and pharmaceuticals	Damages the cell's DNA and affects endocrine activity	[84]
Anthracene		Diluent for wood preservatives and manufacture of dyes and pigments	Carcinogenic and immunotoxic	[85]
Fluoranthene		Manufacture of agrochemicals, dyes and pharmaceuticals	Nephropathy and liver problem	[86]
Fluorene		Manufacture of pharmaceuticals, pigments, dyes, pesticides and thermoset plastic	Carcinogenic	[87]
Phenanthrene		Manufacture of resins and pesticides	Hyperuricemia, and reproductive endocrine disruptor	[88]
Pyrene		Manufacture of pigments	Skin irritation and chronic skin discoloration	[89,90]

wastewater. Oil pollution in the water can be in the form of lubricant, fat, cutting liquids, light hydrocarbons (e.g., kerosene, gasoline, and jet fuel), and heavy hydrocarbons (e.g., grease, tars, diesel oil, and crude oil) [95]. Regular oil pollution in water bodies poses a great threat to the life of birds, animal, humans, and plants. Oil spills are not soluble in water and, being lighter in weight, the oil forms a thick layer of sludge on the top of the water. This sludge layer suffocates the aquatic species by stopping the contact of air with water and also blocks the sunlight, which inhibits the photosynthesis processes of aquatic plants. Many animals and bird consume oil-contaminated water which is poisonous and may kill them. Oil exposure causes many adverse effects to fishes, including enlarged livers, reduced growth, changes in respiration rate, reproduction impairment, and fin erosion. This also generates harmful effects on the eggs and larval survival. According to Webler and Lord, oil spills can harm human health significantly in three major ways: (1) eating seafood containing bioaccumulated oil toxin can directly affect the human health; (2) coastal area human health can deteriorate from breathing in oil vapors; and (3) there can be economic impact on the fishers' lives due to the impact of oil spills on fishes [96]. Inhalation of oil vapors and ingestion of oil toxin-contaminated sea weed can cause lethal effects which can range from slight dizziness and nausea to severe health issues, such as cancer and central nervous system [97–99].

2.9 Heavy metals

The use of HMIs is pervasive in cosmetic products, automobiles, metal plating, electroplating, mining operations, batteries, etc. [100]. The most commonly used and highly toxic heavy metals released in industrial wastewater are Zn, Ni, Pb, Hg, Cd, Cu, Cr, and Pd. A few trace elements (such as Zn, Cu, Mg, and K) are essential to regulate biochemical processes and metabolism in the human body, but only within certain limits [101]. An excess of these HMIs can result in significant health issues. For example, exceeding the recommended concentration of Ni(II) in water causes cancer, infertility, insomnia, and skin, lung, and kidney disorders [101]. Hg and Pb are neurotoxins which directly attack the central nervous system and have other harmful effects, such as kidney, liver, and reproductive system dysfunction, vomiting, dizziness, chest pain, etc. [102,103]. Therefore heavy metal toxicity is an issue of great concern as these are nonbiodegradable, and they also exhibit high potential for bioaccumulation [100,104,105]. Accumulation of HMIs in the water bodies has an adverse effect on flora and fauna and also has a negative impact on the health of animals and humans who are consuming that contaminated food. The World Health Organization (WHO) and the United States Environmental Protection Agency (USEPA) determined some safe permissible limits of HMIs in water, as presented in Table 2.1. Table 2.1 also shows the sources of HMIs and their negative impacts on human health.

2.10 Radioactive substances

Anthropogenic and natural activities mostly cause radioactive or radiological water pollution. Anthropogenic activities include nuclear waste disposal, testing of nuclear weapons, nuclear power plant emissions, spillage from nuclear fuel mining, and manufacturing and use of radioactive sources [106,107]. It is also generated from the mining and processing of ores, which are used in agriculture, industrial activities, research, and medicine. Radiological pollution is an environmental threat due to radioactive decay which emits harmful radioactive radiation, such as alpha, beta, and gamma rays or neutrons. The amount of risk from radiological pollution was determined, based on the concentration of pollutant, the proximity of the pollutant to the body and the type and energy of radiation emission. ^{235}U, ^{128}Pd, ^{137}Cs, ^{89}Sr, ^{65}Zn, ^{59}Fe, and ^{57}Co are common examples of radioactive pollutants in water. Exposure to radioactive radiation might increase the threat of damage to cells, tissues, DNA and other vital molecules. The toxicological effects can lead to genetic mutations, several life-threatening diseases, such as cancer, hemorrhage, anemia, premature aging, premature death, and cardiovascular complications [107,108].

2.11 Inorganic salts

Several inorganic salts or anions, such as nitrate, ammonium, phosphate, and sulfate salts, contribute to the water pollution and deteriorate the water quality. Ammonia and ammonium salts are primarily used in agriculture (as fertilizer), in industries (manufacturing of plastics, papers, rubber etc.), in metal processing (as coolant), and as cleansing agents and food additives [109,110]. Ammonium nitrate is widely used in explosive materials, civil construction, and mining [111]. Excess of ammonium salts in the body results in nervous system dysfunction, kidney damage, reproduction toxicity, embryotoxicity, acidosis, and lung edema [112].

Phosphate and nitrate salts are important nutrients for physiological processes in living bodies. However, their presence in water at levels higher than the recommended limit promotes water pollution. Sewage, chemical fertilizers, waste dumped by industry, and textile and pharmaceutical effluents increase the level of nitrate contamination in water bodies [113,114]. The maximum concentration limit of nitrate-nitrogen was set by the United States and the WHO at 10 and 11.3 mg/L, respectively [115]. Consumption of nitrate-contaminated water is threating to the lives of newborns and pregnant women. It causes a blue baby syndrome of methemoglobinemia and also affects the fetus by creating oxygen deficiency which leads to abortion [116]. Prolonged exposure to nitrate-contaminated water can also cause gastrointestinal and prostate cancer [115]. Similarly, phosphate ion contamination can occur naturally, from the decomposition of rocks, minerals, erosion, and agriculture, or by human activities, such as industrial and sewage discharges. Although it is an essential nutrient up to the permissible limit, it can also cause kidney damage and osteoporosis when the limit in drinking water is exceeded [117]. Sulfate salts discharged from mining, electroplating, nonferrous smelting, and steel pickling into water bodies are also a concerning industrial wastewater pollution issue [118,119]. High sulfate concentration in water increases the acidity of water which affects the taste and causes diarrhea [120]. Additionally, contaminated sulfate water also causes erosion of metal parts of delivery containers, heat exchangers, and boilers [121].

2.12 Conclusion

Water pollution is a severe issue which is adversely affecting every kind of living being on this planet. It stems from different sources and contaminants. Industrial, domestic, pharmaceutical, municipal, and agricultural effluents are the significant sources of water pollution. Water contaminants generally include organic dyes, pharmaceutical and personal care ingredients, pesticides, herbicides, PAHs, oil spills, heavy metals, radioactive substances, and inorganic salts. These pollutants deteriorate the quality of water even in exceptionally low concentration and may have hazardous effects on human health, animals, plants, and aquatic organisms. Thus the removal of these organic and inorganic toxins from wastewater/industrial effluent is of the utmost importance. This chapter has classified all water contaminants along with their hazardous effects.

References

[1] Benelli R, Barbero A, Ferrini S, Scapini P, Cassatella M, Bussolino F, et al. Human immunodeficiency virus transactivator protein (Tat) stimulates chemotaxis, calcium mobilization, and activation of human polymorphonuclear leukocytes: implications for tat-mediated pathogenesis. J Infect Dis 2000;182:1643−51.
[2] Kumar M., Puri A. A review of permissible limits of drinking water, Indian J Occup Environ Med. 2012;16:40−44.

[3] Gautam R., Sharma S., Mahiya S., Chattopadhyaya M. Contamination of heavy metals in aquatic media: transport, toxicity and technologies for remediation, In: Heavy metals in water: presence, removal safety. London: RSC Publishers; 2015;ISBN 978-1-84973-885-9.

[4] Yagub MT, Sen TK, Afroze S, Ang HM. Dye and its removal from aqueous solution by adsorption: a review. Adv Colloid Interface Sci 2014;209:172−84.

[5] Sen TK, Afroze S, Ang H. Equilibrium, kinetics and mechanism of removal of methylene blue from aqueous solution by adsorption onto pine cone biomass of *Pinus radiata*. Water Air Soil Pollut 2011;218:499−515.

[6] Kumar N, Mittal H, Parashar V, Ray SS, Ngila JC. Efficient removal of rhodamine 6G dye from aqueous solution using nickel sulphide incorporated polyacrylamide grafted gum karaya bionano-composite hydrogel. RSC Adv 2016;6:21929−39.

[7] Sokolowska-Gajda J, Freeman HS, Reife A. Synthetic dyes based on environmental considerations. Part 2: Iron complexes formazan dyes. Dye Pigment 1996;30:1−20.

[8] Kumar N, Reddy L, Parashar V, Ngila JC. Controlled synthesis of microsheets of ZnAl layered double hydroxides hexagonal nanoplates for efficient removal of Cr(VI) ions and anionic dye from water. J Environ Chem Eng 2017;5:1718−31.

[9] Sharma VK, Anquandah GA, Yngard RA, Kim H, Fekete J, Bouzek K, et al. Nonylphenol, octyl-phenol, and bisphenol-A in the aquatic environment: a review on occurrence, fate, and treatment. J Environ Sci Health Part A 2009;44:423−42.

[10] Lade HS, Waghmode TR, Kadam AA, Govindwar SP. Enhanced biodegradation and detoxification of disperse azo dye Rubine GFL and textile industry effluent by defined fungal-bacterial consortium. Int Biodeterior Biodegrad 2012;72:94−107.

[11] Kadirvelu K, Kavipriya M, Karthika C, Radhika M, Vennilamani N, Pattabhi S. Utilization of various agricultural wastes for activated carbon preparation and application for the removal of dyes and metal ions from aqueous solutions. Bioresour Technol 2003;87:129−32.

[12] Salleh MAM, Mahmoud DK, Karim WAWA, Idris A. Cationic and anionic dye adsorption by agricultural solid wastes: a comprehensive review. Desalination. 2011;280:1−13.

[13] Gregory P. Classification of dyes by chemical structure. In: Waring DR, Hallas G, editors. The chemistry and application of dyes. Boston, MA: Springer US; 1990. p. 17−47.

[14] Nam S, Renganathan V. Non-enzymatic reduction of azo dyes by NADH. Chemosphere. 2000;40:351−7.

[15] Chen B-Y. Understanding decolorization characteristics of reactive azo dyes by *Pseudomonas luteola*: toxicity and kinetics. Process Biochem 2002;38:437−46.

[16] Wambuguh D, Chianelli RR. Indigo dye waste recovery from blue denim textile effluent: a by-product synergy approach. N J Chem 2008;32:2189−94.

[17] Sousa MM, Miguel C, Rodrigues I, Parola AJ, Pina F, de Melo JSS, et al. A photochemical study on the blue dye indigo: from solution to ancient Andean textiles. Photochem Photobiol Sci 2008;7:1353−9.

[18] Jabs CFI, Drutz HP. The role of intraoperative cystoscopy in prolapse and incontinence surgery. Am J Obstet Gynecol 2001;185:1368−73.

[19] Campos R, Kandelbauer A, Robra KH, Cavaco-Paulo A, Gübitz GM. Indigo degradation with purified laccases from *Trametes hirsuta* and *Sclerotium rolfsii*. J Biotechnol 2001;89:131−9.

[20] Barka N, Assabbane A, Nounah A, Ichou YA. Photocatalytic degradation of indigo carmine in aqueous solution by TiO_2-coated non-woven fibres. J Hazard Mater 2008;152:1054−9.

[21] Jenkins CL. Textile dyes are potential hazards. J Environ Health 1978;40:256−63.

[22] Ikeda K, Sannohe Y, Araki S, Inutsuka S. Intra-arterial dye method with vasomotors (PIAD method) applied for the endoscopic diagnosis of gastric cancer and the side effects of indigo carmine. Endoscopy. 1982;14:119−23.

[23] Allen R.L. Color chemistry, 1st Edn, Springer, Island Press, 1971, ISB 978-1-4615-6665-6.

[24] Raue R, Corbett JF. Nitro and nitroso dyes. In: Ullmann's encyclopedia of industrial chemistry. Weinheim: Wiley-VCH Verlag GmbH & Co. KGaA, 2000. p. 273−290. https://onlinelibrary.wiley.com/doi/epdf/10.1002/14356007.a17_383.

[25] Gordon PF. Nontextile applications of dyes. In: Waring DR, Hallas G, editors. The chemistry and application of dyes. Boston, MA: Springer US; 1990. p. 381−406.

[26] Waring DR, Hallas G. The chemistry and application of dyes. Springer Science & Business Media; 2013.

[27] Daniel J. The excretion and metabolism of edible food colors. Toxicol Appl Pharmacol 1962;4:572—94.

[28] Bachman GB. Nitration of hydrocarbons and other organic compounds (topchiev, A. V.). J Chem Educ 1960;37:A176.

[29] Qin A, Lam JWY, Tang L, Jim CKW, Zhao H, Sun J, et al. Polytriazoles with aggregation-induced emission characteristics: synthesis by click polymerization and application as explosive chemosensors. Macromolecules. 2009;42:1421—4.

[30] Dellarco VL, Prival MJ. Mutagenicity of nitro compounds in *Salmonella typhimurium* in the presence of flavin mononucleotide in a preincubation assay. Environ Mol Mutagenesis 1989;13:116—27.

[31] Roy Choudhury AK. 2 - Dyeing of synthetic fibres. In: Clark M, editor. Handbook of textile and industrial dyeing. Woodhead Publishing; 2011. p. 40—128.

[32] dos Santos AB, Bisschops IAE, Cervantes FJ, van Lier JB. The transformation and toxicity of anthraquinone dyes during thermophilic (55°C) and mesophilic (30°C) anaerobic treatments. J Biotechnol 2005;115:345—53.

[33] Ding F, Zhang L, Diao J-X, Li X-N, Ma L, Sun Y. Human serum albumin stability and toxicity of anthraquinone dye alizarin complexone: an albumin—dye model. Ecotoxicol Environ Saf 2012;79:238—46.

[34] Wang Q, Gao F, Yuan X, Li W, Liu A, Jiao K. Electrochemical studies on the binding of a carcinogenic anthraquinone dye, Purpurin (C.I. 58 205) with DNA. Dye Pigment 2010;84:213—17.

[35] Venturini S, Tamaro M. Mutagenicity of anthraquinone and azo dyes in Ames' *Salmonella typhimurium* test. Mutation Res Genetic Toxicol 1979;68:307—12.

[36] Srivastava S, Sinha R, Roy D. Toxicological effects of malachite green. Aquat Toxicol 2004;66:319—29.

[37] Culp SJ, Beland FA. Malachite green: a toxicological review. J Am Coll Toxicol 1996;15:219—38.

[38] NYEJC. Treating pesticide-contaminated wastewater: development and evaluation of a system. ACS Publications; 1984.

[39] Eskenazi B, Bradman A, Castorina R. Exposures of children to organophosphate pesticides and their potential adverse health effects. Environ Health Perspect 1999;107:409—19.

[40] Garcia FP, Ascencio SYC, Oyarzun JG, Hernandez AC, Alavarado PV. Pesticides: classification, uses toxic measures exposure genotoxic risks. J Res Environ Sci Toxicol 2012;1:279—93.

[41] Pang G-F. Chapter 1 - Introduction. In: Pang G-F, editor. Analytical methods for food safety by mass spectrometry. Academic Press; 2018. p. 1—9.

[42] Story P, Cox M. Review of the effects of organophosphorus and carbamate insecticides on vertebrates. Are there implications for locust management in Australia? Wildl Res 2001;28:179—93.

[43] Jayaraj R, Megha P, Sreedev P. Organochlorine pesticides, their toxic effects on living organisms and their fate in the environment. Interdiscip Toxicol 2016;9:90—100.

[44] Roy JR, Chakraborty S, Chakraborty TR. Estrogen-like endocrine disrupting chemicals affecting puberty in humans—a review. Med Sci Monit 2009;15:RA137—45.

[45] Boldt TS, Jacobsen CS. Different toxic effects of the sulfonylurea herbicides metsulfuron methyl, chlorsulfuron and thifensulfuron methyl on fluorescent pseudomonads isolated from an agricultural soil. FEMS Microbiol Lett 1998;161:29—35.

[46] Saeki M, Toyota K. Effect of bensulfuron-methyl (a sulfonylurea herbicide) on the soil bacterial community of a paddy soil microcosm. Biol Fertil soils 2004;40:110—18.

[47] Lin X-Y, Wang Y, Wang H-L, Chirko T, Ding H-T, Zhao Y-H. Isolation and characterization of a bensulfuron-methyl-degrading strain L1 of bacillus. Pedosphere 2010;20:111—19.

[48] Brown HM. Mode of action, crop selectivity, and soil relations of the sulfonylurea herbicides. Pesticide Sci 1990;29:263—81.

[49] Curren MSS, King JW. Solubility of triazine pesticides in pure and modified subcritical water. Anal Chem 2001;73:740—5.

[50] Jaga K, Dharmani C. The epidemiology of pesticide exposure and cancer: a review. Rev Environ Health 2005;15.

[51] Janke D, Kearney PC, Kaufman DD. Herbicides—chemistry, degradation, and mode of action (Eds) J Basic Microbiol 1989;29:718 Volume 3. XIII + 403 S. New York—Basel 1988. Marcel Dekker Inc. ISBN: 0-8247-7804-9.

[52] Kümmerer K. Pharmaceuticals in the environment. Annu Rev Environ Resour 2010;35:57—75.

[53] Solanki A, Boyer TH. Pharmaceutical removal in synthetic human urine using biochar. Environ Sci Water Res Technol 2017;3:553—65.

[54] Khannanov A, Nekljudov VV, Gareev B, Kiiamov A, Tour JM, Dimiev AM. Oxidatively modified carbon as efficient material for removing radionuclides from water. Carbon. 2017;115:394—401.

[55] Bendz D, Paxeus NA, Ginn TR, Loge FJ. Occurrence and fate of pharmaceutically active compounds in the environment, a case study: Höje River in Sweden. J Hazard Mater 2005;122:195—204.

[56] Ternes TA. Occurrence of drugs in German sewage treatment plants and rivers. Water Res 1998;32:3245—60.

[57] Nautiyal V, Sharma B, Negi V, Aswal R, Singh P, Singh R, et al. Pharmaceutical compounds in drinking water. J Xenobiotics 2016;6.

[58] Juliano C, Magrini GA. Cosmetic ingredients as emerging pollutants of environmental and health concern a mini-review. Cosmetics 2017;4:11.

[59] Brausch JM, Rand GM. A review of personal care products in the aquatic environment: environmental concentrations and toxicity. Chemosphere. 2011;82:1518—32.

[60] Aranaz I, Acosta N, Civera C, Elorza B, Mingo J, Castro C, et al. Cosmetics and cosmeceutical applications of chitin, chitosan and their derivatives. Polym (Basel) 2018;10:213.

[61] Bulloch DN, Nelson ED, Carr SA, Wissman CR, Armstrong JL, Schlenk D, et al. Occurrence of halogenated transformation products of selected pharmaceuticals and personal care products in secondary and tertiary treated wastewaters from Southern California. Environ Sci Technol 2015;49:2044—51.

[62] Song JY, Jhung SH. Adsorption of pharmaceuticals and personal care products over metal-organic frameworks functionalized with hydroxyl groups: quantitative analyses of H-bonding in adsorption. Chem Eng J 2017;322:366—74.

[63] Ternes TA, Joss A, Siegrist H. Peer reviewed: scrutinizing pharmaceuticals and personal care products in wastewater treatment. Environ Sci Technol 2004;38:392A—399AA.

[64] Zulaikha S, Syed Ismail S, Praveena S. Hazardous ingredients in cosmetics and personal care products and health concern: a review. Public Health Res 2015;2015:7—15.

[65] Campo J, Masiá A, Picó Y, Farré M, Barceló D. Distribution and fate of perfluoroalkyl substances in Mediterranean Spanish sewage treatment plants. Sci Total Environ 2014;472:912—22.

[66] Ramos S, Homem V, Alves A, Santos L. A review of organic UV-filters in wastewater treatment plants. Environ Int 2016;86:24—44.

[67] Díaz-Cruz MS, García-Galán MJ, Guerra P, Jelic A, Postigo C, Eljarrat E, et al. Analysis of selected emerging contaminants in sewage sludge. TrAC Trends Anal Chem 2009;28:1263—75.

[68] Chen Y, Vymazal J, Březinová T, Koželuh M, Kule L, Huang J, et al. Occurrence, removal and environmental risk assessment of pharmaceuticals and personal care products in rural wastewater treatment wetlands. Sci Total Environ 2016;566:1660—9.

[69] Fang J-L, Stingley RL, Beland FA, Harrouk W, Lumpkins DL, Howard P. Occurrence, efficacy, metabolism, and toxicity of triclosan. J Environ Sci Health Part C 2010;28:147—71.

[70] Lin D, Zhou Q, Xie X, Liu Y. Potential biochemical and genetic toxicity of triclosan as an emerging pollutant on earthworms (*Eisenia fetida*). Chemosphere. 2010;81:1328—33.

[71] Snyder EH, O'Connor GA, McAvoy DC. Toxicity and bioaccumulation of biosolids-borne triclocarban (TCC) in terrestrial organisms. Chemosphere. 2011;82:460—7.

[72] Halden RU, Paull DH. Co-occurrence of triclocarban and triclosan in US water resources. Environ Sci Technol 2005;39:1420—6.

[73] Babich H, Davis DL. Phenol: a review of environmental and health risks. Regul Toxicol Pharmacology 1981;1:90—109.

[74] Hsiao P-K, Lin Y-C, Shih T-S, Chiung Y-M. Effects of occupational exposure to 1,4-dichlorobenzene on hematologic, kidney, and liver function. Int Arch Occup Environ Health 2009;82:1077—85.

[75] Darbre PD, Byford JR, Shaw LE, Hall S, Coldham NG, Pope GS, et al. Oestrogenic activity of benzylparaben. J Appl Toxicol 2003;23:43−51.

[76] Daghrir R, Dimboukou-Mpira A, Seyhi B, Drogui P. Photosonochemical degradation of butyl-paraben: optimization, toxicity and kinetic studies. Sci Total Environ 2014;490:223−34.

[77] Soni M, Taylor SL, Greenberg N, Burdock G. Evaluation of the health aspects of methyl paraben: a review of the published literature. Food Chem Toxicol 2002;40:1335−73.

[78] Kim S, Choi K. Occurrences, toxicities, and ecological risks of benzophenone-3, a common component of organic sunscreen products: a mini-review. Environ Int 2014;70:143−57.

[79] Molins-Delgado D, Díaz-Cruz MS, Barceló D. Ecological risk assessment associated to the removal of endocrine-disrupting parabens and benzophenone-4 in wastewater treatment. J Hazard Mater 2016;310:143−51.

[80] Baird WM, Hooven LA, Mahadevan B. Carcinogenic polycyclic aromatic hydrocarbon-DNA adducts and mechanism of action. Environ Mol Mutagenesis 2005;45:106−14.

[81] Mukwevho N, Fosso-Kankeu E, Waanders F, Kumar N, Ray SS, Yangkou Mbianda X. Photocatalytic activity of $Gd_2O_2CO_3 \cdot ZnO \cdot CuO$ nanocomposite used for the degradation of phenanthrene. SN Appl Sci 2018;1:10.

[82] Masih J, Masih A, Kulshrestha A, Singhvi R, Taneja A. Characteristics of polycyclic aromatic hydrocarbons in indoor and outdoor atmosphere in the North central part of India. J Hazard Mater 2010;177:190−8.

[83] Abdel-Shafy HI, Mansour MSM. A review on polycyclic aromatic hydrocarbons: source, environmental impact, effect on human health and remediation. Egypt J Pet 2016;25:107−23.

[84] Cavallo D, Ursini CL, Rondinone B, Iavicoli S. Evaluation of a suitable DNA damage biomarker for human biomonitoring of exposed workers. Environ Mol Mutagenesis 2009;50:781−90.

[85] Page TJ, O'Brien S, Holston K, MacWilliams PS, Jefcoate CR, Czuprynski CJ. 7, 12-Dimethylbenz [a] anthracene-induced bone marrow toxicity is p53-dependent. Toxicol Sci 2003;74:85−92.

[86] Šepič E, Bricelj M, Leskovšek H. Toxicity of fluoranthene and its biodegradation metabolites to aquatic organisms. Chemosphere. 2003;52:1125−33.

[87] Peiffer J, Grova N, Hidalgo S, Salquèbre G, Rychen G, Bisson J-F, et al. Behavioral toxicity and physiological changes from repeated exposure to fluorene administered orally or intraperitoneally to adult male Wistar rats: a dose−response study. NeuroToxicology. 2016;53:321−33.

[88] Chen H, Zhang Z, Tian F, Zhang L, Li Y, Cai W, et al. The effect of pH on the acute toxicity of phenanthrene in a marine microalgae Chlorella salina. Sci Rep 2018;8:17577.

[89] Stampfer MR, Bartholomew JC, Smith HS, Bartley JC. Metabolism of benzo[a]pyrene by human mammary epithelial cells: toxicity and DNA adduct formation. Proc Natl Acad Sci 1981;78:6251−5.

[90] Feo F, Pirisi L, Pascale R, Daino L, Frassetto S, Garcea R, et al. Modulatory effect of glucose-6-phosphate dehydrogenase deficiency on benzo(a)pyrene toxicity and transforming activity for in vitro-cultured Human Skin Fibroblasts. Cancer Res 1984;44:3419−25.

[91] Mojiri A, Zhou JL, Ohashi A, Ozaki N, Kindaichi T. Comprehensive review of polycyclic aromatic hydrocarbons in water sources, their effects and treatments. Sci Total Environ 2019;696:133971.

[92] Zhu L, Chen B, Wang J, Shen H. Pollution survey of polycyclic aromatic hydrocarbons in surface water of Hangzhou, China. Chemosphere. 2004;56:1085−95.

[93] Yu H. Environmental carcinogenic polycyclic aromatic hydrocarbons: photochemistry and phototoxicity. J Environ Sci Health Part C 2002;20:149−83.

[94] Petit P, Maître A, Persoons R, Bicout DJ. Lung cancer risk assessment for workers exposed to polycyclic aromatic hydrocarbons in various industries. Environ Int 2019;124:109−20.

[95] Srinivasan A, Viraraghavan T. Oil removal from water using biomaterials. Bioresour Technol 2010;101:6594−600.

[96] Webler T, Lord F. Planning for the human dimensions of oil spills and spill response. Environ Manag 2010;45:723−38.

[97] Jenssen BM. An overview of exposure to, and effects of, petroleum oil and organochlorine pollution in grey seals (Halichoerus grypus). Sci Total Environ 1996;186:109−18.

[98] Davidson CI, Phalen RF, Solomon PA. Airborne particulate matter and human health: a review. Aerosol Sci Technol 2005;39:737−49.

[99] Aguilera F, Méndez J, Pásaro E, Laffon B. Review on the effects of exposure to spilled oils on human health. J Appl Toxicol 2010;30:291–301.

[100] Gumpu MB, Sethuraman S, Krishnan UM, Rayappan JBB. A review on detection of heavy metal ions in water – An electrochemical approach. Sens Actuators B Chem 2015;213:515–33.

[101] Oyaro N, Ogendi J, Murago EN, Gitonga E. The contents of Pb, Cu, Zn and Cd in meat in Nairobi, Kenya. J Food Agric Environ 2007;5(3).

[102] Naseem R, Tahir S. Removal of Pb (II) from aqueous/acidic solutions by using bentonite as an adsorbent. Water Res 2001;35:3982–6.

[103] Song S-T, Saman N, Johari K, Mat H. Removal of Hg (II) from aqueous solution by adsorption using raw and chemically modified rice straw as novel adsorbents. Ind Eng Chem Res 2013;52:13092–101.

[104] Joseph L, Jun B-M, Flora JRV, Park CM, Yoon Y. Removal of heavy metals from water sources in the developing world using low-cost materials: a review. Chemosphere. 2019;229:142–59.

[105] Vilela D, Parmar J, Zeng Y, Zhao Y, Sánchez S. Graphene-based microbots for toxic heavy metal removal and recovery from water. Nano Lett 2016;16:2860–6.

[106] Bo A, Sarina S, Liu H, Zheng Z, Xiao Q, Gu Y, et al. Efficient removal of cationic and anionic radioactive pollutants from water using hydrotalcite-based getters. ACS Appl Mater Interfaces 2016;8:16503–10.

[107] Li J, Wang X, Zhao G, Chen C, Chai Z, Alsaedi A, et al. Metal–organic framework-based materials: superior adsorbents for the capture of toxic and radioactive metal ions. Chem Soc Rev 2018;47:2322–56.

[108] Yang S, Zong P, Hu J, Sheng G, Wang Q, Wang X. Fabrication of β-cyclodextrin conjugated magnetic HNT/iron oxide composite for high-efficient decontamination of U (VI). Chem Eng J 2013;214:376–85.

[109] Timmer B, Olthuis W, van den Berg A. Ammonia sensors and their applications – a review. Sens Actuators B Chem 2005;107:666–77.

[110] Hignett TP. Ammonium salts, nitric acid, and nitrates. In: Hignett TP, editor. Fertilizer manual. Dordrecht: Springer Netherlands; 1985. p. 83–108.

[111] Chaturvedi S, Dave PN. Review on thermal decomposition of ammonium nitrate. J Energetic Mater 2013;31:1–26.

[112] Webb D, Bartley E, Meyer R. A comparison of nitrogen metabolism and ammonia toxicity from ammonium acetate and urea in cattle. J Anim Sci 1972;35:1263–70.

[113] Shrimali M, Singh K. New methods of nitrate removal from water. Environ Pollut 2001;112:351–9.

[114] Singh KP, Singh VK, Malik A, Basant N. Distribution of nitrogen species in groundwater aquifers of an industrial area in alluvial Indo-Gangetic Plains—a case study. Environ Geochem Health 2006;28:473–85.

[115] Ward MH, Jones RR, Brender JD, De Kok TM, Weyer PJ, Nolan BT, et al. Drinking water nitrate and human health: an updated review. Int J Environ Res Public Health 2018;15:1557.

[116] Control CfD, Prevention. Spontaneous abortions possibly related to ingestion of nitrate-contaminated well water—LaGrange County, Indiana, 1991-1994. MMWR Morbidity Mortal Wkly Rep 1996;45:569.

[117] Tchounwou PB, Yedjou CG, Patlolla AK, Sutton DJ. Heavy metal toxicity and the environment. Exp Suppl 2012;101:133–64.

[118] Najib T, Solgi M, Farazmand A, Heydarian SM, Nasernejad B. Optimization of sulfate removal by sulfate reducing bacteria using response surface methodology and heavy metal removal in a sulfido-genic UASB reactor. J Environ Chem Eng 2017;5:3256–65.

[119] Sang P-L, Wang Y-Y, Zhang L-Y, Chai L-Y, Wang H-Y. Effective adsorption of sulfate ions with poly (m-phenylenediamine) in aqueous solution and its adsorption mechanism. Trans Nonferrous Met Soc China 2013;23:243–52.

[120] Silva AM, Lima RMF, Leão VA. Mine water treatment with limestone for sulfate removal. J Hazard Mater 2012;221-222:45–55.

[121] Silva R, Cadorin L, Rubio J. Sulphate ions removal from an aqueous solution: I. Co-precipitation with hydrolysed aluminum-bearing salts. Miner Eng 2010;23:1220–6.

CHAPTER THREE

Water purification using various technologies and their advantages and disadvantages

3.1 Introduction

Water pollution is a serious issue worldwide and is gaining attention due to its severe harmful and adverse effects on the ecosystem and living beings. Availability of fresh and clean water is the major challenge in the growing modern society and is the necessity of all living beings to establish and maintain a healthy life [1]. Due to extensive anthropogenic activities such as at industrial, municipal, domestic households, and agricultural processes, lots of toxic contaminants enter into the groundwater or freshwater bodies, and mixed up in our food chain [2−5]. Consumption of these toxic contaminants, for example, Pb(II), Hg(II), organic dyes, antibiotics, pesticides, polycyclic aromatic, hydrocarbons, etc., even in trace amounts can be carcinogenic, mutagenic, neurotoxin, teratogenic, and damage several organs affecting the reproductive system, respiratory system, gastrointestinal tract, liver, etc. [6−10].

Direct or indirect discharge of polluted effluents into the freshwater reservoirs is therefore deteriorating the quality of freshwater and also contributing to water scarcity. This ongoing challenge of clean water scarcity increases the demand for sustainable, appropriate wastewater management systems and employing the good use of wastewater effluents treatment technologies before discharge. Wastewater treatment will overcome the water shortages by nonpotable and/or potable reuse of water after proper treatment. Various efforts have been carried out to improve the water management, for example, by spreading awareness regarding not wasting water and the discharge of treated effluents into the water streams; and by upgrading wastewater treatment technologies.

Advancement in wastewater treatment technologies is crucial to prevent the entry of any toxic contaminants into the surface water. Several techniques, such as adsorption, photocatalysis, photoelectrocatalysis, reverse osmosis (RO), membrane separation, flocculation, biological precipitation, electrochemical approaches, ion exchange, and desalination, have been used in water remediation [11−22]. Fig. 3.1 depicts a variety of sources of water pollution and current water purification techniques. However, each treatment method has limitations in terms of commercialization. Table 3.1 presents the different methods of water treatment and their advantages and disadvantages.

Carbon Nanomaterial-Based Adsorbents for Water Purification.
DOI: https://doi.org/10.1016/B978-0-12-821959-1.00003-9

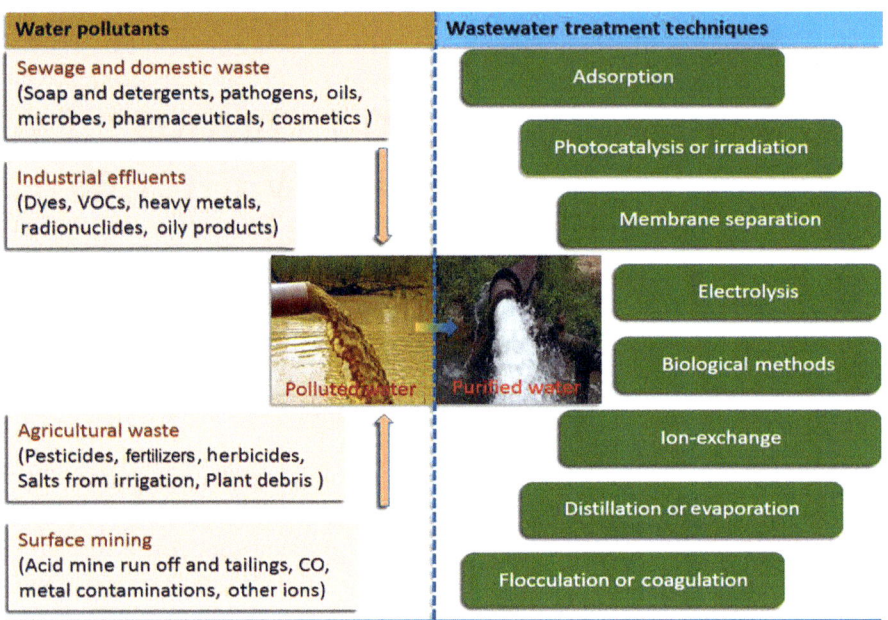

Figure 3.1 Sources of water pollution and wastewater treatment techniques [23]. *Copyright 2020, Elsevier Science Ltd.*

3.2 Wastewater treatment technologies

Sustainable wastewater management and treatment technologies are of growing interest to researchers and economists globally. A variety of conventional and nonconventional wastewater treatment technologies have been considered for the removal of toxic contaminants from water. This chapter primarily focuses on the evaluation and applicability of numerous wastewater technologies. Fig. 3.2 represents the flow chart of classification of wastewater treatment technologies based on physical, chemical, biological, and electrochemical treatment processes, which are further divided into other water purification methods.

3.3 Physical treatment processes

Physical processes are the most straightforward and low-cost methods to treat wastewater effluents. They do not involve the exchange of electrons between the water contaminant and other sacrificial agents. Physical wastewater treatment methods separate the pollutants from the water to produce potable water. Adsorption, filtration,

Table 3.1 Water purification methods and their advantages and disadvantages [23].

Separation methods	Advantages	Disadvantages
Aerobic degradation	Low operation cost, effective for removal of azo dyes	A slow process leads to the growth of microorganisms
Anaerobic degradation	Produced by-product act as energy sources	Require further treatment of by-products under aerobic conditions
Microbial treatment	Eco-friendly	Slow, difficult to standardize, scaling up
Biodegradation	Socially acceptable treatment, Eco-friendly	Slow, high maintenance, the requirement of optimal favourable environments
Advanced oxidation process	Small consumption of chemicals, no sludge production, effective for recalcitrant dyes	Technical constraints, Economically unfeasible, Formation of side-products
Photocatalysis	No sludge formation, foul odor reduces, Economically feasible	Sometimes the formation of toxic by-products, long duration, limited applications
Ozonation	No sludge formation, applied in a gaseous state	High operation cost, short half time
Electrochemical approach	No sludge formation, no or small chemical consumption	High operation cost
Ion-exchange	Effective, low adsorbent loss	Difficult to remove disperse dyes, economic constraints
Membrane filtration	High separation selectivity, efficient	Sludge formation, expensive, applicable at low scale volume, high pressures, complexity
Flocculation/ coagulation	Cost-effective, simple	The high amount of sludge production, usage, and disposal problems
Irradiation	Efficient oxidation at low volume scale	Need dissolved oxygen
Oxidation	Fast and efficient process	Chemicals required, high energy cost
Adsorption	Simple, efficiently remove all kind of pollutants, less energy required, economically feasible	Selectivity, recyclability, nondestructive process

Source: Copyright 2020, Elsevier Science Ltd.

and RO are the popular examples of physical water treatment methods. These processes do not perform any chemical changes in the quality of water.

3.3.1 Reverse osmosis

RO is a widely applied and cost-effective water treatment technology in water desalination and production of potable water. The term osmosis refers to the process in

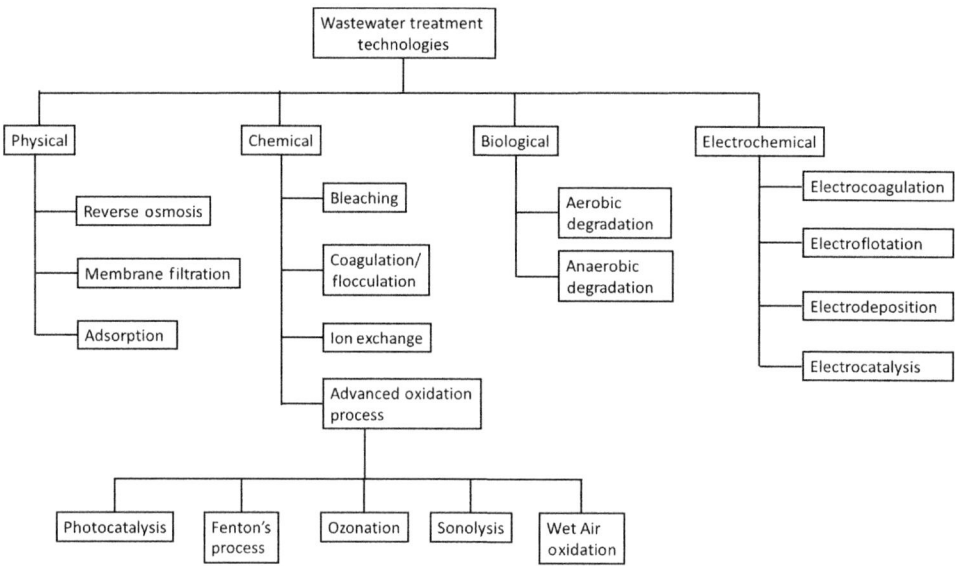

Figure 3.2 Classification of different wastewater technologies.

which water passes through the semipermeable membrane from low solute concentration to high solute concentration. However, RO works in the reverse direction under pressure. It is a pressure-driven technology in which a semipermeable membrane allows only water to pass through, and prevents the dissolved constituents present in the feed water into the pure water [24]. Therefore the recovered clean water is free from particulate matter, solute, endotoxins, and bacteria [25].

RO technology involves a semipermeable membrane to separate the aqueous solution into two phases: (1) concentrate or brine phase, which contains the salts and other compounds which need to be removed; and (2) permeate phase, which is the purified phase of water passing through the semipermeable membrane [25]. It does not require any heating, which does not harm the nutritional value and quality of the water and also does not impart any chemical or pH changes. The inhibition of any constituent to pass through the semipermeable membrane is based on the size and charge of the solute and the physicochemical interaction between the semipermeable membrane, solute, and solvent [26,27].

The RO system consists of prefilters (to remove all the undissolved impurities), pumps, pressure and flow controllers, semipermeable membranes, and monitors. Water is allowed to run in parallel to the membrane with a constant flow rate and pressure is applied to the water to go through the membrane. In a typical RO process, one portion of water is filtered and goes to the product water stream (permeate), whereas the remaining water (concentrate), which does not pass through the membrane is recollected and recycled.

The role of semipermeable membranes in RO technology is crucial to prohibit the passage of unwanted solute into the product water stream. There are different types of semipermeable membranes such as membranes made up of one polymeric material layer or composites with two or more layers. The functional groups present on the semipermeable membranes manipulate the valence and strength of the charge on the membrane, whereas hydrophobicity, roughness, and charge on the membrane decides the degree of adsorption of dissolved constituents present in water [28,29]. The advancement in RO technology is based on the progress in the improvements in the RO membranes, which play an essential role in determining the economic sustainability of the technology and purity of product water.

To improve the selectivity, permeability, mechanical, biological, and chemical stability of RO membrane, the modifications in the composition, structure, and morphology of membranes is acceptable. The current RO membrane in the market is either an asymmetric cellulose type or a thin-film composite (TFC) type of polyamide membranes. TFC polyamide membranes consist of three layers: (1) the first layer is of the polyester web (120−150 μm thick) which act as structural support; (2) a microporous 40 μm thick interlayer; and (3) an ultrathin upper surface layer or 0.2 μm [30]. The ultrathin barrier layer determines the selectivity of the membrane and usually consists of aromatic polyamide via interfacial polymerization [31]. Comparing the cellulose type membranes and TFC type membranes, the latter exhibit the superior qualities of salt rejection, biological stability, thermal stability, and resistance to pressure compaction [32].

Regardless of the many advantages of the RO process and advanced RO membranes, membrane fouling is the main limiting reason for the RO water purification technology [33]. Membrane fouling is caused by the adsorption of solute particles from the feed water on the membrane, which results in pore-clogging and flux decline which directly affect the water quality [34]. Microbial growth, dissolved organic elements (e.g., humic acid), inorganic substances (e.g., salts such as metal hydroxides or carbonates), scaling, and colloidal matter (suspended silica) in feed water are the responsible foulants for membrane fouling. Many efforts have been put forward to mitigate the membrane fouling issue in an RO system, such as pretreatment, designs of new membranes, and development of antifouling RO membranes [35−37]. Among all these, the development of antifouling membranes is the most popular and fundamental method to improve RO efficiency by reducing membrane fouling. A lot of research is going on into the development of manufacturing of antifouling RO membranes [38−41]. By tuning some physicochemical characteristics of RO membranes, such as hydrophilicity, electrostatic charge, steric repulsion, and roughness (Fig. 3.3), the antifouling properties of membranes can be enhanced.

Many foulant molecules are hydrophobic and do not attach to a hydrophilic surface. Therefore on increasing the hydrophilicity of RO membranes, a pure water

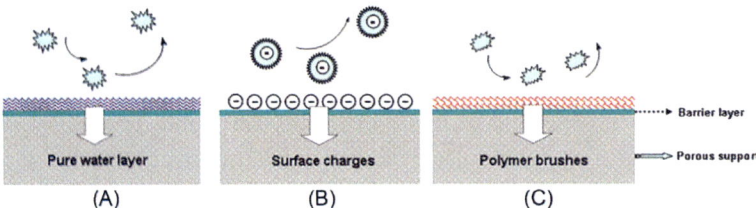

Figure 3.3 Antifouling mechanism of RO membranes by tuning the (A) hydrophilicity, (B) electrostatic charge, and (C) steric repulsion [33]. *Copyrights 2012, Elsevier Science Ltd.*

layer can easily be formed on the surface of highly hydrophilic membranes (Fig. 3.3A). This practice avoids the deposition of hydrophobic foulant molecules on the RO membrane surface, hence preventing the fouling event. However, this exercise allows the deposition of hydrophilic foulant molecules on the surface and causes membrane fouling.

Designing the surface charge of RO membranes according to foulants is another way to improve the antifouling properties of membrane. Electrostatic repulsion force between the foulant and membrane (Fig. 3.3B) is an advantageous technique to reduce the membrane fouling. Nevertheless, these membranes can easily get fouled by the oppositely charged foulants. Another antifouling technique is to modify the surface of membranes with long-chain hydrophilic molecules brushes. This will prevent the deposition of macromolecules, such as protein due to steric repulsion (Fig. 3.3C) [42]. Another technique to reduce membrane fouling is the manufacture of RO membranes with a smoother surface. Such surfaces experience less fouling because it is expected that foulant molecules are more likely to get deposited on the rough topologies than smoother surface [43]. However, smoothening of the RO membrane surface can be disadvantageous to membrane flux [44]. RO is the leading water desalination technology in the current market but the fouling of membranes cannot be avoided, although it can be delayed with the pretreatment and designing of the membranes.

3.3.2 Membrane filtration

A thin physical interface, which controls the passage of certain species from one part of the system to another, depending on their physicochemical properties, is termed as membrane [45]. A membrane filtration method is used to remove the undissolved or suspended particles from the water feed stream using the membranes of specific pore sizes. The difference in the RO system and membrane filtration is that the former separates the dissolved impurities, whereas the latter removes the undissolved particulates from the water [46]. Membrane filtration is usually employed for the pretreatment of water feed before other water treatment [47]. Ultrafiltration (UF) and microfiltration (MF) are the two embodiments of the membrane filtration. UF and MF separate the unwanted particles from the water by sieving through the pores in

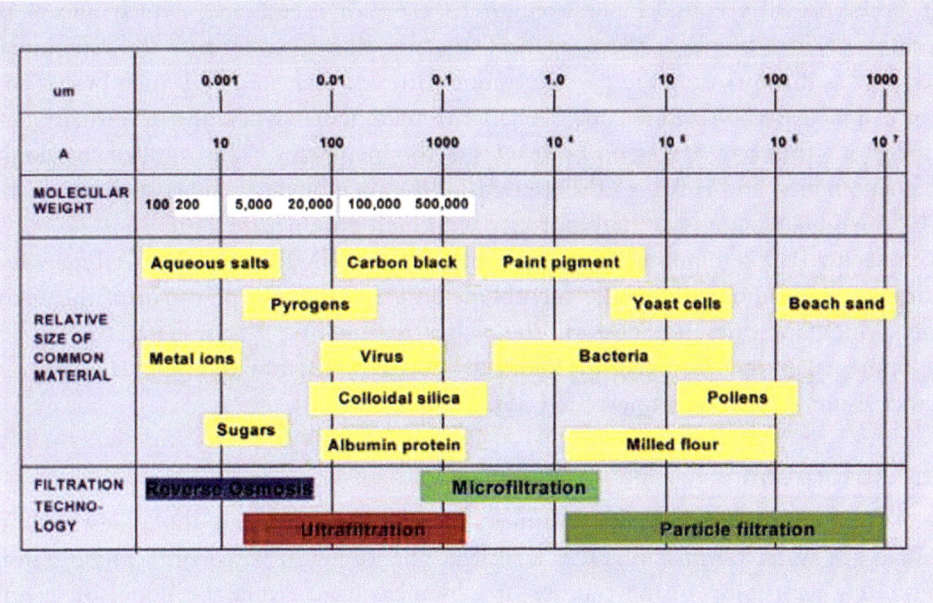

Figure 3.4 Filtration spectrum according to the size of impurities in water feed stream [46]. *Copyrights 2007, Elsevier Science Ltd.*

the membrane. Particles larger than the pore sizes of the membrane are prohibited by the membrane surface from passing through and remain in the feed side of the system.

Fig. 3.4 illustrates the separation spectrum using the membranes of various pore sizes from RO to particle filtration [46]. It expressed the pore sizes of the membranes used in different membrane filtration methods and the targeted species, which can be filtered through those pores. UF membranes exhibit mesopores and generally target the finest particles such as virus, colloidal silica, and albumin proteins to remove from the water. UF has potential application in various membrane separation processes, such as wastewater treatment, recovery of surfactants after use in industries, protein separation, food processing, and gene engineering [48]. On the other hand, MF membranes consist of macropores and are used to filter moderately larger particulates such as microbes, oil emulsion, and bacteria from water [49]. UF and MF membranes generally comprise inorganic materials (ZrO_3, TiO_2, Al_2O_3, etc.), cellulose derivatives, and polymer species [e.g., polyethersulfone (PES), polyacrylonitrile (PAN), polyvinylidene fluoride (PVDF)] [50−52].

The difference in UF and MF membranes is the nature of their porosity and pore size. MF membranes display the homogenous pore structure with a good symmetry of pore sizes and narrow pore size distribution [53]. However, the UF membrane exhibits a high order of asymmetry with a thin layer of active finer pores. An active thin layer of finer pores provides considerable strength to the UF membrane, which allows the supportive layer to be much more open. The active thin layer controls the resistance in water flow as it allows only the passage of fine particles. Generally, UF membranes consist of a thin active layer only one side of the membrane, whereas a few membranes comprise of a thin active layer on both sides of the membrane. This is called the double skin phenomenon and enhances the strength of the membrane at the cost of permeability. UF membranes consist of high porosity with high pore density [46].

Similar to RO semipermeable membranes, UF and MF also suffer fouling which restricts the continuous use of the membrane and decreases the life span of the membrane, which increases the cost of the water treatment process. Therefore several developments in the manufacturing of UF and MF membranes have occurred to enhance fouling-resistant property [54,55].

3.3.3 Adsorption

Overall, adsorption is the most common, cheapest, fastest, and the oldest method employed in water purification. It is a surface phenomenon based on a phase transfer process, that is, it assists in the transfer of substances from either the liquid or gaseous phase onto a solid surface. Generally, it is an exothermic process. In general, the adsorbate (pollutant)-containing solution is shaken with the adsorbent, and the former adheres onto the surface of the adsorbent, via chemical or physicochemical interactions, under specific conditions, until equilibrium is established. Adsorption is further divided into two types, based on the interactions between the adsorbent and adsorbate: chemisorption and physisorption. Chemisorption is an irreversible process, and the driving factor is strong chemical interactions via the exchange of electrons between the adsorbate and the adsorbent surface [56].

In contrast, physisorption is a reversible process, and the forces involved are weak intermolecular physical forces such as $\pi-\pi$ interactions, dipole–dipole interactions, van der Waals forces, and hydrogen bonding, between the adsorbate and adsorbent [22,57]. Several factors, including the surface area of the adsorbent, temperature, pH, interactions between adsorbate and adsorbent, contact time, adsorbent dosage, etc., significantly influence the adsorption efficiency and contribute to the removal of hazardous substances from effluent [22,58]. The details of the adsorption processes and the application of several adsorbents for the removal of toxic water contaminants are described in other forthcoming chapters.

3.4 Chemical treatment processes

Chemical wastewater treatment methods involve the electron exchange between the water contaminants and another introduced chemical into the wastewater.

Such methods not only separate the water pollutants from the water but also aid to mineralize them into CO_2, H_2O, and other innocuous small molecules. Chemical treatment methods are generally combined with physical processes to obtain enhanced water pollutant removal efficiency. For example, adsorption of the water contaminant on the disinfectant chemical or catalyst used in the chemical treatment method is the essential step. Bleaching, coagulation, flocculation, ion-exchange, and advanced oxidation processes (AOPs) are the common examples of chemical wastewater treatment methods.

3.4.1 Bleaching

Bleaching or chlorination is the most commonly used chemical treatment method for the disinfection and sanitization of water at the industrial scale [59]. The efficiency of this process depends on the quality of feed water and the applied chlorination method. After Hurricane Katrine (2005) in the United States, which caused flooding and waterborne diseases, The US Environmental Protection Agency (USEPA) recommended bottled, boiled, and bleached water for consumption [60]. Chlorine (Cl_2) is a popular disinfectant for water treatment. However, the production, transport, and application of Cl_2 in gaseous or solution can be a safety hazard and the by-products formed during the process, such as chlorinated hydrocarbons, can be carcinogenic [61]. Chlorine dioxide (ClO_2) is an alternative preoxidant and bleaching agent for the disinfection of water [62]. The significant advantages of ClO_2 over Cl_2 is the more effectively inactivation of several protozoa (e.g., *Giardia* and *Cryptosporidium*) and ease of transportation and storage. Additionally, it does not produce any hazardous halogenated by-products under proper conditions [62]. In the gaseous state, ClO_2 is highly unstable and forms Cl_2 and O_2, whereas it is stable in an aqueous medium.

Appropriate dosing of ClO_2, to feed water provides a residual concentration, which keeps the drinking water safe from microbiological recontamination [62]. In Europe, the concentration of ClO_2 to treat the water is fixed near $<0.05-0.1$ mg/L, whereas this concentration is increased in the United States from 1 to 1.4 mg/L [63]. The reason for increased dosage of ClO_2 in the United States is that there it is not only used for bleaching activities but also used as a preoxidant for surface water. ClO_2 is a stable free radical and reacts as the electron acceptor. It reacts with the active groups on the contaminants. Huber et al. suggested the use of ClO_2 for the oxidation of pharmaceuticals present in wastewater [62]. The concentration of ClO_2 in feed water should not exceed the recommended dose, as higher dosages are likely to yield high levels of chlorite in water, which is against the USEPA chlorite standard levels (1 mg/L). Chlorite is produced as the reduction product of ClO_2 and is also considered as blood poison [64]. Additionally, ClO_2 has a chlorine-like odor and is irritable to the eyes, nose, and throat. Therefore the safety hazard and generation of halogenated by-products, for example, chloroform and chlorite, on using high concentration of bleaching agents limit the treatment of wastewater through bleaching.

3.4.2 Coagulation/flocculation

Coagulation/flocculation is an established chemical water treatment method using a solid—liquid separation technique to remove the organic, colloidal, suspended, and dissolved solids from wastewater. Fine suspended or colloidal particles do not settle down in the water and/or take an extremely long time to settle down to be removed. This process involves the addition of a chemical (coagulant/flocculant) into the feed water stream, which initiates the agglomeration of suspended or dissolved fine particles into the large particles (flocs) to settle down to the bottom and can easily be removed by sedimentation (Fig. 3.5) [65—67]. Therefore the term coagulation is generally used to describe the initial phase of the process: destabilization of suspension to form aggregations, whereas flocculation is used to express the later phase of the process: the formation of larger aggregate particles of the destabilized particles, which settle down and can be separated easily [68]. It is an easy, effective, and widely used technique to treat industrial wastewater and has already been employed to treat wastewater from various industries such as textile, pulp mill, palm oil mill [69—72]. Besides the main treatment of the feed water stream, coagulation/flocculation is also used for the pre-and post-treatment of water due to its versatility [73].

Two types of chemicals are used in coagulation/flocculation processes, that is, inorganic hydrolyzing metal coagulants, such as aluminum-based coagulants (alum, sodium aluminate, and aluminium chloride) and iron-based coagulants (ferric salts: ferric sulfate, ferrous sulfate, ferric chloride, and ferric chloride sulfate) [74], and organic polymeric flocculants [75]. Inorganic coagulants exhibit the limitations due to the demand of high dosages for effective wastewater treatment and narrow applications [75]. Similarly, synthetic polymeric organic flocculants are nonbiodegradable, which are

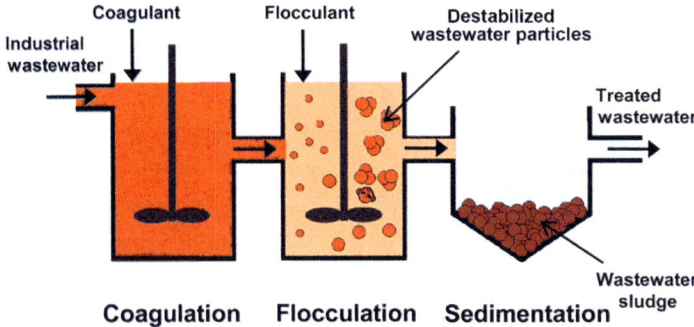

Figure 3.5 Schematic illustration of industrial wastewater treatment following coagulation/flocculation method [67]. *Copyrights 2016, American Chemical Society.*

harmful for the environment after use, and natural polymeric organic flocculants are costly and have a short shelf life [75]. To improve these drawbacks of coagulants and flocculants, nanohybrid materials of inorganic metal and organic polymers are gaining attention [75,76]. Compared to individual materials, synergetic effects of coagulant and flocculant in one hybrid material provide superior performance.

Nevertheless, coagulation/flocculation exhibits significant drawbacks which are sludge formation and the ineffective decolorization of the dissolved textile dyes [77,78]. Further, sludge treatment increases the cost of this process. However, few researchers are pretreating the wastewater to enhance the coagulation potentials and generate less sludge [79,80]. Additionally, the focus has also shifted onto the designing of suitable coagulants/flocculants to produce less sludge as a waste product or reusing the sludge to treat the wastewater [67,81].

3.4.3 Ion exchange

The ion-exchange method involves the reversible exchange of ions between the solid material (ion-exchanger or ion-exchange resin) and liquid phase [82]. In the ion-exchange process, an ionic water contaminant is replaced by a nonobjectionable or less-objectionable ion. The most commonly used ion-exchangers are zeolites, resins, and clay [83−85]. The general classification of ion exchangers is into cationic ion-exchangers, which can exchange cationic moieties, and anionic ion-exchangers, which can exchange anionic moieties. There is also another class of ion-exchangers which can simultaneously exchange the cations and anions and is known as an amphoteric ion-exchanger [86]. Ion-exchangers carry exchangeable ions and insoluble in water and organic solvents, and are added into the wastewater treatment plants to interchange the ions present in the water system to fulfil the purpose of demineralization. Ion exchangers exhibit a unique porous structure, which carries a cross-linked polymer matrix with various functional groups and positively or negatively charged ions [87]. An ideal ion-exchanger should have hydrophilicity, good thermal and chemical stability, be cost-effective, have a high surface area, high speed, and capacity of ion-exchange [82]. The dissolved contaminant in the water should exhibit a similar type of charge so that it can replace the existing ion from the ion-exchanger.

Water softening is the perfect example of ion exchange water treatment to replace Ca^{2+} and Mg^{2+} ions. The ion exchange method has potential in removing toxic ions from chemical industries and producing high-quality water. This process is widely used to remove many ions, such as fluoride, manganese, nitrates, radium, cadmium, lead, chromium, perchlorate, uranium, sulfates, iron, and so on, from the wastewater [88−93]. On introducing an ion-exchanger to a contaminated wastewater tank, the ions in it start to flow freely and are replaced by the counterions in the water with the same charge.

The advantage of this process is that ion-exchange resins are cheap, can easily be regenerated and recycled with high efficiency, and very small amounts of energy are required to carry out the process [82]. However, there are a few disadvantages of this process that limit its application, such as ion exchanger fouling, irreversible adsorption of organic matter on the solid surface of the ion-exchanger, bacterial contamination, and exchanged ion contamination.

3.4.4 Advanced oxidation processes

AOPs are a well-established and effective chemical method to treat wastewater via oxidation to produce potable water. AOPs can be described as an aqueous phase oxidation process and involve the in situ generation of powerful oxidants such as hydroxyl radicals ($\cdot$OH), which help to oxidize the contaminant molecules present in the wastewater [94,95]. It is gaining attention for the efficient removal of several types of pollutants from wastewater by either complete mineralization or transformation into more innocuous products [96−98]. Fast degradation rates and nonselective oxidation of contaminants are the advantages of the AOPs. AOPs include several methods, such as ozonation, photocatalysis, Fenton's process, and sonolysis, for the degradation of water contaminants [95,97,99]. The next subsection will briefly explain several types of AOPs.

3.4.4.1 Photocatalysis

Heterogeneous photocatalysis is most appealing heterogeneous advanced oxidation process and exhibits potential in various application fields such as environmental remediation, selective oxidation and reduction, and medicine [100−103]. It requires sunlight and a semiconductor photocatalyst of the appropriate bandgap to perform the reaction [104]. The difference between the valance band energy level and conduction band energy level in the photocatalytic material is known as an energy gap (E_g). A photocatalyst can absorb the photon, which comprises the energy equal to or higher than the bandgap of it. For efficient photocatalysis, the semiconductor material should absorb the visible spectrum of solar light. Most of the semiconductor materials are either unable to absorb visible light due to a too wide bandgap or show disadvantages of fast recombination of photogenerated charge carriers due to a too small bandgap. Modification of the semiconductor material helps in absorbing the visible light spectrum and delaying the recombination period by tuning the bandgap [105].

In a typical wastewater treatment via photocatalysis process, contaminated water with a suitable catalyst is placed under simulated or genuine solar light irradiation for a sufficient period so that all toxic molecules get mineralized. Natural sunlight is the renewable and reliable light source to excite the photocatalyst to initiate the reaction. With the semiconductor photocatalytic material under light irradiation, the electron gets excited and jumps from the valence band to the conduction band (e_{CB}^-) leaving behind holes (h_{VB}^+) at the valance band. If these photogenerated charge pairs, that is, electrons and holes, get trapped on the surface of the photocatalyst for a more extended recombination period, it can initiate the redox reaction to degrade the water contaminants. The general mechanism of the redox reaction on the surface of the photocatalyst can be expressed as follows (Eqs. 3.1−3.8):

$$\text{Photoexcitation: } h\nu + \text{Photocatalyst} \rightarrow e_{CB}^- + h_{VB}^+ \tag{3.1}$$

$$\text{Recombination:} e_{CB}^- + h_{VB}^+ \rightarrow \text{heat} \tag{3.2}$$

$$\text{Reduction:} O_2 + e_{CB}^- \rightarrow O_2^{-\bullet} \tag{3.3}$$

$$\text{Oxidation:} h_{VB}^+ + H_2O \rightarrow H^+ + HO^- \tag{3.4}$$

$$HO^- + h_{VB}^+ \rightarrow HO^{-\bullet} \tag{3.5}$$

$$h_{VB}^+ + \text{Water Contaminant} \rightarrow \text{Oxidation of water contaminant} \tag{3.6}$$

Photodegradation:

$$O_2^{-\bullet} + \text{Water Contaminant} \rightarrow \text{Intermediate Products} \rightarrow \text{Mineralization} \tag{3.7}$$

$$HO^{-\bullet} + \text{Water Contaminant} \rightarrow \text{Intermediate Products} \rightarrow \text{Mineralization} \tag{3.8}$$

Eq. (3.1) involves the photoexcitation of the photocatalyst and generation of photocharge pairs, that is, e_{CB}^- and h_{VB}^+ at the conduction and valence band, respectively. Early recombination of electron and hole produces the heat and suppresses the photocatalysis process, as photocharge carriers do not take part in the photocatalysis reaction (Eq. 3.2). However, an electron at the conduction band is available for reaction; it reacts with the oxygen molecule to produce a superoxide radical ($O_2^{-\bullet}$) (Eq. 3.3). At the valence band, holes react with a water molecule to produce a hydroxyl anion (Eq. 3.4), which again reacts with holes to produce a hydroxyl radical ($HO^{-\bullet}$) (Eq. 3.5). The photodegradation of the water contaminant is usually affected by the presence of three main active species, that is, h_{VB}^+, $O_2^{-\bullet}$, and $HO^{-\bullet}$ [104]. Holes at the valence band (h_{VB}^+) directly degrade the water contaminant molecule via oxidation. Photocatalysis reaction products ($O_2^{-\bullet}$ and $HO^{-\bullet}$) react with the water pollutants and degrade into mineralized products (CO_2, H_2O, etc.) via the formation of intermediate products. The rate and mechanism of the photocatalytic reaction are also influenced by the various operating parameters.

Fig. 3.6 represents the mechanism (Fig. 3.6A), contribution of various reactive species (Fig. 3.6B), and the effect of several operating parameters (Fig. 3.6C−E) on the rate of photodegradation of metoprolol (MTP) using TiO_2 nanotube arrays [106]. Fig. 3.6A depicts the schematic representation of the photodegradation mechanism and involvement of photogenerated electrons and holes in the degradation of MTP. On photoexcitation the generated holes at the valence band can directly either oxidize

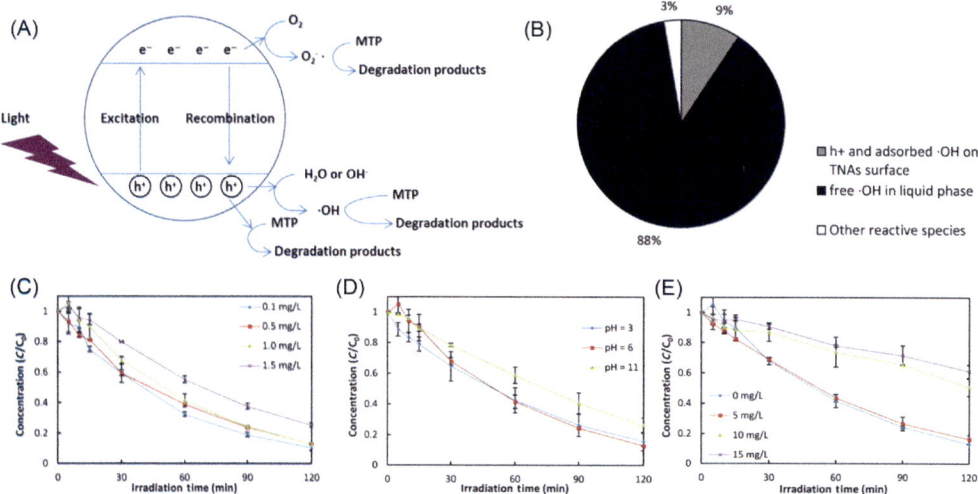

Figure 3.6 (A) Proposed photocatalytic degradation mechanism of metoprolol (MTP) using TiO_2 nanotubes. (B) Various reactive species contribution on photodegradation of MTP, here TNA refers to TiO_2 nanotube arrays. Effect of (C) initial concentration of MTP, (D) pH of the reaction medium, and (E) natural organic matter (NOM) concentration on the rate of MTP photodegradation [106]. *Reproduced and reprint with permission. Copyrights 2018, Elsevier Science Ltd.*

the MTP or generate the hydroxyl radical, which reduces the MTP into the degraded products. Additionally, photoelectrons react with oxygen molecules to produce superoxide radicals, which further react with MTP and produce innocuous degraded products. Out of all the reactive species, that is, h_{VB}^+, $O_2^{-\bullet}$, and $HO^{-\bullet}$, 88% of degradation was achieved by the presence of free hydroxyl radicals, 9% is due to the holes and adsorbed hydroxyl radicals on the TiO_2 nanotubes arrays, and the remaining 3% is due to other reactive species (Fig. 3.6B).

Fig. 3.6C—E presents the effect of the initial concentration of MTP, pH of the reaction medium, and natural organic matter (NOM) concentration on the rate of photodegradation of MTP. Increasing the initial concentration of MTP from 0.1 to 1.0 mg/L does not make a significant impact on the removal efficiency, whereas by further increasing the MTP concentration to 1.5 mg/L the rate of MTP removal decreases (Fig. 3.6C). At a constant concentration of photocatalyst and light irradiation, the number of reactive species was also constant, which can decrease the 1.0 mg/L concentration of MTP efficiently. However, by further increasing the contaminant concentration the shortage of reactive species decreases the degradation of MTP. The effect of pH on degradation rate of MTP was evaluated at acidic condition (pH = 3), moderate condition (pH = 6), and alkaline condition (pH = 11) (Fig. 3.6D). The removal efficiency of MTP decreases at higher pH. The pK_a value of MTP was found to be 9.64 [107], which suggests that at a higher pH (pH = 9), MTP becomes deprotonated and develops a negative charge. This acquires the electrostatic repulsion between the MTP and TiO_2 nanorods arrays, which pose a negative effect on the adsorption of MTP on TiO_2 and reduces photodegradation efficiency. To achieve a high rate of photocatalytic degradation of water pollutant, the adsorption of the pollutant on the catalyst surface should be high. Water bodies also consist of several NOMs, which also interfere in the photodegradation process. Fig. 3.6E shows that the presence of NOMs below 5 mg/L concentration does not make an impact on the rate of removal of MTP, whereas on further increasing to 10 or 15 mg/L, it significantly reduces the removal rate of MTP. NOMs are well-known reactive species quenchers, which decrease the accessibility of reactive species and hence decrease the MTP photodegradation. Therefore the optimization of the operational parameters is needed for enhanced photocatalytic efficiency.

3.4.4.2 Fenton's process

In Fenton's process of wastewater treatment, iron and H_2O_2 are used as catalyst and oxidant, respectively. Fenton's process is a commonly used AOP, which was first discovered by H.J.H. Fenton in 1894 [108]. It contains several advantages such as simple operating conditions (room temperature and atmospheric pressure), high performance, and nontoxicity [109,110]. During the reaction, H_2O_2 can easily break down into harmless species (H_2O and O_2) and does not produce any toxic by-products. This process can be divided into homogenous Fenton's AOP and heterogeneous Fenton's AOP. In homogenous reaction, the catalysis reaction can be performed in the whole liquid medium, whereas in heterogeneous reaction, the catalysis reaction can only be performed on the surface of the catalyst [110]. To perform higher catalytic degradation in heterogeneous Fenton's process, the diffusion and adsorption of oxidant and reactants should be higher on the catalyst surface [110].

The general mechanism of Fenton's process can be expressed as shown in Eq. (3.9) and Fig. 3.7.

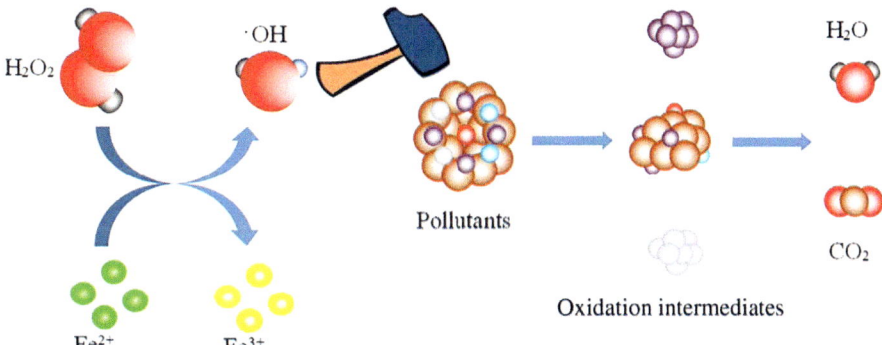

Figure 3.7 General mechanism of mineralization of water pollutants using Fenton's method [111]. *Copyright 2019, Elsevier Science Ltd.*

$$Fe(II) + H^+ + H_2O_2 \rightarrow Fe(III) + H_2O + HO^\bullet \tag{3.9}$$

The reaction of iron (Fe(II)) and H_2O_2, in the acidic medium produces hydroxyl radicals ($^\bullet$OH), which effectively degrade most of the organic pollutants into CO_2 and H_2O [112]. Based on this principle, it has been widely used to degrade a variety of water pollutants such as phenol, organic dyes, pharmaceuticals, pesticides [110,113,114]. However, Fenton's process also exhibits several disadvantages as well, such as high cost, requirements for a large amount of H_2O_2, the formation of iron sludge, limited operating pH range, and difficulty in recycling in the homogenous process [110,115−117]. To improve a few of the limitations, Fenton's process is modified into Fenton's-like process, which includes the replacement of the iron catalyst with another homo/heterogeneous catalysts such as Cu(II), pyrite, schorl, nano zero-valent iron, and metal ion−organic ligand complexes [110].

3.4.4.3 Ozonation
Ozone (O_3) is a powerful oxidizing agent (electrochemical oxidation potential ($E^\circ = 2.07$ V)) which can easily degrade most of the water contaminant species [118]. Oxidation of organic molecules in the presence of O_3 is known as ozonation. In this process, organic contaminants get oxidized in two ways: (1) direct oxidation of organic substance with O_3 molecule under acidic conditions; and (2) indirect oxidation by the reaction of pollutants with $^\bullet$OH, which is produced by the O_3 decomposition in alkaline conditions [119]. Direct oxidation with O_3, selectively oxidizes certain functional groups, anions and C=C bonds, whereas indirect oxidation with OH is faster and nonselective [120,121]. Owing to the capability of selective oxidation, ozonation is also used for the pretreatment of wastewater and combined with other treatment methods such as photocatalysis, adsorption, and so on [118,122,123].

3.4.4.4 Sonolysis
The sonolysis approach or sonochemical technique implies the use of physical impacts of ultrasound waves for the degradation of water contaminants. Sonolysis is a promising AOP, as it does not need any extra chemicals to degrade the organic pollutants. This process can decompose the organic contaminants in water either by direct thermolysis (Eq. 3.10) or by OH radical attack (Eqs. 3.11 and 3.12). Ultrasound sonication generates the compression and rarefaction cycles in the aqueous medium, which follow the formation and violent collapse of cavitation bubbles in a few microseconds. This event of formation and collapse of cavitation bubbles generates the local shock waves ($5000°C$ temperature and 500 atmospheric pressure; for few microseconds) [124,125]. Such incidents help in pyrolytic cleavage of the molecules and generate the radicals at the interface of bubbles and liquid bulk, which takes part in the chemical reactions for degradation of organic contaminants.

$$\text{Organic pollutant} +)))) \rightarrow \text{Degradation products} \tag{3.10}$$

$$H_2O +)))) \rightarrow HO^\bullet + H^\bullet \qquad (3.11)$$

$$HO^\bullet + \text{Organic pollutants} \rightarrow \text{Degradation products} \qquad (3.12)$$

where)))) signifies the ultrasound waves in wastewater.

However, sonolysis exhibits several drawbacks which limit its practical applications, such as high cost and noise pollution [126]. Additionally, to avoid such noise pollution, the additional costs of noise safety, for example, acoustic insulation of the reactors, is an extra barrier during economical treatment processes. Low mineralization efficiency and the high energy consumption are major disadvantages of using sonolysis for wastewater treatment [127]. Therefore the combination of sonolysis with other water treatment methods can promote the faster degradation of pollutants, for example, combination of sonolysis and Fenton's oxidation (sono-Fenton process) [126–128].

3.4.4.5 Wet air oxidation

Wet air oxidation (WAO) or thermal liquid-phase oxidation process is used to treat toxic organic and/or inorganic substances present in wastewater by oxidation in the presence of air or oxygen at high temperature and pressure [129]. In this typical AOP, normal air or gaseous oxygen is introduced into the wastewater at elevated temperature ($150°C - 400°C$) and pressure ($2-40$ MPa) [130]. During this reaction, active reaction species, such as hydroxyl radicals, are generated, which are well-known oxidants for the degradation of toxic substances. In this process, the contaminants present in water are either degraded partially into biodegradable intermediates or mineralized into H_2O, CO_2, and inorganic salts, without the emission of harmful SO_2, furans, NO_x, HCl, fly ashes, etc. [131]. It is a temperature-dependent technique. For example, the improvement in resorcinol removal in wastewater was noted to increase from 27% to 97.5% on increasing the temperature from $150°C$ to $230°C$ [132]. The main disadvantage of this process is that sometimes the complete mineralization of organic substances is not achieved using the WAO technique. This is because of the presence and accumulation of low-molecular-weight oxygenated compounds in the wastewater, which resist the further transformation of intermediates into CO_2. For example, acetic acid mineralization is almost negligible at a working temperature below $300°C$ [133]. Therefore the WAO process is also used as the pretreatment of the wastewater, which later is treated with other wastewater purification techniques [134,135].

3.5 Biological treatment processes

Biological degradation or microbial degradation involves the degradation of organic contaminants in the presence of microbes. It involves the degradation of water contaminants using microbes cultures such as fungal, algae, and enzymes [136]. Microbial communities biodegrade the contaminant molecules into the nutrients for other organisms and hence contribute to the food web [137]. Different microbial groups exhibit different potentials to degrade particular classes of pollutant. For example, mycobacteria are an excellent choice for the bioremediation of polyaromatic hydrocarbons-infected wastewater due to lipophilic surfaces [138]. However, higher doses of pollutant molecules are phytotoxic and can affect the rate of microbial degradation. For example, a high concentration of toluene can disrupt the cell membranes and are toxic to the microorganisms. It can be approached either by in situ or ex situ exercises [139]. In situ microbial degradation involves the degradation of wastewater under natural conditions at the original site to produce CO_2, H_2O, and other innocuous products via the degradation of organic contaminants. It is cost-effective, requires low maintenance, and is environment-friendly. However, ex situ involves the treatment of wastewater somewhere else using microbes under artificial conditionals. Furthermore, on the basis of reaction conditions (presence of oxygen), it can be divided into two types: (1) aerobic and (2) anaerobic degradation.

3.5.1 Aerobic degradation

Aerobic biological degradation of water contaminants depends on the growth of microorganisms in the oxygen-rich environment, which can quickly oxidize the organic compounds into CO_2, H_2O, and other cellular compounds [140]. It is a cost-effective and reliable technique to produce high-quality effluents. Activated sludge (AS), rotating biological reactor (RBR), sequencing batch reactor (SBR), percolating filter, and membrane bioreactor are the common examples of aerobic biological treatment of wastewater [141]. The advantages and disadvantages of these processes are shown in Table 3.2.

3.5.2 Anaerobic degradation process

Anaerobic degradation is carried out by microorganisms in the absence of oxygen. In this process, organic substances are generally (95%) broken down into biogas (a mixture of CH_4 and CO_2). Anaerobic biological degradation exhibits several advantages over aerobic degradation, such as low sludge production, the capability to degrade concentrated wastewater, and low energy input requirements [142]. The effluents after anaerobic degradation can also be used for several other applications such as fertilizer or fish feed [143]. However, the unstable behavior of the anaerobic degradation process with minor changes in the operational parameters limits its application at industrial scale. Anaerobic digestion, anaerobic membrane bioreactor, anaerobic sequencing batch reactor, upflow anaerobic sludge blanket, completely stirred tank reactor, moving bed biofilm reactor, and up-flow anaerobic filter are examples of anaerobic

Table 3.2 Advantages and disadvantages of various aerobic biological degradation processes.

Aerobic degradation process	Advantages	Disadvantages
Activated sludge	A simple and easy method Cost-effective	High sludge production High energy consumption Bulking and foaming Iron and carbonates precipitation Low efficiency in water Low quality of effluents
Rotating biological reactor	High efficiency Cost-effective Low space and maintenance requirement Low sludge production Low power input	Technical skill required to run the reactor Odor Continuous supply of electricity The reactor must be protected from sunlight, rains, winds and cold weather
Sequencing batch reactor	High efficiency Cost-effective Low sludge bulking Low maintenance Flexible Competent method for denitrification and phosphorus removal	High energy consumption Sludge disposal
Percolating Filter	Competent method for ammonia removal Simple and easy process High efficiency	The filter can be blocked by iron and carbonate precipitation Not suitable for high-strength wastewater treatment
Membrane bioreactor	High efficiency Production of high-quality effluents Low sludge production Low energy requirement Compact reactor	Membrane fouling High cost Membrane pollution Aeration Limitations

biological degradation methods of wastewater contaminants [140,144]. Table 3.3 shows the advantages and disadvantages of these anaerobic degradation methods.

3.6 Electrochemical treatment processes

Electrochemical treatment method of wastewater includes the redox reaction at both anode (usually the oxidation of organic pollutants) and the cathode (usually

Table 3.3 Advantages and disadvantages of various anaerobic biological degradation processes.

Anaerobic degradation process	Advantages	Disadvantages
Anaerobic digestion	High removal rate Low sludge production Cost-effective A simple and easy method	High energy requirement Control on the operational parameters Toxic compound
Anaerobic membrane bioreactor	High removal rate Enhance biomass retention	High retention period Membrane fouling
Anaerobic sequencing batch reactor	A simple and easy method Cost-effective	Sludge disposal Cannot tolerate high organic loading rates
Upflow anaerobic sludge blanket	High removal rate No need for sludge separation Flexibility High biomass concentration Cost-effective A simple and easy method Low energy requirements No consumption of support material Can tolerate high organic and hydraulic loading rates	Technical skill requirement Long start-up period
Completely stirred tank reactor	High removal rate A simple and easy method Cost-effective Low maintenance	Low conversion per unit volume Channelling issues

reduction of heavy metal ions) to remove the water contaminants [145]. It generally provides the complete mineralization of large organic contaminant molecules, such as pharmaceuticals, pesticides, and textile dyes by the anodic oxidation [146–148]. Therefore the electrochemical properties of anodic material play a vital role in the removal efficiency of electrochemical wastewater treatment plant [148]. The oxidation of organic contaminants using electrochemical technology can occur via two methods, that is, (1) direct oxidation and (2) indirect oxidation [149]. Direct oxidation involves the mineralization of water contaminants on the anodic surface via direct electron transfer to the anode, without any interference of other active species. However, indirect oxidation involves the participation of physisorbed active oxygen (e.g., $^{\bullet}OH$) or chemisorbed active oxygen (e.g., oxygen in the lattice of a metal oxide anode) [150,151]. Electrochemical treatment approaches are gaining interest and preference over other processes due to their ambient operating conditions, compact system, less space requirements, no secondary waste generation, no additional auxiliary chemical requirements, they are cost-effective, and they can easily be combined with other water treatment processes to produce high-quality effluents [145,149,152]. Several electrochemical treatment-based approaches, such as electrocoagulation, electrodeposition, electrocatalysis, and electrophotocatalysis, have been used to remove water pollutants.

3.6.1 Electrocoagulation

Electrocoagulation (EC) is the simplest electrochemical wastewater treatment technique. It involves an electrochemical cell where voltage is applied to electrodes (generally Fe or Al electrodes) and the wastewater act as electrolyte [153]. This technique is a combination of coagulation, flocculation, and electrochemistry methods. However, the EC method exhibits several advantages over chemical coagulation and flocculation technique such as in situ coagulants generation via electrooxidation of anodic material, less sludge production, stable flocs formations, high removal efficiency, lower cost, and potential to treat oily water [154]. These advantages promote the application of the EC method in treating the polluted water. Generation of in situ coagulants, by the electrooxidation of the sacrificial electrode, destabilizes the suspension of pollutants in the water and forms large flocs which can easily be separated. Destabilization of suspended or colloidal wastewater system and flocs formation is similar to the chemical coagulation and flocculation mechanism. It is highly influenced by the operational parameters, which demand the optimization of reaction conditions [155].

3.6.2 Electroflotation

The electroflotation process is based on the bubble created during the electrochemical reaction and is generally used together with electrocoagulation and other water treatment methods [156]. In this process, during electrochemical treatment oxygen and gaseous hydrogen bubbles are generated via the anodic oxidation and cathodic reduction, which induces the flotation of dirty particles (electroflocs). The size of the gaseous bubbles directly makes an impact on the potential of the treatment process. The gaseous bubbles bind the pollutant particles and bring them to the water surface level, they float and form the froth layer. This layer can easily be removed by means of mechanical devices.

Combining electroflotation with electrocoagulation enhances the removal efficiency by removing those particles that were aggregated during electrocoagulation but not separated [157]. Pretreatment of wastewater using electroflotation removes a significant portion of NOM, silica, and other impurities which also helps in reducing the membrane fouling when using the membrane filtration treatment process [158].

3.6.3 Electrodeposition

Electrodeposition or electroplating technique involves the degradation of organic pollutants at the cathode, while the heavy metals are reduced and simultaneously electroplated or deposited at the cathode [159]. This method is generally used for the removal of inorganic pollutants. The general mechanism of electrodeposition is based on the process that on exposing the wastewater to a direct external current in an electrochemical cell, heavy metal ions present in wastewater starts to get reduced and deposited on the cathode, and thus can be collected. This is a simple and easy electrochemical treatment method with high contaminant removal efficiency [160]. However, this process demands high electrical energy and is not suitable for the low concentration wastewater systems. Therefore, this method is more feasible for the pretreatment of wastewater.

3.6.4 Electrocatalysis

Electrocatalysis involves the redox reaction in an electrochemical cell in the presence of an electrocatalyst, which lowers the overpotential of the electrochemical reactions for driving a specific electrochemical reaction [161]. Active anodes (e.g., RuO_2, Pt) that exhibit low overpotential for oxygen evolution are considered to be good electrocatalysts for oxygen evolution reactions (OER) and selective oxidation of the organic water contaminants. However, nonactive anodes (PbO_2, boron-doped diamond, SnO_2, etc.) with high oxygen evolution overpotential are considered to be poor electrocatalysts for the OER, and generally direct electrochemical oxidation of pollutants takes place on these electrodes [162]. Furthermore, to improve the efficiency of electrocatalysis, it can also be combined with other wastewater treatment methods. Photoelectrocatalysis (electrocatalysis in presence of light), sonoelectrocatalysis (electrocatalysis in the presence of ultrasound waves) and electro-Fenton (a combination of Fenton's process and electrocatalysis) are common examples of upgraded combined electrocatalysis processes [163–165].

3.7 Conclusion

In recent years, the continuous discharge of contaminated water into freshwater bodies has raised the alarm for clean water scarcity. In this context, a significant amount of research has been conducted to develop several technologies to remove emerging contaminants from the wastewater before its discharge into the environment. However, a considerable knowledge gap still exists, which is drawing attention to the safety challenge of the consumption of reused water. Therefore various physical, chemical, biological, and electrochemical wastewater treatment approaches have been discussed in this chapter, along with their advantages and disadvantages. Physical treatment processes are the simplest and most adapted water treatment methods. These methods involve the separation of pollutants from the water by means of adsorption or filtration. RO is one of the physical treatment approaches which has gained a large interest in the market. Adsorption is one of the most studied, economical, and environment-friendly approaches to clean the wastewater. Physical methods have also been considered as pretreatment methods in combination with other methods. However, membrane fouling is the common disadvantage of such processes. Chemical oxidation, especially AOPs, is gaining a large amount of interest of researchers for water treatment. Such processes involve the active oxidation species (e.g., $^\bullet OH$ radicals), to oxidize and mineralize contaminated molecules. However, adsorption (physical method) is the necessary step for AOPs, facilitating the oxidation of pollutants on the surface of catalysts. Therefore, the combination of physical and chemical approaches (physicochemical wastewater methods) is attractive for wastewater treatment processes. It is a good for the removal of organic contaminants.

The biological wastewater treatment method requires the microorganism to degrade the organic water contaminants in the presence (aerobic) and absence (anaerobic) of oxygen/air. It generally produces fertilizers or nutrients for other organisms and thus helps in the food web. It is low-cost, environment-friendly, and requires low maintenance, which makes it accessible for wastewater treatment. However, the relatively long period and culture growth to degrade the organic contaminants are the drawbacks of the biological treatment process. Electrochemical approaches apply electric fields during the wastewater treatment process. Redox reaction at the electrodes and the adsorption of contaminants on the electrodes play a vital role in order to produce clean water. Electrochemical approaches can easily be combined with other processes to enhance the removal efficiency. Photocatalytic wastewater treatment in the presence of an electric field is called photoelectrocatalysis. The hybrid systems, that is, physicochemical, electrochemical, biophysical, and biochemical approaches, can be economical and more effective wastewater treatment methods. However, adsorption of the pollutants on the active species or the catalyst is the crucial step. Therefore, the following the chapters are dedicated to adsorption, the factors affecting the adsorption process, and the adsorption of various contaminants.

References

[1] Jackson RB, Carpenter SR, Dahm CN, McKnight DM, Naiman RJ, Postel SL, et al. Water in a changing world. Ecol Appl 2001;11:1027−45.
[2] Reemtsma T, Weiss S, Mueller J, Petrovic M, González S, Barcelo D, et al. Polar pollutants entry into the water cycle by municipal wastewater: a European perspective. Environ Sci Technol 2006;40:5451−8.
[3] Boroski M, Rodrigues AC, Garcia JC, Sampaio LC, Nozaki J, Hioka N. Combined electrocoagulation and TiO$_2$ photoassisted treatment applied to wastewater effluents from pharmaceutical and cosmetic industries. J Hazard Mater 2009;162:448−54.
[4] Vazquez-Montiel O, Horan NJ, Mara DD. Management of domestic wastewater for reuse in irrigation. Water Sci Technol 1996;33:355.

[5] Cicek N. A review of membrane bioreactors and their potential application in the treatment of agricultural wastewater. Can Biosyst Eng 2003;45(6) 37--6.

[6] Cantor KP. Drinking water and cancer. Cancer Causes Control 1997;8:292−308.

[7] Padhi B. Pollution due to synthetic dyes toxicity & carcinogenicity studies and remediation. Int J Environ Sci 2012;3:940.

[8] Fu F, Wang Q. Removal of heavy metal ions from wastewaters: a review. J Environ Manag 2011;92:407−18.

[9] Gadipelly C, Pérez-González A, Yadav GD, Ortiz I, Ibáñez R, Rathod VK, et al. Pharmaceutical industry wastewater: review of the technologies for water treatment and reuse. Ind Eng Chem Res 2014;53:11571--92.

[10] Rubio-Clemente A, Torres-Palma RA, Penuela GA. Removal of polycyclic aromatic hydrocarbons in aqueous environment by chemical treatments: a review. Sci Total Environ 2014;478:201−25.

[11] Das R, Vecitis CD, Schulze A, Cao B, Ismail AF, Lu X, et al. Recent advances in nanomaterials for water protection and monitoring. Chem Soc Rev 2017;46:6946−7020.

[12] Mittal H, Kumar V, Alhassan SM, Ray SS. Modification of gum ghatti via grafting with acrylamide and analysis of its flocculation, adsorption, and biodegradation properties. Int J Biol Macromol 2018;114:283−94.

[13] Ama OM, Kumar N, Adams FV, Ray SS. Efficient and cost-effective photoelectrochemical degradation of dyes in wastewater over an exfoliated graphite-moo3 nanocomposite electrode. Electrocatalysis 2018;9:623−31.

[14] Nataraj SK, Hosamani KM, Aminabhavi TM. Distillery wastewater treatment by the membrane-based nanofiltration and reverse osmosis processes. Water Res 2006;40:2349−56.

[15] Chen G. Electrochemical technologies in wastewater treatment. Sep Purif Technol 2004;38:11−41.

[16] Luo H, Xu P, Roane TM, Jenkins PE, Ren Z. Microbial desalination cells for improved performance in wastewater treatment, electricity production, and desalination. Bioresour Technol 2012;105:60−6.

[17] Mukwevho N, Fosso-Kankeu E, Waanders F, Kumar N, Ray SS, Yangkou Mbianda X. Photocatalytic activity of $Gd_2O_2CO_3 \cdot ZnO \cdot CuO$ nanocomposite used for the degradation of phenanthrene. SN Appl Sci 2018;1:10.

[18] Kumar N, Mittal H, Alhassan SM, Ray SS. Bionanocomposite hydrogel for the adsorption of dye and reusability of generated waste for the photodegradation of ciprofloxacin: a demonstration of the circularity concept for water purification. ACS Sustain Chem Eng 2018;6:17011−25.

[19] Umukoro EH, Kumar N, Ngila JC, Arotiba OA. Expanded graphite supported p-n MoS_2-SnO_2 heterojunction nanocomposite electrode for enhanced photo-electrocatalytic degradation of a pharmaceutical pollutant. J Electroanalytical Chem 2018;827:193−203.

[20] Kumar N, Sinha Ray S, Ngila JC. Ionic liquid-assisted synthesis of Ag/Ag_2Te nanocrystals via a hydrothermal route for enhanced photocatalytic performance. N J Chem 2017;41:14618−26.

[21] Kumar N, Mittal H, Reddy L, Nair P, Ngila JC, Parashar V. Morphogenesis of ZnO nanostructures: role of acetate $(COOH^-)$ and nitrate (NO_3^-) ligand donors from zinc salt precursors in synthesis and morphology dependent photocatalytic properties. RSC Adv 2015;5:38801−9.

[22] Kumar N, Mittal H, Parashar V, Ray SS, Ngila JC. Efficient removal of rhodamine 6G dye from aqueous solution using nickel sulphide incorporated polyacrylamide grafted gum karaya bionano-composite hydrogel. RSC Adv 2016;6:21929−39.

[23] Gusain R, Kumar N, Ray SS. Recent advances in carbon nanomaterial-based adsorbents for water purification. Coord Chem Rev 2020;405:213111.

[24] Malaeb L, Ayoub GM. Reverse osmosis technology for water treatment: state of the art review. Desalination. 2011;267:1−8.

[25] Bieber S. Chapter 10 - water treatment equipment for in-center hemodialysis. In: Nissenson AR, Fine RN, editors. Handbook of dialysis therapy. 5th ed. Elsevier; 2017. p. 123-43.e1.

[26] Bellona C, Drewes JE. The role of membrane surface charge and solute physico-chemical properties in the rejection of organic acids by NF membranes. J Membr Sci 2005;249:227−34.

[27] Radjenović J, Petrović M, Ventura F, Barceló D. Rejection of pharmaceuticals in nanofiltration and reverse osmosis membrane drinking water treatment. Water Res 2008;42:3601−10.

[28] Gabelich CJ, Ishida KP, Bold RM. Testing of water treatment copolymers for compatibility with polyamide reverse osmosis membranes. Environ Prog 2005;24:410−16.

[29] Mondal S, Hsiao CL, Ranil Wickramasinghe S. Nanofiltration/reverse osmosis for treatment of coproduced waters. Environ Prog 2008;27:173−9.

[30] Petersen RJ, Cadotte JE. Thin film composite reverse osmosis membranes. Handb Ind Membr Technol 1990;307−48.

[31] Cadotte JE. Reverse osmosis membrane. Google Patents; 1977.

[32] Li D, Wang H. Recent developments in reverse osmosis desalination membranes. J Mater Chem 2010;20:4551−66.

[33] Kang G-d, Cao Y-m. Development of antifouling reverse osmosis membranes for water treatment: a review. Water Res 2012;46:584−600.

[34] Van der Bruggen B, Vandecasteele C, Van Gestel T, Doyen W, Leysen R. A review of pressure-driven membrane processes in wastewater treatment and drinking water production. Environ Prog 2003;22:46−56.

[35] Shon H, Vigneswaran S, Kim IS, Cho J, Ngo H. Effect of pretreatment on the fouling of membranes: application in biologically treated sewage effluent. J Membr Sci 2004;234:111−20.

[36] Van Boxtel A, Otten Z, Van der Linden H. Evaluation of process models for fouling control of reverse osmosis of cheese whey. J Membr Sci 1991;58:89−111.

[37] Zhang L, She Q, Wang R, Wongchitphimon S, Chen Y, Fane AG. Unique roles of aminosilane in developing anti-fouling thin film composite (TFC) membranes for pressure retarded osmosis (PRO). Desalination 2016;389:119−28.

[38] Jhaveri JH, Murthy Z. A comprehensive review on anti-fouling nanocomposite membranes for pressure driven membrane separation processes. Desalination 2016;379:137−54.

[39] Ong CS, Goh P, Lau W, Misdan N, Ismail AF. Nanomaterials for biofouling and scaling mitigation of thin film composite membrane: a review. Desalination 2016;393:2−15.

[40] Anand A, Unnikrishnan B, Mao J-Y, Lin H-J, Huang C-C. Graphene-based nanofiltration membranes for improving salt rejection, water flux and antifouling−a review. Desalination 2018;429:119−33.

[41] Zhao X, Zhang R, Liu Y, He M, Su Y, Gao C, et al. Antifouling membrane surface construction: chemistry plays a critical role. J Membr Sci 2018;551:145−71.

[42] Nie F-Q, Xu Z-K, Ye P, Wu J, Seta P. Acrylonitrile-based copolymer membranes containing reactive groups: effects of surface-immobilized poly (ethylene glycol) s on anti-fouling properties and blood compatibility. Polymer 2004;45:399−407.

[43] Sagle AC, Van Wagner EM, Ju H, McCloskey BD, Freeman BD, Sharma MM. PEG-coated reverse osmosis membranes: desalination properties and fouling resistance. J Membr Sci 2009;340:92−108.

[44] Al-Jeshi S, Neville A. An investigation into the relationship between flux and roughness on RO membranes using scanning probe microscopy. Desalination 2006;189:221−8.

[45] Lee A, Elam JW, Darling SB. Membrane materials for water purification: design, development, and application. Environ Sci Water Res Technol 2016;2:17−42.

[46] Pearce G. Introduction to membranes: filtration for water and wastewater treatment. Filtration Sep 2007;44:24−7.

[47] Wilf M, Awerbuch L. The guidebook to membrane desalination technology: reverse osmosis, nanofiltration and hybrid systems: process, design, applications and economics. Balaban Desalination Publications; 2007.

[48] Pellegrin M-L, Aguinaldo J, Arabi S, Sadler ME, Min K, Liu M, et al. Membrane processes. Water Environ Res 2013;85:1092−175.

[49] Pendergast MM, Hoek EM. A review of water treatment membrane nanotechnologies. Energy Environ Sci 2011;4:1946−71.

[50] Yan L, Li YS, Xiang CB, Xianda S. Effect of nano-sized Al2O3-particle addition on PVDF ultrafiltration membrane performance. J Membr Sci 2006;276:162−7.

[51] Laîné J-M, Hagstrom JP, Clark MM, Mallevialle J. Effects of ultrafiltration membrane composition. J AWWA 1989;81:61−7.

[52] Luo M-L, Zhao J-Q, Tang W, Pu C-S. Hydrophilic modification of poly(ether sulfone) ultrafiltration membrane surface by self-assembly of TiO$_2$ nanoparticles. Appl Surf Sci 2005;249:76−84.

[53] Mulder M. Preparation of synthetic membranes. Basic principles of membrane technology. Dordrecht: Springer Netherlands; 1991. p. 54−109.

[54] Zhao W, Su Y, Li C, Shi Q, Ning X, Jiang Z. Fabrication of antifouling polyethersulfone ultrafiltration membranes using Pluronic F127 as both surface modifier and pore-forming agent. J Membr Sci 2008;318:405−12.

[55] Fan X, Su Y, Zhao X, Li Y, Zhang R, Zhao J, et al. Fabrication of polyvinyl chloride ultrafiltration membranes with stable antifouling property by exploring the pore formation and surface modification capabilities of polyvinyl formal. J Membr Sci 2014;464:100−9.

[56] Takijiri K, Morita K, Nakazono T, Sakai K, Ozawa H. Highly stable chemisorption of dyes with pyridyl anchors over TiO$_2$: application in dye-sensitized photoelectrochemical water reduction in aqueous media. Chem Commun 2017;53:3042−5.

[57] Dunne LJ, Manos G. Adsorption and phase behaviour in nanochannels and nanotubes. Springer Science & Business Media; 2009.

[58] Gao Z, Bandosz TJ, Zhao Z, Han M, Qiu J. Investigation of factors affecting adsorption of transition metals on oxidized carbon nanotubes. J Hazard Mater 2009;167:357−65.

[59] Jonnalagadda SB, Nadupalli S. Chlorine dioxide for bleaching, industrial applications and water treatment. Indian Chem Eng 2014;56:123−36.

[60] Lantagne D, Person B, Smith N, Mayer A, Preston K, Blanton E, et al. Emergency water treatment with bleach in the United States: the need to revise EPA recommendations. Environ Sci Technol 2014;48:5093−100.

[61] White GC. Handbook of chlorination. Van Nostrand Reinhold Company; 1986.

[62] Huber MM, Korhonen S, Ternes TA, von Gunten U. Oxidation of pharmaceuticals during water treatment with chlorine dioxide. Water Res 2005;39:3607−17.

[63] Gates DJ. Chlorine dioxide handbook: AWWA; 1998.

[64] Condie LW. Toxicological problems associated with chlorine dioxide. J Am Water Work Assoc 1986;78:73−8.

[65] Verma AK, Dash RR, Bhunia P. A review on chemical coagulation/flocculation technologies for removal of colour from textile wastewaters. J Environ Manag 2012;93:154−68.

[66] Sharma B, Dhuldhoya N, Merchant U. Flocculants—an ecofriendly approach. J Polym Environ 2006;14:195−202.

[67] Teh CY, Budiman PM, Shak KPY, Wu TY. Recent advancement of coagulation−flocculation and its application in wastewater treatment. Ind Eng Chem Res 2016;55:4363−89.

[68] Bratby J. Coagulation and flocculation in water and wastewater treatment. IWA Publishing; 2016.

[69] Ahmad A, Ismail S, Bhatia S. Optimization of coagulation − flocculation process for palm oil mill effluent using response surface methodology. Environ Sci Technol 2005;39:2828−34.

[70] Wong S, Teng T, Ahmad A, Zuhairi A, Najafpour G. Treatment of pulp and paper mill wastewater by polyacrylamide (PAM) in polymer induced flocculation. J Hazard Mater 2006;135:378−88.

[71] Yue Q, Gao B, Wang Y, Zhang H, Sun X, Wang S, et al. Synthesis of polyamine flocculants and their potential use in treating dye wastewater. J Hazard Mater 2008;152:221−7.

[72] Zhong J, Sun X, Wang C. Treatment of oily wastewater produced from refinery processes using flocculation and ceramic membrane filtration. Sep Purif Technol 2003;32:93−8.

[73] Huang H, Schwab K, Jacangelo JG. Pretreatment for low pressure membranes in water treatment: a review. Environ Sci Technol 2009;43:3011−19.

[74] Duan J, Gregory J. Coagulation by hydrolysing metal salts. Adv Colloid Interface Sci 2003;100--102:475−502.

[75] Lee CS, Robinson J, Chong MF. A review on application of flocculants in wastewater treatment. Process Saf Environ Prot 2014;92:489−508.

[76] Lee KE, Morad N, Teng TT, Poh BT. Development, characterization and the application of hybrid materials in coagulation/flocculation of wastewater: a review. Chem Eng J 2012;203:370−86.

[77] Amuda O, Amoo I. Coagulation/flocculation process and sludge conditioning in beverage industrial wastewater treatment. J Hazard Mater 2007;141:778−83.

[78] Anjaneyulu Y, Chary NS, Raj DSS. Decolourization of industrial effluents—available methods and emerging technologies—a review. Rev Environ Sci Bio/Technology 2005;4:245—73.

[79] Jarvis P, Mergen M, Banks J, McIntosh B, Parsons SA, Jefferson B. Pilot scale comparison of enhanced coagulation with magnetic resin plus coagulation systems. Environ Sci Technol 2008;42:1276—82.

[80] Humbert H, Gallard H, Jacquemet V, Croué J-P. Combination of coagulation and ion exchange for the reduction of UF fouling properties of a high DOC content surface water. Water Res 2007;41:3803—11.

[81] Moghaddam SS, Moghaddam MA, Arami M. Coagulation/flocculation process for dye removal using sludge from water treatment plant: optimization through response surface methodology. J Hazard Mater 2010;175:651—7.

[82] Levchuk I, Rueda Márquez JJ, Sillanpää M. Removal of natural organic matter (NOM) from water by ion exchange — a review. Chemosphere. 2018;192:90—104.

[83] Qiu W, Zheng Y. Removal of lead, copper, nickel, cobalt, and zinc from water by a cancrinite-type zeolite synthesized from fly ash. Chem Eng J 2009;145:483—8.

[84] Rengaraj S, Yeon K-H, Moon S-H. Removal of chromium from water and wastewater by ion exchange resins. J Hazard Mater 2001;87:273—87.

[85] Rožić M, Cerjan-Stefanović Š, Kurajica S, Vančina V, Hodžić E. Ammoniacal nitrogen removal from water by treatment with clays and zeolites. Water Res 2000;34:3675—81.

[86] Nesterenko PN, Haddad PR. Zwitterionic ion-exchangers in liquid chromatography. Anal Sci 2000;16:565—74.

[87] Sherrington DC. Preparation, structure and morphology of polymer supports. Chem Commun 1998;2275—86.

[88] Ku Y, Chiou H-M, Wang W. The removal of fluoride ion from aqueous solution by a cation synthetic resin. Sep Sci Technol 2002;37:89—103.

[89] Shin D, Ju K, Cheong S, Rhim J. Removal of radioactive ions from contaminated water by ion exchange resin. Appl Chem Eng 2016;27:633—8.

[90] Erdem E, Karapinar N, Donat R. The removal of heavy metal cations by natural zeolites. J Colloid Interface Sci 2004;280:309—14.

[91] Kumar N, Reddy L, Parashar V, Ngila JC. Controlled synthesis of microsheets of ZnAl layered double hydroxides hexagonal nanoplates for efficient removal of Cr(VI) ions and anionic dye from water. J Environ Chem Eng 2017;5:1718—31.

[92] Gusain R, Kumar N, Fosso-Kankeu E, Ray SS. Efficient removal of Pb(II) and Cd(II) from industrial mine water by a hierarchical MoS_2/SH-MWCNT nanocomposite. ACS Omega 2019;4:13922—35.

[93] Vaaramaa K, Lehto J. Removal of metals and anions from drinking water by ion exchange. Desalination 2003;155:157—70.

[94] Bolton JR, Bircher KG, Tumas W, Tolman CA. Figures-of-merit for the technical development and application of advanced oxidation processes. J Adv Oxid Technol 1996;1:13—17.

[95] Miklos DB, Remy C, Jekel M, Linden KG, Drewes JE, Hübner U. Evaluation of advanced oxidation processes for water and wastewater treatment — a critical review. Water Res 2018;139:118—31.

[96] Poyatos JM, Muñio MM, Almecija MC, Torres JC, Hontoria E, Osorio F. Advanced oxidation processes for wastewater treatment: state of the art. Water Air Soil Pollut 2009;205:187.

[97] Oturan MA, Aaron J-J. Advanced oxidation processes in water/wastewater treatment: principles and applications. a review. Crit Rev Environ Sci Technol 2014;44:2577—641.

[98] Bethi B, Sonawane SH, Bhanvase BA, Gumfekar SP. Nanomaterials-based advanced oxidation processes for wastewater treatment: a review. Chem Eng Processing-Process Intensif 2016;109:178—89.

[99] Dewil R, Mantzavinos D, Poulios I, Rodrigo MA. New perspectives for advanced oxidation processes. J Environ Manag 2017;195:93—9.

[100] Wen J, Xie J, Chen X, Li X. A review on g-C_3N_4-based photocatalysts. Appl Surf Sci 2017;391:72—123.

[101] Gusain R, Kumar P, Sharma OP, Jain SL, Khatri OP. Reduced graphene oxide—CuO nanocomposites for photocatalytic conversion of CO_2 into methanol under visible light irradiation. Appl Catal B: Environ 2016;181:352—62.

[102] Mukwevho N, Gusain R, Fosso-Kankeu E, Kumar N, Waanders F, Ray SS. Removal of naphthalene from simulated wastewater through adsorption-photodegradation by ZnO/Ag/GO nanocomposite. J Ind Eng Chem 2020;81:393—404.

[103] Balli B, Aygun A, Sen F. Medicinal applications of photocatalysts. In: Inamuddin Asiri AM, Lichtfouse E, editors. Nanophotocatalysis and environmental applications: detoxification and disinfection. Cham: Springer International Publishing; 2020. p. 245—65.

[104] Dong S, Feng J, Fan M, Pi Y, Hu L, Han X, et al. Recent developments in heterogeneous photocatalytic water treatment using visible light-responsive photocatalysts: a review. RSC Adv 2015;5:14610—30.

[105] Kumari S, Gusain R, Khatri OP. Tuning the band-gap of h-boron nitride nanoplatelets by covalent grafting of imidazolium ionic liquids. RSC Adv 2016;6:21119—26.

[106] Ye Y, Feng Y, Bruning H, Yntema D, Rijnaarts HHM. Photocatalytic degradation of metoprolol by TiO^2 nanotube arrays and UV-LED: effects of catalyst properties, operational parameters, commonly present water constituents, and photo-induced reactive species. Appl Catal B Environ 2018;220:171—81.

[107] Newton DW, Kluza RB. pKa values of medicinal compounds in pharmacy practice. Drug Intell Clin Pharm 1978;12:546—54.

[108] Fenton H. LXXIII.—Oxidation of tartaric acid in presence of iron. J Chem Soc Trans 1894;65:899—910.

[109] Bigda RJ. Consider Fentons chemistry for wastewater treatment. Chem Eng Prog 1995;91.

[110] Wang N, Zheng T, Zhang G, Wang P. A review on Fenton-like processes for organic wastewater treatment. J Environ Chem Eng 2016;4:762—87.

[111] Zhang M-h, Dong H, Zhao L, Wang D-x, Meng D. A review on Fenton process for organic wastewater treatment based on optimization perspective. Sci Total Environ 2019;670:110—21.

[112] Santos MS, Alves A, Madeira LM. Paraquat removal from water by oxidation with Fenton's reagent. Chem Eng J 2011;175:279—90.

[113] Ikehata K, El-Din MG. Aqueous pesticide degradation by hydrogen peroxide/ultraviolet irradiation and Fenton-type advanced oxidation processes: a review. J Environ Eng Sci 2006;5:81—135.

[114] Babuponnusami A, Muthukumar K. Degradation of phenol in aqueous solution by fenton, sono-fenton and sono-photo-fenton methods. Clean—Soil, Air, Water 2011;39:142—7.

[115] Sun S-P, Zeng X, Lemley AT. Nano-magnetite catalyzed heterogeneous Fenton-like degradation of emerging contaminants carbamazepine and ibuprofen in aqueous suspensions and montmorillonite clay slurries at neutral pH. J Mol Catal A Chem 2013;371:94—103.

[116] Xue X, Hanna K, Abdelmoula M, Deng N. Adsorption and oxidation of PCP on the surface of magnetite: kinetic experiments and spectroscopic investigations. Appl Catal B Environ 2009;89:432—40.

[117] Sun S-P, Lemley AT. p-Nitrophenol degradation by a heterogeneous Fenton-like reaction on nano-magnetite: process optimization, kinetics, and degradation pathways. J Mol Catal A Chem 2011;349:71—9.

[118] Khamparia S, Jaspal DK. Adsorption in combination with ozonation for the treatment of textile waste water: a critical review. Front Environ Sci Eng 2017;11:8.

[119] Broséus R, Vincent S, Aboulfadl K, Daneshvar A, Sauvé S, Barbeau B, et al. Ozone oxidation of pharmaceuticals, endocrine disruptors and pesticides during drinking water treatment. Water Res 2009;43:4707—17.

[120] Alvares A, Diaper C, Parsons S. Partial oxidation by ozone to remove recalcitrance from wastewaters -- a review. Environ Technol 2001;22:409—27.

[121] Umar M, Roddick F, Fan L, Aziz HA. Application of ozone for the removal of bisphenol A from water and wastewater — a review. Chemosphere 2013;90:2197—207.

[122] Agustina TE, Ang HM, Vareek VK. A review of synergistic effect of photocatalysis and ozonation on wastewater treatment. J Photochem Photobiol C Photochem Rev 2005;6:264—73.

[123] Merouani S, Hamdaoui O. Sonolytic ozonation for water treatment: efficiency, recent developments and challenges. Curr Opin Green Sustain Chem, 2019.

[124] Suslick KS. Sonochemistry. Science. 1990;247:1439—45.

[125] Adityosulindro S, Barthe L, González-Labrada K, Jáuregui Haza UJ, Delmas H, Julcour C. Sonolysis and sono-Fenton oxidation for removal of ibuprofen in (waste)water. Ultrason Sonochem 2017;39:889—96.

[126] Bagal MV, Gogate PR. Wastewater treatment using hybrid treatment schemes based on cavitation and Fenton chemistry: a review. Ultrason Sonochem 2014;21:1—14.

[127] Liang J, Komarov S, Hayashi N, Kasai E. Improvement in sonochemical degradation of 4-chlorophenol by combined use of Fenton-like reagents. Ultrason Sonochem 2007;14:201—7.

[128] Neppolian B, Jung H, Choi H, Lee JH, Kang J-W. Sonolytic degradation of methyl tert-butyl ether: the role of coupled fenton process and persulphate ion. Water Res 2002;36:4699—708.

[129] Luck F. A review of industrial catalytic wet air oxidation processes. Catal Today 1996;27:195—202.

[130] Villegas LGC, Mashhadi N, Chen M, Mukherjee D, Taylor KE, Biswas N. A short review of techniques for phenol removal from wastewater. Curr Pollut Rep 2016;2:157—67.

[131] Luck F. Wet air oxidation: past, present and future. Catal Today 1999;53:81—91.

[132] Weber B, Chavez A, Morales-Mejia J, Eichenauer S, Stadlbauer EA, Almanza R. Wet air oxidation of resorcinol as a model treatment for refractory organics in wastewaters from the wood processing industry. J Environ Manag 2015;161:137—43.

[133] Mishra VS, Mahajani VV, Joshi JB. Wet air oxidation. Ind Eng Chem Res 1995;34:2—48.

[134] Rathnayake B, Heponiemi A, Huovinen M, Ojala S, Pirilä M, Loikkanen J, et al. Photocatalysis and catalytic wet air oxidation: degradation and toxicity of bisphenol A containing wastewaters. Environ Technol 2019;1—12.

[135] Saroha AK. Treatment of industrial organic raffinate containing pyridine and its derivatives by coupling of catalytic wet air oxidation and biological processes. J Clean Prod 2017;162:973—81.

[136] Barra Caracciolo A, Topp E, Grenni P. Pharmaceuticals in the environment: biodegradation and effects on natural microbial communities. A review. J Pharm Biomed Anal 2015;106:25—36.

[137] Megharaj M, Ramakrishnan B, Venkateswarlu K, Sethunathan N, Naidu R. Bioremediation approaches for organic pollutants: a critical perspective. Environ Int 2011;37:1362—75.

[138] Bogan B, Lahner L, Sullivan W, Paterek J. Degradation of polycyclic aromatic and straight-chain aliphatic hydrocarbons by a strain of Mycobacterium austroafricanum. J Appl Microbiol 2003;94:230—9.

[139] Aggarwal P, Means J, Hinchee R, Headington G, Gavaskar A. Methods to select chemicals for in situ biodegradation of fuel hydrocarbons. Battelle Columbus Div OH; 1990.

[140] Goli A, Shamiri A, Khosroyar S, Talaiekhozani A, Sanaye R, Azizi K. A review on different aerobic and anaerobic treatment methods in dairy industry wastewater. J Environ Treat Tech 2019;6:113—41.

[141] Ahmed MB, Zhou JL, Ngo HH, Guo W, Thomaidis NS, Xu J. Progress in the biological and chemical treatment technologies for emerging contaminant removal from wastewater: a critical review. J Hazard Mater 2017;323:274—98.

[142] Pavlostathis SG. Anaerobic processes. Water Environ Res 1994;66:342—56.

[143] Bowie CTC, Sneddon DM, Montgomery AR. Considerations in design and operation of a biogas plant. In: Twidell J, Riddoch F, Grainger B, editors. Energy for rural and island communities. Pergamon; 1984. p. 371—7.

[144] Shi X, Leong KY, Ng HY. Anaerobic treatment of pharmaceutical wastewater: a critical review. Bioresour Technol 2017;245:1238—44.

[145] Garcia-Segura S, Ocon JD, Chong MN. Electrochemical oxidation remediation of real wastewater effluents — a review. Process Saf Environ Prot 2018;113:48—67.

[146] Cavalcanti EB, Garcia-Segura S, Centellas F, Brillas E. Electrochemical incineration of omeprazole in neutral aqueous medium using a platinum or boron-doped diamond anode: degradation kinetics and oxidation products. Water Res 2013;47:1803—15.

[147] Samet Y, Agengui L, Abdelhédi R. Anodic oxidation of chlorpyrifos in aqueous solution at lead dioxide electrodes. J Electroanalytical Chem 2010;650:152—8.

[148] Labiadh L, Barbucci A, Cerisola G, Gadri A, Ammar S, Panizza M. Role of anode material on the electrochemical oxidation of methyl orange. J Solid State Electrochem 2015;19:3177—83.

[149] Feng Y, Yang L, Liu J, Logan BE. Electrochemical technologies for wastewater treatment and resource reclamation. Environ Sci Water Res Technol 2016;2:800—31.

[150] Johnson SK, Houk LL, Feng J, Houk R, Johnson DC. Electrochemical incineration of 4-chloro-phenol and the identification of products and intermediates by mass spectrometry. Environ Sci Technol 1999;33:2638−44.

[151] Chang H, Johnson DC. Electrocatalysis of anodic oxygen-transfer reactions activation of electrodes in by addition of Bismuth (III) and Arsenic (III, V). J Electrochem Soc 1990;137:2452−7.

[152] Radjenovic J, Sedlak DL. Challenges and opportunities for electrochemical processes as next-generation technologies for the treatment of contaminated water. Environ Sci Technol 2015;49:11292−302.

[153] Moreno-Casillas HA, Cocke DL, Gomes JAG, Morkovsky P, Parga JR, Peterson E. Electrocoagulation mechanism for COD removal. Sep Purif Technol 2007;56:204−11.

[154] Moussa DT, El-Naas MH, Nasser M, Al-Marri MJ. A comprehensive review of electrocoagulation for water treatment: potentials and challenges. J Environ Manag 2017;186:24−41.

[155] Mollah MY, Morkovsky P, Gomes JA, Kesmez M, Parga J, Cocke DL. Fundamentals, present and future perspectives of electrocoagulation. J Hazard Mater 2004;114:199−210.

[156] Gamage NP, Chellam S. Mechanisms of physically irreversible fouling during surface water micro-filtration and mitigation by aluminum electroflotation pretreatment. Environ Sci Technol 2014;48:1148−57.

[157] Chen G, Chen X, Yue PL. Electrocoagulation and electroflotation of restaurant wastewater. J Environ Eng 2000;126:858−63.

[158] Gamage NP, Rimer JD, Chellam S. Improvements in permeate flux by aluminum electroflotation pretreatment during microfiltration of surface water. J Membr Sci 2012;411:45−53.

[159] Chang J-H, Ellis AV, Yan C-T, Tung C-H. The electrochemical phenomena and kinetics of EDTA−copper wastewater reclamation by electrodeposition and ultrasound. Sep Purif Technol 2009;68:216−21.

[160] Zhu Y, Fan W, Zhou T, Li X. Removal of chelated heavy metals from aqueous solution: a review of current methods and mechanisms. Sci Total Environ 2019;678:253−66.

[161] Li R, Li C. Chapter one - photocatalytic water splitting on semiconductor-based photocatalysts. In: Song C, editor. Advances in catalysis. Academic Press; 2017. p. 1−57.

[162] De Battisti A, Martínez-Huitle CA. Chapter 5 - electrocatalysis in wastewater treatment. In: Martínez-Huitle CA, Rodrigo MA, Scialdone O, editors. Electrochemical water and wastewater treatment. Butterworth-Heinemann; 2018. p. 119−31.

[163] Peleyeju MG, Arotiba OA. Recent trend in visible-light photoelectrocatalytic systems for degradation of organic contaminants in water/wastewater. Environ Sci Water Res Technol 2018;4:1389−411.

[164] Nidheesh P, Gandhimathi R. Trends in electro-Fenton process for water and wastewater treatment: an overview. Desalination 2012;299:1−15.

[165] Martín de Vidales MJ, Sáez C, Cañizares P, Rodrigo MA. Removal of triclosan by conductive-diamond electrolysis and sonoelectrolysis. J Chem Technol Biotechnol 2013;88:823−8.

CHAPTER FOUR

Adsorption in the context of water purification

4.1 Introduction

The term adsorption describes the process of the adhesion of a liquid, gaseous, or solid material onto the surface of another material. The material that is adsorbed on the surface from bulk is termed as the adsorbate and the material surface on which adsorption is being performed is termed as the adsorbent. It is different from absorption, in which the absorbed molecules do not only remain on the surface but also enter into the bulk phase. Adsorption is a surface-based technique, whereas absorption involves the whole volume of the materials. The use of charcoal-based gas masks to prevent dust particles and harmful gases being breathed in from the air is an example of adsorption and ink-soaked cotton is an example of absorption. Adsorption techniques are used in wide range of industrial applications, such as in pharmacy by adsorption of drugs to kill the germs, humidifiers which reduce the humidity by adsorbing moisture, gas masks to avoid inhalation of poisonous gas, chromatographic analysis, gaseous capture (e.g., CO_2 and H_2 adsorption) and storage, water purification, cosmetics, and adsorption chillers [1−8].

Globing warming and discharge of polluted water into freshwater bodies are the alarming issues worldwide. The adsorption technique is being used widely in environmental protection applications, especially for capturing CO_2 and the removal of contaminants from wastewater [5,6]. This chapter will discuss the fundamentals of the adsorption process and mechanisms.

4.2 Adsorption in water purification

Environmental pollution, especially water pollution, is the most critical issue receiving attention worldwide. Several techniques, including coagulation, ion exchange, adsorption, advanced oxidation, membrane treatment, etc., have been utilized in wastewater purification [9]. However, adsorption is the most commonly practiced method for removal of hazardous organic and inorganic pollutants from wastewater. Adsorption is an economical, fast, environmentally benign, and easy to carry out process, and the adsorbent

Carbon Nanomaterial-Based Adsorbents for Water Purification.
DOI: https://doi.org/10.1016/B978-0-12-821959-1.00004-0

can also be reused several times without any deterioration in its performance [10]. Adsorption is a physicochemical interaction on the surface of the adsorbent with the adsorbate. Depending on the nature of the interactions between the adsorbent and adsorbate molecules, the adsorption can be classified as physical adsorption or physisorption and chemical adsorption or chemisorption. Adsorption allows the segregation of toxic substances from dilute solutions.

4.3 Adsorbent

The material that adsorbs the selected compounds from the solution is termed the adsorbent. The physicochemical characteristics of the adsorbent play a significant role in the removal of toxic compounds from wastewater. It should have a high surface area with active sites, specific functional groups, porosity, chemical and thermal stability, defects, unique morphology, ease of functionalization, and mechanical strength [11,12]. A variety of materials have been tested as adsorbents for water contamination remediation (Fig. 4.1). Generally, nanomaterials are preferred for adsorption purposes rather than bulk materials because of their high surface area with active sites. On moving toward the nanoscale, the surface to volume ratio increases along with increased surface energy, which introduces the active sites for the adsorption of water contaminants.

Commonly used nanomaterials for the adsorption of pollutants are metal oxides (i.e., Fe_3O_4, CuO, TiO_2, etc.) [14,15], transition metal chalcogenides (TMCs, i.e., NiS/Ni_3S_4, MoS_2, ZnS, $ZnSe$, etc.) [16−20], h-boron nitride [21], MXenes (transition metal carbides and carbonitrides) [22,23], layered double hydroxides (LDHs) [24], magnetic nanomaterials [25], gels (hydrogels, aerogels, xerogels) [20,25−28], polymer nanocomposites [29−32], zeolites [33], and carbon nanomaterials (CNMs) (graphene and its derivatives, carbon nanotubes (CNTs), fullerenes, nanodiamonds, carbon-rich materials, that is, carbon nanofibers (CNFs), activated coke, nanoporous carbons (NPCs), and graphitic-carbon nitride (g-C_3N_4)) [34−40]. Table 4.1 shows the adsorption capacity of several adsorbents for the removal of various adsorbates from wastewater.

4.4 Types of adsorbents

Adsorbents can be classified either based on their origin (i.e., natural, synthetic, and semisynthetic) or their characteristics (i.e., metal oxide/sulfide-based nanoparticles, layered nanomaterials, gels, polymer–based nanomaterials, carbon–based nanomaterials,

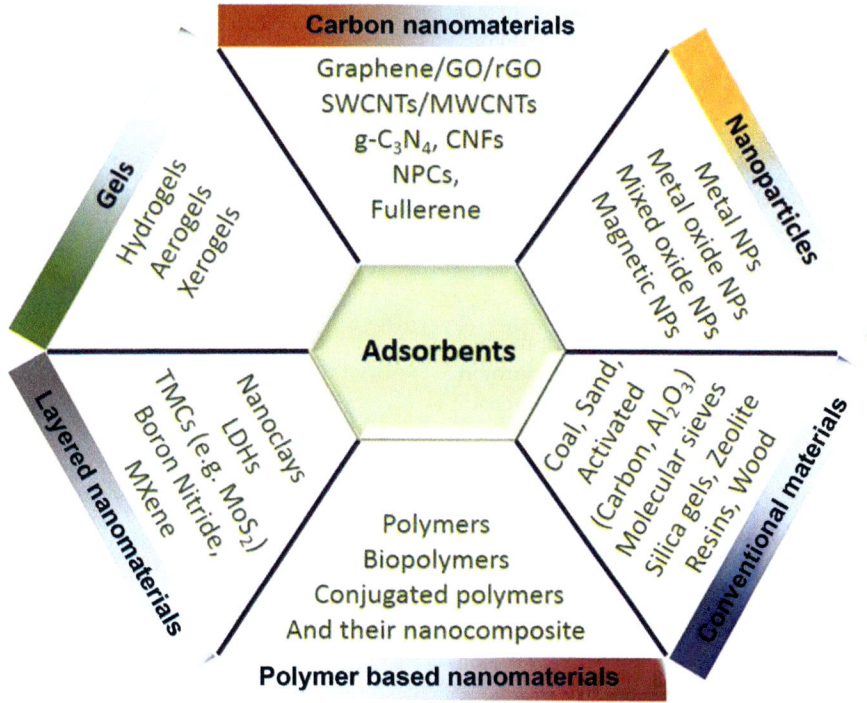

Figure 4.1 Types of adsorbents used for water purification [13]. *Copyright 2020, Elsevier Science Ltd.*

and conventional materials). Herein, in the next section types of adsorbents are discussed based on their characteristics.

4.4.1 Metal oxide and metal sulfide-based nanomaterials

Metal oxide and metal sulfide-based nanomaterials have been investigated for a wide range of applications, such as catalysis, lubrication, pharmaceutical, energy storage, environmental remediation, and adsorbents in water purification [16,54−58]. Transition metals oxides/sulfides are economically favorable, abundantly available, and exhibit remarkable properties for an adsorbent, such as high surface area and active site, which facilitate the adsorption contaminants onto their surface [16,59]. Recently, several transition metal-based adsorbents, such as CuO, MoS_2, WO_3, ZnO, $K_2SnSb_2S_6$, $Cs_2SnSb_2S_6$, $K_2SnSb_2Se_6$, and their nanocomposites with other nanomaterials have been explored as adsorbents to clean wastewater [6,16,60−63]. Kumar et al. [52] synthesized the MoS_2 nanostructures with controlled morphologies (microspheres, microrods, and microrods with hair-like structures) to adsorb Pb(II) ions from industrial mine water. Among them all MoS_2 microrods with hair-like structures (MoS_2-N-H) performed better in Pb(II) adsorption via Pb-S complexation formation and electrostatic interaction (Fig. 4.2A).

Table 4.1 Adsorption capacity of several nanomaterials for various water pollutants in an aqueous medium.

Adsorbent	Adsorbate	Adsorption capacity (mg/g)	References
Graphene sand hybrid (D-GSH)	Methyl Violet (MV)	2564	[41]
	Pb(II)	781	
	Cd(II)	793	
Poly(N-vinylcarbazole)−GO	Pb(II)	982	[42]
Poly(acrylic acid)-based superadsorbent nanocomposite hydrogel (PAA-NC)	Methylene Blue (MB)	2100	[43]
Activated carbon	Naproxen	39.5	[44]
	Ketoprofen	24.7	
	Diclofenac	56.2	
	Ibuprofen	12.6	
Fe_3O_4/MoS_2	Congo Red (CR)	71	[45]
MoS_2 microspheres	MB	297	[16]
	Malachite Green (MG)	204	
	Rhodamine 6G (RhB 6G)	216	
	Fuchsin Acid (FA)	183	
	CR	146	
Polyaniline/TiO_2	Acid Red G (ARG)	454	[46]
Magnetic hierarchical porous carbon	Methyl Orange (MO)	1522	[47]
MWCNTs/iron oxide/β-CD	p-nitrophenol	69.5	[48]
Mg/Al double layered hydroxide	Humic Acid	98.8	[49]
	Fulvic Acid	97.6	
ZrO_2-C	Trichlorophenol	306	[50]
rGO/Nd_2O_3 aerogel	RhB	197	[51]
	Indigo Disulfonate	397	
MoS_2 microrods with hairlike structures (MoS_2-N-H)	Pb(II)	303	[52]
ZnO/Ag/GO	Naphthalene	500	[53]

After adsorbing Pb(II) ions, MoS_2-N-H was converted into $PbMoO_{4-x}S_x$ nanospindles, which were further employed for the photodegradation of ciprofloxacin (CIP). This process reduces the generation of secondary waste and utilizes the spent-adsorbent for the degradation of organic contaminants.

Iron is one of the most abundantly available metals in the Earth's crust. Iron oxide nanomaterials, either standalone or doped and composite forms, are also broadly studied as adsorbents for the removal of organic/inorganic pollutants [66−68]. In an iron oxide-based nanocomposite, iron oxide not only improves the adsorption capacity of the nanocomposite material but also helps in the separation of adsorbents after adsorption, which is a big challenge. Using an external magnetic field, iron oxide-based nanocomposite material can

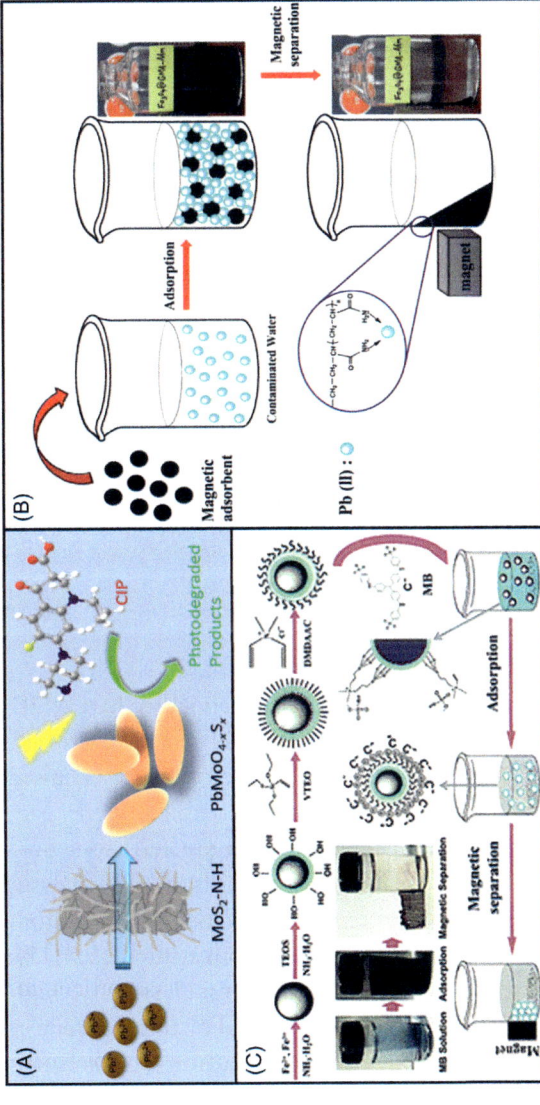

Figure 4.2 (A) Adsorption of Pb(II) ions on MoS$_2$ microrods with hair-like structures (MoS$_2$-N-H) and photodegradation of ciprofloxacin (CIP) using Pb(II) adsorbed MoS$_2$-N-H (PbMoO$_{4-x}$S$_x$) nanospindles [52]. Copyright 2019, American Chemical Society Publishers. (B) Adsorption of Pb(II) on the magnetic Fe$_3$O$_4$@GMA—AAm, followed by the magnetic separation [64]. Copyright 2017, Royal Society of Chemistry Publishers. (C) Schematic illustration of chemical preparation of Fe$_3$O$_4$@SiO$_2$—VTEOS—DMDAAC, adsorption of MB dye on Fe$_3$O$_4$@SiO$_2$—VTEOS—DMDAAC and external magnetic separation [65]. Copyright 2018, Royal Society of Chemistry Publishers.

easily be separated from the aqueous medium after absorbing the contaminant [64]. These characteristics of iron oxide are increasing its demand in research and innovation technologies in wastewater remedies following adsorption. Moradi et al. prepared chelating magnetic nanocomposites by the modification of iron oxide nanoparticles with glycidylmethacrylate and finally grafted them with acrylamide (Fe$_3$O$_4$@GMA−AAm), for the adsorption of Pb(II) ions (Fig. 4.2B) [64]. Within 2 minutes of contact time, the adsorption−desorption equilibrium was established between adsorbate and adsorbent. The adsorption process was followed by chemisorption and Pb(II) adsorbed Fe$_3$O$_4$@GMA−AAm was separated from water by the external magnetic field, as shown in Fig. 4.2B. This process eases the separation of adsorbent after adsorption, without any energy consumption costs, for example, centrifugation.

Similarly the adsorption of selectively anionic organic dye was also performed on magnetic silica modified with triethoxyvinylsilane and diallyl dimethyl ammonium chloride (Fe$_3$O$_4$@SiO$_2$−VTEOS−DMDAAC) [65]. After adsorption, methylene blue (MB) dye-adsorbed Fe$_3$O$_4$@SiO$_2$−VTEOS−DMDAAC nanomaterial was nicely dispersed in the aqueous medium, which was separated using an external magnet (Fig. 4.2C). The collected material (dye-adsorbed Fe$_3$O$_4$@SiO$_2$−VTEOS−DMDAAC), was regenerated in the alkaline medium and again reused several times to remove MB dye from the water. Therefore such a magnetic adsorbent can be effectively recollected, regenerated, and reused for the removal of contaminants with the practical advantages of the magnetic property of the adsorbent.

4.4.2 Layered nanomaterial

Layered nanoadsorbents generally include 2D nanomaterials, such as nanoclays, LDHs, 2D transition metal chalcogenides, MXenes, few-layered graphene, h-BN, etc. [10,69−71]. Metal sulfides are explained in Section 4.4.1 and few-layered graphenes is briefly discussed under carbon-based nanomaterials (Section 4.4.5). 2D nanomaterials exhibit lateral sizes between 10 nm to a few micrometers. Layered nanomaterials with large lateral sizes possess a high surface area, which enhances their applications in adsorption and catalysis. Layered silicate nanomaterials, such as laponite, bentonite, montmorillonite, and hectorite, are common examples of nanoclays materials and are derived from natural clay [72]. The nontoxicity, low cost, small molecular size, structural and chemical stability, high porosity, and large specific surface area of nanoclay favors its potential as an adsorbent in wastewater treatment. The highly porous surface and high potential for ion exchange develop the high adsorption potential of the contaminant on the surface of nanoclay. Fig. 4.3A presents the different surface locations for the adsorption of water pollutants on the surface of nanoclay following different mechanisms, that is, physical adsorption, ion-exchange, and chemisorption [72]. Anirudhan et al. prepared polyamidoxime/organobentonite nanocomposite (OBent-PAN) for the removal of Zn(II), Cd(II), and Cu(II)

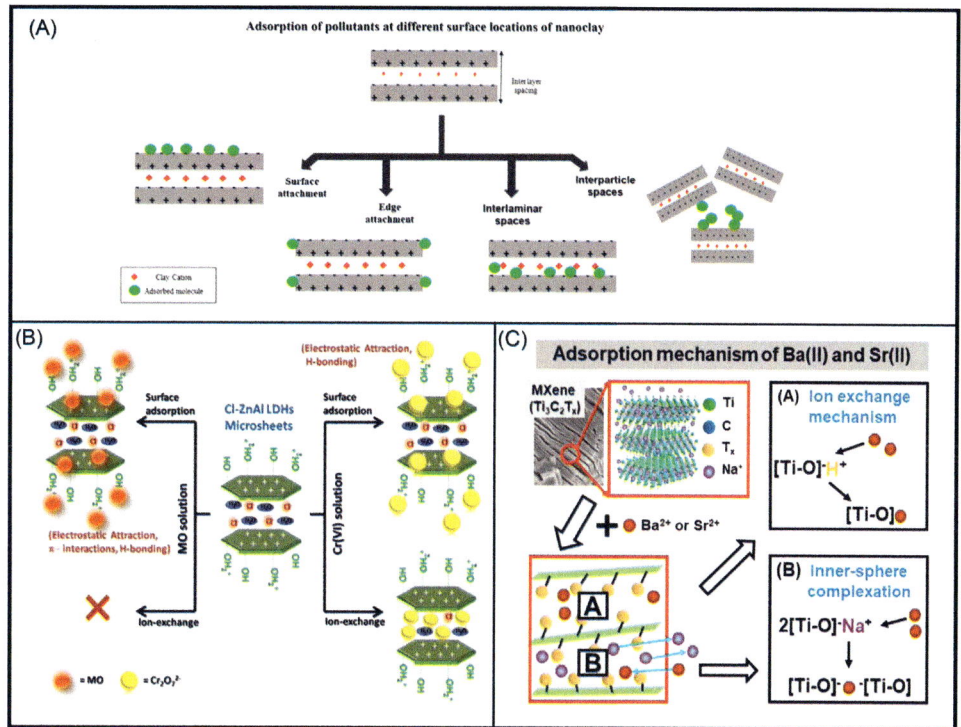

Figure 4.3 (A) Different specific sites for the adsorption of water pollutants on nanoclay [72]. Copyright 2019, Springer Nature Switzerland. (B) Adsorption mechanism of organic dye (MO) and heavy metal ion (Cr(VI)) on Cl intercalated on ZnAl LDHs microsheets [10]. Copyright 2017, Elsevier Science Ltd. (C) Adsorption mechanism of Ba(II) and Sr(II) ions on $Ti_3C_2T_x$ MXenes [74]. Copyright 2020, Elsevier Science Ltd.

ions from industrial wastewater [73]. Ion exchange and metal chelation formation were considered the most favorable mechanisms for the heavy metal ion using polymer−clay (OBent-PAN) nanocomposite with 65.4, 52.6, and 77.43 mg/g adsorption capacity for Zn(II), Cd(II), and Cu(II), respectively.

LDHs nanomaterials or anionic clays are 2D inorganic layered materials with the general formula $[M^{II}_{1-x}M^{III}_x(OH)_2]^{x+}[A^{n-}]_{x/n}.mH_2O$, where M^{II} and M^{III} represent the divalent and trivalent metal ions, respectively, which are carrying excess positive charge in brucite-like layers and A^{n-} is the inorganic or organic interlayer anion [75]. The value of x varies from 0.2 to 0.33. Besides many other applications, LDHs exhibit great potential in wastewater treatment as adsorbent due to their high surface area, low cost, nontoxicity, flexible tunability, exchangeable anionic properties, and surface hydroxyl groups [76]. Hydroxide precipitation, electrostatic interaction, complexation formation, π-π interaction, and chemical bonding are the most common adsorption mechanisms involved in the adsorption of toxic contaminants on the LDHs [77−80]. Fig. 4.3B illustrates the adsorption mechanism of MO dye and Cr(VI) metal ions on the Cl intercalated microsheets of ZnAl LDHs hexagonal nanoplates (Cl-ZnAl LDHs) [10]. Kumar et al. explained the adsorption of MO dye and Cr(VI) ion on Cl-ZnAl LDHs by two plausible mechanisms: (1) physiosorption on the external surface of Cl-ZnAl LDHs, which was associated with the electrostatic, π-π nteraction, and hydrogen bonding between adsorbate and adsorbent; and (2) anion-exchange mechanism. Physiosorption was considered for the adsorption of both the MO and Cr(VI), whereas the anion exchange mechanism was considered only for Cr(VI) adsorption. Cr(VI) adsorption on the Cl-ZnAl LDHs was found to be favorable via anion exchange due to weak electrostatic interaction between the Cl anion and layers, which easily replace the Cl anion by Cr(VI) ions.

MXenes are recent members of the 2D nanomaterials family with the general formula $M_{n+1}X_nT_x$, where M stands for an early transition metal, X is nitrogen and/or carbon, and T_x represents the functional groups present on the surface [71]. The value of n can vary in the range of $1-3$. Due to their nontoxicity, high conductivity, and high thermal and chemical stability, MXenes are gaining attention rapidly worldwide in various applications including adsorption [22]. A very few literatures are available on the applications of various MXenes and, of them all, Ti-based MXenes are the most competent material for environmental remediation applications. Presently, $Ti_3C_2T_x$ is the most investigated MXene in wastewater treatment. Jun et al. employed $Ti_3C_2T_x$ MXene as an adsorbent for the removal of Ba(II) and Sr(II) and explained the adsorption mechanism (Fig. 4.3C) [74]. Ba(II) and Sr(II) are adsorbed by the ion exchange mechanism (Fig. 4.3C (A)) and the inner surface complexation formation (Fig. 4.3C(B)) with the $Ti_3C_2T_x$, which was confirmed by XPS and FTIR analysis.

4.4.3 Polymer-based nanomaterials

Recently, organic polymer-based nanoadsorbents have shown immense potential for water pollutant removal as an efficient alternative to traditional adsorbents due to adjustable surface chemistry, high mechanical strength, high surface area, porosity, ease of functionalization, and regeneration capacity at mild conditions [81]. Biopolymer, conducting polymer, and polymer nanocomposites are the main categories of polymer-based nanoadsorbents. In polymer nanocomposites, the incorporation of inorganic nanomaterial into polymeric adsorbents enhances the mechanical strength, chemical stability, surface area, and selectivity along with the adsorption capacity of the polymer. Polymer nanoadsorbents can readily adsorb various organic dyes, polycyclic aromatic hydrocarbons, inorganic heavy metal ions, phenolic compounds, and their derivatives because of electrostatic, hydrophobic, hydrogen bonding, and ion-exchange interactions [81]. However, the low mechanical strength and thermal and chemical stability are the major drawbacks of polymers, which can be overcome by further modification with other inorganic nanoparticles. Therefore, generally, polymer nanocomposites are attractive adsorbents in this category. Polymer nanocomposites consist of one polymer or copolymer matrix incorporated with inorganic nanomaterials or nanofillers to enhance the physical, chemical, and mechanical properties [82]. The properties of the polymer nanocomposites can be tuned by the nature of the nanomaterials or polymer matrix. These nanocomposites exhibit high thermal stability, biodegradability, mechanical strength, and high surface area, which enhances their capability for environmental remediation applications, including as nanoadsorbents for organic and inorganic water contaminants adsorption [83].

Chitosan, starch, cellulose, dextran, and alginate are common examples of biopolymer-based nanoadsorbents. Most of the used nanoadsorbents are expensive and nonbiodegradable. Therefore biopolymer-based nanoadsorbents are alternatives to the nonbiodegradable adsorbents. They have been widely investigated to adsorb various organic/inorganic water contaminants from wastewater [84–87]. Due to the presence of hydroxyl, carboxyl, amide, and amine functional groups, biopolymer nanoadsorbents can be considered as efficient adsorbents for the physical/chemical adsorption of water pollutants. They are low cost, naturally available, and nontoxic biopolymers with high surface area and adsorption capacity. Biopolymers can easily be functionalized or modified into biopolymer-based nanocomposites due to the presence of reactive functional groups [88]. Chitosan and modified chitosan are good examples of biopolymer-based nanocomposites. Kyzas et al. prepared succinyl–grafted chitosan (CSUC) for the adsorption of organic dye (CR) and heavy metal ion (Zn(II)) in the single and binary component system at different pH (Fig. 4.4A) [89]. On examining the effect of pH, the adsorption of CR and Zn initially increased (pH 2–5) and then the adsorption percentage became constant. Maximum CR removal was observed at pH 5, which might be due to the electrostatic interactions between a positively charged amino group (CSUC) and a negatively charge Cl (CR) group and a negatively charged carboxyl group (CSUC) and a positively charged ammonium group (CR) (Fig. 4.4A). However, for Zn(II) adsorption, two kinds of interaction forces may exist, one might be the chelation

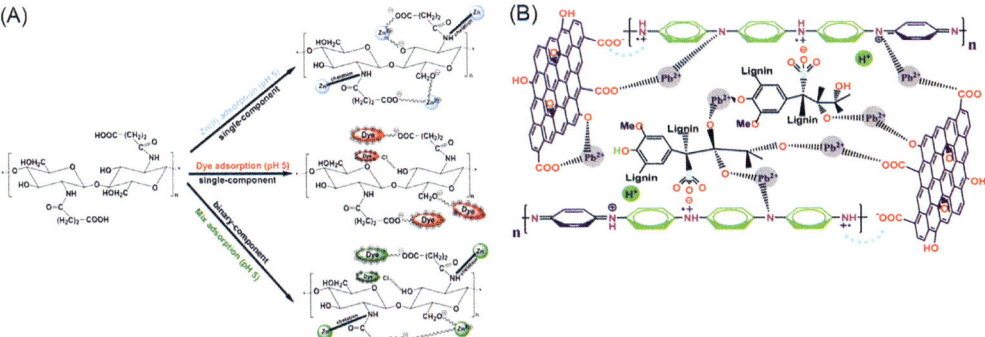

Figure 4.4 (A) Adsorption of CR dye and Zn(II) ion in single-component system and binary component system on CSUC [89]. Reproduced and reprints with permission. Copyrights 2015, Elsevier Science Ltd. (B) Adsorption mechanism of Pb(II) on Lignosulfonate Graphene Oxide Polyaniline (LS-GO-PANI) [90]. Copyrights 2014, American Chemical Society Publishers.

formation between the chitosan and Zn(II) and the other might be the electrostatic interaction between negatively charged carboxyl group (CSUC) and positively charged Zn(II) at pH 5 (Fig. 4.4A). In a binary component (CR and Zn(II)) system, the adsorption of both contaminants was found to be efficient with electrostatic interaction and chelation formation. CSUCs exhibit high reusability potential and only 6% loss in removal ability was observed from the first to the sixth cycle, whereas on further use the decline in removal percentage increases significantly.

Several conductive polymer and conduction polymer-based nanocomposites have also been utilized in wastewater treatment. Conducting polymers and their derived nanomaterials have attracted an enormous amount of attention from researchers in various fields due to their low cost, abundant availability, excellent conductivity, and environmental stability [91,92]. Polypyrrole (Ppy), polyaniline (PANI), and polythiophene (PTh) are common examples of conducting polymers. The availability of massive π-electrons contributes to adsorption through π-π stacking. Also, the heteroatoms, for example, nitrogen atoms in Ppy and PANI and S in PTh, participate in electrostatic interactions and hydrogen bonding for improved adsorption. Lin et al. investigated the adsorption potential of lignosulfonate graphene oxide polyaniline (LS-GO-PANI)-based ternary nanocomposite for Pb(II) ion adsorption (Fig. 4.4B) [90]. In LS-GO-PANI ternary nanocomposite, all three nanomaterials (LS, GO, and PANI), collectively improve the adsorption potential. GO particles improve the specific surface area by avoiding the aggregation of PANI and provide numerous functional groups to the ternary nanocomposite and hence facilitate the interaction between the adsorbent and Pb(II) through electrostatic interaction. The presence of the N atom on the polymeric chain of PANI exhibits strong affinity with Pb(II) and hence contributes to the adsorption process.

Similarly, LS consists of abundant electronegative sulfonic groups which attract the positive Pb(II) significantly. This results in the rapid and efficient removal of Pb(II) from wastewater. The mutual benefits of nanomaterials and polymeric chains make polymer nanocomposites efficient nanoadsorbents with a number of interaction sites.

4.4.4 Gels

Gel-based nanoadsorbents are broadly classified into aerogel, hydrogel, and xerogel. These are three-dimensional (3D) nanoporous polymeric materials, prepared by following the sol—gel process. During preparation, the molecular precursors are swollen in the solvent medium and form a cross-linked polymeric gel-like structure [93]. Based on the process used for the venting out of the liquid medium from the solid network without affecting the skeleton, they are divided into aerogel and xerogel. In aerogel, the wet gel is dried using freeze-drying [94], supercritical drying [95], or ambient pressure drying [96], whereas in xerogel, the solvent is evaporated under standard conditions using conventional methods [97]. When the solvent medium is water during the preparation of nanoporous gel, it is known as a hydrogel [98]. Exceptionally high porosity, large internal pore volume, and a large surface area favors the application of aerogel, xerogel, and hydrogel nanoporous materials in adsorption.

Aerogels are lightweight materials, and exhibit high surface area ($200-1000$ m^2/g), large amount of controlled pore size distribution, high conductivity, transparency, low density, flexibility, low dielectric constant, and high mechanical strength, which makes them potent candidates in a wide range of applications including the adsorption of water contaminants [99,100]. Compared to xerogels, aerogels exhibit a larger surface area and porosity and large pore volume [101]. Graphene-based aerogels, silica-based aerogels, and zeolite-based aerogels are the most common examples of aerogels. They can easily be modified and functionalized in order to improve the selective adsorption via electrostatic interaction, π-π stacking, hydrogen bonding, and chelation. Chang et al. briefly explained the plausible mechanism of MB dye selective adsorption on 4-HIF/NaOH (4-hydroxyindole-formaldehyde/NaOH) aerogel (Fig. 4.5) [102]. Electrostatic and π-π interactions play a major role in the adsorption of MB on 4-HIF/NaOH aerogel. Also, the pretreatment of 4-HIF with NaOH enhanced the adsorption, as the Na cation also interacts with the indole ring of MB dye following the cation$-\pi$ interaction and the hydroxide

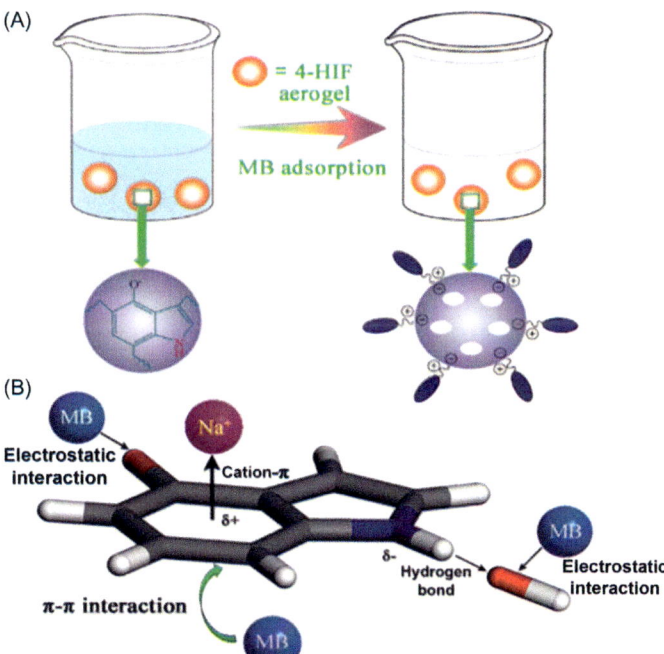

Figure 4.5 Schematic diagram of (A) 4-HIF aerogel structure and adsorption of MB dye on 4-HIF aerogel; (B) mechanism of MB dye adsorption on 4-HIF aerogel. *Copyrights 2019, Royal Society of Chemistry Publications [102].*

anion of NaOH forms hydrogen bonds with the indoleamine, which creates a stable $Na^+-indole-OH^-$ complex and promotes the selective adsorption of MB.

Xerogels are mesoporous materials with high thermal stability, that are nontoxic, cost-effective, biocompatible, with high surface area and high porosity, and can easily be modified [103]. Carbon and silica-based xerogels are mostly studied for water purification following adsorption [103,104]. The pore size of the carbon xerogels (CX) can easily be tuned according to the size of the adsorbates [105]. Zheng et al. prepared an xerogel using polyvinyl alcohol, chiral monomer tartaric acid dihydrazide, and 4-formylphenylboronic acid to adsorb several textile dyes (MB, methyl orange (MO), and Rhodamine B (RhB)) and explained the mechanism (Fig. 4.6) [106]. Electrostatic, π-cation, hydrogen bonding, and π-π interactions are the main interaction forces for the physical adsorption of dye on the xerogels.

Hydrogels are highly porous, economical, and rich in hydroxyl, amide, and carboxyl groups, which can easily be swelled up in water or any biological fluids without dissolving [107]. The swelling characteristic of the hydrogels is a reversible process that occurs through hydration, capillary, and osmotic forces, which are counterbalanced and produce the expanded cross-linked chain network of the polymer. These opposing effects

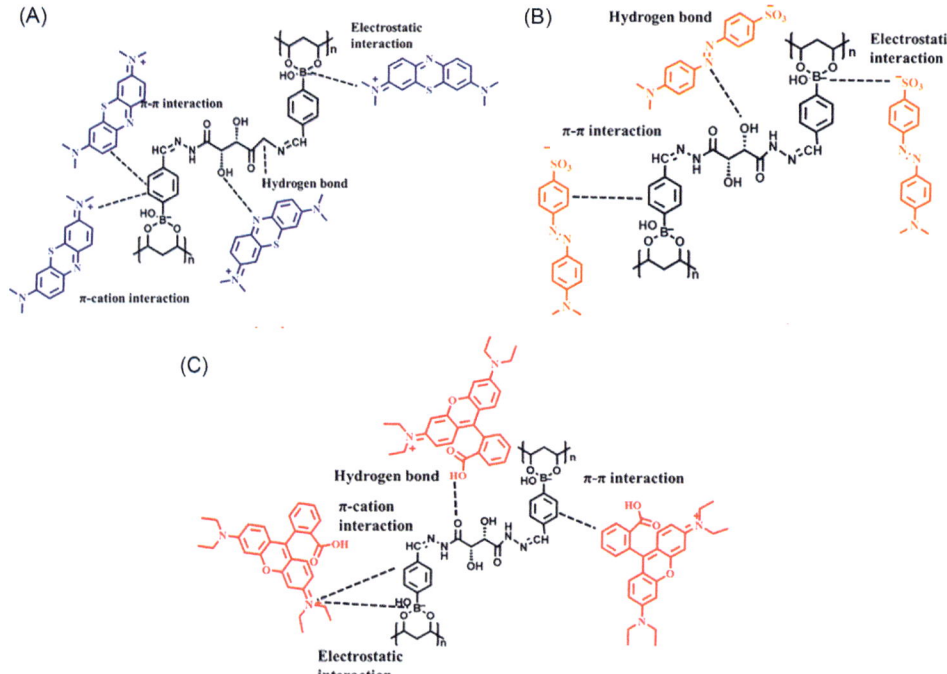

Figure 4.6 Adsorption mechanism of (A) MB, (B) RhB, and (C) MO adsorption on xerogel [106]. *Reproduced and Reprint with slight modifications, Copyrights 2019, Royal Society of Chemistry Publishers.*

regulate the inherent characteristics of a hydrogel such as diffusion, internal transport, and mechanical strength [108]. Several types of a hydrogel, such as graphene-based hydrogels, polyacrylamide-based hydrogel, functionalized copolymer hydrogels, biopolymer-based hydrogels, and hydrogel nanocomposites, have been employed to treat wastewater through adsorption [20,32,109−111]. Electrostatic interaction is the main interaction force of organic contaminant removal using hydrogel. Several functional groups on hydrogel, such as −COOH, −OH, −NH$_2$, and −NH−, interact with the oppositely charged pollutants. For example, − COOH can readily adsorb cationic pollutants via electrostatic interaction, or under optimum pH conditions it will deprotonate to form −COO$^-$ and adsorb the positively charged contaminants [111]. Alternatively, anionic pollutants are preferably adsorbed at acidic conditions. At acidic conditions, the functional groups, such as amino groups, protonate and attract the negatively charged species to be adsorbed [112]. Therefore the optimization of reaction parameters is also necessary to enhance the adsorption efficiency, In addition to electrostatic interaction, ion-exchange, complexation formation, π-π interaction, and hydrogen bonding also participates in the adsorption of water contaminants by hydrogel adsorbent [111].

4.4.5 Carbons nanomaterials

For decades, carbon materials have attracted intense attention from scientists and been employed in a prodigious number of applications [34,113,114]. CNMs are the most common nanoadsorbents, due to their abundant availabilities, excellent adsorption capacities, cost-effectiveness, high chemical and thermal stabilities, high active surface areas, and environment-friendly natures, and thus contribute to wastewater management [115,116]. Being highly porous, with a large surface area, activated carbon (AC) has been the most commonly used adsorbent for years. However, the high cost of AC restricts its use, and more economical alternatives are needed. Therefore different forms of carbon or functionalized carbon such as graphene [15,117], CNTs [118−120], carbon nitride (g-C$_3$N$_4$) [121], carbon nanodiamonds [122], CNFs [123], fullerenes [37], carbon dots [124], and carbon-derived nanocomposites [125,126] have been implemented for adsorption. The high capacity for regeneration of CNMs encourages their application as adsorbents to remove contaminants in wastewater. Numerous studies have been done on the investigation of adsorption characteristics of the carbon-based nanomaterials [118,127]. Carbon nanoadsorbents can be further divided on the basis of their structure (Fig. 4.7) into zero dimensional (0D), one-dimensional (1D), two-dimensional (2D), three-dimensional (3D), and carbon-based nanocomposites.

The discovery of carbon-based 0D nanomaterials started with sp^2 hybridized C$_{60}$ fullerenes [129]. C$_{60}$ fullerene consists of 60 carbon atoms, arranged in a soccer ball structure. Fullerene is also available in other analogous structures, for example, C$_{70}$, C$_{80}$, C$_{76}$, C$_{94}$, etc. [130]. Fullerene exhibits the homogeneous π electron environment all over the structure, which allows a strong π-π interaction with organic

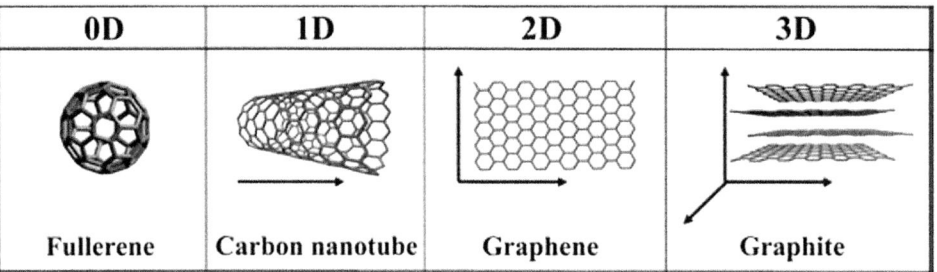

0D	1D	2D	3D
Fullerene	Carbon nanotube	Graphene	Graphite

Figure 4.7 Example of 0D (fullerene), 1D (CNTs), 2D (graphene) and 3D (graphite) CNMs [128]. *Copyrights 2015, Springer Nature.*

contaminants in the adsorption process. The adsorption potential of fullerenes was also found to be higher than activated carbon for the adsorption of organochlorine compounds from aqueous medium [131].

Recently, other 0D carbon nanostructural materials, such as carbon quantum dots (CQDs), nanodiamonds (NDs), fullerene, and carbon black nanoparticles (CBNPs), were introduced with exceptional properties [132,133]. CQDs are quasispherical nanoparticles of less than 10 nm diameter and composed of carbon with hydrogen, oxygen, nitrogen, and other elements. Due to their specific composition and structure, they exhibit unique optical and electrical properties and replaced the use of many traditional metal-based quantum dots in several applications, for example, photocatalysis, pharmaceuticals, and photovoltaics [134,135]. Nanodiamond is a nanocrystalline CNM with a diamond structure and characteristics such as hardness, chemical inertness, and optical transparency [136]. Fullerenes are spheroidal hollow carbon structures in which sp^2 hybridized carbon is directly attached to three neighbors in an arrangement of five-/six-membered rings. Numerous forms of fullerene have been observed and their sizes are noted to be in the range of 30−3000 carbon atoms [137]. CBNPs are produced by the thermal decomposition and combustion of the carbonaceous material [138]. They are inexpensive, exhibit good conductivity but are carcinogenic, which can be suppressed by biological activities [139,140]. These 0D CNMs have been used as nanoadsorbents in water purification but are still not much explored [141−144].

CNTs and CNFs are the most popular examples of 1D CNMs. CNTs are aromatic sp^2 hybridized carbon nanosheets, which are rolled to form hollow tubular graphitic structures. With a unique structure, 1D CNMs exhibit remarkable chemical and mechanical properties with high aspect ratio, ease of functionalization, fast water transport, and most important high surface area [145] These properties of 1D CNMs make them potent candidates for various applications, such as energy storage, pharmaceuticals, catalysis, and adsorption in environmental remediation [146−149]. CNTs can be of two types: single-walled CNTs (SWCNTs) and multiwalled CNTs (MWCNTs) (Fig. 4.8).

Along with the interstitial spaces and external surfaces, CNTs also exhibit pores and channels between the adjacent tubes for the adsorption of contaminants. In CNTs

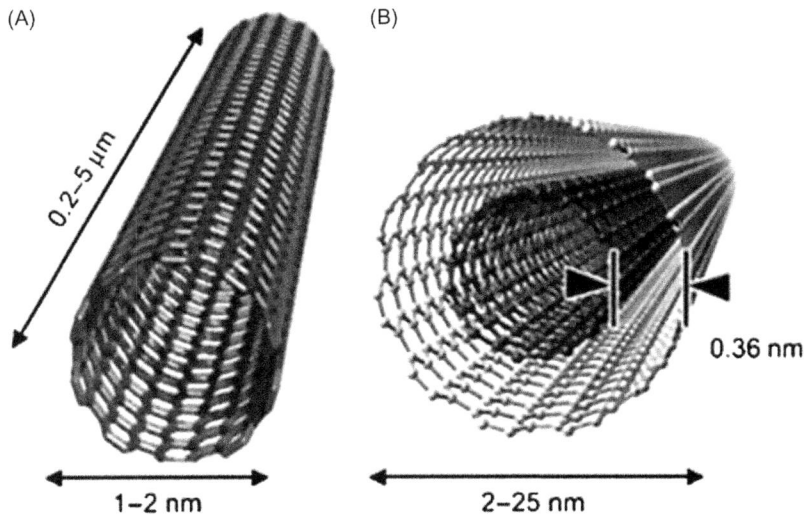

(A)

0.2-5 μm

1-2 nm

(B)

0.36 nm

2-25 nm

Figure 4.8 The conceptual structure of (A) SWCNTs and (B) MWCNTs with their typical dimensions in length, width, and the distance between the layers [152]. *Copyright 2009, Taylor and Francis.*

adsorption can occur at the external and internal surface of tubes, the groove between the contact of adjacent tubes, inside the open end pore of tubes and interstitial channels between the tubes [128]. CNTs can easily interact and adsorb the water contaminant molecules through electrostatic forces, π-π interactions, hydrogen bonding, hydrophobic interaction, and van der Waal forces. Similarly, CNFs exhibit a graphite platelets structure, arranged in a well-ordered manner, which grow in all possible directions in a fiber axis orientation [150]. The availability of an extensive number of edges for adsorption purposes promotes the adsorption through physical and/or chemical interactions. Highly ordered crystalline nature provide a high surface area ($100-250$ m^2/g) which can also be tailored by modification up to 700 or more m^2/g and deliver enhanced adsorption efficiency [150]. The adsorption selectivity and efficiency of CNTs can also be altered by introducing the different functional groups, which play a major role in changing the adsorption properties of CNT, including hydrogen-bonding potential and polarity [101,151].

Graphene is a popular 2D material because of its extraordinary properties which are due to exceptional high specific surface area, chemical inertness, and unique morphology [153]. Graphene is a two-dimensional carbon nanosheet, arranged in a honeycomb lattice. Theoretically, the surface area of graphene is exceptionally high (2600 m^2/g) which encourages its application in adsorption and it delivers an attractive and promising alternative to various traditional employed adsorbents, such as activated carbons. Graphene and its derivatives, such as graphene oxide (GO), reduced graphene oxide (rGO), functionalized graphene oxide (fGO), and nanocomposites with other materials, easily adsorb the pollutants primarily via complexation formation, electrostatic, π-π, hydrophobic, and van der Waals interactions [127]. Fig. 4.9 illustrates the various interaction forces involved in

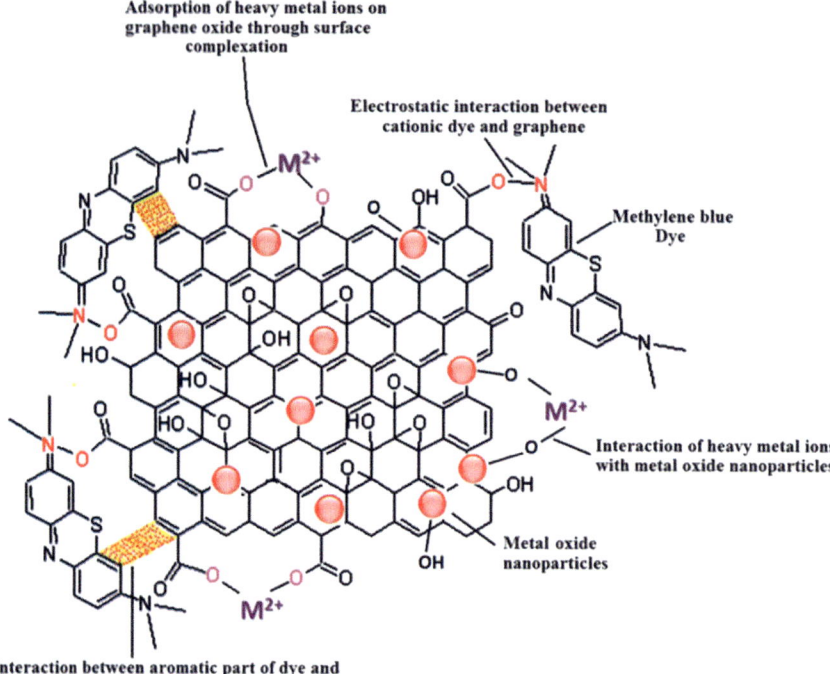

Figure 4.9 Different interaction forces that facilitate the adsorption of organic dye (MB) on the metal oxide/graphene oxide nanocomposite material [15]. *Copyright 2014, Royal Society of Chemistry Publications.*

the adsorption of MB dye on a metal oxide/graphene oxide nanocomposite [15]. Functionalization and/or incorporation of graphene can significantly improve the sensitivity and selectivity of graphene-based nanoadsorbents, which supports its use in industrial applications [154]. Furthermore, the separation of graphene-based nanomaterials after adsorption is again a challenging task, which can be overcome by the magnetization of graphene-based nanoadsorbents which can be separated by an external magnetic field [155]. Graphitic carbon nitride (g-C$_3$N$_4$) is another type of layered 2D CNM that is widely studied as a photocatalyst rather than adsorbent.

Graphite is the most stable 3D CNM. It grows in all three axes and has been explored as an adsorbent to remove water contaminants [156]. However, due to a small surface area in comparison to other CNMs, it is less popular as an adsorbent in water purification systems [157]. 3D CNMs can also be constructed using graphene into graphene aerogels, sponges, foams, and hydrogels, and investigated for adsorption applications [158]. Generally, electrostatic and π-π interactions are the driving forces of adsorption of contaminants on the 3D CNMs.

Among all the forms of carbon, graphene (2D) and CNTs (1D)-based nanomaterials and nanocomposites have received a huge amount of attention as nanoadsorbents [159,160]. The adsorption potential of all forms of CNMS (0D, 1D, 2D, 3D and carbon nanocomposites) is explained in Chapters 8−13.

4.4.6 Conventional adsorbents

Activated carbon, ion-exchange resins, alumina, zeolites, and silica gel are the most popular examples of the conventional adsorbent materials. Classification of conventional adsorbents can further be divided into three categories (Fig. 4.10): (1) commercially activated carbon; (2) inorganic materials which include zeolites, silica gel, activated alumina, and molecular sieves; and (3) ion-exchange resins [161]. The fundamental characteristics of all kinds of conventional adsorbents are high porosity and high surface area.

Indeed, the carbon-based adsorbents are the most widely used and the oldest adsorbents used municipally and in industries. Charcoal is one of the finest examples of the oldest used carbon-based adsorbents. The low cost, nontoxicity, porous texture, and high surface area enhance activated carbon's use as an adsorbent to adsorb contaminants. It can efficiently remove the organic (pesticides, textile dyes, pharmaceuticals, aromatic, and phenolic derivatives) and inorganic (heavy metal ions) contaminants from wastewater through adsorption. Activated carbon can be prepared from a variety of industrial and agricultural waste, such as coke, wood, coconut shells, sawdust, and rice hulls, by catalytic activation of pyrolyzed char [162]. Activated carbon exists in two forms: (1) powdered activated carbon (PAC) and (2) granular activated carbon (GAC) [163]. Between these two forms of activated carbons, GAC is much more adaptable as an adsorbent for the removal of water contaminants. Fig. 4.11A shows the highly porous nature of activated carbon from corn straw

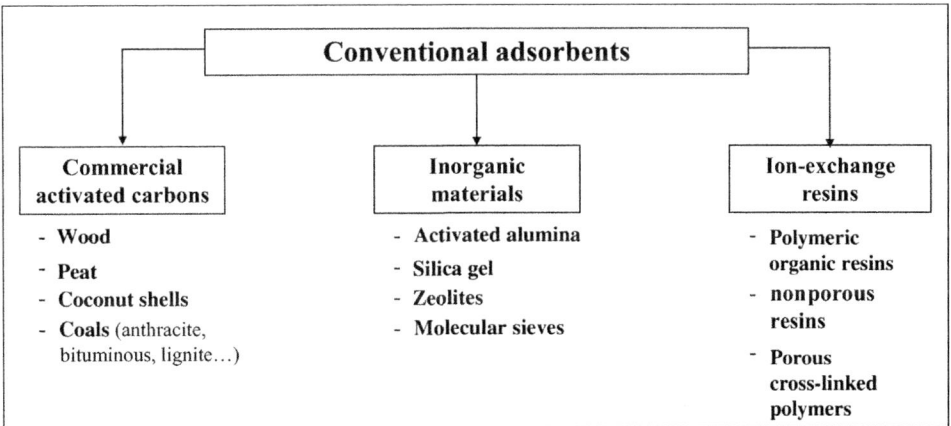

Figure 4.10 Classification of conventional adsorbents [161]. *Copyright 2019, Springer Nature Switzerland.*

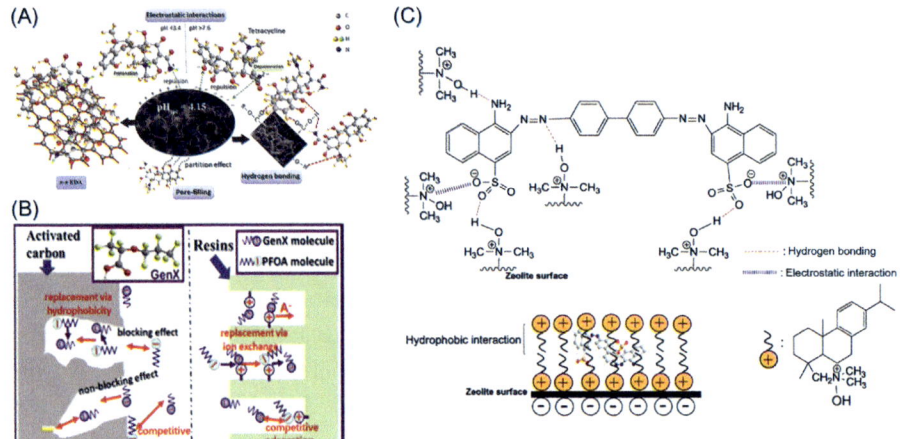

Figure 4.11 (A) Possible interaction forces during the adsorption of tetracycline on the activated carbon [164]. Copyrights 2019, Elsevier Science Ltd. (B) Adsorption mechanism of GenX and replacement by co-existing pollutant PFOA on activated carbon and anion exchange resin [165]. The internal letters represent the attraction forces during the adsorption of adsorbate on adsorbent which signifies as follows: a: hydrophobic interaction; b: electrostatic repulsion; replacement of GenX by PFOA by hydrophobic interactions; d: anion exchange, e: electrostatic interaction and f: replacement of GenX by coexisting PFOA by anion-exchange. Copyrights 2019, Elsevier Science Ltd. (C) A plausible mechanism of CR dye on DAAO modified zeolite by electrostatic, hydrophobic and hydrogen bonding [166]. Copyrights 2014, Elsevier Science Ltd.

and the adsorption of organic contaminant, that is, tetracycline through pore filling, electrostatic interaction, hydrogen bonding, and π-π electron donor–acceptor interactions [164]. The graphitized surface of activated carbon supports the π-π interaction while the reduced the volume of mesopores and micropores after the adsorption confirms the adsorption via pore filling. FTIR and XPS results confirm the hydrogen bonding between the hydroxyl group, carboxylic group of activated carbon, and heteroatoms (O and N) on the tetracycline surface. Therefore the highly porous nature and functional groups on the activated carbon mutually contribute to the efficient adsorption of water pollutant.

Similarly, activated alumina, zeolites (crystalline alumina silicates), and polymer resins are highly porous materials with a large surface area which promotes the adsorption efficiency. Fig. 4.11B displays the adsorption mechanism of perfluoro-2-propoxypropanoic acid (GenX) on activated carbon and anion exchange resin [165]. Hydrophobic interaction between GenX and activated carbon is the key adsorption interaction force for the adsorption of GenX on the activated carbon and overcomes the negative effect caused due to electron repulsion between negatively charged adsorbent and adsorbate. In the presence of coexisting perfluorooctanoic acid (PFOA), the adsorption of GenX is decreased on activated carbon as more hydrophobic PFOA significantly replaces the GenX on the activated carbon surface following hydrophobic interaction. However, anion exchange is the main driving force for the adsorption of GenX on anion-exchange resin. Unlike activated carbon, anion-exchange resin also performs electrostatic interactions between negatively charged GenX and positively charged resin. Adsorbed GenX on ion-exchange resin surface can also be replaced by coexisting PFOA via ion exchange. Fig. 4.11C presents the detailed mechanism of adsorption of Congo red (CR) dye on the N,N-dimethyl dehydroabietyla-mine oxide (DAAO) modified zeolites [166]. Hydrogen bonding between the heteroatoms (O and N) in functional groups of CR dye and H atoms of quaternary ammonium of DAAO modified zeolite layers, electrostatic interaction between the positively charged DAAO modified zeolite surface and negative groups ($-SO_3^-$) on CR dye, and hydrophobic interactions are the leading influential forces for the high rate of adsorption. Therefore, plenty of active sites with functional groups on the conventional adsorbents contribute efficiently to the adsorption of water contaminants on the surface.

4.5 Types of adsorption processes

Contact between the solid adsorbent and the pollutant solution is vital to achieving effective adsorption for both laboratory and industrial-scale treatments [167]. A variety of contact systems, including batch methods, continuous moving bed, fixed-bed-, fluidized bed, and pulsed bed processes have been utilized to collect experimental data and for practical applications [168]. However, each method has its own advantages and disadvantages, which are listed in Table 4.2. The most commonly used methods for liquid/solid adsorption are a continuous process (fixed-bed-type or columns) and discontinuous process (batch process) methods (Fig. 4.12).

4.5.1 Discontinuous process/batch process

Batch process methods are often used for laboratory-scale optimization of adsorption conditions, and small- and medium-sized water applications. They are favored by researchers due to their low-cost, simplicity (need only a mixing tank), and smooth operation. Parameters, including pH, temperature, contact time, and ionic strength, are easily adjustable in these processes. In this mode of operation, a known quantity of adsorbent is mixed into the wastewater until the concentration of pollutant is decreased and the adsorption—desorption equilibrium is reached.

4.5.2 Continuous process/column process

Continuous process methods are principally exploited in large-sized/industrial applications. They are expensive processes, as they require more equipment. Continuous process methods allow the contact of an almost constant concentration of adsorbate with

Table 4.2 Advantage and disadvantages of several adsorption processes.

Types of adsorption method	Description of adsorption method	Advantages	Disadvantages
Batch adsorption	A known quantity of adsorbent is mixed in the solution (consisting adsobate) of constant volume and stirred/shaken at constant rate for the sufficient time to reach adsorption-desorption equilibrium.	Cheap and simple to operate; Most of the adsorption reaction parameters such as pH, contact time, temperature and ionic strength can be monitored/controlled.	Only applicable for small size adsorption application and not suitable for industrial scale.
Continuous moving bed adsorption	It is a steady-state system; in which both adsorbent and adsorbate are in continuous motion and the bed of adsorbent is fixed.	The adsorbent can be collected and regenerated.	Complex and expensive technique A large amount of adsorbent is needed in this process
Continuous fixed-bed adsorption	It is an unsteady state rate-controlling process; Adsorbate is continuously flowed at a constant rate over the fixed bed of adsorbent; Adsorption of adsorbate follows at a specific region of bed known as Mass transfer zone (MTZ).	Simple and inexpensive method; Using for industrial purposes with a high amount of contaminated water.	After a particular period, the MTZ particles do not take participate in adsorption; To conduct quantitative adsorption approach it is difficult to design and optimize fixed-bed column
Continuous fluidized bed adsorption	In this method, gas has passed through to fluidizing the adsorbents particles and adsorbate is in contact with the fluidizing adsorbent bed.	Applicable for industrial applications; Employed for high amount of contaminated water.	Complicated and costly method; Sufficient and insufficient flow of adsorbent bed, does not provide constant contact of adsorbate and adsorbent molecules
Pulsed bed adsorption	In this method, the column consists of number of adsorbent beds and the bed from the bottom can be replaced with new/fresh bed from the top.	Simple and cheap technique; Required low amount of adsorbent.	Used for small amount of wastewater with minimum concentration of contaminant;

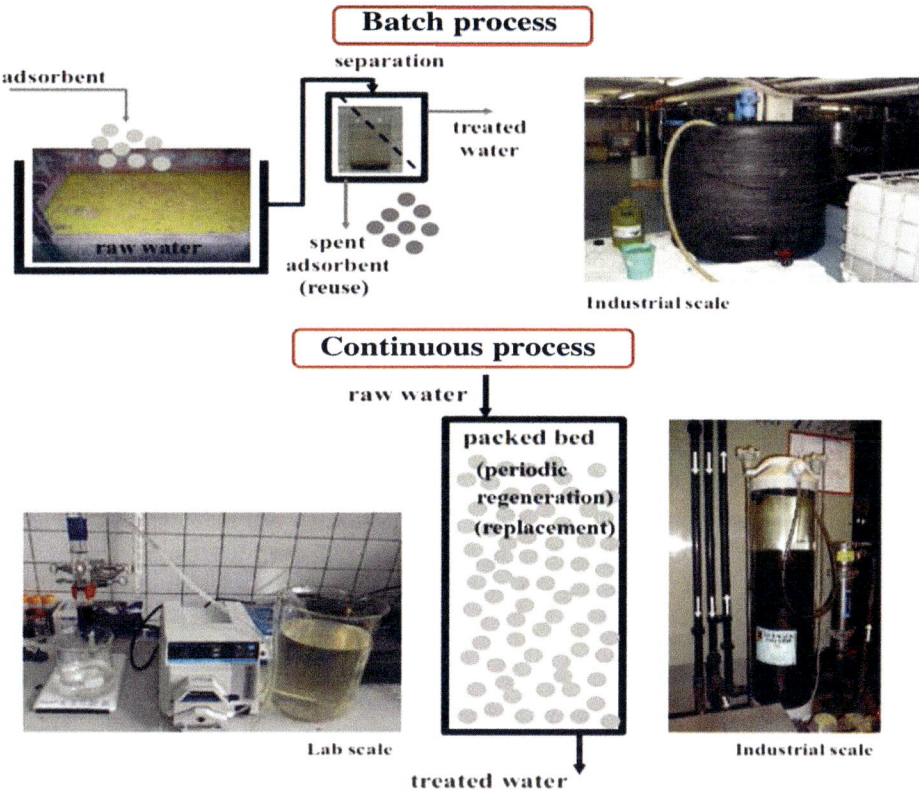

Figure 4.12 Two significant processes used for adsorption: (A) batch processes, (B) continuous processes. *Reproduced with permission from Crini G, Lichtfouse E, Wilson LD, Morin-Crini N. Conventional and non-conventional adsorbents for wastewater treatment. Environ Chem Lett 2018:1—19 [169], with slight modifications. Copyright 2018, Springer International Publishing.*

adsorbent, longer residence times, and better mass and heat transfer behavior [169]. In the column method, the adsorbent is in continuous contact with the adsorbate molecules and provides efficient adsorption.

4.6 Adsorption mechanism

Understanding the mechanism of the adsorption process is an important and challenging step, which can directly suggest improvements in adsorption feasibility through advances in adsorbent design and adsorption/desorption conditions. Improved feasibility motivates further improvement in economic sustainability, reaction environment, regeneration, recyclability of the adsorbent, and overall practicality of the adsorption process. Therefore determination and study of the adsorption mechanism are essential. The adsorbent exhibits higher energy at the surface rather than in bulk. The degree of adsorption is enhanced with the increase in surface energy of the adsorbent at given reaction conditions. Adsorption mechanisms can be broadly divided into three categories, that is, physisorption, chemisorption, and ion exchange (Fig. 4.13) [169]. The most described mechanisms are the physisorption and chemisorption mechanisms.

4.6.1 Physisorption adsorption mechanism

Physisorption is the most extensively studied mechanism that involves reversible weak intermolecular physical interactions. It refers to surface adsorption without disruption of the electronic orbitals of the adsorbent and adsorbate, involving van der Waals interactions, electrostatic interactions, hydrogen bonding, diffusion, and hydrophobic interactions (π-π interactions and Yoshida interactions) [10,16,170]. Physisorption is an exothermic phenomenon, and the rate of reaction decreases with increase in temperature [171]. It follows the formation of multilayer of the adsorbed molecule on the

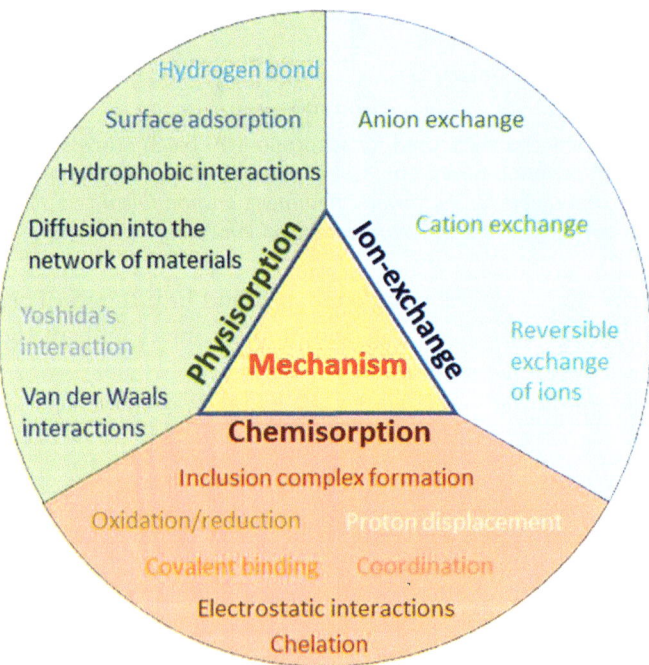

Figure 4.13 Schematic showing the most common adsorption mechanisms [13]. *Copyright 2020, Elsevier Science Ltd.*

surface of the adsorbent. However, to study the role of physisorption in the adsorption process, the foremost challenge is to determine the role of the surface functional groups of the adsorbate and adsorbent, which participate in each interaction. Once the role of the surface functional groups in the adsorption mechanism is clarified, modification of the surface functional groups and adsorption conditions is possible. Bayrak et al. investigated the adsorption efficiency of MB dye on activated carbon produced by watermelon rind (WAC) and concluded that the adsorption is based on physisorption [172]. Fig. 4.14 displays the weak physical forcers that is, electrostatic interaction and π-π stacking between the adsorbate (MB) and adsorbent (WAC) responsible for the adsorption.

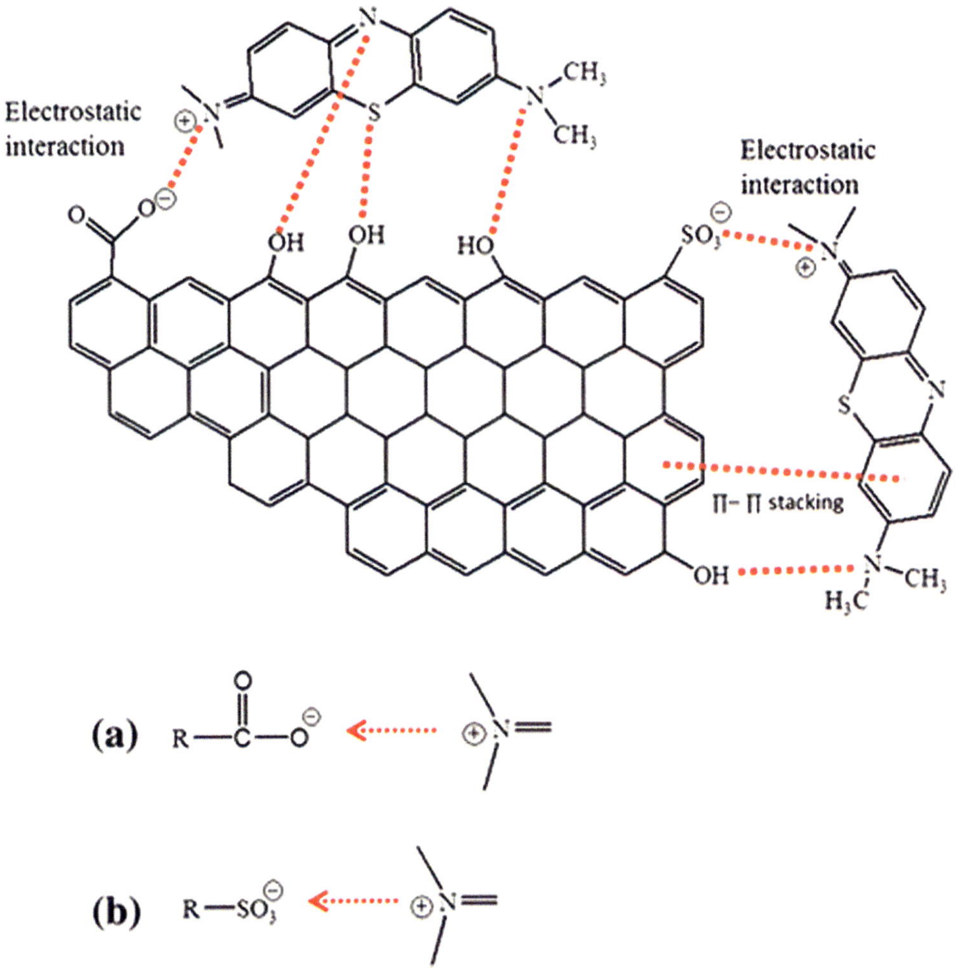

Figure 4.14 Graphical representation of possible adsorption mechanism of methylene blue dye on WAC [172]. *Copyright 2016, Springer International Publishing Switzerland.*

4.6.2 Chemisorption adsorption mechanism

In contrast, chemisorption involves chemical interactions, that is, the involvement of electronic orbitals and valence forces, between absorbent and adsorbate. It generates chemical bonds to the surface of the adsorbent and is irreversible. The mechanism of chemisorption processes can involve complex formation, chelation, covalent bonding, redox reactions, and proton displacement. As a consequence of chemisorption, it changes in the electronic state of adsorbent can be observed using analytical techniques, for example, FTIR, Raman, XPS, magnetic susceptibility, etc. As chemisorption is a chemical process, it is associated with the activation energy of the reactants, and the rate of chemisorption depends on the activation energy of the process, that is, processes, with higher activation energy take more time reach equilibrium, and hence the rate of adsorption is lower. For a process to be energetically favorable, the adsorbate should interact efficiently with the active sites of the adsorbent. Chemisorption is a temperature-dependent process; therefore, adsorption processes that are governed by chemisorption can be controlled by altering the temperature [173]. Chemisorption process is endothermic and generally increases with increase in temperature. Generally, adsorption of inorganic contaminants follows chemisorption mechanism [52,174]. Ray et al. examined the effect of temperature on the adsorption efficiency of molybdenum sulfide/thiol-functionalized multiwalled CNTs (MoS_2/SH-MWCNT) towards the removal of heavy metal ions (Pb(II) and Cd(II)) from industrial wastewater. The adsorption process was following chemisorption mechanism and the rate of adsorption was found to be increased with temperature [174].

However, the adsorption of crystal violet (CV) dye on the woody biochar is controlled by both physisorption and chemisorption [175]. Fig. 4.15 shows that the physical forces such as van der Waals interaction, electrostatic interaction, H-bonding, π-π interaction which helps for the physisorption of CV on Gliricidia biochar which was derived at $700°C$ temperature (GBC700). In contrast for chemisorption, interaction between the negatively charged surface of GBC700 and positively charged CV dye takes place. At pH > 7, due to the formation of carbonate and phenolate anions, the GBC700 surface becomes negatively charged and interact positively charged CV. Therefore, the collective physical and chemical forces tend for higher adsorption of CV dye molecules.

4.6.3 Ion-exchange adsorption mechanism

Ion exchange is an adsorption phenomenon which is reversible, stoichiometric, and electrostatic [176]. In this process, the ions on the surface of the adsorbent are replaced by adsorbate ions, for example, demineralization of water, in which cations are replaced by H^+, and anions are replaced by OH^-. Ion-exchange resins are of two types: cation exchangers, which possess positively charged ions and replace the cations; and anion exchangers, which possess negatively charged ions and replace the anions. Ion exchange is a complex process, as when the targeted material is adsorbed, the exchanged ion is released

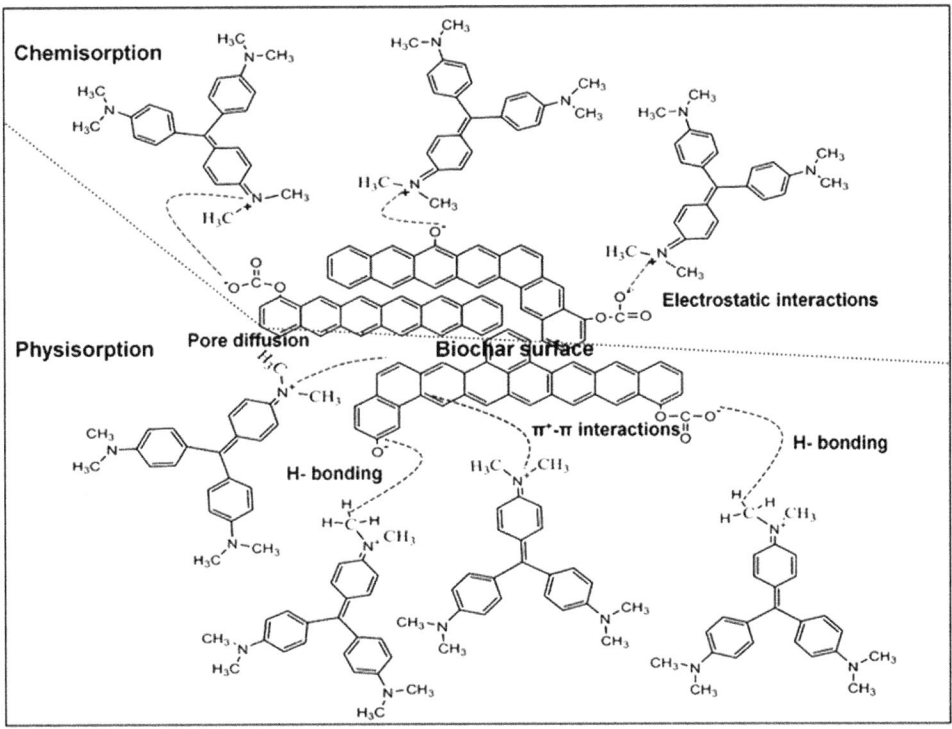

Figure 4.15 Depicting the possible adsorption mechanism of crystal violet dye on Gliricidia biochar (GBC700) derived at 700°C [175]. *Copyright 2017, Springer Nature.*

into the surrounding environment, and the electroneutrality of the environment has to be maintained throughout the process. This is significantly affected by the presence of coions and their transport through the exchange matrix. The particle diffusion, co-ion presence, conditions of the process, and selectivity of the targeted adsorbate can be improved on understanding the adsorbent—adsorbate interaction mechanism.

Chemisorption (surface complexation formation, precipitation, chemical reduction), ion-exchange (cation exchange, anion exchange), and physisorption (mostly electrostatic interactions) mechanisms are collectively explained by Ma et al. [177]. Fig. 4.16 illustrates the primarily considered adsorption forces for the heavy metal ions adsorption on biochar [177]. These five modes of adsorption can be described as follows:

1. Heavy metal ions react with the electron–rich domain and functional groups of the adsorbent for surface complexation formation: Chemisorption.
2. Sometimes heavy metal ions react with the adsorbent and form some insoluble salts to get precipitate: Chemisorption.
3. Reduction of the heavy metal ion and the adsorption of the reduced species is known as chemical reduction adsorption: Chemisorption.

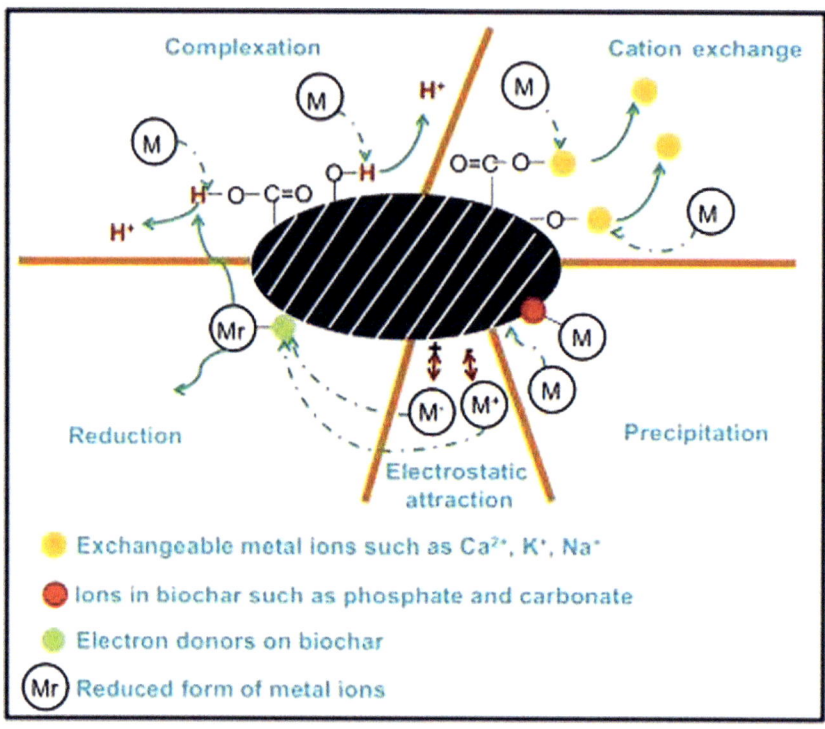

Figure 4.16 Proposed illustration of adsorption of heavy metal ions on biochar with possible interaction forces [177]. *Copyright 2017, Elsevier Science Ltd.*

4. Cation exchange occurs between the metals ions and protons on biochar surface: Ion exchange.
5. Electrostatic interaction is constructed by the interaction between oppositely charged species. Positively charged heavy metal ions attracted towards the negatively charged biochar surface and got adsorbed: Physisorption and Chemisorption.

4.7 Conclusion

Among the various known technologies to treat wastewater, adsorption is one of the most popular, low cost, simple, and easy to scale up technologies. It can remove both organic and inorganic contaminants to generate effluents-free water after treatment. In this process, a foreign substance (adsorbent) with high surface active area, porosity, and functional groups is introduced into the polluted water, which interacts with the water contaminants (adsorbate) in the aqueous phase and adsorbs them to clean the water. For efficient adsorption, the chemistry between the adsorbate and adsorbent should be comparable to produce several interactions, such as π-π stacking, electrostatic, hydrophobic, hydrogen bonding, ion-exchange, and pore filling. The design and modification of the adsorbent is the key step to performing the transfer of adsorbate from the aqueous phase to the surface. A detailed description of several types of adsorbents is explained along with the adsorption mechanisms. Keeping in view the advancements in nanotechnology, the main focus is on the nanoadsorbents. Additionally, nanoadsorbents exhibit a higher surface area than bulk phases, which is the fundamental property for a high rate of adsorption. Batch and column adsorption processes were also discussed to understand the advantages and disadvantages of each process. Furthermore, different adsorption mechanism, that is, physisorption, chemisorption, and ion exchange are also explained in detail, which directly suggest the required improvements in the adsorbent design and adsorption/desorption condition to enhance the adsorption feasibility.

References

[1] Wakshlak RB-K, Pedahzur R, Avnir D. Antibacterial activity of silver-killed bacteria: the "zombies" effect. Sci Rep 2015;5:9555.

[2] Umprayn K, Mendes RW. Hygroscopicity and moisture adsorption kinetics of pharmaceutical solids: a review. Drug Dev Ind Pharm 1987;13:653−93.

[3] Von Blucher H, De Ruiter E. Activated charcoal filter layer for gas masks. Google Patents; 1991.

[4] Macko T, Pasch H. Separation of linear polyethylene from isotactic, atactic, and syndiotactic polypropylene by high-temperature adsorption liquid chromatography. Macromolecules 2009;42:6063−7.

[5] Ghanbari T, Abnisa F, Daud WMAW. A review on production of metal organic frameworks (MOF) for CO_2 adsorption. Sci Total Environ 2019;135090.

[6] Gusain R, Gupta K, Joshi P, Khatri OP. Adsorptive removal and photocatalytic degradation of organic pollutants using metal oxides and their composites: a comprehensive review. Adv Colloid Interface Sci 2019;272:102009.

[7] Petry T, Bury D, Fautz R, Hauser M, Huber B, Markowetz A, et al. Review of data on the dermal penetration of mineral oils and waxes used in cosmetic applications. Toxicol Lett 2017;280:70−8.

[8] Askalany AA, Salem M, Ismail IM, Ali AHH, Morsy MG. A review on adsorption cooling systems with adsorbent carbon. Renew Sustain Energy Rev 2012;16:493−500.

[9] Santhosh C, Velmurugan V, Jacob G, Jeong SK, Grace AN, Bhatnagar A. Role of nanomaterials in water treatment applications: a review. Chem Eng J 2016;306:1116−37.

[10] Kumar N, Reddy L, Parashar V, Ngila JC. Controlled synthesis of microsheets of ZnAl layered double hydroxides hexagonal nanoplates for efficient removal of Cr(VI) ions and anionic dye from water. J Environ Chem Eng 2017;5:1718−31.

[11] Carmalin Sophia A, Lima EC, Allaudeen N, Rajan S. Application of graphene based materials for adsorption of pharmaceutical traces from water and wastewater- a review. Desalination Water Treat 2016;57:27573−86.

[12] Ihsanullah, Abbas A, Al-Amer AM, Laoui T, Al-Marri MJ, Nasser MS, et al. Heavy metal removal from aqueous solution by advanced carbon nanotubes: critical review of adsorption applications. Sep Purif Technol 2016;157:141−61.

[13] Gusain R, Kumar N, Ray SS. Recent advances in carbon nanomaterial-based adsorbents for water purification. Coord Chem Rev 2020;405:213111.

[14] Hu JS, Zhong LS, Song WG, Wan LJ. Synthesis of hierarchically structured metal oxides and their application in heavy metal ion removal. Adv Mater 2008;20:2977−82.

[15] Upadhyay RK, Soin N, Roy SS. Role of graphene/metal oxide composites as photocatalysts, adsorbents and disinfectants in water treatment: a review. RSC Adv 2014;4:3823−51.

[16] Massey AT, Gusain R, Kumari S, Khatri OP. Hierarchical microspheres of MoS_2 nanosheets: efficient and regenerative adsorbent for removal of water-soluble dyes. Ind Eng Chem Res 2016;55:7124—31.

[17] Wang Z, Mi B. Environmental applications of 2D molybdenum disulfide (MoS_2) nanosheets. Environ Sci Technol 2017;51:8229—44.

[18] Pala IR, Brock SL. ZnS nanoparticle gels for remediation of Pb^{2+} and Hg^{2+} polluted water. ACS Appl Mater Interfaces 2012;4:2160—7.

[19] Taghizadeh F, Ghaedi M, Kamali K, Sharifpour E, Sahraie R, Purkait M. Comparison of nickel and/or zinc selenide nanoparticle loaded on activated carbon as efficient adsorbents for kinetic and equilibrium study of removal of Arsenazo (III) dye. Powder Technol 2013;245:217—26.

[20] Kumar N, Mittal H, Parashar V, Ray SS, Ngila JC. Efficient removal of rhodamine 6G dye from aqueous solution using nickel sulphide incorporated polyacrylamide grafted gum karaya bionano-composite hydrogel. RSC Adv 2016;6:21929—39.

[21] Lei W, Portehault D, Liu D, Qin S, Chen Y. Porous boron nitride nanosheets for effective water cleaning. Nat Commun 2013;4:1777.

[22] Zhang Y, Wang L, Zhang N, Zhou Z. Adsorptive environmental applications of MXene nanomaterials: a review. RSC Adv 2018;8:19895—905.

[23] Zhang Q, Teng J, Zou G, Peng Q, Du Q, Jiao T, et al. Efficient phosphate sequestration for water purification by unique sandwich-like MXene/magnetic iron oxide nanocomposites. Nanoscale 2016;8:7085—93.

[24] Chubar N, Gilmour R, Gerda V, Mičušík M, Omastova M, Heister K, et al. Layered double hydroxides as the next generation inorganic anion exchangers: synthetic methods versus applicability. Adv Colloid Interface Sci 2017;245:62—80.

[25] Ambashta RD, Sillanpää M. Water purification using magnetic assistance: a review. J Hazard Mater 2010;180:38—49.

[26] Lee B, Lee S, Lee M, Jeong DH, Baek Y, Yoon J, et al. Carbon nanotube-bonded graphene hybrid aerogels and their application to water purification. Nanoscale 2015;7:6782—9.

[27] Ye S, Liu Y, Feng J. Low-density, mechanical compressible, water-induced self-recoverable graphene aerogels for water treatment. ACS Appl Mater Interfaces 2017;9:22456—64.

[28] Yang Y, Deng Y, Tong Z, Wang C. Renewable lignin-based xerogels with self-cleaning properties and superhydrophobicity. ACS Sustain Chem Eng 2014;2:1729—33.

[29] Yin J, Deng B. Polymer-matrix nanocomposite membranes for water treatment. J Membr Sci 2015;479:256—75.

[30] Unuabonah EI, Taubert A. Clay—polymer nanocomposites (CPNs): adsorbents of the future for water treatment. Appl clay Sci 2014;99:83—92.

[31] Zhao G, Huang X, Tang Z, Huang Q, Niu F, Wang X-K. Polymer-based nanocomposites for heavy metal ions removal from aqueous solution: a review. Polym Chem 2018.

[32] Kumar N, Mittal H, Alhassan SM, Ray SS. Bionanocomposite hydrogel for the adsorption of dye and reusability of generated waste for the photodegradation of ciprofloxacin: a demonstration of the circularity concept for water purification. ACS Sustain Chem Eng 2018;6:17011—25.

[33] Wang S, Peng Y. Natural zeolites as effective adsorbents in water and wastewater treatment. Chem Eng J 2010;156:11—24.

[34] Mauter MS, Elimelech M. Environmental applications of carbon-based nanomaterials. Environ Sci Technol 2008;42:5843—59.

[35] Kumar S, Nair RR, Pillai PB, Gupta SN, Iyengar MAR, Sood AK. Graphene oxide—$MnFe_2O_4$ magnetic nanohybrids for efficient removal of lead and arsenic from water. ACS Appl Mater Interfaces 2014;6:17426—36.

[36] Gupta K, Khatri OP. Reduced graphene oxide as an effective adsorbent for removal of malachite green dye: plausible adsorption pathways. J Colloid Interface Sci 2017;501:11—21.

[37] Gupta VK, Saleh TA. Sorption of pollutants by porous carbon, carbon nanotubes and fullerene- an overview. Environ Sci Pollut Res 2013;20:2828—43.

[38] Si Y, Ren T, Ding B, Yu J, Sun G. Synthesis of mesoporous magnetic Fe_3O_4@carbon nanofibers utilizing in situ polymerized polybenzoxazine for water purification. J Mater Chem 2012;22:4619—22.

[39] Fu Y, Qin L, Huang D, Zeng G, Lai C, Li B, et al. Chitosan functionalized activated coke for Au nanoparticles anchoring: green synthesis and catalytic activities in hydrogenation of nitrophenols and azo dyes. Appl Catal B: Environ 2019;255:117740.

[40] Fu Y, Xu P, Huang D, Zeng G, Lai C, Qin L, et al. Au nanoparticles decorated on activated coke via a facile preparation for efficient catalytic reduction of nitrophenols and azo dyes. Appl Surf Sci 2019;473:578—88.

[41] Khan S, Achazhiyath Edathil A, Banat F. Sustainable synthesis of graphene-based adsorbent using date syrup. Sci Rep 2019;9:18106.

[42] Musico YLF, Santos CM, Dalida MLP, Rodrigues DF. Improved removal of lead (II) from water using a polymer-based graphene oxide nanocomposite. J Mater Chem A 2013;1:3789—96.

[43] Hu X-S, Liang R, Sun G. Super-adsorbent hydrogel for removal of methylene blue dye from aqueous solution. J Mater Chem A 2018;6:17612—24.

[44] Baccar R, Sarrà M, Bouzid J, Feki M, Blánquez P. Removal of pharmaceutical compounds by activated carbon prepared from agricultural by-product. Chem Eng J 2012;s211 − 212 : 310 − 7.

[45] Song HJ, You S, Jia XH, Yang J. MoS$_2$ nanosheets decorated with magnetic Fe$_3$O$_4$ nanoparticles and their ultrafast adsorption for wastewater treatment. Ceram Int 2015;41:13896—902.

[46] Wang N, Li J, Lv W, Feng J, Yan W. Synthesis of polyaniline/TiO$_2$ composite with excellent adsorption performance on acid red G. RSC Adv 2015;5:21132—41.

[47] Siyasukh A, Chimupala Y, Tonanon N. Preparation of magnetic hierarchical porous carbon spheres with graphitic features for high methyl orange adsorption capacity. Carbon 2018;134:207—21.

[48] Liu W, Jiang X, Chen X. A novel method of synthesizing cyclodextrin grafted multiwall carbon nanotubes/iron oxides and its adsorption of organic pollutant. Appl Surf Sci 2014;320:764—71.

[49] Fang L, Hou J, Xu C, Wang Y, Li J, Xiao F, et al. Enhanced removal of natural organic matters by calcined Mg/Al layered double hydroxide nanocrystalline particles: adsorption, reusability and mechanism studies. Appl Surf Sci 2018;442:45—53.

[50] Tan Y, Zhu L, Niu H, Cai Y, Wu F, Zhao X. Synthesis of flower-shaped ZrO$_2$—C composites for adsorptive removal of trichlorophenol from aqueous solution. RSC Adv 2015;5:77175—83.

[51] Pan L, Liu S, Oderinde O, Li K, Yao F, Fu G. Facile fabrication of graphene-based aerogel with rare earth metal oxide for water purification. Appl Surf Sci 2018;427:779—86.

[52] Kumar N, Fosso-Kankeu E, Ray SS. Achieving controllable MoS$_2$ nanostructures with increased interlayer spacing for efficient removal of Pb(II) from aquatic systems. ACS Appl Mater Interfaces 2019;11:19141—55.

[53] Mukwevho N, Gusain R, Fosso-Kankeu E, Kumar N, Waanders F, Ray SS. Removal of naphthalene from simulated wastewater through adsorption-photodegradation by ZnO/Ag/GO nanocomposite. J Ind Eng Chem 2020;81:393—404.

[54] Kumar N, Mittal H, Reddy L, Nair P, Ngila JC, Parashar V. Morphogenesis of ZnO nanostructures: role of acetate (COOH −) and nitrate (NO$_3$ −) ligand donors from zinc salt precursors in synthesis and morphology dependent photocatalytic properties. RSC Adv 2015;5:38801—9.

[55] Gusain R, Singhal N, Singh R, Kumar U, Khatri OP. Ionic-liquid-functionalized copper oxide nanorods for photocatalytic splitting of water. ChemPlusChem 2016;81:489—95.

[56] Kumari S, Mungse HP, Gusain R, Kumar N, Sugimura H, Khatri OP. Octadecanethiol-grafted molybdenum disulfide nanosheets as oil-dispersible additive for reduction of friction and wear. FlatChem 2017;3:16—25.

[57] Kumar N, George BPA, Abrahamse H, Parashar V, Ray SS, Ngila JC. A novel approach to low-temperature synthesis of cubic HfO$_2$ nanostructures and their cytotoxicity. Sci Rep 2017;7:9351.

[58] Gusain R, Kumar P, Sharma OP, Jain SL, Khatri OP. Reduced graphene oxide—CuO nanocomposites for photocatalytic conversion of CO$_2$ into methanol under visible light irradiation. Appl Catal B Environ 2016;181:352—62.

[59] Chen L, Zhao X, Pan B, Zhang W, Hua M, Lv L, et al. Preferable removal of phosphate from water using hydrous zirconium oxide-based nanocomposite of high stability. J Hazard Mater 2015;284:35—42.

[60] Raul PK, Senapati S, Sahoo AK, Umlong IM, Devi RR, Thakur AJ, et al. CuO nanorods: a potential and efficient adsorbent in water purification. RSC Adv 2014;4:40580—7.

[61] Luo JY, Lin YR, Liang BW, Li YD, Mo XW, Zeng QG. Controllable dye adsorption behavior on amorphous tungsten oxide nanosheet surfaces. RSC Adv 2015;5:100898—904.

[62] Yuvaraja G, Prasad C, Vijaya Y, Subbaiah MV. Application of ZnO nanorods as an adsorbent material for the removal of As(III) from aqueous solution: kinetics, isotherms and thermodynamic studies. Int J Ind Chem 2018;9:17—25.

[63] Hassanzadeh Fard Z, Islam SM, Kanatzidis MG. Porous amorphous chalcogenides as selective adsorbents for heavy metals. Chem Mater 2015;27:6189—92.

[64] Moradi A, Najafi Moghadam P, Hasanzadeh R, Sillanpää M. Chelating magnetic nanocomposite for the rapid removal of Pb(ii) ions from aqueous solutions: characterization, kinetic, isotherm and thermodynamic studies. RSC Adv 2017;7:433—48.

[65] Chen J, Chen H. Removal of anionic dyes from an aqueous solution by a magnetic cationic adsorbent modified with DMDAAC. N J Chem 2018;42:7262—71.

[66] Mac Rae I. Removal of pesticides in water by microbial cells adsorbed to magnetite. Water Res 1985;19:825—30.

[67] Saha B, Das S, Saikia J, Das G. Preferential and enhanced adsorption of different dyes on iron oxide nanoparticles: a comparative study. J Phys Chem C 2011;115:8024—33.

[68] Wang Y, Cheng R, Wen Z, Zhao L. Facile preparation of Fe_3O_4 nanoparticles with cetyltrimethylammonium bromide (CTAB) assistant and a study of its adsorption capacity. Chem Eng J 2012;181:823—7.

[69] Almasri DA, Rhadfi T, Atieh MA, McKay G, Ahzi S. High performance hydroxyiron modified montmorillonite nanoclay adsorbent for arsenite removal. Chem Eng J 2018;335:1—12.

[70] Manos MJ, Kanatzidis MG. Sequestration of heavy metals from water with layered metal sulfides. Chemistry—A Eur J 2009;15:4779—84.

[71] Rasool K, Pandey RP, Rasheed PA, Buczek S, Gogotsi Y, Mahmoud KA. Water treatment and environmental remediation applications of two-dimensional metal carbides (MXenes). Mater Today 2019;30:80—102.

[72] Awasthi A, Jadhao P, Kumari K. Clay nano-adsorbent: structures, applications and mechanism for water treatment. SN Appl Sci 2019;1:1076.

[73] Anirudhan TS, Ramachandran M. Synthesis and characterization of amidoximated polyacrylonitrile/organobentonite composite for Cu(II), Zn(II), and Cd(II) adsorption from aqueous solutions and industry wastewaters. Ind Eng Chem Res 2008;47:6175—84.

[74] Jun B-M, Park CM, Heo J, Yoon Y. Adsorption of Ba^{2+} and Sr^{2+} on $Ti_3C_2T_x$ MXene in model fracking wastewater. J Environ Manag 2020;256:109940.

[75] Zubair M, Daud M, McKay G, Shehzad F, Al-Harthi MA. Recent progress in layered double hydroxides (LDH)-containing hybrids as adsorbents for water remediation. Appl Clay Sci 2017;143:279—92.

[76] Wang J, Zhang T, Li M, Yang Y, Lu P, Ning P, et al. Arsenic removal from water/wastewater using layered double hydroxide derived adsorbents, a critical review. RSC Adv 2018;8:22694—709.

[77] Gong J, Liu T, Wang X, Hu X, Zhang L. Efficient removal of heavy metal ions from aqueous systems with the assembly of anisotropic layered double hydroxide nanocrystals@carbon nanosphere. Environ Sci Technol 2011;45:6181—7.

[78] Wu X-L, Wang L, Chen C-L, Xu A-W, Wang X-K. Water-dispersible magnetite-graphene-LDH composites for efficient arsenate removal. J Mater Chem 2011;21:17353—9.

[79] Zhang F, Song Y, Song S, Zhang R, Hou W. Synthesis of magnetite—graphene oxide-layered double hydroxide composites and applications for the removal of Pb(II) and 2,4-dichlorophenoxyacetic acid from aqueous solutions. ACS Appl Mater Interfaces 2015;7:7251—63.

[80] Zhang H, Huang F, Liu D-L, Shi P. Highly efficient removal of Cr(VI) from wastewater via adsorption with novel magnetic Fe_3O_4@C@MgAl-layered double-hydroxide. Chin Chem Lett 2015;26:1137—43.

[81] Zare EN, Motahari A, Sillanpää M. Nanoadsorbents based on conducting polymer nanocomposites with main focus on polyaniline and its derivatives for removal of heavy metal ions/dyes: a review. Environ Res 2018;162:173—95.

[82] Singh PP. Ambika. 10 - Environmental remediation by nanoadsorbents-based polymer nanocomposite. In: Hussain CM, Mishra AK, editors. New polymer nanocomposites for environmental remediation. Elsevier; 2018. p. 223—41.

[83] Bushra R. 11 - Nanoadsorbents-based polymer nanocomposite for environmental remediation. In: Hussain CM, Mishra AK, editors. New polymer nanocomposites for environmental remediation. Elsevier; 2018. p. 243—60.

[84] Lu T, Xiang T, Huang X-L, Li C, Zhao W-F, Zhang Q, et al. Post-crosslinking towards stimuli-responsive sodium alginate beads for the removal of dye and heavy metals. Carbohydr Polym 2015;133:587—95.

[85] Mosaferi M, Nemati S, Khataee A, Nasseri S, Hashemi AA. Removal of Arsenic (III, V) from aqueous solution by nanoscale zero-valent iron stabilized with starch and carboxymethyl cellulose. J Environ Health Sci Eng 2014;12:74.

[86] Srinivasan S, Chelliah P, Srinivasan V, Stantley AB, Subramani K. Chitosan and reinforced chitosan films for the removal of Cr (VI) heavy metal from synthetic aqueous solution. Orient J Chem 2016;32:671−80.

[87] Yu X, Tong S, Ge M, Wu L, Zuo J, Cao C, et al. Adsorption of heavy metal ions from aqueous solution by carboxylated cellulose nanocrystals. J Environ Sci 2013;25:933−43.

[88] Vakili M, Rafatullah M, Salamatinia B, Abdullah AZ, Ibrahim MH, Tan KB, et al. Application of chitosan and its derivatives as adsorbents for dye removal from water and wastewater: a review. Carbohydr Polym 2014;113:115−30.

[89] Kyzas GZ, Siafaka PI, Pavlidou EG, Chrissafis KJ, Bikiaris DN. Synthesis and adsorption application of succinyl-grafted chitosan for the simultaneous removal of zinc and cationic dye from binary hazardous mixtures. Chem Eng J 2015;259:438−48.

[90] Yang J, Wu J-X, Lü Q-F, Lin T-T. Facile preparation of lignosulfonate−graphene oxide−polyaniline ternary nanocomposite as an effective adsorbent for Pb(II) ions. ACS Sustain Chem Eng 2014;2:1203−11.

[91] Naseri M, Fotouhi L, Ehsani A. Recent progress in the development of conducting polymer-based nanocomposites for electrochemical biosensors applications: a mini-review. Chem Rec 2018;18:599−618.

[92] Zhan C, Yu G, Lu Y, Wang L, Wujcik E, Wei S. Conductive polymer nanocomposites: a critical review of modern advanced devices. J Mater Chem C 2017;5:1569−85.

[93] Rivas Murillo JS, Bachlechner ME, Campo FA, Barbero EJ. Structure and mechanical properties of silica aerogels and xerogels modeled by molecular dynamics simulation. J Non-Crystalline Solids 2010;356:1325−31.

[94] Chen H-B, Liu B, Huang W, Wang J-S, Zeng G, Wu W-H, et al. Fabrication and properties of irradiation-cross-linked poly (vinyl alcohol)/clay aerogel composites. ACS Appl Mater Interfaces 2014;6:16227−36.

[95] Maleki H, Durães L, Portugal A. Synthesis of lightweight polymer-reinforced silica aerogels with improved mechanical and thermal insulation properties for space applications. Microporous Mesoporous Mater 2014;197:116−29.

[96] Leventis N, Palczer A, McCorkle L, Zhang G, Sotiriou-Leventis C. Nanoengineered silica-polymer composite aerogels with no need for supercritical fluid drying. J Sol-Gel Sci Technol 2005;35:99−105.

[97] Zanto EJ, Al-Muhtaseb SA, Ritter JA. Sol − gel-derived carbon aerogels and xerogels: design of experiments approach to materials synthesis. Ind Eng Chem Res 2002;41:3151−62.

[98] Ahmed EM. Hydrogel: preparation, characterization, and applications: a review. J Adv Res 2015;6:105−21.

[99] Maleki H. Recent advances in aerogels for environmental remediation applications: a review. Chem Eng J 2016;300:98−118.

[100] Fricke J, Tillotson T. Aerogels: production, characterization, and applications. Thin Solid Films 1997;297:212−23.

[101] Khajeh M, Laurent S, Dastafkan K. Nanoadsorbents: classification, preparation, and applications (with emphasis on aqueous media). Chem Rev 2013;113:7728−68.

[102] Zhang L, Yang L, Xu Y, Chang G. Renewable 4-HIF/NaOH aerogel for efficient methylene blue removal via cation−π interaction induced electrostatic interaction. RSC Adv 2019;9:29772−8.

[103] Sriram G, Uthappa UT, Kigga M, Jung H-Y, Altalhi T, Brahmkhatri V, et al. Xerogel activated diatoms as an effective hybrid adsorbent for the efficient removal of malachite green. N J Chem 2019;43:3810−20.

[104] Álvarez S, Ribeiro RS, Gomes HT, Sotelo JL, García J. Synthesis of carbon xerogels and their application in adsorption studies of caffeine and diclofenac as emerging contaminants. Chem Eng Res Des 2015;95:229−38.

[105] Mahata N, Silva AR, Pereira MFR, Freire C, de Castro B, Figueiredo JL. Anchoring of a [Mn (salen)Cl] complex onto mesoporous carbon xerogels. J Colloid Interface Sci 2007;311:152−8.

[106] Ren S, Sun P, Wu A, Sun N, Sun L, Dong B, et al. Ultra-fast self-healing PVA organogels based on dynamic covalent chemistry for dye selective adsorption. N J Chem 2019;43:7701−7.

[107] Varaprasad K, Raghavendra GM, Jayaramudu T, Yallapu MM, Sadiku R. A mini review on hydrogels classification and recent developments in miscellaneous applications. Mater Sci Eng C 2017;79:958−71.

[108] Buwalda SJ, Boere KW, Dijkstra PJ, Feijen J, Vermonden T, Hennink WE. Hydrogels in a historical perspective: from simple networks to smart materials. J Controlled Rel 2014;190:254−73.

[109] Chen Y, Chen L, Bai H, Li L. Graphene oxide−chitosan composite hydrogels as broad-spectrum adsorbents for water purification. J Mater Chem A 2013;1:1992−2001.

[110] Yang S, Fu S, Liu H, Zhou Y, Li X. Hydrogel beads based on carboxymethyl cellulose for removal heavy metal ions. J Appl Polym Sci 2011;119:1204−10.

[111] Van Tran V, Park D, Lee Y-C. Hydrogel applications for adsorption of contaminants in water and wastewater treatment. Environ Sci Pollut Res 2018;25:24569−99.

[112] İnal M, Erduran N. Removal of various anionic dyes using sodium alginate/poly(N-vinyl-2-pyrrolidone) blend hydrogel beads. Polym Bull 2015;72:1735−52.

[113] Allen MJ, Tung VC, Kaner RB. Honeycomb carbon: a review of graphene. Chem Rev 2009;110:132−45.

[114] Wang J. Carbon-nanotube based electrochemical biosensors: a review. Electroanalysis 2005;17:7−14.

[115] Zhang X, Gao B, Creamer AE, Cao C, Li Y. Adsorption of VOCs onto engineered carbon materials: a review. J Hazard Mater 2017;338:102−23.

[116] Yang K, Xing B. Adsorption of organic compounds by carbon nanomaterials in aqueous phase: Polanyi theory and its application. Chem Rev 2010;110:5989−6008.

[117] Kyzas GZ, Deliyanni EA, Bikiaris DN, Mitropoulos AC. Graphene composites as dye adsorbents. Chem Eng Res Des 2017.

[118] Ren X, Chen C, Nagatsu M, Wang X. Carbon nanotubes as adsorbents in environmental pollution management: a review. Chem Eng J 2011;170:395−410.

[119] Ghaedi M, Hassanzadeh A, Kokhdan SN. Multiwalled carbon nanotubes as adsorbents for the kinetic and equilibrium study of the removal of alizarin red S and morin. J Chem Eng Data 2011;56:2511−20.

[120] Sadegh H, Ghoshekandi RS, Masjedi A, Mahmoodi Z, Kazemi M. A review on Carbon nanotubes adsorbents for the removal of pollutants from aqueous solutions. Int J Nano Dimens 2016;7:109.

[121] Cai X, He J, Chen L, Chen K, Li Y, Zhang K, et al. A 2D-g-C_3N_4 nanosheet as an eco-friendly adsorbent for various environmental pollutants in water. Chemosphere 2017;171:192−201.

[122] Wang H-D, Yang Q, Hui Niu C, Badea I. Adsorption of azo dye onto nanodiamond surface. Diam Relat Mater 2012;26:1−6.

[123] Teng M, Qiao J, Li F, Bera PK. Electrospun mesoporous carbon nanofibers produced from phenolic resin and their use in the adsorption of large dye molecules. Carbon 2012;50:2877−86.

[124] Wang L, Cheng C, Tapas S, Lei J, Matsuoka M, Zhang J, et al. Carbon dots modified mesoporous organosilica as an adsorbent for the removal of 2, 4-dichlorophenol and heavy metal ions. J Mater Chem A 2015;3:13357−64.

[125] Xiong Z, Zhang LL, Ma J, Zhao XS. Photocatalytic degradation of dyes over graphene−gold nanocomposites under visible light irradiation. Chem Commun 2010;46:6099−101.

[126] Yao Y, Miao S, Liu S, Ma LP, Sun H, Wang S. Synthesis, characterization, and adsorption properties of magnetic Fe_3O_4@graphene nanocomposite. Chem Eng J 2012;184:326−32.

[127] Baig N, Ihsanullah Sajid M, Saleh TA. Graphene-based adsorbents for the removal of toxic organic pollutants: a review. J Environ Manag 2019;244:370−82.

[128] Machado F, Fagan S, Zanella da Silva I, Andrade M. Carbon Nanoadsorvents. 2015.

[129] Kroto HW, Heath JR, O'Brien SC, Curl RF, Smalley RE. C60: buckminsterfullerene. Nature 1985;318:162−3.

[130] Shanbogh P, Sundaram N. Fullerenes revisited. Resonance 2015;20:123−35.

[131] Berezkin V, Viktorovskii I, Golubev L, Petrova V, Khoroshko L. A comparative study of the sorption capacity of activated charcoal, soot, and fullerenes for organochlorine compounds. Tech Phys Lett 2002;28:885−8.

[132] Yang F, Ren X, LeCroy GE, Song J, Wang P, Beckerle L, et al. Zero-dimensional carbon allotropes—carbon nanoparticles versus fullerenes in functionalization by electronic polymers for different optical and redox properties. ACS Omega 2018;3:5685—91.

[133] Raja IS, Song S-J, Kang MS, Lee YB, Kim B, Hong SW, et al. Toxicity of zero- and one-dimensional carbon nanomaterials. Nanomaterials 2019;9:1214.

[134] Li X, Rui M, Song J, Shen Z, Zeng H. Carbon and graphene quantum dots for optoelectronic and energy devices: a review. Adv Funct Mater 2015;25:4929—47.

[135] Molaei MJ. A review on nanostructured carbon quantum dots and their applications in biotechnology, sensors, and chemiluminescence. Talanta 2019;196:456—78.

[136] Zhang Y, Rhee KY, Hui D, Park S-J. A critical review of nanodiamond based nanocomposites: synthesis, properties and applications. Compos Part B: Eng 2018;143:19—27.

[137] Malhotra BD, Ali MA. Chapter 2 - Functionalized carbon nanomaterials for biosensors. In: Malhotra BD, Ali MA, editors. Nanomaterials for biosensors. William Andrew Publishing; 2018. p. 75—103.

[138] Chaudhuri I, Fruijtier-Pölloth C, Ngiewih Y, Levy L. Evaluating the evidence on genotoxicity and reproductive toxicity of carbon black: a critical review. Crit Rev Toxicol 2018;48:143—69.

[139] Lindner K, Ströbele M, Schlick S, Webering S, Jenckel A, Kopf J, et al. Biological effects of carbon black nanoparticles are changed by surface coating with polycyclic aromatic hydrocarbons. Part Fibre Toxicol 2017;14:8.

[140] Li F, Qi L, Yang J, Xu M, Luo X, Ma D. Polyurethane/conducting carbon black composites: structure, electric conductivity, strain recovery behavior, and their relationships. J Appl Polym Sci 2000;75:68—77.

[141] Tadesse A, RamaDevi D, Hagos M, Battu G, Basavaiah K. Synthesis of nitrogen doped carbon quantum dots/magnetite nanocomposites for efficient removal of methyl blue dye pollutant from contaminated water. RSC Adv 2018;8:8528—36.

[142] Ma W, Fanqing M, Guo L, Wu L, Shibo D, Chen Z, et al. A carbon quantum dot synthesizing method and its application of modifying functional water purification material. Google Patents; 2018.

[143] Raeiszadeh M, Hakimian A, Shojaei A, Molavi H. Nanodiamond-filled chitosan as an efficient adsorbent for anionic dye removal from aqueous solutions. J Environ Chem Eng 2018;6.

[144] Ballesteros E, Gallego M, Valcárcel M. Analytical potential of fullerene as adsorbent for organic and organometallic compounds from aqueous solutions. J Chromatogr A 2000;869:101—10.

[145] Lee J, Jeong S, Liu Z. Progress and challenges of carbon nanotube membrane in water treatment. Crit Rev Environ Sci Technol 2016;46:999—1046.

[146] Zhang Q, Huang J-Q, Qian W-Z, Zhang Y-Y, Wei F. The road for nanomaterials industry: a review of carbon nanotube production, post-treatment, and bulk applications for composites and energy storage. Small 2013;9:1237—65.

[147] Vairavapandian D, Vichchulada P, Lay MD. Preparation and modification of carbon nanotubes: review of recent advances and applications in catalysis and sensing. Analytica Chim Acta 2008;626:119—29.

[148] Eatemadi A, Daraee H, Karimkhanloo H, Kouhi M, Zarghami N, Akbarzadeh A, et al. Carbon nanotubes: properties, synthesis, purification, and medical applications. Nanoscale Res Lett 2014;9:393.

[149] Sarkar B, Mandal S, Tsang YF, Kumar P, Kim K-H, Ok YS. Designer carbon nanotubes for contaminant removal in water and wastewater: a critical review. Sci Total Environ 2018;612:561—81.

[150] Park C, Engel ES, Crowe A, Gilbert TR, Rodriguez NM. Use of carbon nanofibers in the removal of organic solvents from water. Langmuir 2000;16:8050—6.

[151] ALOthman ZA, Habila M, Yilmaz E, Soylak M. Solid phase extraction of Cd (II), Pb (II), Zn (II) and Ni (II) from food samples using multiwalled carbon nanotubes impregnated with 4-(2-thiazolylazo) resorcinol. Microchimica Acta 2012;177:397—403.

[152] Singh I, Rehni A, Kumar P, Kumar M. Carbon nanotubes: synthesis, properties and pharmaceutical applications. Fuller Nanotubes Carbon Nanostructures - Fuller Nanotub Carbon Nanostr 2009;17:361—77.

[153] Kim S, Park CM, Jang M, Son A, Her N, Yu M, et al. Aqueous removal of inorganic and organic contaminants by graphene-based nanoadsorbents: a review. Chemosphere 2018;212:1104—24.

[154] Chen M, Huo C, Li Y, Wang J. Selective adsorption and efficient removal of phosphate from aqueous medium with graphene—lanthanum composite. ACS Sustain Chem Eng 2016;4:1296—302.

[155] Bao S, Yang W, Wang Y, Yu Y, Sun Y. One-pot synthesis of magnetic graphene oxide composites as an efficient and recoverable adsorbent for Cd(II) and Pb(II) removal from aqueous solution. J Hazard Mater 2020;381:120914.

[156] Gao W, Majumder M, Alemany LB, Narayanan TN, Ibarra MA, Pradhan BK, et al. Engineered graphite oxide materials for application in water purification. ACS Appl Mater Interfaces 2011;3:1821−6.

[157] Pérez-Ramírez EE, De La Rosa-Álvarez G, Salas P, Velasco-Santos C, Martínez-Hernández AL. Comparison as effective photocatalyst or adsorbent of carbon materials of one, two, and three dimensions for the removal of reactive red 2 in water. Environ Eng Sci 2015;32:872−80.

[158] Hiew BYZ, Lee LY, Lee XJ, Thangalazhy-Gopakumar S, Gan S, Lim SS, et al. Review on synthesis of 3D graphene-based configurations and their adsorption performance for hazardous water pollutants. Process Saf Environ Prot 2018;116:262−86.

[159] Coroş M, Pogăcean F, Măgeruşan L, Socaci C, Pruneanu S. A brief overview on synthesis and applications of graphene and graphene-based nanomaterials. Front Mater Sci 2019;13:23−32.

[160] Thostenson ET, Ren Z, Chou T-W. Advances in the science and technology of carbon nanotubes and their composites: a review. Compos Sci Technol 2001;61:1899−912.

[161] Crini G, Lichtfouse E, Wilson LD, Morin-Crini N. Conventional and non-conventional adsorbents for wastewater treatment. Environ Chem Lett 2019;17:195−213.

[162] Dias JM, Alvim-Ferraz MCM, Almeida MF, Rivera-Utrilla J, Sánchez-Polo M. Waste materials for activated carbon preparation and its use in aqueous-phase treatment: a review. J Environ Manag 2007;85:833−46.

[163] Dąbrowski A, Podkościelny P, Hubicki Z, Barczak M. Adsorption of phenolic compounds by activated carbon—a critical review. Chemosphere 2005;58:1049−70.

[164] Yang Q, Wu P, Liu J, Rehman S, Ahmed Z, Ruan B, et al. Batch interaction of emerging tetracycline contaminant with novel phosphoric acid activated corn straw porous carbon: adsorption rate and nature of mechanism. Environ Res 2019;108899.

[165] Wang W, Maimaiti A, Shi H, Wu R, Wang R, Li Z, et al. Adsorption behavior and mechanism of emerging perfluoro-2-propoxypropanoic acid (GenX) on activated carbons and resins. Chem Eng J 2019;364:132−8.

[166] Liu S, Ding Y, Li P, Diao K, Tan X, Lei F, et al. Adsorption of the anionic dye Congo red from aqueous solution onto natural zeolites modified with N,N-dimethyl dehydroabietylamine oxide. Chem Eng J 2014;248:135−44.

[167] Ali I. Water treatment by adsorption columns: evaluation at ground level. Sep Purif Rev 2014;43:175−205.

[168] Patel H. Fixed-bed column adsorption study: a comprehensive review. Appl Water Sci 2019;9:45.

[169] Crini G, Lichtfouse E, Wilson LD, Morin-Crini N. Conventional and non-conventional adsorbents for wastewater treatment. Environ Chem Lett 2018;1−19.

[170] Dąbrowski A. Adsorption—from theory to practice. Adv Colloid Interface Sci 2001;93:135−224.

[171] Sing KSW. 10 - Adsorption by active carbons. In: Rouquerol F, Rouquerol J, Sing KSW, Llewellyn P, Maurin G, editors. Adsorption by powders and porous solids. 2nd ed. Oxford: Academic Press; 2014; p. 321−91.

[172] Üner O, Geçgel Ü, Bayrak Y. Adsorption of methylene blue by an efficient activated carbon prepared from citrullus lanatus rind: kinetic, isotherm, thermodynamic, and mechanism analysis. Water Air Soil Pollut 2016;227:247.

[173] Herrero E, Feliu JM, Blais S, Radovic-Hrapovic Z, Jerkiewicz G. Temperature dependence of CO chemisorption and its oxidative desorption on the Pt (111) electrode. Langmuir 2000;16:4779−83.

[174] Gusain R, Kumar N, Fosso-Kankeu E, Ray SS. Efficient removal of Pb(II) and Cd(II) from industrial mine water by a hierarchical MoS_2/SH-MWCNT nanocomposite. ACS Omega 2019;4:13922−35.

[175] Wathukarage A, Herath I, Iqbal MCM, Vithanage M. Mechanistic understanding of crystal violet dye sorption by woody biochar: implications for wastewater treatment. Environ Geochem Health 2019;41:1647−61.

[176] Kammerer J, Carle R, Kammerer DR. Adsorption and ion exchange: basic principles and their application in food processing. J Agric Food Chem 2010;59:22−42.

[177] Li H, Dong X, da Silva EB, de Oliveira LM, Chen Y, Ma LQ. Mechanisms of metal sorption by biochars: biochar characteristics and modifications. Chemosphere 2017;178:466−78.

CHAPTER FIVE

Adsorption equilibrium isotherms, kinetics and thermodynamics

5.1 Introduction

Adsorption is a successful separation technique and has been employed in various applications, including wastewater treatment by adsorbing contaminants from water [1,2]. However, adsorption efficiency and characteristics of the adsorbents, for example, adsorption capacity, feasibility, etc. are examined by different adsorption isotherms (Henry's, Langmuir, Dubinin−Radushkevich, Freundlich, Temkin, Redlich−Peterson, Koble−Carrigan and so on), kinetic models (first order, pseudo-first-order, second-order, pseudo-second-order, intraparticle diffusion, etc.) and thermodynamic parameters (activation energy, entropy, free energy, etc.) [3,4]. This chapter briefly explains the various models of adsorption isotherms and chemical kinetics and elaborates on the use of thermodynamic parameters during the adsorption process.

5.2 Adsorption equilibrium isotherms

Adsorption isotherms assist in the prediction of the adsorption mechanism and interactions between the adsorbate and adsorbent during the adsorption process, at a constant temperature and specific solution pH. The amount of adsorbed substance on the adsorbent is calculated based on remaining concentration (mostly in ppm) in the solution. Each adsorption isotherm has mathematical equations to calculate the other parameters, such as adsorption capacity, and to predict the mechanism. Based on initial concentration (C_o), remaining concentration (C_t) at time t, and the remaining concentration at equilibrium (C_e) of dye in the solution, values are placed in the respective mathematical equations and the best-fitted model provides information about the mechanism of the adsorption and the surface properties of the adsorbent material. Several adsorption models, including the Henry, Langmuir, Dubinin−Radushkevich, Freundlich, Temkin, Redlich−Peterson, Koble−Corrigan, Jovanovic, and Halsey isotherms, have been developed to fit experimental data to predict the mechanism of adsorption (monolayer/multilayers or homogenous/heterogeneous). The most commonly used adsorption isotherm models for the

Carbon Nanomaterial-Based Adsorbents for Water Purification.
DOI: https://doi.org/10.1016/B978-0-12-821959-1.00005-2

removal of contaminants from wastewater are the Langmuir and Freundlich adsorption isotherms, which help to calculate the maximum adsorption capacity (Q_m) of adsorbents. All isotherms are based on linear mathematical ($y = mx + c$) equations and a straight line is observed in the plot. The linear regression constant (R^2) value aids in finding out which adsorption isotherm is suitable to explain the adsorption behavior. The value of R^2 near unity promotes the accessibility of the isotherm model to study the adsorption [5].

The R^2 value is the most reliable tool to define the best-fitted model as it analyzes the adsorption system, quantifies the adsorbate distribution, and verifies the adsorption isotherm theoretical assumptions. In this chapter, we are discussing the basics of a few adsorption isotherms.

5.2.1 Henry's adsorption isotherm

Henry's adsorption isotherm is the simplest isotherm, and states that the amount of adsorbate adsorbed is proportional to the partial pressure of the gas (adsorbate) [6]. This model is generally applicable at low concentration of adsorbate. At equilibrium condition of adsorption the adsorbate amount and the adsorbed phases can be linearly related by Henry's adsorption isotherm linear expression (Eq. 5.1) as follows:

$$Q_e = K_{HE} C_e \tag{5.1}$$

where Q_e (mg/g) and C_e (mg/L) are the amount of adsorbate at equilibrium on the adsorbent and in the solution, respectively, and K_{HE} is the Henry's isotherm constant.

5.2.2 Langmuir adsorption isotherms

This is the most regularly studied adsorption isotherm model, based on the assumption of the formation of a monolayer of adsorbate homogeneously on the adsorbent. The adsorption process that follows the Langmuir adsorption isotherm model indicates that the used adsorbent exhibits finite numbers of identical active sites available for the interaction with adsorbate, and there is no lateral interaction and struggle owing to steric hindrance between the adsorbed molecules on the surface [7,8]. Therefore it follows that there is homogeneous distribution of the adsorbate onto the surface of the adsorbent, with constant enthalpy and adsorption activation energy [9]. Once all available active surface sites are occupied by the adsorbate molecule/ion, no further adsorption onto the surface will take place. The mathematical representation of the Langmuir adsorption isotherm is expressed as:

$$Q_e = \frac{Q_m.K_L.C_e}{1 + K_L.C_e} \tag{5.2}$$

Here, C_e (mg/g) represents the amount of adsorbate remains in the solution at equilibrium, Q_e (mg/g) represents the amount of adsorbate (mg) adsorbed per unit of adsorbent (g), Q_m (mg/g) signifies the maximum adsorption capacity, and K_L (L/mg) is the

Langmuir isotherm constant. The value of K_L correlates with the variation in surface characteristics of the adsorbent, such as specific surface area and porosity.

Eq. (5.2) can be transformed into two linear equations to calculate the Langmuir isotherm parameters. Linear Langmuir isotherm-1 can be expressed as Eq. (5.3):

$$\frac{C_e}{Q_e} = \frac{1}{Q_m} C_e + \frac{1}{Q_m.K_L} \qquad (5.3)$$

The Langmuir isotherm graph is plotted by C_e/Q_e (Y-axis) against C_e (X-axis) and a straight line is observed with intercept $1/(Q_m.K_L)$ and slope $1/Q_m$ values. The regression constant (R^2) value from the graph justifies the applicability of this model. Linear Langmuir isotherm-2 can be written as Eq. (5.4):

$$\frac{1}{Q_e} = \frac{1}{K_L.Q_m} . \frac{1}{C_e} + \frac{1}{Q_m} \qquad (5.4)$$

Using this equation, Q_m and K_L can be calculated using the intercept and slope values from the graph plotted between $1/Q_e$ versus $1/C_e$.

Langmuir plot also helps to calculate the dimensionless constant, which is known as the separation factor (R_L) following Eq. (5.5):

$$R_L = \frac{1}{1 + K_L.C_o} \qquad (5.5)$$

The R_L value indicates whether the adsorption method is favorable ($0 < R_L < 1$) or unfavorable ($R_L > 1$), irreversible ($R_L = 0$) or linear ($R_L < 1$) using K_L and the initial concentration (C_o) in ppm of adsorbate in solution. Adsorption behavior of a number of organic dyes, heavy metal ions, and other water pollutants can be explained following the Langmuir adsorption isotherm model system [10,11].

5.2.3 Freundlich adsorption isotherm

The Freundlich isotherm is highly accepted for heterogeneous adsorption systems in which the adsorbent exhibits dissimilar active surface sites with nonuniform distribution of energies [10,12]. This isotherm is not limited to the construction of a monolayer of adsorbate onto adsorbent but follows a multilayer trend. The mathematical representation of the Freundlich adsorption isotherm can be written as follows:

$$Q_e = K_f.(C_e)^{1/n} \qquad (5.6)$$

Eq. (5.6) can further be modified into a linear form as expressed in Eq. (5.7)

$$\log Q_e = \log K_f + \frac{1}{n} \log C_e \qquad (5.7)$$

where Q_e signifies the amount of adsorbate (mg/g) adsorbed onto the adsorbent at equilibrium, C_e represents the amount of adsorbate (mg/g) remains into the solution at equilibrium, K_f is the adsorption capacity (mg/L) of adsorbent, and n signifies the adsorption intensity.

The Freundlich adsorption graph is plotted between log Q_e and log C_e to help calculate the adsorption capacity and adsorption intensity using intercept and slope, respectively. A higher value of $1/n$ ($1/n > 1$) indicates that the adsorbent works well for high concentration solutions, however, the low value of $1/n$ ($1/n < 1$) shows the potential of the adsorbent adsorption capacity for low concentration solutions as well [13]. Generally, layered materials or hybrid materials follow the Freundlich adsorption isotherm, as a different intrastructure model in hybrid materials restricts the formation of an identical monolayer of adsorbate onto the surface [6].

5.2.4 Temkin adsorption isotherm

According to the Temkin adsorption isotherm, the heat of adsorption (ΔH_{ads}) of the adsorbate molecules' surface drops with the exposure of the adsorbent surfaces [14]. It follows the indirect interaction between the adsorbent and adsorbate and is characterized by the equally distributed binding energies. The mathematical linear equation (Eq. 5.8) for the Temkin isotherm is derived into linear form as follows:

$$Q_e = \frac{RT}{b} \ln K_T + \frac{RT}{b} \ln C_e \tag{5.8}$$

where R is the universal gas constant which is equal to 8.314 J/K/mol, T is the temperature (K), K_T is the Temkin isotherm constant (L/g), linked to maximum binding energy, and b represents the Temkin constant (J/mol), related to the heat of adsorption. The Temkin plot (Q_e vs ln C_e) provides information about K_T and b via the intercept and slope, respectively. A positive b value indicates that the adsorption process is endothermic in nature and vice versa.

5.2.5 Dubinin−Radushkevich adsorption isotherm

The Dubinin−Radushkevich (D−R) adsorption isotherm is applicable for heterogeneous surfaces and, generally, the adsorption mechanism is expressed using Gaussian energy distribution on heterogeneous surfaces [15]. This isotherm is considered for the microporous adsorbents and follows a pore-filling mechanism [8]. It is also a temperature-dependent isotherm and all suitable data can be obtained by plotting the D-R adsorption isotherm graphs at different temperatures [16]. It helps to distinguish the physisorption and chemisorption of adsorbate on the adsorbent with the help of the calculation of mean free energy (E) [8]. E value less than 8 kJ/mol ($E < 8$) signifies that the adsorption is followed by physical interactions; however, if it is more than

8 kJ/mol ($E > 8$) it signifies that the adsorption is chemisorption [17]. Eq. (5.9) is used to calculate the E:

$$E = \frac{1}{\sqrt{2\beta}} \tag{5.9}$$

where β signifies the Dubinin−Radushkevich isotherm (D−R) constant (mol^2/kJ^2) and can be calculated following the D−R isotherm model.

The equilibrium relation of adsorbate−adsorbent during the adsorption process following D−R adsorption isotherm can be mathematically represented as follows:

$$\varepsilon = RT\ln\left(1 + \frac{1}{C_e}\right) \tag{5.10}$$

where ε is the Polanyi potential or adsorption potential at temperature T (K).

The D−R adsorption isotherm following a Gaussian-type distribution on heterogeneous surfaces can be expressed by Eq. (5.11):

$$\ln Q_e = \ln Q_s - \beta\varepsilon^2 \tag{5.11}$$

where Q_s represents the theoretical isotherm saturation capacity (mg/g).

Following the D−R adsorption isotherm graph between ln Q_e and ε^2, β can be calculated using the slope, which can further be used to determine the mechanism of adsorption.

5.2.6 Jovanovic adsorption isotherm

Similar to the Langmuir adsorption isotherm, the Jovanovic adsorption isotherm is also applied to the homogeneous surface and monolayer adsorption. But, the Jovanovic isotherm model also considers the probability of mechanical contacts between the adsorbed and bulk phase [18]. The Jovanovic adsorption isotherm model can be expressed (Eq. 5.12) as follows:

$$Q_e = Q_{max}(1 - e^{-K_j C_e}) \tag{5.12}$$

Eq. (5.12) can be linearized as

$$\ln Q_e = \ln Q_{max} - K_j C_e \tag{5.13}$$

where Q_e (mg/g) and Q_{max} (mg/g) are the amounts of adsorbate adsorbed on the adsorbent at equilibrium and the maximum uptake of adsorbate by adsorbent, respectively, and K_J is the Jovanovic isotherm constant. The values of Q_{max} and K_J can be calculated using the slope and intercept of the Jovanovic isotherm graph, which can be obtained on plotting ln Q_e against C_e.

5.2.7 Halsey adsorption isotherm

This adsorption isotherm is applied for the heterogeneous systems following multilayer adsorption at a relatively large distance from the surface [19]. The Halsey isotherm equation can be represented as follows:

$$Q_e = \left(\frac{K_H}{C_e}\right)^{1/n_H} \tag{5.14}$$

This equation can be linearized into Eq. (5.15)

$$\ln Q_e = \frac{1}{n_H}\ln K_H - \frac{1}{n_H}\ln C_e \tag{5.15}$$

where n_H and K_H are the Halsey isotherm constants and can be calculated using the slope and intercept of the Halsey adsorption plot (ln Q_e vs ln C_e), respectively.

5.2.8 Hill adsorption isotherm

The Hill adsorption isotherm model suggests the binding energy of various adsorbing species on the homogeneous adsorbent surface. This model postulates that the Hill adsorption process is a cooperative phenomenon as the ligand-binding ability at one site on macromolecule may encourage different sites of the macromolecule for ligand binding [14]. The Hill isotherm equation and the linearized form of the Hill equation through logarithmic can be represented as Eqs. (5.16) and (5.17), respectively.

$$Q_e = \frac{Q_H(C_e)^{n_H}}{K_D + (C_e)^{n_H}} \tag{5.16}$$

$$\ln\left(\frac{Q_e}{Q_H - Q_e}\right) = n_H\ln C_e - \ln K_D \tag{5.17}$$

where Q_H (mg/g) is the maximum uptake of the adsorbate to the adsorption sites saturation (mg/g), K_D is the Hill constant and n_H is the Hill cooperative coefficient which represents the binding interactions. $n_H > 1$ represents positive cooperative binding interaction, $n_H < 1$ represents negative cooperative binding interaction, and $n_H = 1$ represents noncooperative binding. n_H and K_D can be calculated using the Hill isotherm, whereas the values of n_H and K_D further help to calculate the dissociation constant per site (K_d, mg/L) (Eq. 5.18) and association constant ($K_a = 1/K_d$).

$$K_d = (K_D)^{1/n_H} \tag{5.18}$$

5.2.9 Elovich adsorption isotherm

The Elovich isotherm is based on the kinetic principle and follows multilayer chemisorption. This model suggests that adsorption sites on the adsorbent increase exponentially with the adsorption [20]. The Elovich isotherm model can be illustrated as Eq. (5.19).

$$\frac{Q_e}{Q_m} = K_E C_e e^{-\left(\frac{Q_e}{Q_m}\right)}$$

(5.19)

This equation can be linearized using the logarithmic as follows:

$$\ln\frac{Q_e}{C_e} = \ln K_E Q_m - \frac{Q_e}{Q_m}$$

(5.20)

where K_E belongs to the Elovich constant.

5.2.10 Redlich−Peterson adsorption isotherm (R−P isotherm)

The R−P adsorption isotherm model has incorporated three parameters in the empirical equation which include elements from the Langmuir isotherm and Freundlich isotherm models [21]. Therefore this isotherm does not follow monolayer adsorption ideally. This isotherm can be expressed as follows:

$$Q_e = \frac{K_{RP} C_e}{1 + a_{RP} C_e^{\beta}}$$

(5.21)

where K_{RP} (L/g) and a_{RP} (L/mg) are the Redlich−Peterson isotherm constants. β is the exponent value which lies in between 0 and 1. When the value of $\beta = 1$, the adsorption process behaves like the Langmuir adsorption isotherm [22]. Therefore in this condition, β is the Langmuir adsorption constant and relates to the energy of adsorption and α is equal to the multiplication of β and Q_m.

$$Q_e = \frac{K_{RP} C_e}{1 + a_{RP} C_e}$$

(5.22)

When the value of β approaches to 0, it follows Henry's isotherm as follows [22]:

$$Q_e = \frac{K_{RP} C_e}{1 + a_{RP}}$$

(5.23)

At high liquid phase adsorbate concentrations, Eq. (5.21) reduces to Eq. (5.24) and follows the Freundlich isotherm. In this case the value of $a_{RP} C_e^{\beta}$ is more than 1.

$$Q_e = \frac{K_{RP}}{a_{RP}} C_e^{1-\beta}$$

(5.24)

where K_{RP}/a_{RP} illustrates the Freundlich isotherm constant, K_F and $1 - \beta = 1/n$.

The Eq. (5.24) can be solved into the linear equation as Eqs. (5.25) and (5.26)

$$\frac{C_e}{Q_e} = \frac{a_{RP}}{K_{RP}} C_e^{\beta} + \frac{1}{K_{RP}} \tag{5.25}$$

$$\ln\left(K_{RP}\frac{C_e}{Q_e} - 1\right) = \ln a_{RP} + \beta \ln C_e \tag{5.26}$$

Eq. (5.26) is the most accepted R−P isotherm linear equation, and the graph is plotted between $\ln\left(K_{RP}\frac{C_e}{Q_e} - 1\right)$ against C_e.

5.2.11 Sips adsorption isotherm

The Sips adsorption isotherm model is also the combined isotherm of the Langmuir isotherm and Freundlich isotherm that is used to predict the adsorption over heterogeneous surfaces [23,24]. It explains the adsorption energy distribution on the surface of the adsorbent. This isotherm also overcomes the drawback of the Freundlich isotherm model, which is the continuous increase in the amount of adsorbed adsorbate with an increase in concentration. This equation can be expressed as follows:

$$Q_e = \frac{Q_m K_s C_e^{1/n}}{1 + K_s C_e^{1/n}} \tag{5.27}$$

where K_s (mg/L) is the Sips adsorption isotherm constant and $1/n$ relates to the surface heterogeneity. If the value of $1/n$ is near to 0 it indicates that the adsorbent surface is heterogeneous and follows the Freundlich model, whereas if the value of $1/n$ is close to 1 it indicates that the surface exhibits homogenous binding sites and the isotherm model is reduced to the Langmuir model [24,25]. The pseudo-linear derivation of the above equation can be derived into Eq. (5.28) [26]:

$$\frac{1}{Q_e} = \frac{1}{Q_m K_s}\left(\frac{1}{C_e}\right)^n + \frac{1}{Q_m} \tag{5.28}$$

The pseudo-linear plot of Sips adsorption isotherm can be benefited to calculate three isotherm constants, that is, K_S, $1/n$, and Q_m, following the trial and error method.

5.2.12 Koble−Carrigan isotherm

The Koble−Carrigan adsorption isotherm is also a three-parameter isotherm equation, which is a resemblance of the Sips adsorption isotherm. Similar to the Sips isotherm, the Koble−Carrigan model also incorporates the Langmuir and Freundlich isotherms [27].

It is usually studied for heterogeneous surfaces. The Koble—Carrigan isotherm equation can be represented as follows:

$$Q_e = \frac{A_{KC} B_{KC} C_e^{n_{KC}}}{1 + B_{KC} C_e^{n_{KC}}} \tag{5.29}$$

where A_{KC} (mg/g, B_{KC} (L/g), and n_{KC} are the Koble—Carrigan isotherm constants. At a high concentration of adsorbate in solution with the value of $n_{KC} = 1$, this isotherm approximates to the Freundlich isotherm. However, with a high concentration of adsorbate with $n_{KC} < 1$, this model is not applicable for providing experimental data with low error values.

The above equation can be linearized into Eq. (5.30) as follows:

$$\frac{1}{Q_e} = \frac{1}{A_{KC} B_{KC} C_e^{n_{KC}}} + \frac{1}{A_{KC}} \tag{5.30}$$

The Koble—Carrigan isotherm constants can be calculated using the linear plot of the above equation using a trial and error optimization.

5.3 Adsorption chemical kinetics

To investigate the chemical rate of adsorption and rate-determining step of the adsorption process, the chemical kinetics models of the reactions are explored. A number of kinetic models have been proposed to study the reaction order of the adsorption process, such as reversible first-order, reversible second-order, irreversible first-order, irreversible second-order, pseudo-first-order, pseudo-second-order, intraparticle diffusion model, and so on [28—32]. Generally, the chemical kinetics models of adsorption can be classified into two classes: (1) adsorption reaction kinetics models; and (2) adsorption diffusion kinetic models (Fig. 5.1). Adsorption reaction kinetic models are derived from the chemical reaction kinetics. However, adsorption diffusion kinetic models are based on the following three steps: (1) liquid film diffusion or external diffusion: diffusion of adsorbate across the liquid film around adsorbent particles; (2) intraparticle diffusion: diffusion through pores; and (3) double exponential model: consequently adsorption and desorption between the adsorbate molecule and active sites of the adsorbent. Usually, three different models are chosen for the study of chemical kinetics of contaminant removal from wastewater, that is, (1) pseudo-first-order, (2) pseudo-second-order, and (3) intraparticle diffusion models. Chemical kinetics are employed to determine the rate-controlling step, selection of material as adsorbent, and also provide information on how other factors, such as pH and time, are affecting the reaction. In the next subsection, there is a brief description of several adsorption chemical kinetic models:

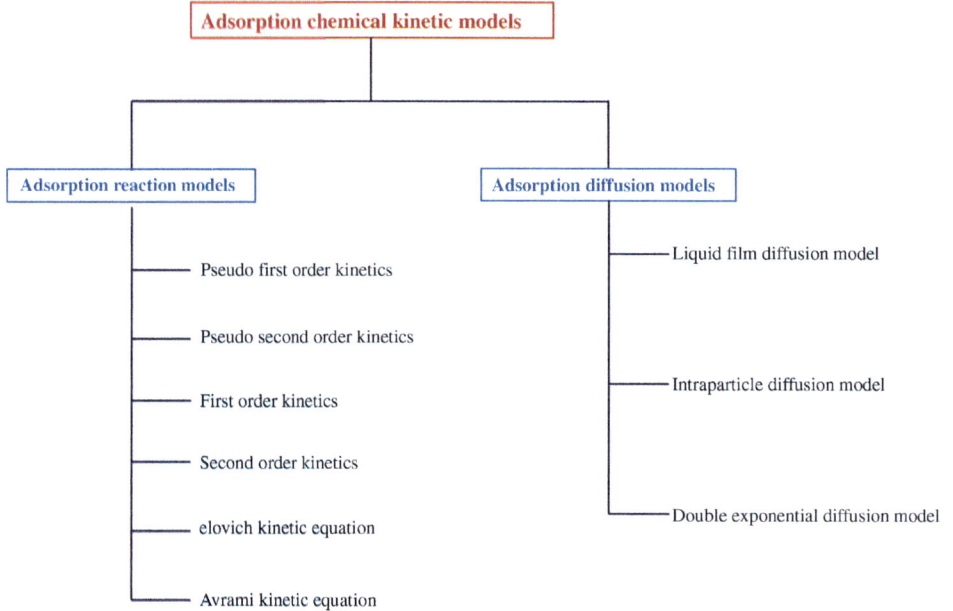

Figure 5.1 Classification of adsorption chemical kinetic models based on the adsorption reaction model and adsorption diffusion models.

5.3.1 Pseudo-first-order kinetics or Lagergren model

Lagergren proposed a pseudo-first-order kinetic model in 1898 which is widely applied and accepted for solid–liquid systems and the equation can be expressed as follows:

$$Q_t = Q_e(1 - e^{(-k_1 . t)}) \qquad (5.31)$$

Eq. (5.11) can further be modified into a linear form

$$\log(Q_e - Q_t) = \log Q_e - \frac{k_1}{2.303} t \qquad (5.32)$$

where Q_e (mg/g) and Q_t signify the adsorption capacity of adsorbent at equilibrium and time t (min), respectively; and k_1 is the rate constant (min^{-1}). k_1 can be determined from the slope of the plot between $\log(Q_e - Q_t)$ versus t. This plot also helps to check the predicted Q_e from the intercept. Many adsorption systems for the adsorption of acid dyes on the biopolymer chitosan were studied using pseudo-first-order kinetics [33]. The pseudo-first-order model can successfully be applied for the high concentration of adsorbate in the solution.

5.3.2 Pseudo-second-order kinetics or Ho and McKay model

The pseudo-second-order model has the advantage of investigating the adsorption kinetics for low concentrated solutions. Ho described the pseudo-second-order kinetic model in 1995 for divalent metal ions adsorption onto peat [34]. It is assumed that in a pseudo-second-order reaction, the rate-limiting step is chemisorption and the adsorption capacity mainly depends on the active surface sites on the adsorbent. The differential equation for the pseudo-second-order model is represented as:

$$\frac{dQ_t}{dt} = k_2(Q_e - Q_t)^2 \tag{5.33}$$

where k_2 (g/mg/min) represents the pseudo-second-order rate constant. This equation can further be integrated for boundary conditions ($q_t = 0 - q_t$ and $t = 0 - t$) and can be easily linearized into Eq. (5.34):

$$\frac{t}{Q_t} = \frac{1}{k_2 \cdot Q_e^2} + \frac{1}{Q_e} t \tag{5.34}$$

Eq. (5.14) is practiced to calculate the k_2 and Q_e from the intercept and slope of the plot between t/Q_t and t. The adsorption rate of contaminants from waste effluents containing low ppm metal ions and other pollutants are studied using pseudo-second-order kinetics [2].

5.3.3 First-order kinetic model

A first-order reversible reaction is considered for the surface reaction, and the rate of the forward reaction is equal to the rate of the backward reaction. This model can be simply represented as follows:

$$A \underset{k_2}{\overset{k_1}{\longleftrightarrow}} B \tag{5.35}$$

$$\frac{dC_B}{dt} = -\frac{dC_A}{dt} = k_1 C_A - k_2 C_B \tag{5.36}$$

where k_1 and k_2 are the rate constants of adsorption and desorption, and C_A (mg/L) and C_B (mg/L) are the remaining concentration of adsorbate in the solution and on the adsorbent, respectively. C_{A0} is the initial concentration (mg/L) of the adsorbate in the solution and equal to the sum of C_A and C_B. At equilibrium the rate of adsorption and the rate of desorption are equal and can be expressed as:

$$k_1 C_{A_{eq}} = k_2 C_{B_{eq}} \tag{5.37}$$

or

$$\frac{k_1}{k_2} = \frac{C_{B_{eq}}}{C_{A_{eq}}} = K \tag{5.38}$$

where K is the rate constant of the adsorption reaction.

The above equation can be transformed into the following equations:

$$\ln\left(1 - \frac{C_{A_0} - C_A}{C_{A_0} - C_{A_{eq}}}\right) = -(k_1 + k_2)t \tag{5.39}$$

$$\ln\left(\frac{C_{A_{eq}}}{C_A - C_{A_{eq}}}\right) = (k_1 + k_2)t - \ln K \tag{5.40}$$

A straight line is observed on plotting $\ln\left(\frac{C_{A_{eq}}}{C_A - C_{A_{eq}}}\right)$ versus time and $(k_1 + k_2)$ and K can be calculated using the slope and intercept, respectively. On using Eq. (5.40) and the slope and intercept values, the values of k_1 and k_2 can also be calculated.

5.3.4 Second-order kinetic model

The second-order adsorption kinetic model is widely studied for the adsorption of heavy metal ions on the adsorbent [35]. This model can be represented as:

$$2A \xrightarrow{k_2} B \tag{5.41}$$

This equation can further be rearranged as Eqs. (5.42) and (5.43):

$$-\frac{dC_A}{dt} = k_2 C_A^2 \tag{5.42}$$

$$-\frac{dC_A}{C_A^2} = k_2 dt \tag{5.43}$$

Further the linearized form of the second-order kinetic model can be generated on integrating the above equations on applying boundary limits (i.e., $t = 0$ to t and $C_A = C_{A0}$ to C_{At}) such as:

$$\frac{1}{C_{A_t}} - \frac{1}{C_{A_0}} = k_2 t \tag{5.44}$$

where C_{At} and C_{A0} are the remaining concentration of adsorbate in the solution after time t and initial concentration of adsorbate, respectively.

A second-order kinetic rate graph is observed on plotting the straight line of $1/C_{At}$ versus t and the second-order rate constant (k_2, $dm^3/mg/min$) can be calculated using the slope.

5.3.5 Elovich kinetic equation

The Elovich kinetic adsorption model or "Roginsky–Zeldovich" kinetic model is more compatible with chemisorption and was proposed by Roginsky and Zeldovich (1934) [36]. They applied this model to explain the adsorption of CO on MgO. Later, Elovich and Zhabrova explored the adsorption of ethylene and hydrogen following the same equation. Thus the Elovich equation explains the chemical nature of adsorption kinetics without desorption and can be expressed as follows Eq. (5.45) [16,36]:

$$\frac{dQ_t}{dt} = \alpha \exp(-\beta Q_t) \tag{5.45}$$

where α and β are the initial adsorption rate (mg/g/min) and the Elovich constant (g/mg).

The linearized form of the Elovich equation can be represented as:

$$Q_t = \frac{1}{\beta}\ln(\alpha\beta) + \frac{1}{\beta}\ln t \tag{5.46}$$

Any adsorption process following the Elovich kinetic model provides a straight line on plotting the adsorption data with Q_t versus $\ln t$. The value of β is produced from the slope of the graph and the value of α can be calculated using the value of β and the intercept of the graph.

5.3.6 Avrami kinetic equation

The Avrami adsorption kinetic model has been recently applied on the adsorption of various materials on adsorbents [37,38]. This model is reaction time- and temperature-dependent and the Avrami equation is expressed as Eq. (5.47):

$$\frac{dQ}{dt} = k_A^{n_A} n t^{n-1}(Q_e - Q) \tag{5.47}$$

where k_A is the Avrami kinetic constant (s^{-1}) and n_A is the Avrami model constant.

This equation can be integrated and further transformed into a linear equation as follows [39]:

$$\ln\left(\frac{Q_e}{Q_e - Q}\right) = n_A \ln k_A + n_A \ln t \tag{5.48}$$

Any adsorption process following the Avrami kinetic model provides a straight line on plotting the data between $\ln\left(\frac{Q_e}{Q_e - Q}\right)$ versus $\ln t$. The values of n_A and k_A can be calculated using the slope and intercept, respectively.

5.3.7 Liquid film diffusion kinetic model (McKay kinetic model)

This model suggests that the rate of adsorption of adsorbate onto adsorbent in solution is equal to the rate of adsorbate diffusion across the liquid film. Therefore in this kinetic model liquid film diffusion is the rate-determining step. The liquid film diffusion kinetic model can be expressed by the linear equation as follows [40]:

$$\ln(1 - F) = -k_{fd}t \tag{5.49}$$

where F is fractional attainment ($F = Q_t/Q_e$) and k_{fd} (min^{-1}) is the constant of liquid film diffusion kinetics. If the diffusion of adsorbate controls the adsorption process through the liquid film (boundary layer), it provides a straight line on plotting the adsorption data between $\ln(1-F)$ and t with zero intercepts. The value of k_{fd} can be obtained from the slope of the graph which further helps to calculate the effective liquid film diffusion coefficient (D_{efd}) using Eq. (5.50).

$$k_{fd} = \frac{3D_{efd}}{r_0 \Delta r_0 k} \tag{5.50}$$

where r_0 (cm) and Δr_0 are the radii of adsorbent and thickness of the liquid film. K is the equilibrium constant of the adsorption process.

5.3.8 Intraparticle diffusion kinetic model

The intraparticle diffusion rate model is usually applied for the porous materials and the diffusion of the adsorbate into the pores of different size is the physisorption phenomenon. This kinetic model was proposed by Weber and Morris in 1962 [41]. The rate-limiting step during the transport of adsorbate molecules/ions from the bulk solution to the adsorbent solid surface is diffusion via intraparticle diffusion [42]. The diffusion through the intraparticle diffusion model is represented by the following equation:

$$q_t = k_p.t^{1/2} + C \tag{5.51}$$

where k_p signifies the intraparticle diffusion rate constant (mg.min$^{1/2}$/g) and C is the intercept constant.

The graph plotted between q_t and $t^{1/2}$ is utilized to calculate the k_p and understand the mechanism.

All the kinetic models described here should provide linear relationship data and any kinetic model is applied to the adsorption process when the linear regression constant (R^2) is nearly unity [43].

5.3.9 The double exponential diffusion model

This model was proposed by Wilczak and Keinath and exhibits a two-step mechanism: (1) rapid adsorption due to external and internal diffusion, (2) followed by slow adsorption controlled by intraparticle diffusion [44]. This two-step mechanism model can be fairly described by the following equation:

$$Q_t = Q_e - \frac{D_1}{m_a}\exp(-K_1 t) - \frac{D_2}{m_a}\exp(-K_2 t) \tag{5.52}$$

where D_1 and D_2 (mmol/L) are the adsorption rate parameters and K_1, and K_2 (min^{-1}) demonstrate the rate constants for fast and slow steps, respectively. m_a represents the adsorbent mass.

If $K_1 \gg K_2$, then the effect of the first or fast step on the overall kinetics is supposed to be negligible and the above reaction can be rearranged to the simplified Eq. (5.53) and then linearized into Eq. (5.54).

$$Q_t = Q_e - \frac{D_2}{m_a}\exp(-K_2 t) \tag{5.53}$$

$$\ln(Q_e - Q_t) = \ln\frac{D_2}{m_a} - K_2 t \tag{5.54}$$

The values of D_2 and K_2 can be calculated using the intercept and slope of the graph on plotting $\ln(Q_e - Q_t)$ versus t, respectively.

5.4 Thermodynamics parameters

Adsorption rate and adsorption quantity are the two major factors to check the potential of the adsorption process. However, thermodynamic parameters are required to study whether the adsorption process is spontaneous or nonspontaneous [45]. Changes in Gibb's free energy ($\Delta G°$), entropy ($\Delta S°$), and enthalpy ($\Delta H°$) are the standard thermodynamic parameters to provide the insights into adsorption feasibility, based on the temperature-dependent adsorption conditions. The kinetic energy of the solute in the solution is directly related to the temperature. With an increase in the temperature, the simultaneous increase in the diffusion rate of the adsorbate is also observed. Therefore temperature significantly alters the equilibrium state of the adsorption process and thus changes the thermodynamic parameters [46]. $\Delta G°$ at each temperature can be determined using the given equation:

$$\Delta G° = -\mathrm{RT}\,\ln K_a \tag{5.55}$$

where K_a is the apparent equilibrium adsorption constant at temperature T (K) and R is the universal gas constant. K_a can be calculated by Q_e/C_e at a particular temperature.

Further, the value of ΔG° for each temperature is given by Eq. (5.56):

$$\Delta G^{\circ} = \Delta H^{\circ} - T \Delta S^{\circ} \tag{5.56}$$

The minus value of ΔG° and plus value of ΔS° supports the spontaneity of the adsorption process. For a spontaneous process, the plot of ΔG° versus temperature (T) is always linear and entropy (ΔS°) and enthalpy (ΔH°) can be calculated using the slope and intercept of the graph. The positive ΔH° value shows the endothermic nature of the adsorption process.

Changes in Gibbs free energy can also be measured using the Van't Hoof equation, which can be depicted as follows [47]:

$$\Delta G^{\circ} = RTlnK_e^{\circ} \tag{5.57}$$

where K_e° is the dimensionless thermodynamic equilibrium constant.

Therefore the thermodynamic equilibrium constant and entropy and enthalpy can also be related to each other as Eq. (5.58):

$$lnK_e^{\circ} = -\frac{\Delta H^{\circ}}{RT} + \frac{\Delta S^{\circ}}{R} \tag{5.58}$$

Similarly, on plotting the graph between ln K_e° and $1/T$, a straight line is observed and the intercept and slope are used to calculate entropy change and enthalpy change.

5.5 Conclusion

The study of various adsorption isotherms, kinetic models, and thermodynamic parameters is necessary to speculate on the mechanism of the adsorption process, adsorption rate, and spontaneity of the process, respectively. This helps to predict whether the adsorption is physisorption or chemisorption, monolayer or multilayer, and how the diffusion is taking part in the adsorption process. To scale up the adsorption process, the thorough knowledge of adsorption equilibrium isotherms to evaluate adsorption capacity, chemical kinetics to calculate the rate of adsorption, and thermodynamic parameters to check the feasibility of the adsorption process is required. Based on the above-described models, it would be helpful in predicting the mechanism of the adsorption process according to the followed adsorption isotherm and kinetics. The analysis of various adsorption isotherm models is also essential to optimize adsorbents usage during the process. The value of linear regression constant (R^2) is used to analyze the best-fitted model.

References

[1] Zhu H-Y, Jiang R, Xiao L, Li W. A novel magnetically separable γ-Fe_2O_3/crosslinked chitosan adsorbent: preparation, characterization and adsorption application for removal of hazardous azo dye. J Hazard Mater 2010;179:251−7.

[2] Reemtsma T, Weiss S, Mueller J, Petrovic M, González S, Barcelo D, et al. Polar pollutants entry into the water cycle by municipal wastewater: a European perspective. Environ Sci Technol 2006;40:5451−8.

[3] Han R, Zhang J, Han P, Wang Y, Zhao Z, Tang M. Study of equilibrium, kinetic and thermodynamic parameters about methylene blue adsorption onto natural zeolite. Chem Eng J 2009;145:496−504.

[4] Kavitha D, Namasivayam C. Experimental and kinetic studies on methylene blue adsorption by coir pith carbon. Bioresour Technol 2007;98:14−21.

[5] Kumar N, Fosso-Kankeu E, Ray SS. Achieving controllable MoS_2 nanostructures with increased interlayer spacing for efficient removal of pb(ii) from aquatic systems. ACS Appl Mater Interfaces 2019;11:19141−55.

[6] Ayawei N, Ebelegi AN, Wankasi D. Modelling and Interpretation of Adsorption Isotherms. J Chem 2017;2017:11.

[7] Mittal H, Mishra SB. Gum ghatti and Fe_3O_4 magnetic nanoparticles based nanocomposites for the effective adsorption of rhodamine B. Carbohydr Polym 2014;101:1255−64.

[8] Vijayaraghavan K, Padmesh TVN, Palanivelu K, Velan M. Biosorption of nickel(II) ions onto *Sargassum wightii*: application of two-parameter and three-parameter isotherm models. J Hazard Mater 2006;133:304−8.

[9] Kundu S, Gupta AK. Arsenic adsorption onto iron oxide-coated cement (IOCC): regression analysis of equilibrium data with several isotherm models and their optimization. Chem Eng J 2006;122:93−106.

[10] Kumar N, Reddy L, Parashar V, Ngila JC. Controlled synthesis of microsheets of ZnAl layered double hydroxides hexagonal nanoplates for efficient removal of Cr(VI) ions and anionic dye from water. J Environ Chem Eng 2017;5:1718−31.

[11] Santhosh C, Velmurugan V, Jacob G, Jeong SK, Grace AN, Bhatnagar A. Role of nanomaterials in water treatment applications: a review. Chem Eng J 2016;306:1116−37.

[12] Kumar N, Mittal H, Parashar V, Ray SS, Ngila JC. Efficient removal of rhodamine 6G dye from aqueous solution using nickel sulphide incorporated polyacrylamide grafted gum karaya bionanocomposite hydrogel. RSC Adv 2016;6:21929−39.

[13] Yan H, Yang L, Yang Z, Yang H, Li A, Cheng R. Preparation of chitosan/poly (acrylic acid) magnetic composite microspheres and applications in the removal of copper (II) ions from aqueous solutions. J Hazard Mater 2012;229:371−80.

[14] Ringot D, Lerzy B, Chaplain K, Bonhoure J-P, Auclair E, Larondelle Y. In vitro biosorption of ochratoxin A on the yeast industry by-products: comparison of isotherm models. Bioresour Technol 2007;98:1812−21.

[15] Çelebi O, Üzüm Ç, Shahwan T, Erten HN. A radiotracer study of the adsorption behavior of aqueous Ba^{2+} ions on nanoparticles of zero-valent iron. J Hazard Mater 2007;148:761−7.

[16] Günay A, Arslankaya E, Tosun I. Lead removal from aqueous solution by natural and pretreated clinoptilolite: adsorption equilibrium and kinetics. J Hazard Mater 2007;146:362−71.

[17] Kogo B, Biamah E, Langat P. Optimized design of a hybrid biological sewage treatment system for domestic wastewater supply. J Geosci Environ Prot 2017;05:14−29.

[18] Jaroniec M. Statistical interpretation of the Jovanović adsorption isotherms. Colloid Polym Sci 1976;254:601−5.

[19] Halsey G. Physical adsorption on non-uniform surfaces. J Chem Phys 1948;16:931−7.

[20] Gubernak M, Zapala W, Kaczmarski K. Analysis of amylbenzene adsorption equilibria on an RP-18e chromatographic column. Acta Chromatogr 2003;13.

[21] Foo KY, Hameed BH. Insights into the modeling of adsorption isotherm systems. Chem Eng J 2010;156:2−10.

[22] Ng J, Cheung W, McKay G. Equilibrium studies of the sorption of Cu (II) ions onto chitosan. J Colloid Interface Sci 2002;255:64−74.

[23] Anirudhan TS, Senan P. Adsorption characteristics of cytochrome C onto cationic Langmuir monolayers of sulfonated poly(glycidylmethacrylate)-grafted cellulose: mass transfer analysis, isotherm modeling and thermodynamics. Chem Eng J 2011;168:678−90.

[24] Papageorgiou SK, Katsaros FK, Kouvelos EP, Nolan JW, Le Deit H, Kanellopoulos NK. Heavy metal sorption by calcium alginate beads from *Laminaria digitata*. J Hazard Mater 2006;137:1765—72.

[25] Ngah WSW, Fatinathan S, Yosop NA. Isotherm and kinetic studies on the adsorption of humic acid onto chitosan-H_2SO_4 beads. Desalination 2011;272:293—300.

[26] Nanta P, Kasemwong K, Skolpap W. Isotherm and kinetic modeling on superparamagnetic nanoparticles adsorption of polysaccharide. J Environ Chem Eng 2018;6:794—802.

[27] Koble RA, Corrigan TE. Adsorption isotherms for pure hydrocarbons. Ind Eng Chem 1952;44:383—7.

[28] Saiers JE, Hornberger GM, Liang L. First- and second-order kinetics approaches for modeling the transport of colloidal particles in porous media. Water Resour Res 1994;30:2499—506.

[29] McCoy M, Liapis A. Evaluation of kinetic models for biospecific adsorption and its implications for finite bath and column performance. J Chromatogr A 1991;548:25—60.

[30] Venkata Mohan S, Chandrasekhar Rao N, Karthikeyan J. Adsorptive removal of direct azo dye from aqueous phase onto coal based sorbents: a kinetic and mechanistic study. J Hazard Mater 2002;90:189—204.

[31] Chu K, Hashim M. Modeling batch equilibrium and kinetics of copper removal by crab shell. Sep Sci Technol 2003;38:3927—50.

[32] O'Shannessy DJ, Winzor DJ. Interpretation of deviations from pseudo-first-order kinetic behavior in the characterization of ligand binding by biosensor technology. Anal Biochem 1996;236:275—83.

[33] Wong YC, Szeto YS, Cheung WH, McKay G. Pseudo-first-order kinetic studies of the sorption of acid dyes onto chitosan. J Appl Polym Sci 2004;92:1633—45.

[34] Ho YS, McKay G. Sorption of dye from aqueous solution by peat. Chem Eng J 1998;70:115—24.

[35] Kundu S, Gupta AK. Sorption kinetics of As(V) with iron-oxide-coated cement—a new adsorbent and its application in the removal of arsenic from real-life groundwater samples. J Environ Sci Health Part A 2005;40:2227—46.

[36] Low M. Kinetics of chemisorption of gases on solids. Chem Rev 1960;60:267—312.

[37] Serna-Guerrero R, Sayari A. Modeling adsorption of CO_2 on amine-functionalized mesoporous silica. 2: kinetics and breakthrough curves. Chem Eng J 2010;161:182—90.

[38] Lopes EC, dos Anjos FS, Vieira EF, Cestari AR. An alternative Avrami equation to evaluate kinetic parameters of the interaction of Hg (II) with thin chitosan membranes. J Colloid Interface Sci 2003;263:542—7.

[39] Tan KL, Hameed BH. Insight into the adsorption kinetics models for the removal of contaminants from aqueous solutions. J Taiwan Inst Chem Eng 2017;74:25—48.

[40] Boyd G, Adamson A, Myers Jr L. The exchange adsorption of ions from aqueous solutions by organic zeolites. II. Kinetics1. J Am Chem Soc 1947;69:2836—48.

[41] Morris JC, Weber jr WJ. Removal of biologically-resistant pollutants from waste waters by adsorption. Advances in water pollution research. Elsevier; 1964. p. 231—66.

[42] McKay G. The adsorption of dyestuffs from aqueous solutions using the activated carbon adsorption model to determine breakthrough curves. Chem Eng J 1984;28:95—104.

[43] Gerente C, Lee VKC, Cloirec PL, McKay G. Application of chitosan for the removal of metals from wastewaters by adsorption—mechanisms and models review. Crit Rev Environ Sci Technol 2007;37:41—127.

[44] Tosun İ. Ammonium removal from aqueous solutions by clinoptilolite: determination of isotherm and thermodynamic parameters and comparison of kinetics by the double exponential model and conventional kinetic models. Int J Environ Res Public Health 2012;9:970—84.

[45] Kumar N, Mittal H, Alhassan SM, Ray SS. Bionanocomposite hydrogel for the adsorption of dye and reusability of generated waste for the photodegradation of ciprofloxacin: a demonstration of the circularity concept for water purification, ACS Sustainable Chem Eng 2018;6:17011—25.

[46] Nouri L, Ghodbane I, Hamdaoui O, Chiha M. Batch sorption dynamics and equilibrium for the removal of cadmium ions from aqueous phase using wheat bran. J Hazard Mater 2007;149:115—25.

[47] Lima EC, Hosseini-Bandegharaei A, Moreno-Piraján JC, Anastopoulos I. A critical review of the estimation of the thermodynamic parameters on adsorption equilibria. Wrong use of equilibrium constant in the Van't Hoof equation for calculation of thermodynamic parameters of adsorption. J Mol Liq 2019;273:425—34.

CHAPTER SIX

Effect of reaction parameters on the adsorption

Adsorption is an easy and economical approach to clean wastewater. The adsorption capacity of the adsorbent determines the effectiveness of the adsorbent to remove the contaminant. However, the rate of adsorption of water pollutants on the adsorbent depends significantly on several reaction parameters, such as initial pollutant concentration, the amount of adsorbent added, contact time, temperature, pH, and size and surface characteristics of adsorbent [1,2]. To achieve the maximum adsorption rate, the optimum values of any parameters that are influencing the adsorption process should be analyzed. In-depth knowledge and optimization of all these reaction parameters will help to build up the industrial-scale water treatment process through adsorption. Determination of the establishment of the adsorption—desorption equilibrium is one of the parameters which reveals the complete utilization of all the active sites on adsorbent. The pH of the aqueous medium changes the surface charges of the adsorbent and hence the degree of adsorption is significantly influenced following electrostatic interaction. Thus the effect of these operational parameters on the adsorption of water contaminants is essential and briefly discussed in this chapter. Table 6.1 represents the optimum calculated values of several factors affecting the adsorption of various pollutants using different adsorbents, as described in the available literature.

6.2 Effect of adsorbent dosage

Optimization of adsorbent dosage is one of the most important actions, to determine the adsorption capacity of adsorbent under specific operating conditions. This is also valuable from an economical point of view as it provides the idea of maximum adsorption of dye using a minimum dosage of the adsorbent. Generally, with an increase in the amount of adsorbent in the wastewater, the rate of adsorption of the pollutant is also increased due to the increase in the number of available active sites on the adsorbent. To optimize the ideal concentration of the adsorbent, different concentrations of adsorbent are

Carbon Nanomaterial-Based Adsorbents for Water Purification.
DOI: https://doi.org/10.1016/B978-0-12-821959-1.00006-4

added into the solution with a fixed amount of water pollutant and shaken until the equilibrium is established between the adsorption and desorption processes of adsorbate [14]. For example, Wu et al. performed the adsorption studies of methylene blue (MB) organic dye using various amounts of Fe_3O_4-graphene@mesoporous SiO_2 (MG@m-SiO_2) (Fig. 6.1A) [4]. They reported the increase in the adsorption percentage from 39.95% to 100% on increasing the MG@m-SiO_2 amount from 2.5 to 10 mg/L. However, Shrivastava et al. demonstrated in their findings that the adsorption of dye (malachite green (MG)) increased on increasing the dosage of adsorbent maize cob powder (MCP) (Fig. 6.1B) but the percentage of the dye adsorbed per unit mass of the adsorbent decreased significantly [15]. This might be due to the available adsorption sites of adsorbents that remain unsaturated during adsorption. Shrestha et al. also studied the effect of the adsorbent magnetic Fe_3O_4/sugarcane bagasse activated carbon (Fe_3O_4/AC) composite dosage on the uptake capacity of Arsenic (As (III)) (Fig. 6.1C) from wastewater [16].

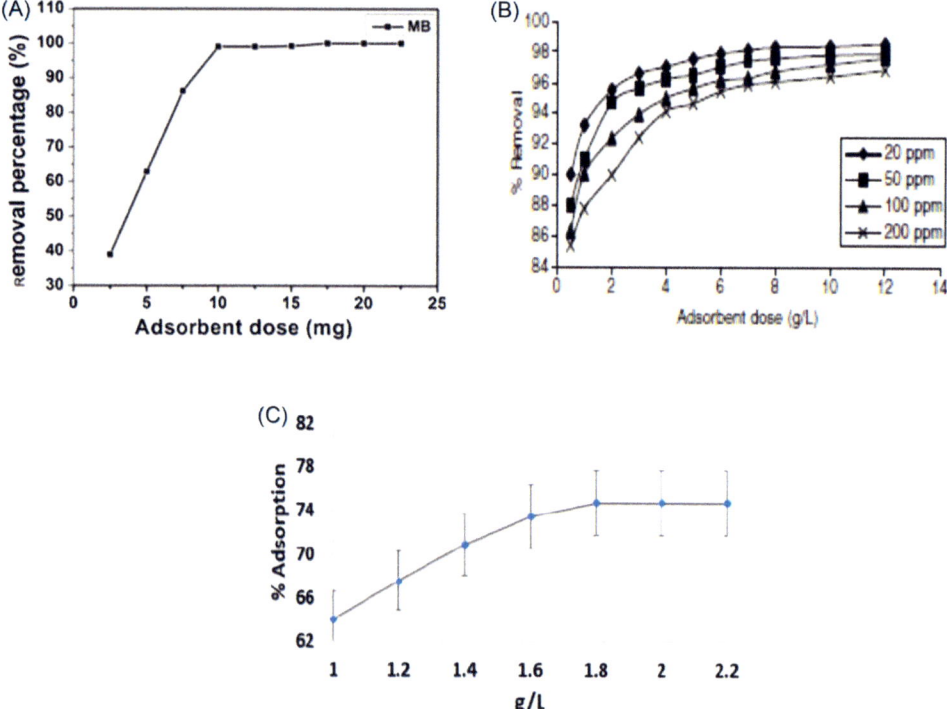

Figure 6.1 Effect of adsorbent dosage on the adsorption of (A) MB dye on magnetic graphene@-mesoporous SiO_2 (MG@m-SiO_2) [4]. Copyright 2016, Elsevier Science Ltd. (B) MG dye on maize cob powder (MCP) [15]. Copyright 2009, Elsevier Science Ltd. (C) As(III) on Fe_3O_4/sugarcane bagasse activated carbon (Fe_3O_4/AC) [16]. Copyright 2019, MDPI publishers.

Table 6.1 Values of several influencing factors affecting adsorption process.

Pollutant	Adsorbent	Adsorbent dosage, (mg/mL)	Pollutant initial concentration (mg/L)	Temperature (K)	Contact time (min)	pH	Reference
Malachite Green (MG) dye	Graphene oxide (GO) Reduced GO (rGO)	1	25	298	100	3	[3]
MB	Magnetic graphene@mesoporous SiO_2 nanocomposites (MG@m-SiO_2)	0.4	–	–	180	11	[4]
Pb(II)	Chitosan and carboxy-methyl cellulose cross-linked beads using arginine (CS-ag-CM)	0.1	325	315	40	6.5	[5]
Cd(II)		0.1	300	310	40	6.5	
Diclofenac	amine-functionalized chitosan coated with Fe_3O_4 nanoparticles (AmCS@Fe_3O_4)	0.5	1000	–	60	4.5	[6]
As(V)	Modified saxaul ash	1.5	0.25	323	60	7	[7]
Cephalexin	Alligator weed-activated carbon (AWAC)	1.6	35	303	240	6	[8]
Neutral Red Dye	N-doped porous carbon nanosheet (N-PCNS)	0.25	150	308	180	1	[9]
Cu(II)	Montmorillonite (Mt),	4	8	–	70	2	[10]
Cu(II)	Trioctylamine modified Montmorillonite (Mt-TOA)	0.04	8		70		
Organophosphorus pesticides	Magnetic graphene oxide–based silica nanoparticles Fe_3O_4@SiO_2@GO–PEA	1.5	–	298	15	7	[11]
Bentazon	Activated carbon	0.5	80	293	90	3.5	[12]
Organophosphorus profenofos	Fe/Ni bimetallic nanoparticles	0.8	1.2	318	16	7	[13]

They checked the effect of the Fe_3O_4/AC dosage from 1 to 2.2 g/L. The maximum adsorption of As(III) was observed with the 1.8 g/L dosage of Fe_3O_4/AC and on further increasing the amount of Fe_3O_4/AC it did not make a significant impact in the percentage of As(III) adsorption ($\sim$75%). This is due to the aggregation or overlapping of active adsorption sites onto the adsorbent with an increase in the adsorbent amount. Similar results were also observed during the adsorption of aniline on graphene oxide [17]. Therefore the optimization of the adsorbent is necessary to investigate the maximum uptake of adsorbate per unit mass of adsorbent.

6.3 Effect of initial concentration of pollutant

The initial concentration of the water pollutant in the solution can alter the pollutant removal efficiency of an adsorbent that relies on the combination of pollutant concentration and available binding sites on the surface of the adsorbent. The initial pollutant concentration can provide the required driving force to repress the mass transfer resistance of pollutant between the solid (adsorbent) and liquid (contaminated water) phases [18]. To study the effect of initial pollutant concentration in an aqueous medium, different solutions of polluted water with various amounts of pollutant and a fixed amount of adsorbent are mixed and stirred. This solution mixture is shaken until equilibrium is approached. Ikram et al. analyzed the effect of the initial amount of heavy metal ions (Pb(II) and Cd(II)) (Fig. 6.2A) on the removal efficiency of chitosan (CS) and carboxymethyl cellulose (CM) cross-linked beads in which arginine (ag) was used as a cross-linker [5]. The uptake of heavy metal ions was first noticed to be increased on the regular increase in the concentration of Pb(II) and Cd(II) from 50 to 300 mg/L, whereas on further increasing the concentration there is no effect on the adsorption of Pb(II) and Cd(II). This might be because at the 300 mg/L concentration of pollutant, all active sites of the adsorbent are occupied and no more other active sites are available for the adsorption activity. Therefore the optimum initial concentration of Pb(II) and Cd(II) was observed to be 300 and 325 mg/L, respectively, using a fixed amount of adsorbent. This is attributed to the fact that at lower concentrations of heavy metal ions the number of collisions occurring between the adsorbent and adsorbate are less and these increase with an increase in the initial pollutant concentration. This increased collision leads to a sharp increase in the adsorption of Pb(II) and Cd(II) on the adsorbent. This also optimized the maximum amount of adsorbed adsorbate per unit mass of adsorbent [21]. However, after a specific concentration (300 mg/L), no significant increase in adsorption was noted, due to the establishment of a dynamic equilibrium between the adsorbed

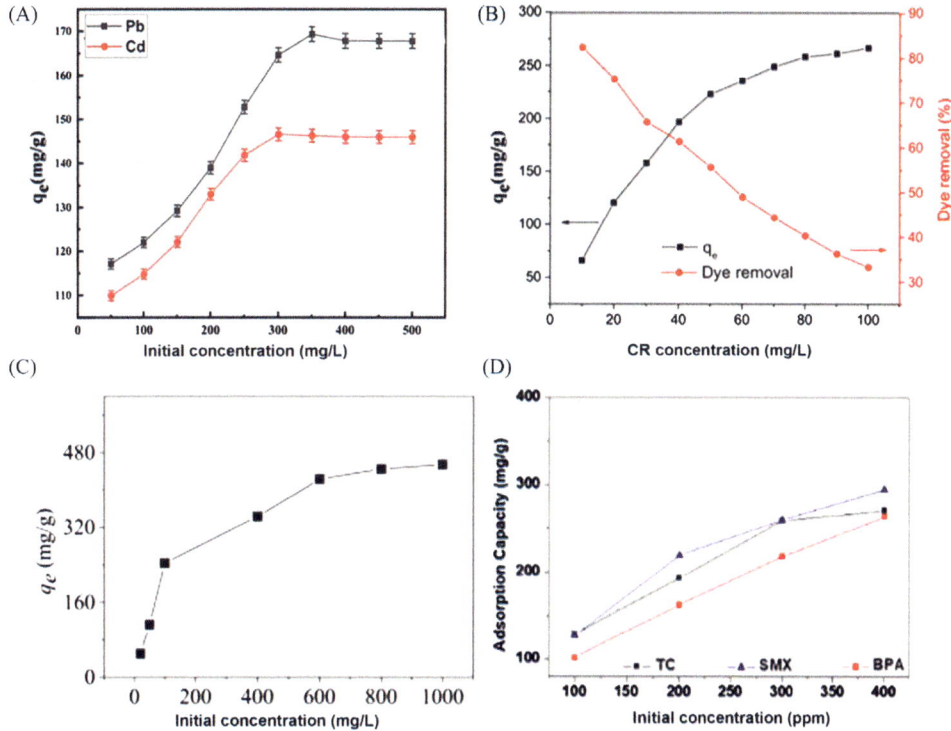

Figure 6.2 Effect of the initial concentration of water pollutants on the adsorption efficiency of adsorbent. Adsorption of (A) Pb(II) and Cd(II) onto CS-ag-CM [5]. Copyright 2019, Royal Society of Chemistry. (B) CR dye on S-doped Fe_2O_3/Carbon nanocomposite [19]. Copyright 2018, Springer Nature. (C) Diclofenac sodium on AmCS@Fe_3O_4 nanocomposite [6]. Copyright 2019, Elsevier Science Ltd. (D) Tetracycline (TC), Sulfamethoxazole (SMX) and Bisphenol A (BPA) on sawdust-derived functionalized graphitic carbon [20]. Copyright 2018, Elsevier Science Ltd.

metal ion and free metal ions (adsorbate) in the solution. Also, at a higher concentration of adsorbate, the available sites for adsorption become less numerous and all the active adsorbent sites get saturated with the adsorbate [22,23]. Similar findings are also observed for the other pollutants such as organic dyes, pharmaceutical ingredients, and personal care products (Fig. 6.1B—D) [6,19,20]. Khoshsang et al. studied the effect of the initial concentration of Congo red (CR) dye (10—100 mg/L) (Fig. 6.2B) on the adsorption capacity of a fixed amount of S-doped Fe_2O_3/carbon nanocomposite (5 mg in 40 mL wastewater) [19]. The driving force for increased adsorption capacity with the increasing concentration of CR might be the high mass transfer of CR with high initial concentration. The adsorption of pharmaceutical ingredients also responds similarly to increases in the initial concentration (Fig. 6.2C and D) [6,20].

6.4 Effect of temperature

Depending on the nature of the adsorption process, that is, exothermic or endothermic, the temperature of the reaction also alters the adsorption capacity of the adsorbent. Generally, if the adsorption process follows endothermic behavior, the adsorption capacity of the adsorbent will also increase with an increase in the temperature and vice versa. This might be due to the increase in the mobility of the pollutant and also in the increase in the number of active sites for adsorption due to the swelling effect [24]. On the contrary, in exothermic adsorption processes, the adsorption capacity of the adsorbent will decrease with an increase in the temperature. Usually, in exothermic reactions, with an increase in temperature, the desorption of pollutant from adsorbent surface dominates, which is due to the weakening of the adsorptive forces between the adsorbent and adsorbate [25]. Also, the thermodynamic parameters (using Eqs. 6.1 and 6.2), such as Gibbs free energy (ΔG°), entropy (ΔS°), and enthalpy (ΔH°) can be also evaluated to check the spontaneity [26].

$$\Delta G^\circ = -\mathrm{RT} \ln K_a \qquad (6.1)$$

$$\ln K_a = \frac{\Delta S^\circ}{R} - \frac{\Delta H^\circ}{\mathrm{RT}} \qquad (6.2)$$

where R (J/Mol/K), T (K), and K_a are universal gas constant, the temperature of the reaction medium, and equilibrium constant, respectively.

Nawawi et al. applied the above equation to calculate the nature of the adsorption process of reactive red 120 dye on chitosan beads [27]. The positive value of ΔH° confirms the endothermic behavior of the adsorption process followed by chemisorption. Furthermore, the negative value of ΔG° and the positive value of ΔS° show the feasibility of adsorption. Similarly, the negative value of ΔH° supports the exothermic behavior of the adsorption process [28]. As is evident, in exothermic processes the adsorption capacity of the adsorbent decreases with an increase in temperature. Therefore in exothermic reactions the value of ΔG increases with an increase in temperature [29].

Xu et al. investigated the adsorption potential of polydopamine (PDA) microspheres toward nine organic dyes, that is, anionic dyes: eosin-B (EB), methyl orange (MO), eosin-Y (EY), acid chrome blue K (ACBK); cationic dyes: malachite green (MG), rhodamine B (RhB), safranine T (ST), methylene blue (MB); and neutral dye: neutral red (NR) [30]. However, the effect of temperature was explored on the three representative dyes, that is, MB, MG, and NR (Fig. 6.3A). The adsorption process was found to be favorable at higher temperature due to the rapid diffusion of dyes at a raised temperature and providing an opportunity for the enhanced interaction of adsorbate at the active sites of adsorbents. A similar trend of the effect of temperature was also observed for the adsorption of inorganic pollutants from the wastewater [31,34]. An adsorption study of heavy metal ions (Co(II), Pb(II), Ni(II), and Cd(II)) was performed using a geopolymer which was derived

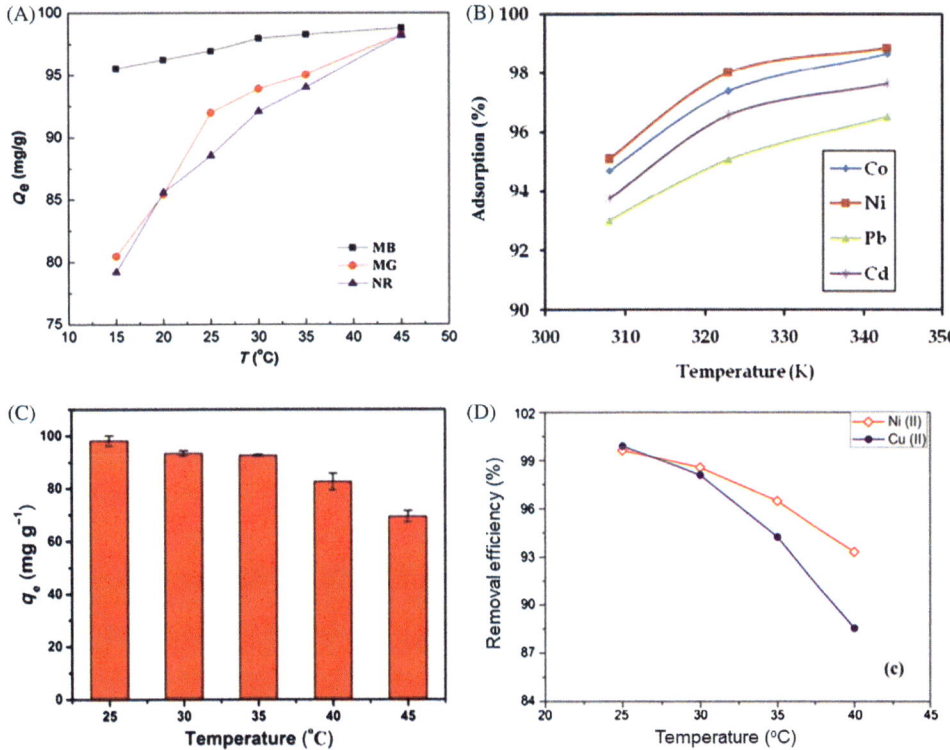

Figure 6.3 Effect of temperature on the adsorption of (A) MB, MG, and NR on PDA microspheres [30]. Copyright 2016, Elsevier Science Ltd. (B) Co(II), Ni(II), Pb(II) and Cd(II) on geopolymer derived from pyrophyllite mine waste [31]. Copyright 2018, Elsevier Science Ltd. (C) TC on p-BN [32]. Copyright 2017, Elsevier Science Ltd. (d) Ni(II) and Cu(II) on cross-linked magnetic chitosan [33]. Copyright 2018, Taylor & Francis Group.

from pyrophyllite mine waste [31]. They have also explored the effect of several parameters including temperature (Fig. 6.3B). The adsorption process was noted to be endothermic and the adsorption of heavy metal ions was found increase with an increase in temperature. Again, the higher collision rate of metals ions onto the surface of the adsorbent and the diffusion rate into the pores of the adsorbent were considered as the driving forces for the enhanced adsorption with increased temperature.

However, an opposite trend of decreased adsorption capacity of various adsorbents with an increase in temperature was also observed for the adsorption of organic and inorganic pollutants [32,35–38]. These adsorption processes are exothermic. The adsorption of tetracycline (TC) on the porous hexagonal BN (p-BN) inhibited with an increase in temperature (Fig. 6.3C) [32]. The reason for the poorer adsorption capacity of p-BN toward TC at higher temperature is due to the improved activity of TC with a raise in temperature of the medium that disturbs the stable adsorption and consequently impedes the adsorption process.

Similarly, the exothermic adsorption of Ni(II) and Cu(II), reduces the adsorption capacity of cross–linked magnetic chitosan at a higher temperature (Fig. 6.3D) [33].

6.5 Effect of contact time

Generally, the rate of pollutant removal via adsorption increases with an increase in the contact time of the adsorption reaction as it provides maximum opportunities for inter-action between the functional groups of the adsorbent and adsorbate [39,40]. However, once the adsorption−desorption equilibrium is achieved, no further uptake of a pollutant from the wastewater could be observed on further increasing the contact time. At this equilibrium time, the rate of desorption of the adsorbate from the surface of the adsorbent is equal to the rate of adsorption and both processes remain in a state of dynamic equilibrium. Therefore the optimization of contact time or equilibrium time is necessary to ensure that the adsorption process is complete and attained the maximum adsorption capacity.

Zhao et al. observed the effect of contact time on the removal percentage of Pb(II), Cd (II), and Cu(II) (Fig. 6.4A) through adsorption on the surface of the carboxyl functionalized magnetite nanoparticles (CMNPs) [41]. It can clearly be seen from Fig. 6.4A that the adsorption of heavy metal ions increased rapidly for the initial 30 minutes, and after that it remains constant. This reveals that the equilibrium time for the adsorption of Pb(II), Cd(II), and Cu(II) on CMNPs was approached in 30 minutes and all the active sites get saturated. The rapid adsorption of heavy metal ions during initial contact time might be due to the availability of a larger number of active sites on the adsorbent for adsorption. As the active sites start to get saturated the adsorption efficiency of the adsorbent also becomes lower and when all the active sites are saturated it can adsorb no more pollutant. The effect of contact

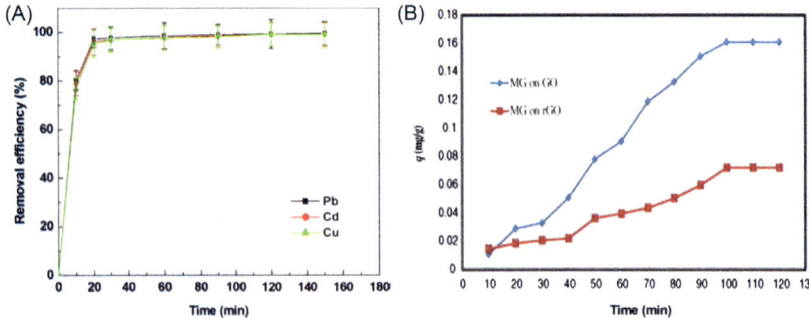

Figure 6.4 Effect of reaction contact time on the adsorption of (A) heavy metal ions (Pb(II), Cd(II), and Cu(II)) on CMNPs [41]. Copyright 2015, American Chemical Society. (B) MG organic dye on GO and rGO [3]. Copyright 2016, Elsevier Science Ltd.

time (10−120 minutes) on graphene oxide (GO) and reduced graphene oxide (rGO) for the adsorptive removal of malachite green (MG) organic dye from wastewater was also studied [3]. For the first 100 minutes the adsorption efficiency of both GO and rGO (Fig. 6.4B) significantly increased with the increase in the contact time but later at 110 and 120 minutes, no adsorption was observed. Therefore 100 minutes was considered as the optimum time to attain a stable adsorption−desorption equilibrium phase.

However, the optimum contact times for MG removal using different adsorbents such as acid-modified sphagnum peat moss (SPM), carboxylate group functionalized multiwalled carbon nanotubes (MWCNT-COOH), polycarboxylic magnetic polydopamine (Fe_3O_4@PDA-COOH), ZnO nanorod-loaded activated carbon, and ZnS:Cu nanoparticles loaded on activating carbon were found to be 45, 50, 600, 3, and 3 min, respectively, under specific conditions [42−46]. Therefore the maximum adsorption capacity is dependent on the maximum interaction of each adsorbate and active sites on the adsorbent that required the optimization of contact time.

6.6 Effect of pH

The pH of the solution is one of the prime influencing factors in the adsorption process. Adsorption efficiency of the adsorbent changes remarkably with the change in the pH of the solution. Modification of the pH of the wastewater makes a significant impact on the surface charge and thus the degree of ionization of the adsorbent that controls the electrostatic interaction between the pollutant and adsorbent and hence the adsorption [47]. On increasing the pH of the medium, beyond the pH_{pzc} (zero point charge) value of the adsorbent the surface-induced negative charge attracts the positive pollutants from the medium to be adsorbed [48,49]. The value of pH_{pzc} is the pH value at which the surface of the adsorbent acquires a neutral charge and is stated as the point of zero charges [50]. Therefore, at a higher pH value (pH $>$ pH_{pzc}) the adsorption of cationic pollutants is enhanced by increased electrostatic interaction between the adsorbate and adsorbent, whereas the adsorption of the anionic pollutants reduces, due to increased electrostatic repulsions [5,19].

Li et al. investigated the effect of pH on the adsorption of cationic (MB dye) and anionic (MO dye) dyes zirconium−metalloporphyrin modified metal−organic frameworks (PCN-222/MOF-545) [51]. The pH_{pzc} value of the PCN-222/MOF-545 was found to be pH 8, which signifies the slightly basic nature of the adsorbent surface. Additionally, this pH_{pzc} value also supports the efficient removal of the cationic (MB) dye. The maximum removal of MO dye was observed at pH 5 (pH $<$ pH_{pzc}) (Fig. 6.5A), and in this state the surface of the adsorbent provides positive charge density and prefers the adsorption of anionic pollutants. However, the regular reduction in adsorption of MO

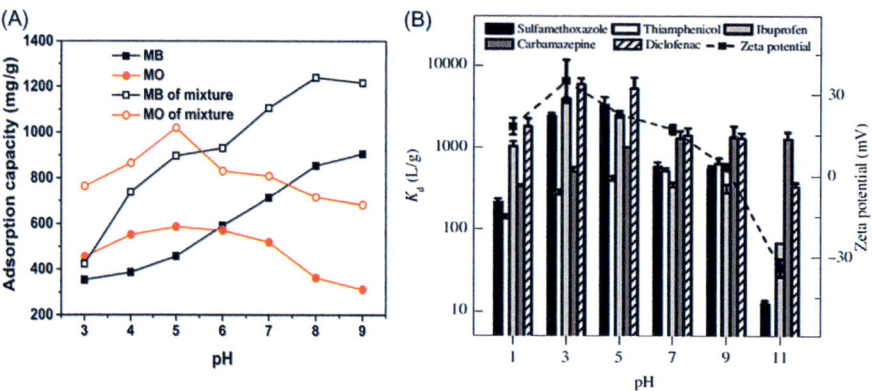

Figure 6.5 Effect of pH of the aqueous medium on the adsorption efficiency of (A) MB, MO and mixtures of MB and MO dyes on PCN-222/MOF-545 [51]; Copyright 2017, Royal Society of Chemistry Publishers (B) sulfamethoxazole, diclofenac, ibuprofen, thiamphenicol, and carbamazepine on multiwalled carbon nanotubes (MWCNTs) [52]. Copyright 2016, Elsevier Science Ltd.

with an increase in pH revealed the unfavorable condition that might be due the start of an electrostatic repulsion between the anionic dye and negative surface. In contrast, a higher pH medium (pH $<$ pH$_{pzc}$) is a favorable condition for the MB dye (Fig. 6.5A) due to the enhanced interaction between the cationic dye and the negatively charged surface of the adsorbent. Furthermore, on mixing both dyes (MB and MO), PCN-222/MOF-545 showed enhanced adsorption capacity toward both dyes. A high pH medium improved the selectivity toward the MB dye.

However, the adsorption of neutral pollutants is also influenced by changing the pH of the medium. Fig. 6.5B represents the effect of pH (pH range 1−11) on the adsorption of several pharmaceutical ingredients (sulfamethoxazole, diclofenac, ibuprofen, thiamphenicol, and carbamazepine) on multiwalled carbon nanotubes (MWCNTs) through the values of adsorbent-to-solution distribution coefficients (K_d) and the zeta potential of MWCNT with changes in pH [52]. The adsorption potential of MWCNTs toward sulfamethoxazole, diclofenac, and ibuprofen is consistent with the trend of zeta potential of MWCNT. However, the adsorption of thiamphenicol and carbamazepine was increased with an increase in pH. In this event, the enhanced adsorption of thiamphenicol and carbamazepine was attributed to the hydrogen bonding and π-hydrogen bonding and degree of ionization. The functional groups of the pharmaceutical ingredients (NH$_2$, NH, and OH groups) interact with the oxygen functionalities on the surface of MWCNT following hydrogen bonding. At higher pH, the hydrogen bonding donors (oxygen–containing functional group) on the MWCNT surface get ionized. Therefore the interaction between the hydrogen bonding donors of pharmaceuticals and hydrogen bonding acceptors or π–donors on the MWCNT is facilitated which promotes the adsorption.

Therefore high pH values (pH $>$ pH$_{pzc}$) facilitate the adsorption of cationic pollutants, whereas anionic pollutants are favorably adsorbed at low pH values (pH $<$ pH$_{pzc}$), and pH changes also influence the adsorption rate by the degree of ionization of the pollutant and adsorbent [53−55].

6.7 Effect of competing ions

The modification of the removal percentage of a pollutant in the presence of other ions is known as the effect of competing or coexisting ions. The presence of other ions significantly affects the removal efficiency as they can easily compete with the adsorbate for the active adsorption sites on the surface on the adsorbent. Also, it can promote the removal efficiency of water pollutant. CO_3^{2-}, HCO_3^-, NO_2^-, NO_3^-, SO_4^{2-}, Ca^{2+}, and Mg^{2+} are the most commonly found ions in the drinking water and could participate in the competition with the water pollutant to get adsorbed [56]. Saad et al. studied the effect of competitive ions (Cl^-. SO_4^{2-}, HCO_3^-, Ca^{2+}, and Mg^{2+}) on the uptake amount of nitrate (NO_3^-) and phosphate ($H_2PO_4^-$) anions using ammonium-functionalized Mobil Composite Material No. 48 (MCM-48) [57]. Table 6.2 displays the effect of the presence of competing anions on the removal percentage of NO_3^- and $H_2PO_4^-$ anions. The presence of Cl^-. SO_4^{2-}, HCO_3^-, Ca^{2+}, and Mg^{2+} ions reduced the amount of NO_3^- adsorption by 43%, 45%, 15%, 38%, and 18%, respectively. Similarly, the presence of the same ions, that is, Cl^-. SO_4^{2-}, HCO_3^-, Ca^{2+}, and Mg^{2+}, makes a negative

Table 6.2 Effects of coexisting ions on the amount of adsorbed nitrate and phosphate anions on ammonium-functionalized Mobil Composite Material No. 48 (MCM-48) [57].

Coexisting ions	The amount adsorbed (mg/g)
NO_3^- alone	12.2 ± 0.3
$NO_3^- + Cl^-$	6.9 ± 1.3
$NO_3^- + SO_4^{2-}$	6.7 ± 1.0
$NO_3^- + HCO_3^-$	10.3 ± 0.4
$NO_3^- + Ca^{2+}$	7.5 ± 0.6
$NO_3^- + Mg^{2+}$	10.0 ± 1.4
$H_2PO_4^-$ alone	16.5 ± 0.6
$H_2PO_4^- + Cl^-$	14.9 ± 0.1
$H_2PO_4^- + SO_4^{2-}$	9.5 ± 0.7
$H_2PO_4^- + HCO_3^-$	17.3 ± 0.1
$H_2PO_4^- + Ca^{2+}$	15.1 ± 0.1
$H_2PO_4^- + Mg^{2+}$	$12.7 \pm 0.$

Source: Copyright 2007, Elsevier Science Ltd.

impact on $H_2PO_4^-$ ion adsorption by 9%, 42%, 8%, and 23%, respectively. The slight increase in the adsorption amount of $H_2PO_4^-$ ion in the presence of HCO_3^- might be due to the modification in pH to 4.0, which was found to be the optimum pH for the adsorption of $H_2PO_4^-$ ions. The effect of the presence of major ions (HCO_3^-, Cl^-. SO_4^{2-}, and Ca^{2+}) and minor ions (PO_4^{3-}, SiO_3^{2-}, humic acid (HA)) in local groundwater was also investigated during the removal of As(V) on reclaimed iron oxide-coated sands (RIOCS) [58]. On examining the multiion system and single-ion system, the competitive effect on As(V) removal was found to be in the following order $PO_4^{3-} > SiO_3^{2-} > HA > Cl^-$ and $PO_4^{3-} > SiO_3^{2-} > HCO_3^- > HA > SO_4^{2-} > Cl^-$, respectively, at all initial pH conditions (pH 5–8). The presence of Ca^{2+} acts as a promotor for the As(V) adsorption both in the multiion system and the single-ion system, whereas SO_4^{2-} slightly decreases the As(V) adsorption in the single-ion system and promotes it in the multiion system. The increased adsorption of As(V) in the presence of Ca^{2+} might be due to either the formation of insoluble $Ca_3(AsO_4)_2$ or Ca^{2+} might interact with RIOCS and neutralize the negative surface charge which helps the efficient adsorption of negatively charged As(V) [59,60]. Similarly, the precipitation of FeAsS on coexisting with SO_4^{2-} anions promotes the As(V) removal from the water [61]. Thus the existence of competitive ions significantly changes the removal percentage of water pollutants.

6.8 Effect of morphology and surface characteristics of adsorbent

Adsorption is a surface phenomenon and is significantly influenced by the morphology and surface characteristics, such as surface area, pore-volume, pore size, functional group, and defects, of the adsorbent. A larger surface area of an adsorbent provides more active sites to take part in adsorption. Nanosized adsorbent material exhibits a higher surface area and higher degree of surface activity than the bulk material and unique physical and chemical properties that promote adsorption [62,63]. Therefore, nanostructural materials, also termed nanoadsorbents, are favored in the modern era of nanotechnology for wastewater treatment. Zeng et al. have revealed that g–C_3N_4 nanobelts comprise more significant surface area than bulk g–C_3N_4 and hence perform better surface applications [64].

Watanabe et al. evaluated the surface properties of five samples of nano-Boehmites of different sizes and also found out that the nano-Boehmites with highest surface area showed the highest adsorption capacity for phosphate ion removal [65]. Liu et al. prepared activated carbon from risk husk (P-AC), which was neutralized (P95-AC) and further activated with different amounts of KOH (PK-AC and K-AC) for dye (Rhodamine B

(RhB)) adsorption [66]. The various K-ACs exhibited the highest surface area in the following order: K-AC (2516 m^2/g) > PK-AC (1543 m^2/g) > P95-AC (1803 m^2/g) > P-AC (892 m^2/g). The adsorbing trend of RhB dye on activated carbon was also found to be consistent with the trend of BET. Furthermore, the selectivity of pollutant removal is also dependent on the surface characteristics of the adsorbent.

Chen et al. prepared mesoporous carbon nanocomposites and employed these for the removal of organic dyes and selective heavy metal ions [67]. Carbon nanocomposites have five times more adsorption efficiency than pure carbon for dye removal. Carbon nano-composites exhibit more affinity toward MB dye removal than MO dye, even though the hydrodynamic radius of the former is bigger than the latter. However, the enhanced MB dye adsorption might be due to the π-π interactions as MB exhibit a higher order of con-jugation with the domain of six-membered aromatic rings. These carbon nanocomposites also achieved selective adsorption toward Cr(II) ion in the presence of Zn(II), Ni(II), Cu (II), and Pb(II) ions [67]. After 24 hours of the adsorption process, no Cr(II) was detected in the wastewater. The pore sizes in the adsorbent are more significant than the metal ion hydrodynamic radius, which allowed the passage of all metal ions through the pores. However, the selectivity of adsorbent toward Cr(II) can be explained based on electrostatic interactions and the ion–exchange mechanism.

Ge et al. studied the morphology-dependent adsorption characteristics of CeO_2 to remove fluoride ions from aqueous solution [68]. CeO_2 nanorods, CeO_2 octahedrons, and CeO_2 nanocubes have shown 71.5, 28.3, and 7.0 mg/g adsorption capacity for fluo-ride adsorption that is not consistent with the trend surface area of the nanoadsorbent. This might be due to the adsorption of fluoride being more efficient on the [1 1 0] facets on the CeO_2 nanorods than on the [1 1 1] and [1 0 0] facets of the CeO_2 octahedrons and CeO_2 nanocubes, respectively. The presence of the Ce^{3+}-O defect, ion–exchange, monolayer surface adsorption, and pore-filling adsorption mechanisms was responsible for the efficient removal of fluoride.

Therefore, the competent removal of pollutants in wastewater significantly depends on the morphology and surface characteristics of the adsorbent, which should be studied carefully.

6.9 Conclusion

This chapter provides an insight into the influence of several experimental condi-tions, such as pH, an initial amount of adsorbate, adsorbent dosage, reaction temperature, presence of competent ions, morphology, and surface characteristics of the adsorbent, on the adsorption of various organic and inorganic water pollutants. For example, to achieve maximum removal of pollutant the pH of the aqueous medium should be tuned according to the properties of the pollutant. The adsorption efficiency of the adsorbent decreases with an increase in the initial concentration of adsorbate materials due to the rapid engagement of all the active sites. From an economical outlook, the optimization of adsorbent concentration is necessary to attain the highest adsorption capacity. The optimization of temperature is a crucial parameter to provide a favorable environment for the adsorption process. At higher temperatures, the adsorption efficacy of adsorbents is preferable for endothermic reactions, whereas the adverse effects are applicable for an exothermic reaction. The optimization of contact time of the adsorption reaction process helps to identify the ideal reaction for the complete saturation of all active sites with pollutant moieties. The presence of coexisting/competitive ions affects the removal percentage of water contaminants through adsorption. Competitive ions either reduce the removal efficiency by getting adsorbed onto the active surface site or enhance the removal efficacy by precipitation of the contaminant molecules. Furthermore, the design of the nanocomposite to accomplish the desired characteristics, such as a high surface area with maximum active sites that requires the functional group to interact with pollutants, defects, and perfect morphology, is also one of the major contributions to improving the adsorption characteristics of nanoadsorbents. Therefore, the optimization of each adsorption reaction parameter is a necessary step and these parameters should be carefully optimized before the scaling-up process of water treatment to the industrial level.

References

[1] Yagub MT, Sen TK, Afroze S, Ang HM. Dye and its removal from aqueous solution by adsorption: a review. Adv Colloid Interface Sci 2014;209:172−84.

[2] Abas SNA, Ismail MHS, Kamal ML, Izhar S. Adsorption process of heavy metals by low-cost adsorbent: a review. World Appl Sci J 2013;28:1518−30.

[3] Robati D, Rajabi M, Moradi O, Najafi F, Tyagi I, Agarwal S, et al. Kinetics and thermodynamics of malachite green dye adsorption from aqueous solutions on graphene oxide and reduced graphene oxide. J Mol Liq 2016;214:259−63.

[4] Wu X-L, Shi Y, Zhong S, Lin H, Chen J-R. Facile synthesis of Fe_3O_4-graphene@ mesoporous SiO_2 nanocomposites for efficient removal of Methylene Blue. Appl Surf Sci 2016;378:80−6.

[5] Manzoor K, Ahmad M, Ahmad S, Ikram S. Removal of Pb(ii) and Cd(ii) from wastewater using arginine cross-linked chitosan−carboxymethyl cellulose beads as green adsorbent. RSC Adv 2019;9:7890−902.

[6] Liang XX, Omer A, Hu Z-h, Wang Yg, Yu D, Ouyang X-k. Efficient adsorption of diclofenac sodium from aqueous solutions using magnetic amine-functionalized chitosan. Chemosphere. 2019;217:270−8.

[7] Rahdar S, Taghavi M, Khaksefidi R, Ahmadi S. Adsorption of arsenic (V) from aqueous solution using modified saxaul ash: isotherm and thermodynamic study. Appl Water Sci 2019;9:87.

[8] Miao M-S, Liu Q, Shu L, Wang Z, Liu Y-Z, Kong Q. Removal of cephalexin from effluent by activated carbon prepared from alligator weed: kinetics, isotherms, and thermodynamic analyses. Process Saf Environ Prot 2016;104:481−9.

[9] Wang Z, Wang K, Wang Y, Wang S, Chen Z, Chen J, et al. Large-scale fabrication of N-doped porous carbon nanosheets for dye adsorption and supercapacitor applications. Nanoscale. 2019;11:8785−97.

[10] Datta D, Uslu H, Kumar S. Adsorptive separation of Cu^{2+} from an aqueous solution using trioctylamine supported montmorillonite. J Chem Eng Data 2015;60:3193−200.

[11] Wanjeri VWO, Sheppard CJ, Prinsloo ARE, Ngila JC, Ndungu PG. Isotherm and kinetic investigations on the adsorption of organophosphorus pesticides on graphene oxide based silica coated magnetic nanoparticles functionalized with 2-phenylethylamine. J Environ Chem Eng 2018;6:1333−46.

[12] Omri A, Wali A, Benzina M. Adsorption of bentazon on activated carbon prepared from Lawsonia inermis wood: equilibrium, kinetic and thermodynamic studies. Arab J Chem 2016;9:S1729–39.

[13] Mansouriieh N, Sohrabi MR, Khosravi M. Adsorption kinetics and thermodynamics of organophosphorus profenofos pesticide onto Fe/Ni bimetallic nanoparticles. Int J Environ Sci Technol 2016;13:1393–404.

[14] Salleh MAM, Mahmoud DK, Karim WAWA, Idris A. Cationic and anionic dye adsorption by agricultural solid wastes: a comprehensive review. Desalination. 2011;280:1–13.

[15] Sonawane G, Shrivastava V. Kinetics of decolourization of malachite green from aqueous medium by maize cob (Zea maize): an agricultural solid waste. Desalination. 2009;247:430–41.

[16] Joshi S, Sharma M, Kumari A, Shrestha S, Shrestha B. Arsenic removal from water by adsorption onto iron oxide/nano-porous carbon magnetic composite. Appl Sci 2019;9:3732.

[17] Fakhri A. Adsorption characteristics of graphene oxide as a solid adsorbent for aniline removal from aqueous solutions: kinetics, thermodynamics and mechanism studies. J Saudi Chem Soc 2017;21:S52–7.

[18] Malkoc E, Nuhoglu Y. Investigations of nickel (II) removal from aqueous solutions using tea factory waste. J Hazard Mater 2005;127:120–8.

[19] Khoshsang H, Ghaffarinejad A, Kazemi H, Wang Y, Arandiyan H. One-pot synthesis of S-doped Fe$_2$O$_3$/C magnetic nanocomposite as an adsorbent for anionic dye removal: equilibrium and kinetic studies. J Nanostruct Chem 2018;8:23–32.

[20] Ahsan MA, Islam MT, Hernandez C, Castro E, Katla SK, Kim H, et al. Biomass conversion of saw dust to a functionalized carbonaceous materials for the removal of Tetracycline, Sulfamethoxazole and Bisphenol A from water. J Environ Chem Eng 2018;6:4329–38.

[21] Ghaedi M, Tavallali H, Sharifi M, Kokhdan SN, Asghari A. Preparation of low cost activated carbon from Myrtus communis and pomegranate and their efficient application for removal of Congo red from aqueous solution. Spectrochim Acta Part A Mol Biomol Spectrosc 2012;86:107–14.

[22] Namasivayam C, Muniasamy N, Gayatri K, Rani M, Ranganathan K. Removal of dyes from aqueous solutions by cellulosic waste orange peel. Bioresour Technol 1996;57:37–43.

[23] Ghaedi M, Amirabad SZ, Marahel F, Kokhdan SN, Sahraei R, Nosrati M, et al. Synthesis and characterization of Cadmium selenide nanoparticles loaded on activated carbon and its efficient application for removal of Muroxide from aqueous solution. Spectrochim Acta Part A Mol Biomol Spectrosc 2011;83:46–51.

[24] Senthilkumaar S, Kalaamani P, Subburaam CV. Liquid phase adsorption of Crystal violet onto activated carbons derived from male flowers of coconut tree. J Hazard Mater 2006;136:800–8.

[25] Ofomaja AE, Ho Y-S. Equilibrium sorption of anionic dye from aqueous solution by palm kernel fibre as sorbent. Dye Pigment 2007;74:60–6.

[26] Bilgiç A, Çimen A. Removal of chromium(vi) from polluted wastewater by chemical modification of silica gel with 4-acetyl-3-hydroxyaniline. RSC Adv 2019;9:37403–14.

[27] Mubarak NSA, Jawad AH, Nawawi WI. Equilibrium, kinetic and thermodynamic studies of Reactive Red 120 dye adsorption by chitosan beads from aqueous solution. Energy Ecol Environ 2017;2:85–93.

[28] Xue F, Wang F, Chen S, Ju S, Xing W. Adsorption equilibrium, kinetics, and thermodynamic studies of cefpirome sulfate by using macroporous resin. J Chem Eng Data 2017;62:4266–72.

[29] Shikuku VO, Zanella R, Kowenje CO, Donato FF, Bandeira NMG, Prestes OD. Single and binary adsorption of sulfonamide antibiotics onto iron-modified clay: linear and nonlinear isotherms, kinetics, thermodynamics, and mechanistic studies. Appl Water Sci 2018;8:175.

[30] Fu J, Xin Q, Wu X, Chen Z, Yan Y, Liu S, et al. Selective adsorption and separation of organic dyes from aqueous solution on polydopamine microspheres. J Colloid Interface Sci 2016;461:292–304.

[31] Panda L, Rath SS, Rao DS, Nayak BB, Das B, Misra PK. Thorough understanding of the kinetics and mechanism of heavy metal adsorption onto a pyrophyllite mine waste based geopolymer. J Mol Liq 2018;263:428–41.

[32] Song Q, Fang Y, Liu Z, Li L, Wang Y, Liang J, et al. The performance of porous hexagonal BN in high adsorption capacity towards antibiotics pollutants from aqueous solution. Chem Eng J 2017;325:71–9.

[33] Thuan LV, Chau TB, Ngan TTK, Vu TX, Nguyen DD, Nguyen M-H, et al. Preparation of cross-linked magnetic chitosan particles from steel slag and shrimp shells for removal of heavy metals. Environ Technol 2018;39:1745–52.

[34] Malik R, Dahiya S, lata S. An experimental and quantum chemical study of removal of utmostly quantified heavy metals in wastewater using coconut husk: a novel approach to mechanism. Int J Biol Macromolecules 2017;98:139—49.

[35] Tirtom VN, Dinçer A, Becerik S, Aydemir T, Çelik A. Removal of lead (II) ions from aqueous solution by using crosslinked chitosan-clay beads. Desalination Water Treat 2012;39:76—82.

[36] Al-Ghouti MA, Da'ana D, Abu-Dieyeh M, Khraisheh M. Adsorptive removal of mercury from water by adsorbents derived from date pits. Sci Rep 2019;9:15327.

[37] Kubilay Ş, Gürkan R, Savran A, Şahan T. Removal of Cu(II), Zn(II) and Co(II) ions from aqueous solutions by adsorption onto natural bentonite. Adsorption 2007;13:41—51.

[38] Muhammad A, Shah A-u-HA, Bilal S, Rahman G. Basic blue dye adsorption from water using polyaniline/magnetite (Fe_3O_4) composites: kinetic and thermodynamic aspects. Materials 2019;12:1764.

[39] Massey AT, Gusain R, Kumari S, Khatri OP. Hierarchical microspheres of MoS_2 nanosheets: efficient and regenerative adsorbent for removal of water-soluble dyes. Ind Eng Chem Res 2016;55:7124—31.

[40] Charpentier TVJ, Neville A, Lanigan JL, Barker R, Smith MJ, Richardson T. Preparation of magnetic carboxymethylchitosan nanoparticles for adsorption of heavy metal ions. ACS Omega 2016;1:77—83.

[41] Shi J, Li H, Lu H, Zhao X. Use of carboxyl functional magnetite nanoparticles as potential sorbents for the removal of heavy metal ions from aqueous solution. J Chem Eng Data 2015;60:2035—41.

[42] Hemmati F, Norouzbeigi R, Sarbisheh F, Shayesteh H. Malachite green removal using modified sphagnum peat moss as a low-cost biosorbent: kinetic, equilibrium and thermodynamic studies. J Taiwan Inst Chem Eng 2016;58:482—9.

[43] Sadegh H, Shahryari-ghoshekandi R, Agarwal S, Tyagi I, Asif M, Gupta VK. Microwave-assisted removal of malachite green by carboxylate functionalized multi-walled carbon nanotubes: kinetics and equilibrium study. J Mol Liq 2015;206:151—8.

[44] Pan X, Zuo G, Su T, Cheng S, Gu Y, Qi X, et al. Polycarboxylic magnetic polydopamine submicrospheres for effective adsorption of malachite green. Colloids Surf A Physicochem Eng Asp 2019;560:106—13.

[45] Nasiri Azad F, Ghaedi M, Dashtian K, Hajati S, Goudarzi A, Jamshidi M. Enhanced simultaneous removal of malachite green and safranin O by ZnO nanorod-loaded activated carbon: modeling, optimization and adsorption isotherms. N J Chem 2015;39:7998—8005.

[46] Dastkhoon M, Ghaedi M, Asfaram A, Goudarzi A, Langroodi SM, Tyagi I, et al. Ultrasound assisted adsorption of malachite green dye onto ZnS:Cu-NP-AC: equilibrium isotherms and kinetic studies — response surface optimization. Sep Purif Technol 2015;156:780—8.

[47] Gusain R, Kumar N, Fosso-Kankeu E, Ray SS. Efficient removal of Pb(II) and Cd(II) from industrial mine water by a hierarchical MoS_2/SH-MWCNT nanocomposite. ACS Omega 2019;4:13922—35.

[48] Kumar N, Mittal H, Alhassan SM, Ray SS. Bionanocomposite hydrogel for the adsorption of dye and reusability of generated waste for the photodegradation of ciprofloxacin: a demonstration of the circularity concept for water purification, ACS Sustainable Chem Eng 2018;6:17011—25.

[49] Kumar N, Mittal H, Parashar V, Ray SS, Ngila JC. Efficient removal of rhodamine 6G dye from aqueous solution using nickel sulphide incorporated polyacrylamide grafted gum karaya bionanocomposite hydrogel, RSC Adv 2016;6:21929—39.

[50] Krishna Kumar AS, Jiang S-J, Tseng W-L. Effective adsorption of chromium(vi)/Cr(iii) from aqueous solution using ionic liquid functionalized multiwalled carbon nanotubes as a super sorbent. J Mater Chem A 2015;3:7044—57.

[51] Li H, Cao X, Zhang C, Yu Q, Zhao Z, Niu X, et al. Enhanced adsorptive removal of anionic and cationic dyes from single or mixed dye solutions using MOF PCN-222. RSC Adv 2017;7:16273—81.

[52] Zhao H, Liu X, Cao Z, Zhan Y, Shi X, Yang Y, et al. Adsorption behavior and mechanism of chloramphenicols, sulfonamides, and non-antibiotic pharmaceuticals on multi-walled carbon nanotubes. J Hazard Mater 2016;310:235—45.

[53] Zhan W, Xu C, Qian G, Huang G, Tang X, Lin B. Adsorption of Cu(ii), Zn(ii), and Pb(ii) from aqueous single and binary metal solutions by regenerated cellulose and sodium alginate chemically modified with polyethyleneimine. RSC Adv 2018;8:18723—33.

[54] Bautista-Toledo I, Ferro-García MA, Rivera-Utrilla J, Moreno-Castilla C, Vegas Fernández FJ. Bisphenol A removal from water by activated carbon. effects of carbon characteristics and solution chemistry. Environ Sci Technol 2005;39:6246–50.

[55] Lazo-Cannata JC, Nieto-Márquez A, Jacoby A, Paredes-Doig AL, Romero A, Sun-Kou MR, et al. Adsorption of phenol and nitrophenols by carbon nanospheres: effect of pH and ionic strength. Sep Purif Technol 2011;80:217–24.

[56] Kumar N, Fosso-Kankeu E, Ray SS. Achieving controllable MoS_2 nanostructures with increased interlayer spacing for efficient removal of pb(ii) from aquatic systems. ACS Appl Mater Interfaces 2019;11:19141–55.

[57] Saad R, Belkacemi K, Hamoudi S. Adsorption of phosphate and nitrate anions on ammonium-functionalized MCM-48: effects of experimental conditions. J Colloid Interface Sci 2007;311:375–81.

[58] Hsu J-C, Lin C-J, Liao C-H, Chen S-T. Evaluation of the multiple-ion competition in the adsorption of As(V) onto reclaimed iron-oxide coated sands by fractional factorial design. Chemosphere. 2008;72:1049–55.

[59] McNeill LS, Edwards M. Arsenic removal during precipitative softening. J Environ Eng 1997;123:453–60.

[60] Bothe JV, Brown PW. Arsenic immobilization by calcium arsenate formation. Environ Sci Technol 1999;33:3806–11.

[61] Sun H, Wang L, Zhang R, Sui J, Xu G. Treatment of groundwater polluted by arsenic compounds by zero valent iron. J Hazard Mater 2006;129:297–303.

[62] Khan I, Saeed K, Khan I. Nanoparticles: properties, applications and toxicities. Arab J Chem 2019;12:908–31.

[63] Marei NN, Nassar NN, Vitale G. The effect of the nanosize on surface properties of NiO nanoparticles for the adsorption of Quinolin-65. Phys Chem Chem Phys 2016;18:6839–49.

[64] Zeng Y, Liu C, Wang L, Zhang S, Ding Y, Xu Y, et al. A three-dimensional graphitic carbon nitride belt network for enhanced visible light photocatalytic hydrogen evolution. J Mater Chem A 2016;4:19003–10.

[65] Watanabe Y, Kasama T, Fukushi K, Ikoma T, Komatsu Y, Tanaka J, et al. Synthesis of nano-sized boehmites for optimum phosphate sorption. Sep Sci Technol 2011;46:818–24.

[66] Ding L, Zou B, Gao W, Liu Q, Wang Z, Guo Y, et al. Adsorption of rhodamine-B from aqueous solution using treated rice husk-based activated carbon. Colloids Surf A Physicochem Eng Asp 2014;446:1–7.

[67] Chen L, Ji T, Mu L, Shi Y, Brisbin L, Guo Z, et al. Facile synthesis of mesoporous carbon nanocomposites from natural biomass for efficient dye adsorption and selective heavy metal removal. RSC Adv 2016;6:2259–69.

[68] Kang D, Yu X, Ge M. Morphology-dependent properties and adsorption performance of CeO_2 for fluoride removal. Chem Eng J 2017;330:36–43.

CHAPTER SEVEN

Carbon nanomaterials: synthesis, functionalization, and properties

7.1 Introduction

Nanotechnology is a revolutionary technology of the 21st century which is helping to significantly advance and modify the numerous conventionally existing technologies such as smart textiles [1,2], smart packaging [3], sensors [4], papermaking [5−7], energy production [8], batteries [9], energy storage [10,11], energy conversion [12], composites [13], biomedical [14], and environmental protection [15]. The nano-materials used in nanotechnology have at least one measurement element in the nano-sized range, that is, from 1 to 100 nm. The properties, such as electrical conductivity, thermal conductivity, chemical reactivity, melting point, photoluminescence, magnetic permeability, and specific surface area, are dependent on the size of the nanomaterials. The size and shape regulate the physicochemical properties of nanostructures enabling them to be utilized in diverse applications. Numerous nanomaterials (inorganic/organic and carbon-based) have been synthesized and applied in various areas. Basically, nanomaterials can be classified into four parts based on their dimensionality: zero-dimensional nanomaterials (0D-NMs), one-dimensional nanomaterials (1D-NMs), two-dimensional nanomaterials (2D-NMs), and three-dimensional nanomaterials (3D-NMs). 0D-NMs (e.g., nanoparticles) have all their dimensions at the nanoscales, whereas 1D-NMs (e.g., rods, belts, fibers, wires, and tubes) are carrying two dimensions at the nanoscale. In the case of 2D-NMs (e.g., sheet, plates, thin films, flakes and coat-ings), the material's thickness is at the nanoscale. Graphene sheets are the first synthesized 2D nanomaterials. 3D-NMs encompass powder, fibrous, multilayers, and polycrystalline materials that are formed using the basic units of 2D-NMs, 1D-NMs, and 0D-NMs. These materials are multifunctional in nature and have a high surface area, low density, and versatile surface properties.

Carbon is one of the most abundant and versatile elements on the planet and has wide-ranging applications [16]. Carbon-based nanomaterials (CNMs) have unique and outstanding characteristics, such as excellent mechanical strength, chemical, and ther-mal stability, low density, resistance to corrosion, and hardness, which make them competitive in a variety of commercial fields, including drug delivery, energy storage, microelectronics, environmental remediation, biotechnology, packaging, and coating

Carbon Nanomaterial-Based Adsorbents for Water Purification.
DOI: https://doi.org/10.1016/B978-0-12-821959-1.00007-6

[17−21]. Based on their structural dimensions, CNMs are classified into three categories: (1) 0D-CNMs, which include nanodiamonds, onion-like carbon structures, fullerenes, and carbon-dots (C-dots); (2) 1D CNMs, such as carbon nanotubes (CNTs), carbon nanofibers and carbon nanohorns; and (3) 2D CNMs, that is, graphene, graphene oxides (GO), reduced graphene oxide (rGO), carbon nanoribbons, and graphite–carbon nitride (g-C_3N_4) [22]. Fig. 7.1 depicts the large family of carbon nanostructures. The synthesis of CNMs for targeted applications is fascinating due to their unique properties, for example, high surface area, superior directionality,

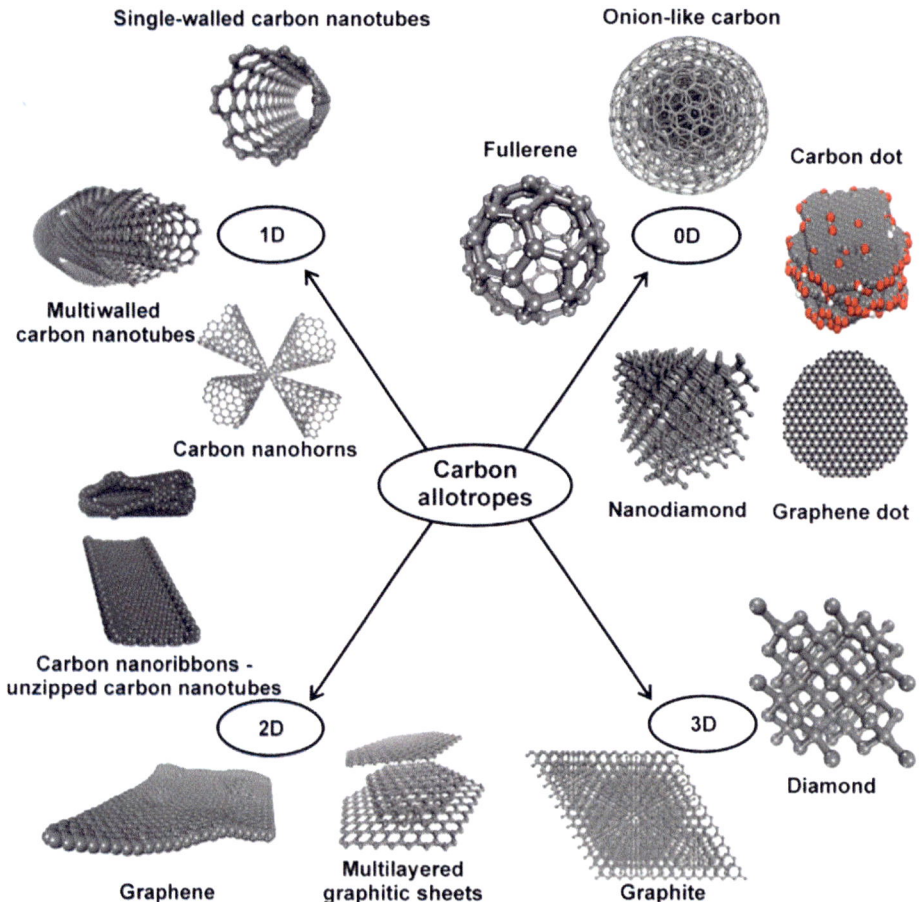

Figure 7.1 Classification of carbon nanomaterials based on structural dimension: 0D, 1D, 2D, and 3D (although graphite and diamond are considered to be bulk materials). *Reproduced with permission from Georgakilas V, Perman JA, Tucek J, Zboril R. Broad family of carbon nanoallotropes: classification, chemistry, and applications of fullerenes, carbon dots, nanotubes, graphene, nanodiamonds, and combined superstructures. Chem Rev 2015;115:4744−822 [22]. Copyright 2015, American Chemical Society.*

remarkable optical properties, and flexibility, which makes their synthesis different and more challenging than that of bulk materials for advanced applications [23,24]. This chapter highlights the overview of CNMs, including chemical structures, synthesis routes, functionalization approaches, and their properties.

7.2 Fabrication methods for carbon nanomaterials

CNMs encompass a variety of materials with different nanostructural forms, including graphene, CNTs, g–C$_3$N$_4$, nanoporous carbon (NPC) materials, carbon dots, carbon nanofibers, fullerenes, and nanodiamonds. Since the 1900s, several synthetic routes have been established to fabricate different CNMs (Fig. 7.2). Despite excellent advancements in fabrication methods, further development of synthetic

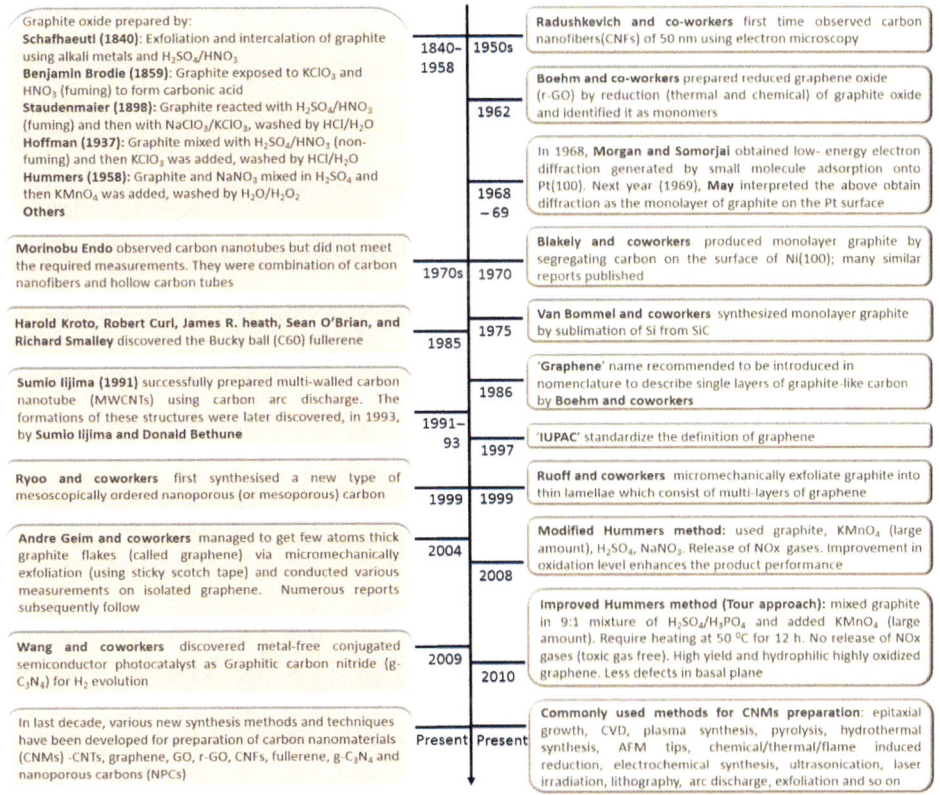

Figure 7.2 Schematic illustrating the history of CNM synthesis. *Reproduced with permission from Gusain R, Kumar N, Ray SS. Recent advances in carbon nanomaterial-based adsorbents for water purification. Coord Chem Rev 2020;405:213111 [25]. Copyright 2020, Elsevier Science Ltd.*

approaches is still required to make them less toxic and more cost-efficient and facile. The following section concisely explains the commonly used synthesis protocols of common CNMs: graphene (G), GO, rGO, CNTs, graphitic carbon nitride (g-C_3N_4), nanodiamonds (NDs), fullerenes, NPCs, and carbon nanofibers (CNFs).

7.2.1 Graphene-based nanomaterials (GNMs—G, GO, and rGO)

Graphene is the thinnest known material and has excellent mechanical stability, thermal conductivity, and optical and transport properties. It is made up of a single layer of sp^2 hybridized carbon atoms arranged in a honeycomb structure [26,27]. It is the primary 2D building block of carbon materials. It can be wrapped into fullerenes (0D), rolled into CNTs (1D), and stacked into graphite (3D) (Fig. 7.3). Exfoliation and chemical cleavage of bulk graphite into few-layer graphene (FLG) or single-layer graphene (SLG) are the principal synthetic routes to graphene. Geim and coworkers received the Nobel Prize in 2010 for the discovery of single-layer graphene (G) through mechanical exfoliation of stacked graphite using Scotch tape [29]. The synthesis of planar FLG has been achieved by pyrolysis of camphor using chemical vapor deposition (CVD) on Ni foils [30]. Overall, graphene synthesis methods can be divided into two types: (1) top-down approaches (using graphite as a starting material) such as mechanical exfoliation, liquid-phase exfoliation, chemical synthesis (using oxidation−exfoliation−reduction), solid exfoliation, intercalation exfoliation, CNTs unzipping, and electrochemical exfoliation; and (2) bottom-up approaches (using nongraphitic/hydrocarbons as starting materials) such as CVD, epitaxial growth on SiC,

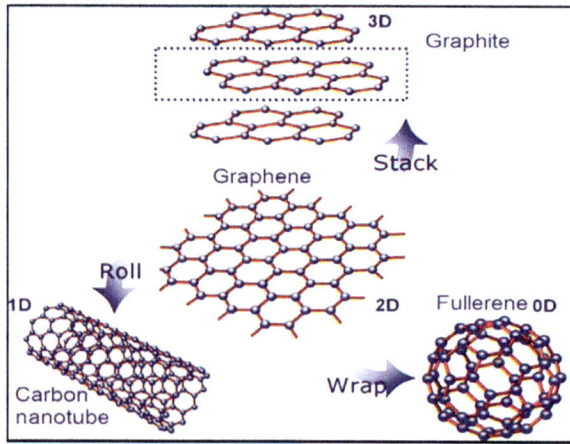

Figure 7.3 Graphene: the basic 2D building block of carbon materials. It can be used to form fullerenes (0D), carbon nanotubes (1D), and graphite (3D). *Reproduced with permission from Wan X, Huang Y, Chen Y. Focusing on energy and optoelectronic applications: a journey for graphene and graphene oxide at large scale. Acc Chem Res. 2012;45:598−607 [28]. Copyright 2012, American Chemical Society.*

pyrolysis, and laser-assisted synthesis. The important fabrication methods using bottom-up and top-down approaches have been illustrated in Fig. 7.4.

However, the difficulties of the controlled synthesis of graphene with limited numbers of layers, small yields, minimization of folds, etc. are issues which remain to be addressed. The reported large-scale synthesis methods for graphene are oxidation—exfoliation—reduction, CVD, and exfoliation.

7.2.1.1 Oxidation—exfoliation—reduction

Graphene is usually employed as a derivative, for example, GO, rGO, or functionalized graphene oxide. All graphene derivatives are prepared by modification of GO. The most common manner for GO synthesis is the oxidation—exfoliation—reduction approach. Well-known approaches for GO synthesis using graphite include the Staudenmaier method [32], the Brodie method [33], the Hummers method [34], and the modified Hummers method [26,35]. All of these methods involve harsh oxidation of graphite using acids ($H_2SO_4/H_3PO_4/HNO_3$), followed by alkali or water intercalation, to form graphite oxide. The graphite oxide is exfoliated to form graphene oxide using various exfoliation strategies (Fig. 7.5). The resulting GO can be further modified into rGO via thermal [37], chemical [38], photocatalytic [39], hydrothermal [40],

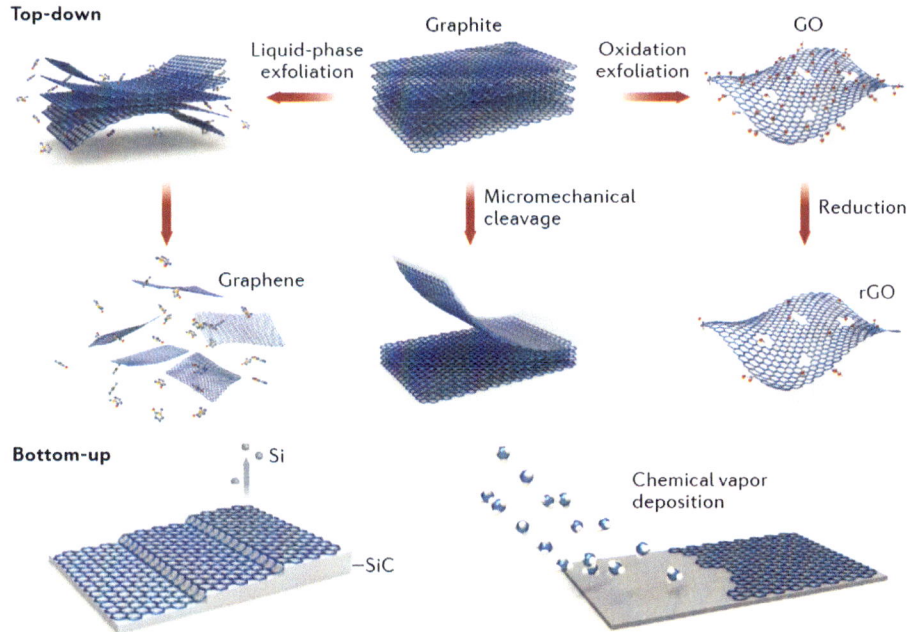

Figure 7.4 Important graphene fabrication techniques. *Reproduced with permission from Wang X-Y, Narita A, Müllen K. Precision synthesis versus bulk-scale fabrication of graphenes. Nat Rev Chem 2017;2:0100 [31] with slight modifications. Copyright 2017, Springer-Nature.*

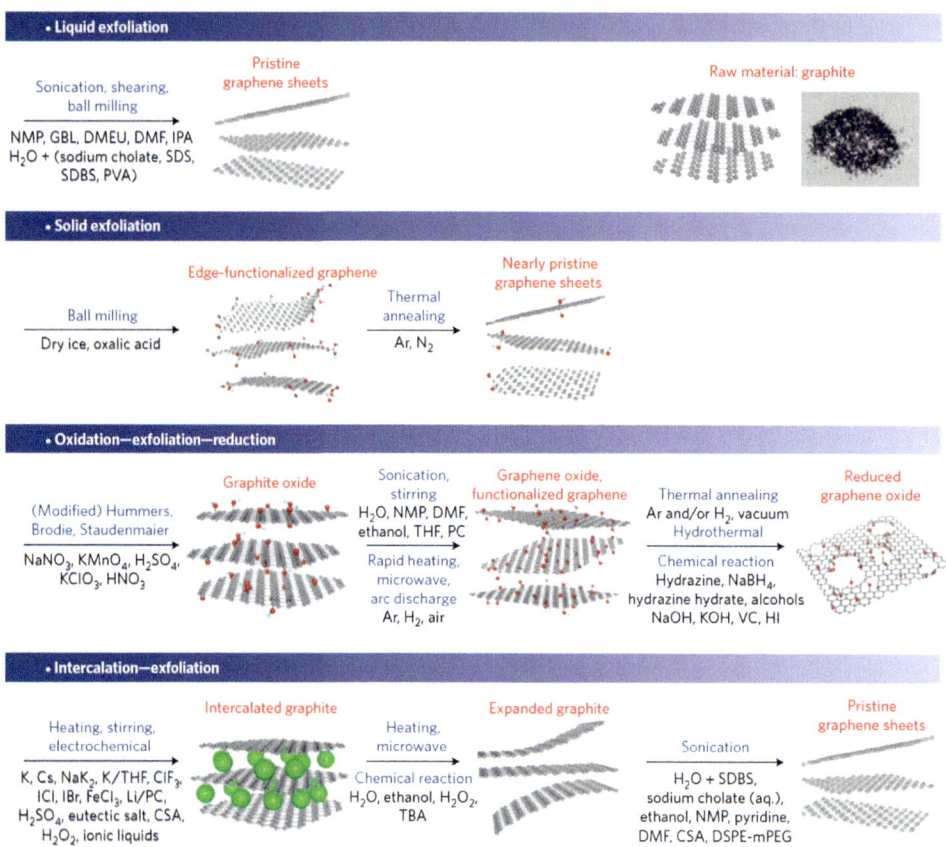

Figure 7.5 Preparation of graphene sheets using various exfoliation methods with graphite as starting material. *Reproduced with permission from Ren W, Cheng H-M. The global growth of graphene. Nat Nanotechnol. 2014;9:726—30 [36]. Copyrights 2014, Springer-Nature.*

or microwave [41] reduction. The most common process for reducing GO to rGO is a chemical reduction, using a variety of organic and inorganic reducing agents, such as sodium borohydride [42], hydrazine hydrate [23], amino acids [43], pyrrole [44], ascorbic acid [45], and urea [46].

7.2.1.2 Chemical vapor deposition

In 2006, for the first time, a few-layer graphene was reported using the CVD technique (bottom-up approach) [30]. Since then, significant research has been done in this area to develop a scaled-up graphene method. In the CVD approach, a wide variety of nongraphite carbons or hydrocarbons are decomposed at a very high temperature on metallic substrates in the presence of an inert atmosphere. Various types of substrates such as Ni, Cu, Co, Ir, Au, Pt, Pd, Ge, Re, Rh, Ru, alloys, SiO_x, and hexagonal boron nitride (h-BN) are used for large area graphene-sheets growth [47–53]. Appropriate carbon sources can be gases (CH_4, C_2H_4, C_2H_2), liquids (C_2H_5OH, C_6H_6, C_6H_{14}, oils), or solids (polymers, camphor, sucrose). The graphene growth mechanism in CVD depends on a substrate material which may involve several steps such as decomposition of carbon molecules, adsorption, diffusion, dehydrogenation, dissolution, nucleation, separation, and precipitation [54,55]. Moreover, the quality of graphene can be regulated by changing the conditions such as carbon source, substrate, partial pressure, and mass transport of gas. For example, carbon is easily soluble in a Ni substrate at high temperatures, thus carbon dissolves, diffuses, and precipitates to create a few- to multilayer graphene [55]. Cu substrate is most famous because of its low-cost, low carbon solubility, and the ability to create a single layer of graphene with high quality using a gas precursor (viz., CH_4) [56]. Solid precursors have also been used in CVD growth of graphene, but they lead to multilayers graphene.

The CVD process results in a high surface area and high-quality graphene, which fulfils the necessities of electronic and photonic applications. Although this method offers scalability and production potential, it is an energy-intensive, lengthy, and expensive process and requires complicated infrastructure, compared with exfoliation and mechanical approaches. Thus various modifications have been introduced in this process for an overall reduction in the cost of CVD-developed graphene. For instance, to reduce the temperature of the process, a plasma-assisted CVD process has been developed, which helps the fast dissociation of gases at low temperatures [57]. A few of the most common modifications are surface wave plasma CVD (SWP-CVD), microwave-assisted surface wave plasma CVD (MW-SWP-CVD), low-pressure CVD (LP-CVD), remote plasma-enhanced CVD (r-PE-CVD), ultrahigh vacuum CVD (UHV-CVD), microwave plasma CVD (MP-CVD), radiofrequency plasma–enhanced CVD (RF-PE-CVD), atmospheric pressure CVD (AP-CVD), and inductively-coupled plasma CVD (ICP-CVD) [58,59]. Furthermore, the developed roll–roll automated method can yield graphene films continuously with high productivity [60]. The aforementioned improved methods basically produce lower-quality graphene, including high defects, thick sheets, and with high sheet resistance. Overall, the improved methods have led to significant advances toward the ultimate aim of the affordable production of graphene, but there is still a need for improvements. Various kinds of catalysts/substrates and carbon sources have been employed in CVD techniques for the development of single/multilayer graphene. Table 7.1 highlights the different CVD techniques and synthesis conditions.

7.2.1.3 Exfoliation

Exfoliation of graphite is the most popular approach for scaling up production of graphene sheets. Exfoliation of graphite can be achieved by various approaches: mechanical exfoliation (viz., Scotch-tape method), solid exfoliation (using ball milling), liquid

Table 7.1 Different CVD techniques for graphene growth and related synthesis parameters.

Technique	Carbon source	Temperature (°C)	Substrate/catalyst	Graphene domain	Graphene layers	Reference
LP-CVD	CH_4	1000	Cu	Coverage of substrate	Monolayer	[61]
	CH_4	950	Cu	cm-sized	Monolayer	[62]
	C_6H_6	300	Cu	micron-sized	Monolayer	[63]
	CH_3OH, C_2H_5OH, C_3H_5OH	650–850	Cu	670, 130, and 168 nm	Mono- and few-layer	[64]
	CH_4	1000	Ni	–	Multilayer	[65]
	C_2H_4	600–900	Ni	Coverage of substrate	Few-layer	[66]
AP-CVD	CH_4	1050	Cu	Micron-sized	Mono- and few-layer	[49]
	C_2H_5OH	650–850	Cu	Micron-sized	Monolayer	[67]
	Naphthalene	300–600	Cu	Micron-sized	Monolayer	[68]
	Naphthalene and Urea	900–1000	Cu	Nanosized	Mono- and few-layer	[69]
	C_2H_2, C_6H_6	600	Ni	Coverage of substrate	Monolayer	[70]
	CH_4	1000	Ga (liquid)	Several-microns	Multilayer	[71]
PE-CVD	CH_4	500	Cu	14 and 26 nm	Multilayer	[72]
	CH_4, C_2H_4, C_2H_2	435–650	Sapphire, SiO_2, pyrolytic graphite, h-BN	~ 200 nm	N-doped monolayer	[73]
UHV-CVD	CH_4	800	Cu	Nanosized	Monolayer	[74]
	C_2H_4	~ 877	Pt	Micron-sized	Mono- and few-layer	[75]
SWP-CVD	CH_4	300–400	Cu	cm-sized	Few-layer	[76]
ICP-CVD	C_2H_2	600–750	CuNi, AuNi	–	Mono- and few-layer	[77]
	CH_4	700–1000	Fe_2O_3	Nanosized	Few-layer	[78]
MW-SWP-CVD	$C_{10}H_{16}O$, CH_4	550	Cu	cm-sized	–	[79]
MP-CVD	CH_4	500	Stainless steel	Micron-sized	Mono- and few-layer	[80]
RFPE-CVD	CH_4	650	Ni	cm-sized	Mono- and few-layer	[81]

exfoliation, and intercalation exfoliation (Fig. 7.5). Due to the low-yield and poor control over graphene layers, mechanical exfoliation of graphite is not an appropriate method for the production of graphene for its projected applications. Liquid-phase exfoliation (LPE) has emerged as a promising method for the synthesis of FLG and graphene nanoplatelets directly from graphite. This method involves the intercalation/insertion of small molecules or atoms in-between graphitic layers, such as water, organic solvents (N-Methyl-2-Pyrrolidone, azobenzene, NH_3, ethanol, 1,2-Dichlorobenzene) [82], surfactants (Triton X-100), polymer (polyvinylpyrrolidone), supercritical CO_2 [83], NaOH, iron acetate, and potassium, that can reduce the interlayer van der Waals interactions and thereby easily facilitate the exfoliation process using shearing force, pressure, and thermal treatment to segregate the graphene layers [84] (Fig. 7.5). The LPE techniques comprise a sonication process [85], microfluidization [86], jet cavitation [87], and high-shear mixing [88]. No method is perfect; all carry their own advantages and restrictions. These methods can lead to defect-rich graphene by causing little damage to the aromatic moieties. The LPE methods produce high-purity FLG and GNPs in the form of inks/dispersions and liquid suspensions, which is very useful for printed electronics, composite fillers, and conductive coatings.

7.2.2 Carbon nanotubes

Radushkevich et al. first reported CNTs in 1952 [89], and in 1991 Iijima described the synthesis of multiwalled carbon nanotubes (MWCNTs) using the arc evaporation method [90]. CNTs can be compared with rolled sheets of graphene. They can be synthesized by arc discharge, laser ablation, CVD, plasma–enhanced CVD (PECVD), and pyrolysis methods [91] (Table 7.2). The arc discharge method, with graphite electrodes, is the most frequently employed and the best way to produce high-quality CNTs. Current is passed through graphite electrodes in the He environment at a high temperature ($>1700°C$) which causes the graphite to vaporize and recondense onto

Table 7.2 Comparative descriptions of commonly used synthesis methods for CNTs.

Method	CVD	Arc-discharge	Laser ablation
Yield rate	$>75\%$	30%	70%
Mass production	Successfully achieved	Possible	Possible
SWCNTs or MWCNTs	Both	Both	Both
Advantages	Low temperature, high purity, aligned growth possible, simple	Simple, low cost, high-quality nanotubes	Room temperature synthesis, relatively high purity
Disadvantages	Nanotubes with defects, mostly used for MWCNTs	Tangled nanotubes, high temperature, purification required	Highly pure precursors needed, limited to lab-scale

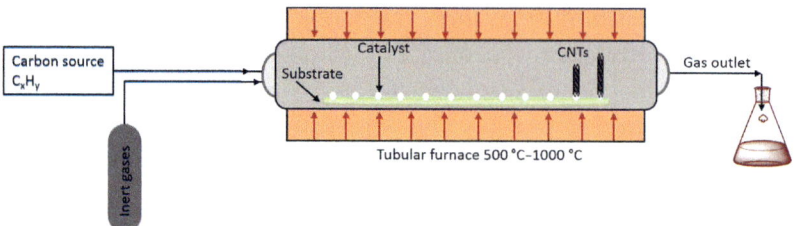

Figure 7.6 Schematic representation of chemical vapor deposition (CVD) reactor for CNTs preparation.

the walls of the reaction vessel and cathode. The condensed carbon on the cathode is composed of CNTs. Generally, MWCNTs are produced using the arc discharge method when the anode comprises solely graphite; however, to selectively produce SWCNTs, the graphite anode is doped with a transition metal [91,92]. Another standard procedure for producing CNTs is the CVD method, which involves the catalytic decomposition of carbon monoxide or a hydrocarbon feedstock in the presence of a transition metal catalyst [93] (Fig. 7.6). The key advantage of the CVD method is the high purity of the resulting CNTs. In the laser ablation method, pure graphite in a quartz tube is vaporized by heating with a high-powered laser inside a furnace ($\geq 1200°$C), under Ar atmosphere [94]. Similar to the arc discharge method, for the synthesis of SWCNTs the addition of a transition metal to graphite is crucial in the laser ablation method. The laser pulse power influences the diameter of the nanotubes; increased pulse power produces thinner CNTs. The most commonly used way for the production of high-purity SWCNTs is the high-pressure carbon monoxide (HiPCO) process [95]. It uses CO as a C source and $Ni(CO)_5/Fe(CO)_5$ as a catalyst at high temperature ($900°$C$-1000°$C) and high pressure ($30-50$ atm). This method could produce high-purity (97 mol% purity) SWCNT at a large scale (10 g/day).

7.2.3 Fullerene

Fullerenes, one of the allotropic forms of carbon, were first introduced by Smalley et al. (1985) in the carbon family via graphite vaporization using laser irradiation under an inert gaseous environment at low pressure [96]. Among several forms of fullerenes, C_{60} is the most common form and exhibits 20 hexagonal and 12 pentagonal rings of sp^2 hybridized carbon atoms arranged in a symmetrical icosahedral closed cage structure. Huffman et al. (1990) were the first to produce C_{60} Fullerene in macroscopic quantities following the arc discharge method using pure graphitic carbon soot [97]. The fullerene synthesis route via this method is similar to that described in the synthesis of CNTs. Since 1990 the arc discharge method is quite popular for achieving a large yield of fullerene under an inert atmosphere. This method involves electric arc generation between graphite rods under an inert atmosphere, and as a result condensed soot is collected and extracted in solvent. Since its discovery several other fullerene synthesis routes, such as laser ablation, [98] diffusion flame, [99] chemical route, [100] ion beam sputtering [101] and electron beam evaporation [102], have been discovered and adopted.

7.2.4 Nanodiamonds

Nanodiamonds (NDs) are new members of the nanocarbon family, with structure and properties of diamond in the nanorange [103]. NDs were discovered accidentally in 1963 by the shock compression of carbon black and graphite in blast chambers [104]. Generally, a mixture of hydrocarbons (CH_4/H_2) [105], CNTs [106], carbon film containing Si [107], graphite [108], carbon NPs [109], ethanol [110] and other carbon sources [111] are employed as precursors for the synthesis of NDs. Several methods, such as hydrothermal [112], ion bombardment [113], laser bombarding [114], microwave plasma chemical vapor deposition techniques [115], detonation [116], ball milling [117], high-temperature/high-pressure method [118], and ultrasound assistance [119], have been explored for the synthesis of NDs.

However, detonation and CVD routes are considered as the most popular and frequently used methods to synthesize high-quality nanodiamonds. Microplasma-CVD (MWCVD) routes are gaining major attention, as these do not only provide the desired characteristics to the material [120], but also provide an advantage for surface and interface engineering. [121,122] Gaseous precursor decomposition under microplasma generates radicals that can further nucleate the nanoparticle [123]. Gaseous flow in a small volume and high pressure influences the particle size, growth, and agglomeration with short residence times. Noble gases are used to stabilize the microplasma system during the experiment. An argon-rich environment supports the synthesis of small-sized nanodiamonds by controlling the nucleation process [124]. In a typical MWCVD process, Ar gas is bubbled at atmospheric pressure into the chamber along with reactive gaseous precursors. Subsequently, microplasma is ignited with a high voltage, direct, and constant current supply, which is stabilized with a power resistor. The precursors dissociate in the microplasma and can be collect at the exit of the reactor in acetone [125].

The detonation technique for the synthesis of nanodiamonds is the most convenient and accepted at an industrial level [116]. In this typical exercise, the explosion of oxygen-deficient explosives (a usual mixture of TNT with other explosives) in a closed chamber results in the formation of nanodiamond particles (diameter 5−20 nm). Detonation waves completely decompose the explosive molecules into atoms in several microseconds (0.2−0.5 µs). The condensed amorphous carbon atom phase experiences 0%−100% phase transformation into nanodiamonds in the pressure range of 17−23 MPa [126]. Another challenge to synthesize NDs using the detonation

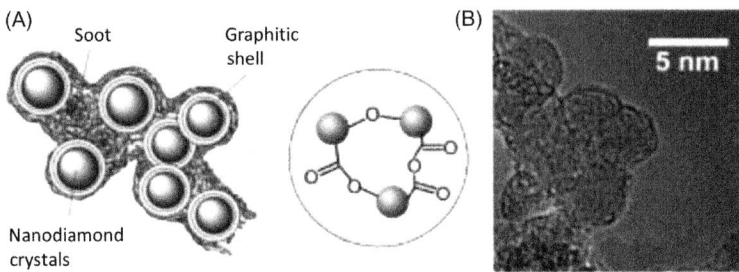

Figure 7.7 (A) Schematic shows the morphology and (B) TEM image of NDs clusters prepared by the detonation method. *Reproduced with permission from Krueger A. Diamond nanoparticles: jewels for chemistry and physics. Adv Mater 2008;20:2445−9 [127]. Copyright 2008, Wiley-VCH.*

technique is purification. Acid contamination (viz., soot) is generally removed by continuous water washing, which is done by treatment in ammonia water and then thermolysis at high temperatures ($200°C−240°C$) to enhance their stability in aqueous suspension. Fig. 7.7 describes the schematic morphology and surface chemical moieties, and TEM image of NDs synthesized by detonation method.

7.2.5 Nanoporous carbon

Nanoporous materials can be classified by pore size, that is, microporous (pore size ≤ 2 nm), mesoporous (pore size $2−50$ nm), and macroporous (pore size $50−1000$ nm). Generally, NPCs are synthesized via the pyrolysis and physical and/or chemical activation of organic precursors such as wood, coal, fruit peel, or polymers, at high temperatures [128−130]. However, the utility of these methods is limited by the low conductivity, defects, graphitization at higher temperatures, and slow transport of the resulting NPCs. The synthetic methods can be improved by the use of hard or soft templates [131,132], carbonization of organic precursors with a thermosetting and/or thermally unstable component [133,134], catalyst-supported activation of organic precursors [135,136], or carbonization of aerogels or cryogels [137]. The use of hard or soft templates facilitates the synthesis of NPCs with well-controlled pore sizes [138]. Hard-template synthesis involves the use of presynthesized organic or inorganic templates, whereas soft-template synthesis relies on the generation of nanostructures via the self-assembly of organic molecules. The most common hard-template synthetic route for mesoporous carbon materials was first reported by Knox et al. using a spherical solid gel as the template [139]. Highly ordered NPCs with oriented mesoporous structures can be obtained using the hard-template method. This method includes the following steps (1) preparation of a solid gel with a controlled pore structure as a template, (2) impregnation of the template into the precursor, (3) crosslinking and carbonization of the precursor, and (4) dissolution of the template. However, the sacrifice of the solid template and mesoporous structures of the NPC during extraction from the template limits the utility of hard-template synthesis. These limitations can be overcome using soft-template synthesis. It is thermodynamically feasible and depends on the chemical interactions between self-assembled templates and organic precursors. Being amphiphilic in nature, surfactants and block copolymers are utilized as templates in this method. For successful soft-template driven synthesis of NPCs, the precursor should exhibit excellent interaction with a pore-forming component for required self-assembled nanostructure. The pore-forming component should also have high thermal stability at experimental conditions but be readily decomposed by carbonization or readily extractable. Furthermore, the carbon yielding component should form a highly cross-linked polymer so that it can sustain the nanostructure throughout the carbonization or extraction of the pore-forming component.

7.2.6 Carbon nanofibers

Carbon nanofibers are carbon filaments with diameters at the nanoscale. They have high surface to volume ratios, excellent mechanical stabilities, high aspect ratios, and nanoscale-driven properties, with fascinating applications. Electrospinning [140], CVD [141], and templating are the common synthetic routes to carbon nanofibers [142]. Electrospinning is the most well-established way to synthesize good-quality carbon nanofibers and is cost-effective. This method requires a sol−gel or polymer solution (positively charged) in a syringe pump to be stretched under high voltage, which results in fine filaments of carbon on a conductive electrode-collector (Fig. 7.8). The solution is pumped through a spinneret at a high voltage and a constant rate, which forms a Taylor cone at the tip of the nozzle using a synergic combination of charge repulsion mechanism and surface tension of solution drops. The solvent is evaporated in the process, the growth of Taylor cone stops, and the stretched material solidifies on the electrode-collector as nanofibers of a few nanometers in diameter. Electrospinning can be done in two ways: vertically and horizontally. Various types of collectors (e.g., wheels and rods) are used for electrospinning. The nature of fibers alignment is controlled by the rotation speed of the electrode-collector [144]. Generally, cellulose, polyacrylonitrile, polyvinyl alcohol, poly(vinyl pyrrolidone), phenolic resins, polybenzimidazole, etc., are electrospun to form carbon nanofibers. Fig. 7.9 depicts the morphology of carbonized polyacrylonitrile (PAN) nanofibers synthesized by electrospinning.

7.2.7 Graphitic carbon nitride (g-C₃N₄)

The most common way to synthesize g-C_3N_4 is by thermal polymerization of a nitrogen-rich material or prebonded C-N core structure [146−148]. Several precursors, such as dicyandiamide, cyanamide, urea [147], thiourea [149], and melamine [146,150−152], have been used. Generally, in a typical thermal polymerization

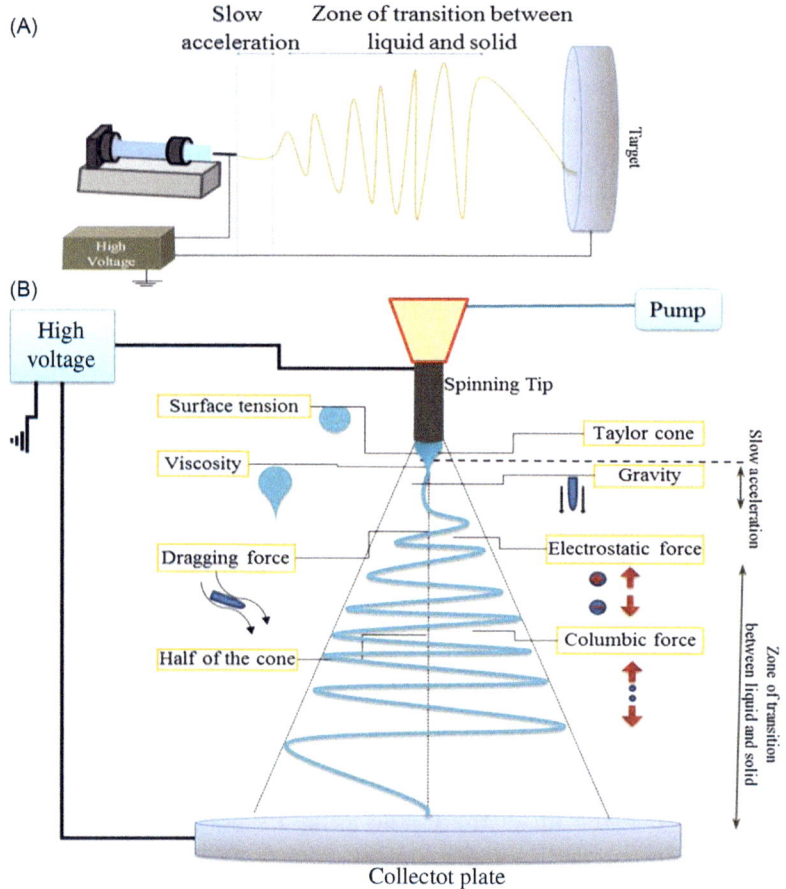

Figure 7.8 Schematics show needle electrospinning: (A) vertical and (B) horizontal electrospinning with a Taylor cone at the tip of the spinneret. *Reproduced with permission from Khalf A, Madihally SV. Recent advances in multiaxial electrospinning for drug delivery. Eur J Pharm Biopharm 2017;112:1−17 [143]. Copyright 2017, Elsevier.*

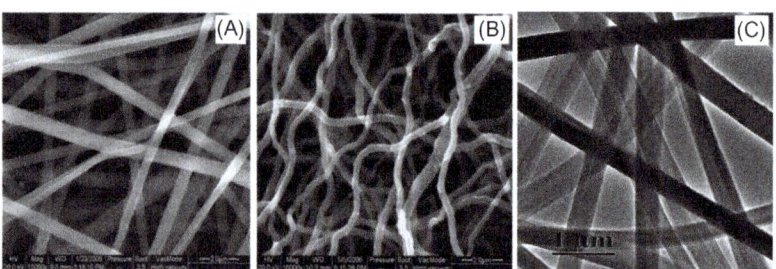

Figure 7.9 SEM (A, B) and TEM (C) images of carbonized polyacrylonitrile (PAN) nanofibers. *Reproduced with permission from Guo Q, Zhou X, Li X, Chen S, Seema A, Greiner A, et al. Supercapacitors based on hybrid carbon nanofibers containing multiwalled carbon nanotubes. J Mater Chem. 2009;19:2810−6 [145]. Copyright 2009, Royal Society of Chemistry.*

reaction to synthesize g-C_3N_4, precursors are first converted to melam and further undergo condensation by eliminating ammonia to prepare g-C_3N_4 at around $500°C-530°C$ [153]. However, on increasing the temperature slightly above to $600°C$, g-C_3N_4 becomes highly unstable, and at $700°C$ it breaks into nitrogen and cyano fragments. The use of different materials and different experimental conditions can modify properties such as the porosity, surface area, morphology, photoluminescence properties, and C/N ratio of g-C_3N_4.

7.3 Surface functionalization

Owing to advanced properties, such as excellent electrical and mechanical properties, and large surface area, CNMs are considered as potential building blocks for advanced nanoscale functionalized materials. Increased dispersions of CNMs in different media and their interactions with other inorganic/organic and polymeric molecules is indispensable to achieve their full potential in various applications such as water purification, sensing, electronics, nanocomposite, biomedical, and energy applications. Surface functionalization/modification of CNMs can enhance the interactions. Functionalization of CNMs improves the dispersion, colloidal stability, selectivity, and multifunctional assembly of nanostructures. For instance, CNMs can easily be converted to have a hydrophobic, hydrophilic, anticorrosive, anionic, cationic, and zwitterion surface just by the introduction of the required surface functionalities. Based on the bonding characteristics between CNMs and functional molecules, functionalized approaches are divided into two categories: covalent and noncovalent functionalization.

7.3.1 Covalent functionalization

In covalent functionalization, different chemical species attach to CNMs by forming chemical bonds. This functionalization in CNMs is governed by oxygen/nitrogen-carrying groups that link to the $\pi-\pi$ conjugated skeleton of the carbon nanostructures. Covalent functionalization is a commonly used strategy for the modification of CNMs to optimize their full-scale potential in a wide range of applications. The functionalization of GO can be accomplished by covalent or noncovalent functionalization, or by doping. In covalent functionalization, oxygen-containing groups, such as carboxylic acids, hydroxyl groups, and epoxy groups, which are present on the GO surface, actively participate. Generally, this approach involves the reaction of the GO with a silane linker, such as (3-chloropropyl) trimethoxysilane, or (3-aminopropyl) trimethoxysilane, which reacts covalently with the GO on one side, while the tail reacts with another group

[154–158]. Fig. 7.10 demonstrates the synthesis strategies for functionalization of GO through the hydroxyl, epoxy, and carboxylic groups. Hu et al. exhibited silanization of GO and rGO using 3-methacryloxypropyltrimethoxysilane (MPS), and further copolymerization with poly(methyl methacrylate) (PMMA) as a process for modification of GO to enhance the solubility and tensile strength (Fig. 7.10A). The functionalized (GO-MPS, and rGO-MPS) materials had shown better solubility than GO and rGO. Moreover, in situ copolymerization of silanized-GO/rGO improved the mechanical strength of the polymer [159]. GO can also be functionalized via modification of epoxy groups that are present on the basal surface, considering a nucleophilic reaction at the α-C of epoxide, and using the $-NH_2$ group-carrying molecule to catalyze the epoxide ring-opening reaction (Fig. 7.10B). Yang et al. reported the synthesis of polydispersed chemically converted graphene (p-CCG) using ionic liquid (1-(3-aminopropyl)-3-methylimidazolium bromide ($R-NH_2$)) for epoxide ring-opening reaction, and to form functionalized GO. The prepared p-CCG demonstrated good solubility in polar solutions. These materials were used as chemical-modified electrodes in electrochemical applications [160]. Similarly, carboxyl groups of GO can be modified/functionalized via esterification, amidation, and activation with polymer-linkers or small chemical molecules. Zhang et al. prepared graphene–C_{60} nanohybrids using pyrrolidine fullerene and GO for nonlinear optical applications. This hybrid material was formed by the esterification reaction between the hydroxyl group of pyrrolidine fullerene and carboxyl groups of GO (Fig. 7.10C) [161]. In another study, researchers exhibited reductive covalent functionalization of graphene via the reaction between potassium graphenides and aryl diazonium salts [162].

Similar to GO, covalent functionalization of other CNMs can be achieved by adding carboxylic groups on the surface via an oxidation process using concentrated acids, HNO_3, H_2SO_4, H_3SO_4, H_3PO_4, HCl, H_2O_2, and an acid mixture [163]. The organic and inorganic molecules can be anchored to oxidized CNMs through the carboxyl groups using covalent functionalization to improve the dispersion of CNMs in different media. There are numerous protocols for covalent functionalization of CNTs, which are listed in Fig. 7.11.

Functionalization of fullerene is a necessary step, due to its limited solubility in many solvents and to enhance its potential for various applications. Thus, the properties of fullerene are tailored and targeted to specific functions via functionalization. Functionalization of fullerenes can be done either by covalent or noncovalent means (hydrogen bonding, metal ion coordination, or $\pi-\pi$ interactions) [165,166]. Britz et al. modified fullerenes with ester groups using a chemical route (nucleophilic reaction) and further functionalized them with SWCNTs for sensor applications [167]. Fullerenes are hydrophobic in behavior but could also be turned hydrophilic and amphiphilic on functionalization [168,169]. Similar to fullerenes, nanodiamonds are

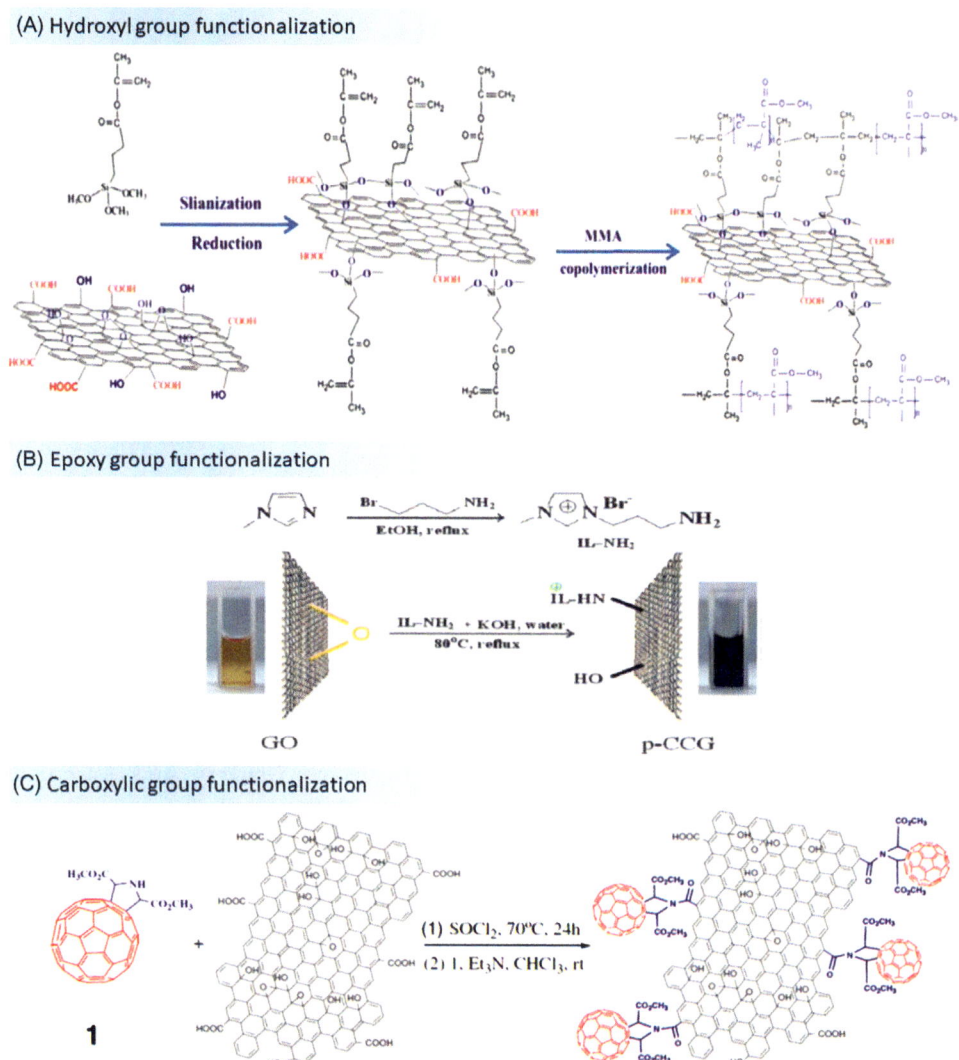

Figure 7.10 (A) An example of silanization of GO through hydroxyl (−OH) using MPS, and followed by copolymerization with PMMA. (B) Epoxy group functionalization of GO using IL-NH₂ to synthesis p-CCG. (C) Carboxylic group functionalization of GO using pyrrolidine fullerene to prepare graphene-C₆₀ nanohybrids. *(A) Reproduced with permission from Hu X, Su E, Zhu B, Jia J, Yao P, Bai Y. Preparation of silanized graphene/poly(methyl methacrylate) nanocomposites in situ copolymerization and its mechanical properties. Compos Sci Technol. 2014;97:6−11 [159]. Copyrights 2014, Elsevier. (B) Reproduced with permission from Yang H, Shan C, Li F, Han D, Zhang Q, Niu L. Covalent functionalization of polydisperse chemically-converted graphene sheets with amine-terminated ionic liquid. Chem Commun 2009:3880−2 [160]. Copyright 2009, Royal Society of Chemistry. (C)Reproduced with permission from Zhang X, Huang Y, Wang Y, Ma Y, Liu Z, Chen Y. Synthesis and characterization of a graphene−C60 hybrid material. Carbon. 2009;47:334−7 [161]. Copyrights 2009, Elsevier.*

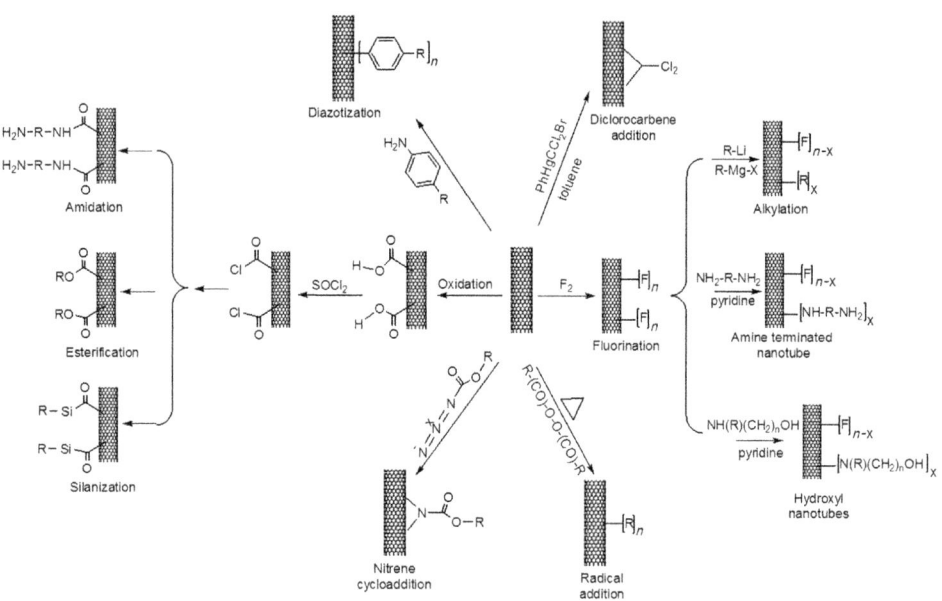

Figure 7.11 Various synthesis protocols for covalent functionalization of CNTs. *Reproduced with permission from Cheng HKF, Basu T, Sahoo NG, Li L, Chan SH. Current advances in the carbon nanotube/thermotropic main-chain liquid crystalline polymer nanocomposites and their blends. Polymers. 2012;4:889–912 [164]. Copyrights 2012, MDPI.*

also functionalized covalently using excess hydroxyl (−OH) and carboxylic (−COOH) groups on the surface. After activation of NDs, amidation and esterification of carboxylic groups are achieved to graft various molecules, whereas silanization and esterification strategies are used for hydroxyl group modification of NDs (Fig. 7.12). The activation process is generally carried out using thionyl chloride ($SOCl_2$), which reacts with a carboxyl group and converts it into acyl chloride. $SOCl_2$ is not specific, it can also convert hydroxyl groups into chlorides by nucleophilic substitution. For mild activation, 1-ethyl-3-(3-dimethylaminopropyl)carbodiimide/N-hydroxysuccinimide (EDS/NHS) combinations have been studied, in which the reaction can be carried out in water/buffer at room temperature [170]. Various modification methodologies allow easy grafting of a wide range of functional groups including small molecules, proteins, monomers, polymers, and dyes, and enhance the demand of NDs in several applications such as biomedical, environmental remediation, electrochemical field and energy.

7.3.2 Noncovalent functionalization

Noncovalent functionalization occurs via weak physicochemical interactions, for example, $\pi-\pi$ interactions, hydrogen bonding, hydrophobic interactions, electrostatic

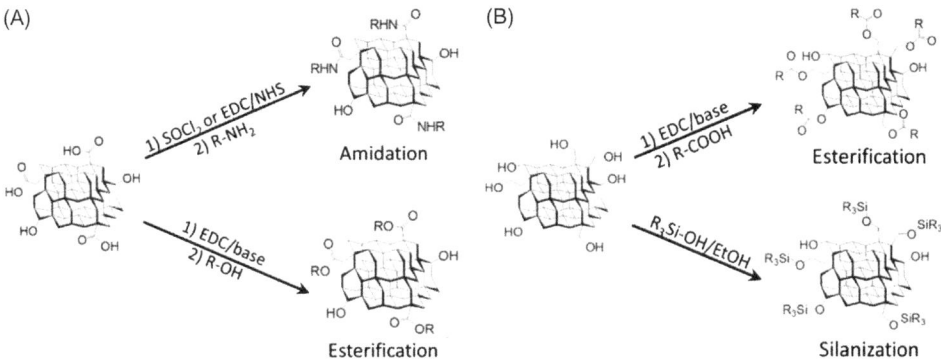

Figure 7.12 Covalent functionalization of NDs using carboxylic groups (A) and hydroxyl groups (B) of the surface (R = proteins, peptides, dyes, amino-alkyl, glycidyl moieties). *Reproduced with permission from Reina G, Zhao L, Bianco A, Komatsu N. Chemical functionalization of nanodiamonds: opportunities and challenges ahead. Angew Chem Int Ed. 2019;58:17918—29 [118]. Copyright 2019, Wiley-VCH.*

interactions, or van der Waals forces [158]. It is noted that repulsive forces and attraction forces are critical for noncovalent interactions and supramolecular assembly. Foreign molecules can be functionalized non-covalently on CNMs using physisorption. Besides, in the presence of weak interactions, noncovalent functionalization can also occur in forms of polymer wrapping, $\pi-\pi$ interactions, and e^- donor—acceptor ligand systems [171]. The noncovalent functionalization is carried out by grinding in a mortar, blending, extruding from a syringe, sonication, adsorption, etc. [171]. The overall energies of noncovalent interactions are less than the covalent interactions, but the effects on the surface are almost similar. The plus point of noncovalent interactions is their reversible nature, which makes it an autocorrect process. From covalent interactions, functional nanomaterials can be obtained with improved response toward external factors such as pH and temperature. This concept has been applied in the drug delivery process using hydrogels [172]. The weak interactions that govern the noncovalent functionalization are active in close vicinity and achievable at ambient conditions. These interactions lead to a change in surface properties (hydrophobic/hydrophilic) and solubility of CNMs. For instance, the excellent dispersion of CNMs such as G, GO, and rGO can be obtained using ultrasonication in solvents N-methyl pyrrolidone (NMP), phenyl ethyl alcohol, and N,N-dimethylformamide. These solvent molecules could enter into sheets of 2D materials and easily adsorb to the available surface via van der Waals forces during the ultrasonication process. At a later stage, these solvent can be detached by a heating process at an appropriate temperature [173].

Moreover, noncovalent functionalization has a drawback of reproducibility as the adsorption of guest molecules can differ during the process. In another method, non-covalent functionalization sustains the sp^2 hybridized carbon network in CNMs, which results in functionalized nanomaterials with new characteristics while keeping most of the intrinsic properties. Thus noncovalent functionalization can yield advanced CNMs with elevated biocompatibility, reactivity, catalytic activity, dispersibility, adsorption, and sensing properties. Using noncovalent functionalization, graphene and GO can be functionalized using polymers, biomolecules, drugs, 2D analogues, and carbon nanoallotropes. For example, to make highly hydrophilic graphene hybrids, tetrasulfonated copper phthalocyanine (TSCuPC) was noncovalent functionalized on graphene via $\pi-\pi$ stacking (Fig. 7.13A). The $\pi-\pi$ stacking was confirmed by UV–visible absorbance spectrum of graphene hybrids, in which absorbance bands related to TSCuPC disappear due to interactions with G (Fig. 7.13B); the Q band (663 nm) of the monomer unit moves to 705 nm [174]. Hsiao et al. reported the synthesis of water-soluble polyurethane (WPU)/rGO nanocomposite by considering noncovalent functionalization for electromagnetic shielding application [175]. In this work, stearyl trimethyl ammonium chloride (STAC, a surfactant) was used for functionalization of GO in order to enhance the dispersibility and for the addition of nitrogen functionalities. Here, polyurethane was functionalized with sulfonate groups to make it soluble and used as a matrix for the preparation of nanocomposite. The excellent dispersion of STAC-modified graphene nanosheets (S-GNS) in the matrix was attributed to extremely well-matched electrostatic interaction between amine moieties of S-GNS and sulfonate groups of WPU.

Similar to G and GO, CNTs can also be noncovalently functionalized using polymer wrapping, surfactant loading, $\pi-\pi$ interactions (need aromatic molecules), and e^- donor–acceptor ligand systems. Fig. 7.14 summarizes the functionalization of CNTs using covalent (sidewalls or end groups), noncovalent, defect-group, and endohedral functionalization or doping to enhance their properties for various applications. Noncovalent functionalization can be done either exohedrally (surfactants, drugs and polymer loading on the surface) or endohedrally (insertion of C_{60} inside CNTs) [176]. Haddad et al. investigated the noncovalent functionalization of CNTs with a water-soluble vitamin (biotin) using $\pi-\pi$ interactions and electropolymerization for preparation of biosensors [177]. For SWCNTs modification, two strategies were used: $\pi-\pi$ interactions using biotin–pyrene and electropolymerization of biotin–pyrrole monomers. The noncovalently biotin-modified SWCNTs exhibited an improved permeability for enzymatically produced H_2O_2.

Zhan et al. reported a new method for the immobilization of myoglobin (Mb) on noncovalently poly(methacrylic acid-*co*-acrylamide) (P(MAA-*co*-AAM)-functionalized MWCNTs via hydrophobic interactions, H-bonding, and electrostatic attraction for the construction of highly effective amperometric sensors [178]. Here, copolymer P(MAA-*co*-AAM) was attached to MWCNTs through noncovalent functionalization using hydrophobic interactions, which led to an increase in stability and solubility of MWCNTs. Moreover, Mb was added on P(MAA-*co*-AAM)−MWCNTs surface via

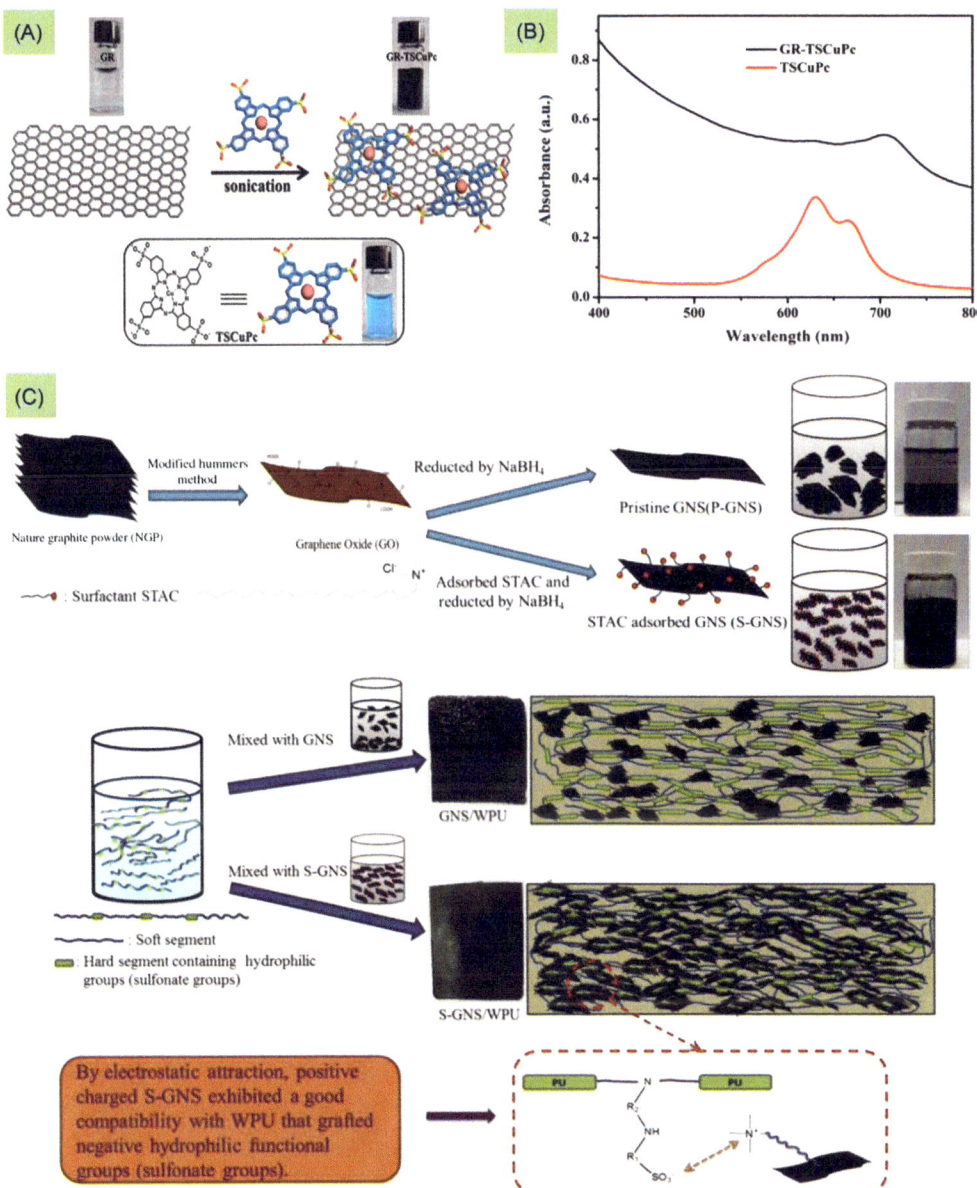

Figure 7.13 (A) Noncovalent functionalization of graphene (GR in image) with phthalocyanine derivative (TSCuPc) via π–π stacking. (B) The UV-vis absorbance spectra of TSCuPc and hybrids (GR-TSCuPc). (C) Preparation of water-soluble polyurethane (WPU) nanocomposite with surfactant (STAC) functionalized rGO using a noncovalent functionalization. *(B) Reprinted with permission from Jiang B-P, Hu L-F, Wang D-J, Ji S-C, Shen X-C, Liang H. Graphene loading water-soluble phthalocyanine for dual-modality photothermal/photodynamic therapy via a one-step method. J Mater Chem B 2014;2:7141–8 [174]. Copyrights 2014, Royal Society of Chemistry. (C) Reproduced with permission*
(Continued)

electrostatic attraction and/or H–bonding. Likewise, other CNMs, such as NDs (Fig. 7.15B), fullerene, g–C$_3$N$_4$, CNFs, and NPCs, can also be non–covalently functionalized with various molecules including drugs, polymers, DNA, dyes, proteins, active aromatic moieties, using hydrophobic interactions, electrostatic attraction, H–bonding, weak van der Waals forces, and π–stacking.

7.3.3 Functionalization with nanoparticles

Surface functionalization/modification of CNMs using various nanostructures including metallic, metal oxides, metal sulfides, and many more, with controlled morphology and crystallinity have been widely researched for the nanotechnology applications in numerous areas, such as electronics, biomedicine, energy, catalysis, and environmental remediation [25,163,179–181]. Modification of CNMs results in the formation of nanocomposite, heterostructures, nanohybrids, and multifunctionalized

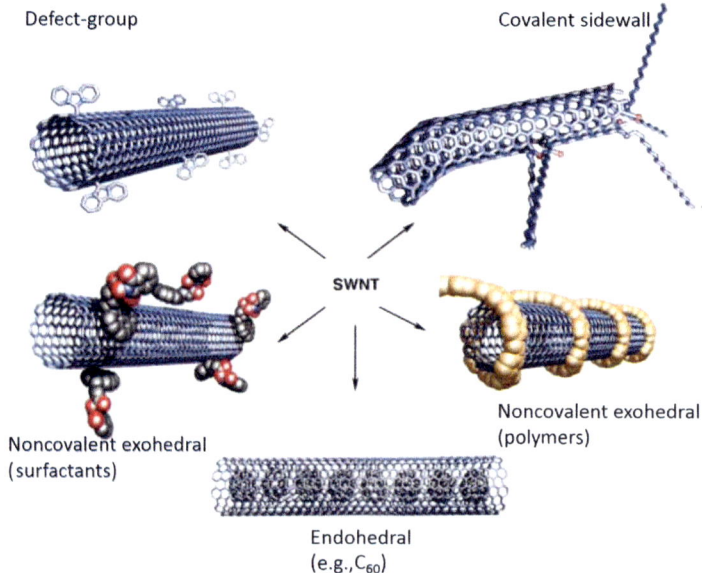

Figure 7.14 Methods for CNT functionalization: defect-group functionalization, covalent sidewall functionalization, noncovalent exohedral functionalization using polymers and surfactants, and endohedral functionalization using small molecules, for example, C$_{60}$. *Reproduced with permission from Hirsch A. Functionalization of Single-Walled Carbon Nanotubes. Angew Chem Int Ed. 2002;41:1853–9 [176]. Copyright 2015, Wiley-VCH.*

◀ *from Hsiao S-T, Ma C-CM, Tien H-W, Liao W-H, Wang Y-S, Li S-M, et al. Using a non-covalent modification to prepare a high electromagnetic interference shielding performance graphene nanosheet/waterborne polyurethane composite. Carbon. 2013;60:57–66 [175]. Copyrights 2013, Elsevier Science Ltd.*

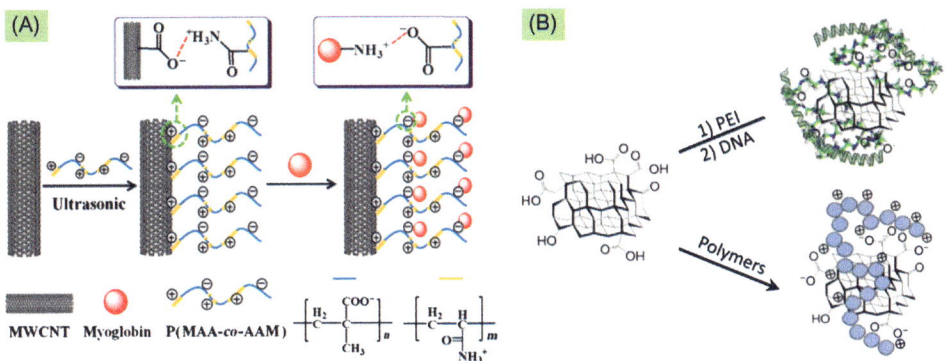

Figure 7.15 (A) Schematic representation of preparation method for Mb−P(MAA-*co*-AAM)−MWCNTs nanocomposites using hydrophobic interactions, electrostatic attraction, and/or H-bonding. (B) Noncovalent functionalization DNA and polymers molecules on NDs via electrostatic interactions (here, PEI is polyethyleneimine). *(A) Reproduced with permission from Zhan K, Liu H, Zhang H, Chen Y, Ni H, Wu M, et al. A facile method for the immobilization of myoglobin on multiwalled carbon nanotubes: poly(methacrylic acid-co-acrylamide) nanocomposite and its application for direct bio-detection of H2O2. J Electroanalytical Chem 2014;724:80−6 [178]. Copyrights 2014, Elsevier Science Ltd. (B) Reproduced with permission from Reina G, Zhao L, Bianco A, Komatsu N. Chemical functionalization of nanodiamonds: opportunities and challenges ahead. Angew Chem Int Ed 2019;58:17918−29 [118]. Copyright 2019, Wiley-VCH.*

materials, which have better properties than the pristine CNMs. The functionalization of CNMs with NPs can be realized by wet chemistry approaches such as sol−gel synthesis, coprecipitation, microwave synthesis, and hydrothermal synthesis. Some researchers have also studied atomic layer deposition, physical vapor deposition, CVD, and template methods for CNMs modification [182]. Among these synthesis methods, the hydrothermal reaction is most popular for the modification of CNMs as it allows the strict control on morphology, particle size, crystallinity, and surface reaction by adjusting reaction temperature, pressure, solution composition, solvent and capping agent selection, and aging time [158,183−186]. In hydrothermal, a reaction occurs at high temperature in the presence of pressure of vapor saturation. Alternatively, microwave methods offer a fast reaction using microwave thermal energy, considering concentrated regional heating [186].

Liu et al. demonstrated bovine serum albumin (BSA)-modified GO/rGO as ideal templates for successful assembly of diverse NPs (Au, Ag, Pt and Pd) with different shape, size, structural composition, and properties (Fig. 7.16A) [187]. The fabricated NPs on rGO were fully dispersed and had controlled morphologies. In another study, the authors demonstrated a novel strategy for the encapsulation of metal oxides (Co_3O_4, SiO_2) by considering electrostatic attraction between positive charged NPs and the negatively charged GO surface. Oxides NPs were first modified by aminopropyltrimethoxysilane (APS) to make the surface positive [188]. The surface-modified

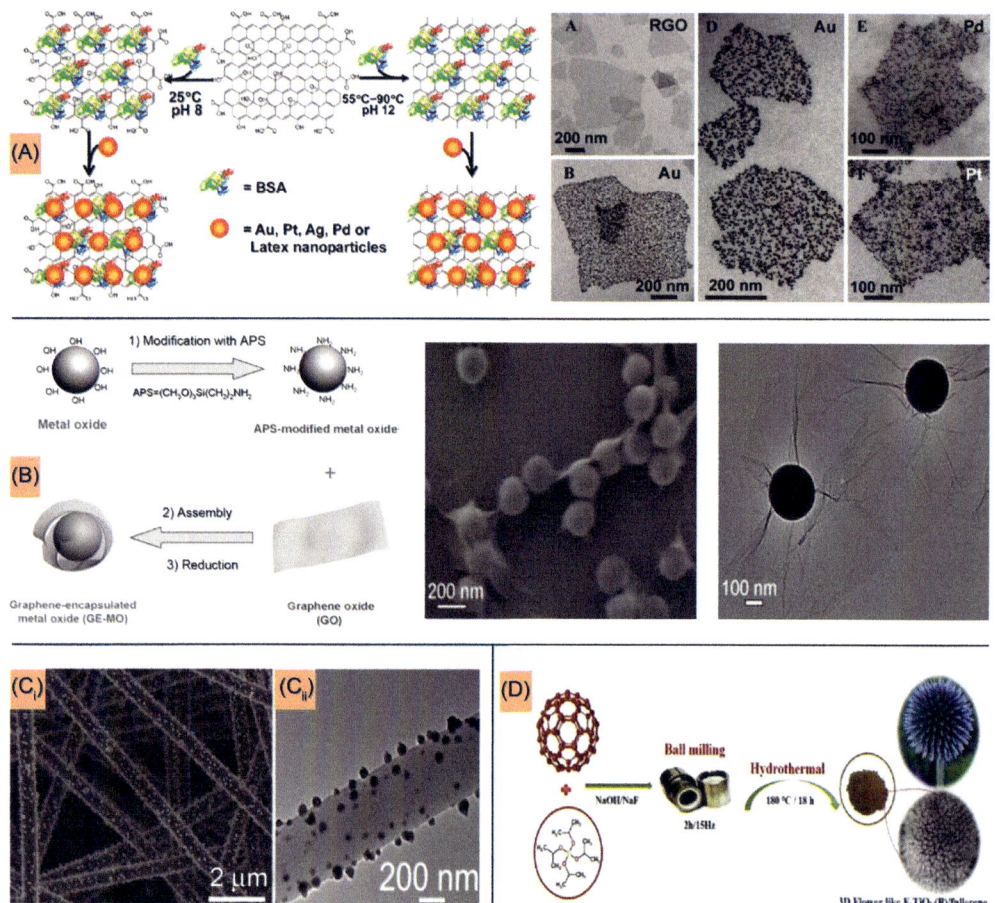

Figure 7.16 (A) Immobilization of bovine serum albumin (BSA) protein on GO, resulting in ideal templates for successful assembly of diverse NPs (Au, Ag, Pt, and Pd); TEM images of different NPs. (B) Modification of negatively charged graphene with metal oxide (SiO_2) by considering electrostatic attraction; SEM images SiO_2 anchored GO. (Ci-Cii) FE-SEM and TEM image of core-shell catalysts ($Pd_{16}S_7/MoS_2$)-modified CNFs by the S vapor-assisted CVD and electrospinning. (D) Strategy to fabricate 3D flower-like F-TiO_2-B/fullerene nanocomposite. *(A) Reproduced with permission from Liu J, Fu S, Yuan B, Li Y, Deng Z. Toward a universal "adhesive nanosheet" for the assembly of multiple nanoparticles based on a protein-induced reduction/decoration of graphene oxide. J Am Chem Soc 2010;132:7279—81 [187]. Copyrights 2010, American Chemical Society. (B) Reproduced with permission from Yang S, Feng X, Ivanovici S, Müllen K. Fabrication of graphene-encapsulated oxide nanoparticles: towards high-performance anode materials for lithium storage. Angew Chem Int Ed 2010;49:8408—11 [188]. Copyright 2010, Wiley-VCH. (C) Reproduced with permission from Wen Y, Zhu H, Zhang L, Hao J, Wang C, Zhang S, et al. Beyond colloidal synthesis: nanofiber reactor to design self-supported core—shell Pd16S7/MoS2/CNFs electrode for efficient and durable hydrogen evolution catalysis. ACS Appl Energy Mater 2019;2:2013—21 [189]. Copyrights 2019, American Chemical Society. (D) Reproduced with permission from Panahian Y, Arsalani N, Nasiri R. Enhanced photo and sono-photo degradation of crystal violet dye in aqueous solution by 3D flower like F-TiO2(B)/fullerene under visible light. J Photochemistry Photobiology A: Chem 2018;365:45—51 [190]. Copyrights 2018, Elsevier.*

metal oxides NPs were amassed on GO (carrying negative charge) by electrostatic attraction and the formed composites were ultimately reduced by hydrazine (Fig. 7.16B). Recently, Wen et al. reported a new strategy for the preparation of catalysts ($Pd_{16}S_7$/MoS_2)-modified CNFs electrode by the S vapor-assisted CVD and electrospinning, for electrocatalytic water splitting [189] (Fig. 7.16C). Nasiri et al. exhibited the synthesis of three-dimensional (3D) flower-like F-doped titanium dioxide bronze and fullerene (F-TiO_2-B/fullerene) nanocomposite using initially ball milling and later a hydrothermal method, considering titanium tetraisopropoxide as a titanium source [190]. The prepared nanocomposite was utilized for sonophotodegradation of a crystal violet dye (Fig. 7.16D).

Furthermore, CNMs can also be modified by atomic-level crystal engineering with heteroatoms addition/doping or defects introduction. Doping can be introduced by adding electron-donating and electron-withdrawing elements in the system, which affect the structural, electronic, and catalytic properties. Dopants make changes in the crystal lattice and bond angles, causing more active sites [191]. Numerous methods have been studied for the introduction of heteroatoms in CNMs using in situ, and postsynthesis methods. Doping of CNMs is facile, can be attained using CVD, ball milling, melt-mixing, electrothermal or electrochemical techniques, annealing/pyrolysis, and by chemical means, or by electrostatic field-tuning techniques [192].

7.4 Diverse properties

CNMs, especially 1D and 2D, are popular for their exceptional mechanical strength, and their flexible rigidity is comparable with carbon fibers and high-tensile steel. They also carry superb optical, electrical, and thermal properties. Many topical research studies have been dedicated to combining CNMs with different materials for gaining maximum advantages of multifunctional properties (e.g., thermal, electrical, optical, and surface characteristics), rather than focusing only on improving the mechanical properties [193].

7.4.1 Surface properties

Surface features/chemistry of CNMs are very crucial to realize a wide range of applications related to adsorption, absorption, hydrophilicity, hydrophobicity, chemical reactivity, corrosivity, porosity, and surface area. The surface properties of G and GO/rGO can be manipulated by the presence of defects/edges, dopants, and oxygen functional groups in the structure. Defects and edges are inexorable in GO and derivatives and appear during the preparation process. Unlike CNTs, fullerenes, and h-BN,

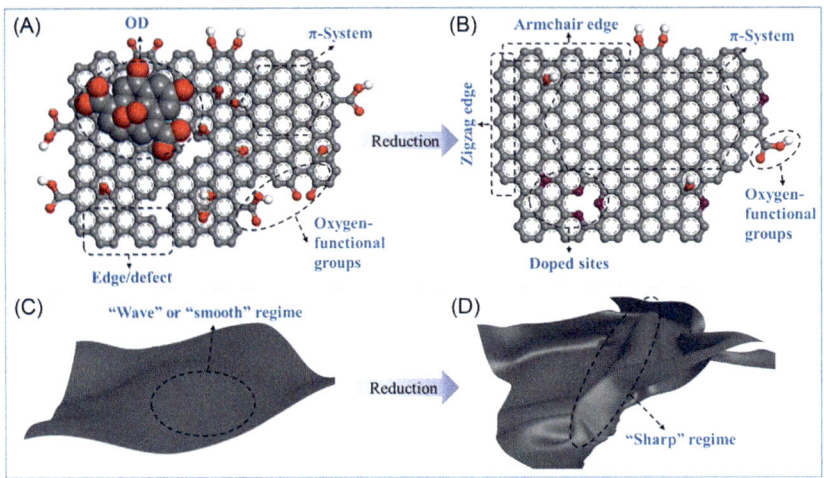

Figure 7.17 The surface structures (A and B) and morphologies (C and D) of GO and rGO/G nanosheets. *Reproduced with permission from Yang K, Wang J, Chen X, Zhao Q, Ghaffar A, Chen B. Application of graphene-based materials in water purification: from the nanoscale to specific devices. Environ Sci Nano. 2018;5:1264−97 [194]. Copyright 2018, Royal Chemical Society.*

GO/rGO have open edges with either zigzag or armchair structures (Fig. 7.17A and B). Theoretical studies show that the electronic structures at edges/defects vary from the π–system structures and result in unique chemical reactivity [195]. Zigzag edges have remarkable chemical, magnetic, and electronic properties due to their interesting flat band characteristics and edge localized states. The surplus electronic states at zigzag edges in rGO/G make them suitable for catalysis and interactions with small atoms and molecules [195,196]. The presence of defects/edges obstructs electron passage on the surface and decreases thermal transport [197]. Due to their hydrophilic nature, GO nanosheets are well-dispersed and have flat laminar structures (Fig. 7.17C). Smooth, wavy, and sharp surface regimes can be obtained during the reduction of GO (Fig. 7.17D) [198]. These structural regimes exhibit dissimilar chemical activities. Sharp curvatures have better reactivity than wavy and smooth regimes due to the protruding unpaired sp^3-like orbitals. These crumpled surfaces avoid π−π stacking and provide appropriate modification of nanoparticles within the composite or self-assembly. Moreover, sharp regimes (crumpled regions) have high surface energies and provide reactive adsorption sites for the elimination of toxic water pollutants [199].

CNMs generally offer a high surface area and porosity for various surface-related applications. SLG is a rich stage for surface reactions as it has a very large theoretical surface area of ~2600 m²/g. The experimental BET surface area of G/rGO was recorded in the range of 270−1550 m²/g, some reports have even shown its value to be near to SLG [200,201]. High surface area graphene is an excellent material for the uptake of different gaseous molecules (CO_2 and H_2) and various pollutants from the environment. The removal of CO_2 using FLG was found to be 35 wt.% at 195K and 1 atm, which was approximately equal to the theoretical value of SLG (37.9 wt.%) [201]. The CO_2 attached alternatively to graphene rings in a parallel manner. As per the literature, a theoretical surface area of closed SWCNTs was calculated to be 1315 m^2/g [202], while an experimental BET surface area of SWCNTs is obtained in the range of 400–900 m^2/g, and a surface area of MWCNTs is usually observed in between 200 and 400 m^2/g [203]. Depending on the surface area and pore size, the characteristic pore volumes of MWCNTs are noted to be in the range of 0.025– >1.67 cm^3/g [204]. The surface area of CNFs is found to be in the range 50–350 m^2/g, which is much lower than in activated carbon fibers and activated carbons [205]. During activation of CNFs using KOH, the surface area can be increased to 400–1200 m^2/g on heating at 800°C–1000°C [206]. Loutfy et al. demonstrated the synthesis of NPCs from fullerenes using commonly used porous carbon synthesis methods. The porous carbon exhibited exceptional high surface area (3142 m^2/g) with a high pore volume of 0.92 cm^3/g, and narrow pore size distribution (4–30 Å). Owing to these outstanding properties, fullerene-based porous carbon may be suitable materials for gas adsorption and storage, highly active catalyst support, purification of gas or liquid stream, and water treatments [207]. The oxygenic surface group carrying NDs have a surface area over 300 m^2/g. The carboxylic group terminated

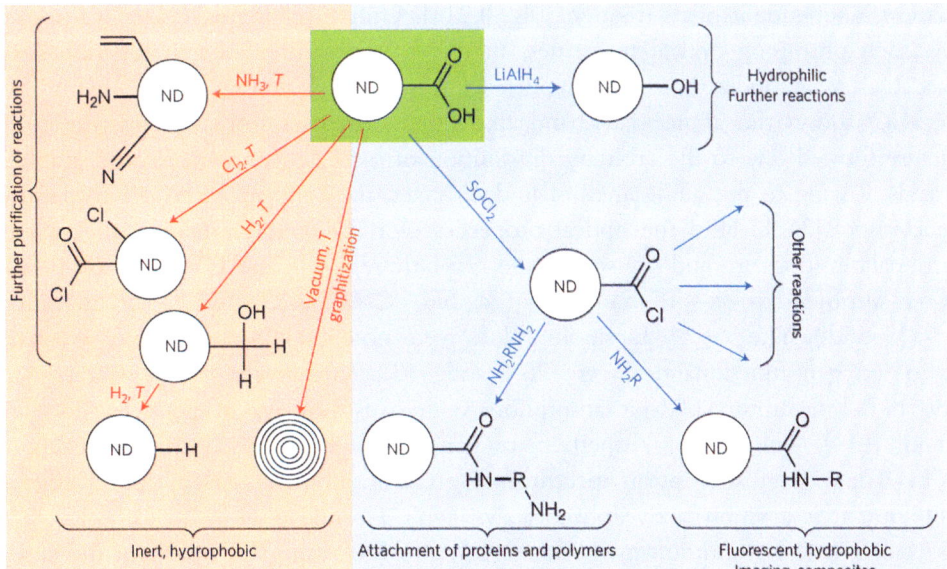

Figure 7.18 Surface modification of NDs using wet chemistry techniques (shown in blue-colored arrows) or high-temperature gas reactions, (shown in red-colored arrows). *Reproduced with permission from Mochalin VN, Shenderova O, Ho D, Gogotsi Y. The properties and applications of nanodiamonds. Nat Nanotechnol 2012;7:11–23 [208]. Copyrights 2012, Nature.*

NDs can be tailored using wet chemistry techniques (shown in blue-colored arrows) or high-temperature gas reactions, (shown in red-colored arrows) to make the surface inert, hydrophilic, hydrophobic, and to utilize them in bioimaging, composites catalytic, and fluorescent applications (Fig. 7.18) [208]. Similarly, other CNMs can also be modified to introduce different surface properties.

7.4.2 Optical properties

Among the various CNMs, graphene has shown superb optical properties, as well as having high conductivity and high strength. The optical properties of graphene are dependent on the number of layers. A single layer of graphene transmits 97.7% of light and merely absorbs $\sim 2.3\%$ of light with 0.1 reflections from its initial path [209]. The light absorption increases with an increase in a number of layers, with a consequent decrease in transmittance. According to the literature, the relationship between absorption and graphene layers is approximately linear. The graphene has the ability to absorb radiation from different regions of the electromagnetic spectrum, which is attributed to its band arrangement (interband and intraband), absence of bandgap, and interaction between the Dirac fermions of graphene and electromagnetic radiation. Due to the absence of discrete energy levels, graphene absorbs the electromagnetic radiation independent of its frequency [210,211]. Unlike graphene, GO/rGO doesn't show such promising optical properties. In the GO structure, oxygen functionalities-linked C is sp^3 hybridized, while unoxidized C in the graphitic region is sp^2 hybridized. The rGO shows the appropriate bandgap, which originates from the unsymmetrical structure formed due to the arbitrary distribution of oxidized sp^3 domains and graphitic domains. Owing to prevailing areas of midgap states, the bandgap of rGO has transport characteristics [212]. Thus the optical properties of rGO are governed by the oxygen functionalities' domains and the size of the graphitic regions. The UV–vis absorption of rGO is attributed to $\pi - \pi^\star$ transitions (graphitic C-C bonds) and $n - \pi^\star$ transitions (C $=$ O bonds). The $\pi - \pi^\star$ transitions values were noted at 4.8 and 5.2 eV, when the sizes of oxygen functionalities were 25% and 75%, respectively [213]. Whereas, for MWCNTs, maximum UV–vis absorption value was noticed in between 200 and 250 nm [214]. The optical properties are generally studied to check the quality of CNTs. The UV–vis absorption spectrum of g-C_3N_4 exhibited an absorption edge at ~ 440 nm. It is a visibly active photocatalyst with a bandgap of ~ 2.7 eV [215]. The NDs show optical absorption in the range of $300 - 1000$ nm. The nature of the absorbance spectrum depends on the size and purity of NDs [216]. The absorption spectrum of pure C_{60} fullerene showed two peaks at 342 and 445 nm (broad) [217]. The optical properties also include Raman features, photoluminescence, and infrared spectroscopic characteristics. The optical properties of CNMs can easily be tailored by adjusting the dopants types, doping density, defects concentration, local excitonic effect, surface feature, and charge transfer.

7.4.3 Mechanical properties

CNMs, in particularly 1D and 2D, have shown excellent mechanical properties. The monolayer graphene (defect-free) is known to be the strongest material with a break strength of 42 N/m, an intrinsic tensile strength of 130 GPa, and a Young's modulus of 1.0 TPa [218]. The fracture toughness of graphene was reported to be ~ 4.0 MPa m$^{1/2}$, indicating the low fracture tendency of graphene [219]. CNTs with 1D geometry have different properties in radial and axial directions. CNTs are extremely strong in the axial axis with a tensile strength of 11−63 GPa and Young's modulus of 270−950 GPa. SWCNTs with different geometries (armchair, chiral, zig-zag) have better mechanical properties than MWCNTs. The mechanical properties of CNMs originate from the continuous covalent sp^2 carbon network present throughout the structures [220]. Owing to superb mechanical properties, CNMs have been employed in advanced composite materials. In the last decades, polymer nanocomposite with CNMs has garnered much attention because of the huge increase in mechanical properties with a small loading amount of CNMs. The mechanical properties of composites are influenced by dispersions of CNMs and the surface interactions between CNMs and the polymer matrix. CNTs, being 1D, can easily be aligned/dispersed in the polymers by use of external forces via various techniques including electrical fields, spinning processes, and mechanical stretching [221]. Whereas other CNMs need to be functionalized to enhance the dispersion in composites and to assure the better interfacial interactions between CNMs and polymer matrixes. Li et al. investigated the enhancement in mechanical properties of polyvinyl alcohol

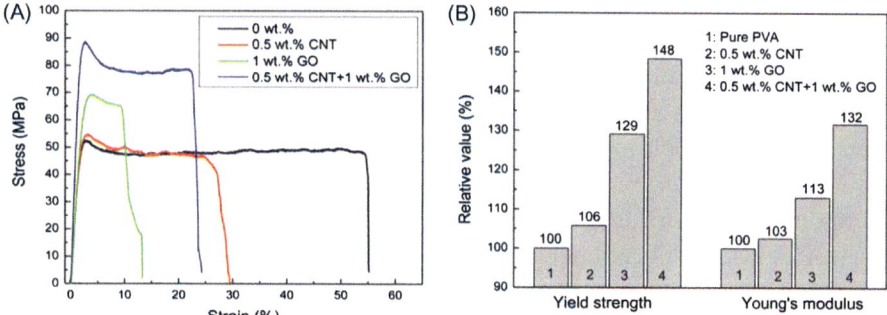

Figure 7.19 (A) Stress—strain graphs of PVA, GO/PVA, CNT/PVA, and GO-CNT/PVA nanocomposites; (B) Young's modulus and yield strength of GO/PVA, CNT/PVA, and GO-CNT/PVA nanocomposites. *Reproduced with permission from Li Y, Yang T, Yu T, Zheng L, Liao K. Synergistic effect of hybrid carbon nantube—graphene oxide as a nanofiller in enhancing the mechanical properties of PVA composites. J Mater Chem. 2011;21:10844—51 [222]. Copyrights, 2011, Royal Chemical Society.*

(PVA) matrix by addition of GO, CNTs, and combined GO-CNT [222]. They observed that the ternary polymer nanocomposite GO/CNT/PVA demonstrated substantially better mechanical properties than the binary composites GO/PVA and CNT/PVA (Fig. 7.19). The Young's modulus and tensile strength of GO (1 wt.%)/CNT (1 wt.%)/PVA nanocomposite improved by 31% and 48%, respectively, which was attributed to the hydrophilic character of GO that might be helping in the good dispersion of CNT in the polymer matrix [222].

7.4.4 Electrical conductivity

The high aspect ratio CNMs (viz., graphene and CNTs) exhibit high electrical conductivity due to the large accessibility of highly mobile pi (π) electrons that are situated above and below the surface of the sp^2 carbon-rich atomic thick sheets. Graphene is a highly electrically conductive material with electrical conductivity of $6500 \, S/m$ [223], and electron mobility of $24 \, m^2/(V \, s^2)$ [224]. Whereas GO exhibits electrical resistivity of $1.64 \times 10^4 \, \Omega m$, due to the interruption of the sp^2 hybridized C orbitals and presence of oxygen functionalities [225]. In order to introduce electrical conductivity, GO can be reduced using various routes including chemical, thermal, microwave, and laser techniques. After reduction, the electrical conductivity of GO significantly improves and is observed to be in the range of $\sim 0.1 - 2.98 \times 10^4 \, S/m$ [226,227]. The electrical conduction mechanism in rGO is different from graphene as it follows the electron hopping process. Another side, the electrical conductivity of pure CNTs can be noted to be in the range of $10^6 - 10^7 \, S/m$ [228]. The electrical conductivity of fullerene is in the order of $10^{-6} \, S/cm$ [229], whereas the electrical conductivity of detonation nanodiamonds is extremely low $\sim 10^{-11} \, S/cm$ [230]. The electrical properties of CNMs depend on impurities, dopants concentration, defects, and surrounding chemical environments.

The high aspect ratio CNMs can be used as nanofillers to produce conductive composites. The percolation thresholds of nanomaterial fillers are important as it decides the amount of fillers at which the composite itself becomes electrically conductive. The percolation thresholds of CNMs in composites are governed by aspect ratio, alignment, dispersion, degree of surface modification of fillers, polymer types, and nanocomposites processing techniques [231]. For instance, Zhang et al. studied the influence of individual fillers (graphene and CNT) and mixed fillers CNT-graphene on the enhancement of the electrical conductivity of polyether sulfones (PES) composites [232]. The electrical conductivity of the graphene (5 wt.%)/PES nanocomposite was higher than CNTs/PES nanocomposite. When mixed fillers (CNT−graphene, 1:1) were introduced in composites, the percolation threshold value reduced to 0.22 vol.%, which was much lower than the graphene/PES composites and the electrical conductivity was 8.9 and 2.2 times higher than the individual PES composite of

CNTs and graphene. This increase in electrical conductivity was accredited to the synergistic effect of mixed nanofillers, and results in an effective 3D percolation network and ideal codispersion of CNTs and graphene [232].

7.4.5 Thermal conductivity

Similar to the electrical conductivity, the in-plane thermal conductivity of graphene ($\sim 3000-5000$ W/(m K)) is significantly higher than GO (0.5–1 W/(m K)) [233,234]. On the other hand, rGO can exhibit thermal conductivity in the range of $\sim 3-61$ W/(m K), which is less than graphene [233]. The longitudinal thermal conductivity of SWCNTs is observed in the range of 2800–6000 W/(m K) at room temperature, which is higher than CNFs, carbon black, and graphite, and comparable to diamond [235,236]. Compared with 1D and 2D CNMs, fullerene C_{60} has shown lower thermal conductivity (~ 0.4 W/(m K)) at room temperature [237]. Among CNMs, CNTs and graphene have been commonly used to improve the thermal properties of nanocomposites due to their high thermal stability, extremely high thermal conductivity, and configurational influence on the thermal contact resistance [221]. The thermal conductivity of nanocomposites depends on filler loading, the type of polymer matrix, filler dispersion, the aspect ratio of the filler, and thermal resistance at the interface between polymer and filler [238,239]. For instance, in one study, the thermal conductivity of the epoxy polymer was enhanced by making a composite with pristine multigraphene platelets (MGP) and MWCNTs, and mixed-filler MGPs/MWCNTs [240]. The thermal conductivity of MGP (1 wt.%)/epoxy composite was improved by 23.8%, while the thermal conductivity of MWCNT (1 wt.%)/epoxy

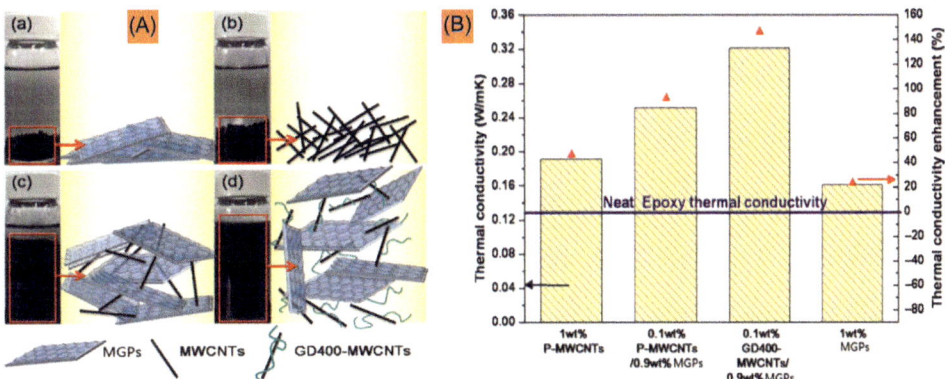

Figure 7.20 (A) Snaps and schemes exhibit the dispersion of MGPs and MWCNTs in THF; (B) thermal conductivity and % thermal conductivity enhancement of various epoxy composites. *Reproduced with permission from Yang S-Y, Lin W-N, Huang Y-L, Tien H-W, Wang J-Y, Ma C-CM, et al. Synergetic effects of graphene platelets and carbon nanotubes on the mechanical and thermal properties of epoxy composites. Carbon 2011;49:793–803 [240]. Copyrights 2011, Elsevier.*

composite was improved by 62%, which was ascribed to the low aspect ratio of MGPs, resulting in poor interfacial contact between epoxy matrix and MGPs, leading to phonon scattering. Moreover, on the addition of the mixed-filler, the thermal conductivity of P-MWCNT/MGP/epoxy composites was significantly increased to 93%, compared to individual MGPs and MWCNTs composites (Fig. 7.20). This was ascribed to improved surface contact between the epoxy matrix and mixed-filler, as well as proper dispersion as long and twisted MWCNTs can interconnect the neighboring MGPs [240].

7.5 Conclusion

Over the last three decades, CNMs have been one of the most researched materials for the development of various advanced technological and industrial-related applications. Owing to their extraordinary characteristics, such as their electrical, mechanical, thermal, optical, and surface properties, CNMs have garnered huge attention in numerous research areas including electrochemical sensors, electronic devices, energy production and storage, biomedicine, water treatment, catalysis, environment protection, composites, and coatings. There is still a need to develop simple and cost-effective large-scale production routes for CNMs in order to achieve their full potential. Functionalization (i.e., covalent and noncovalent) of CNMs is crucial to improve their solubility, dispersion in different media, and to introduce hydrophilic and hydrophobic surfaces for numerous applications. CNMs can also be modified by various NPs such as metallic, metal oxides, metal sulfides, and many more, and to improve the intrinsic properties of CNMs. The activation process is used to enhance surface area and porosity of CNMs. Among the CNMs, 1D and 2D nanostructures have shown excellent thermal, electrical, mechanical, and optical properties. As well as the considerable research advances in CNMs, there is still a requirement for organized toxicology assessments to determine the toxicity of CNMs. Despite a lot of research on CNMs for various applications, CNMs are projected to be explored greatly in the near future for the development of devices and consumer products.

References

[1] Rehan M, Barhoum A, Van Assche G, Dufresne A, Gätjen L, Wilken R. Towards multifunctional cellulosic fabric: UV photo-reduction and in-situ synthesis of silver nanoparticles into cellulose fabrics. Int J Biol Macromolecules 2017;98:877−86.
[2] Rehan M, Khattab TA, Barohum A, Gätjen L, Wilken R. Development of Ag/AgX (X = Cl, I) nanoparticles toward antimicrobial, UV-protected and self-cleanable viscose fibers. Carbohydr Polym 2018;197:227−36.

[3] Barhoum A, Li H, Chen M, Cheng L, Yang W, Dufresne A. Emerging applications of cellulose nanofibers. In: Barhoum A, Bechelany M, Makhlouf ASH, editors. Handbook of nanofibers. Cham: Springer International Publishing; 2019. p. 1131−56.

[4] Quesada-González D, Merkoçi A. Nanomaterial-based devices for point-of-care diagnostic applications. Chem Soc Rev 2018;47:4697−709.

[5] Samyn P, Barhoum A, Öhlund T, Dufresne A. Review: nanoparticles and nanostructured materials in papermaking. J Mater Sci 2018;53:146−84.

[6] Barhoum A, Samyn P, Öhlund T, Dufresne A. Review of recent research on flexible multifunctional nanopapers. Nanoscale 2017;9:15181−205.

[7] El-Sherbiny S, El-Sheikh SM, Barhoum A. Preparation and modification of nano calcium carbonate filler from waste marble dust and commercial limestone for papermaking wet end application. Powder Technol 2015;279:290−300.

[8] Chen J-J, Symes MD, Cronin L. Highly reduced and protonated aqueous solutions of [P2W18O62] 6 − for on-demand hydrogen generation and energy storage. Nat Chem 2018;10:1042−7.

[9] Li Y, Lu J. Metal−air batteries: will they be the future electrochemical energy storage device of choice? ACS Energy Lett 2017;2:1370−7.

[10] Ali GA, Divyashree A, Supriya S, Chong KF, Ethiraj AS, Reddy M, et al. Carbon nanospheres derived from Lablab purpureus for high performance supercapacitor electrodes: a green approach. Dalton Trans 2017;46:14034−44.

[11] Pumera M. Graphene-based nanomaterials for energy storage. Energy Environ Sci 2011;4:668−74.

[12] Zhang Q, Uchaker E, Candelaria SL, Cao G. Nanomaterials for energy conversion and storage. Chem Soc Rev 2013;42:3127−71.

[13] Pastoriza-Santos I, Kinnear C, Pérez-Juste J, Mulvaney P, Liz-Marzán LM. Plasmonic polymer nanocomposites. Nat Rev Mater 2018;3:375−91.

[14] Yan J, Lu Y, Chen G, Yang M, Gu Z. Advances in liquid metals for biomedical applications. Chem Soc Rev 2018;47:2518−33.

[15] Kumar N, Mittal H, Alhassan SM, Ray SS. Bionanocomposite hydrogel for the adsorption of dye and reusability of generated waste for the photodegradation of ciprofloxacin: a demonstration of the circularity concept for water purification. ACS Sustain Chem Eng 2018;6:17011−25.

[16] Zhang Y, Yin Q-Z. Carbon and other light element contents in the Earth's core based on first-principles molecular dynamics. Proc Natl Acad Sci 2012;109:19579−83.

[17] Bianco A, Kostarelos K, Prato M. Applications of carbon nanotubes in drug delivery. Curr Opin Chem Biol 2005;9:674−9.

[18] De Volder MF, Tawfick SH, Baughman RH, Hart AJ. Carbon nanotubes: present and future commercial applications. Science 2013;339:535−9.

[19] Notarianni M, Liu J, Vernon K, Motta N. Synthesis and applications of carbon nanomaterials for energy generation and storage. Beilstein J Nanotechnol 2016;7:149.

[20] Mauter MS, Elimelech M. Environmental applications of carbon-based nanomaterials. Environ Sci Technol 2008;42:5843−59.

[21] Huang Y, Liang J, Chen Y. An overview of the applications of graphene-based materials in supercapacitors. Small 2012;8:1805−34.

[22] Georgakilas V, Perman JA, Tucek J, Zboril R. Broad family of carbon nanoallotropes: classification, chemistry, and applications of fullerenes, carbon dots, nanotubes, graphene, nanodiamonds, and combined superstructures. Chem Rev 2015;115:4744−822.

[23] Stankovich S, Dikin DA, Piner RD, Kohlhaas KA, Kleinhammes A, Jia Y, et al. Synthesis of graphene-based nanosheets via chemical reduction of exfoliated graphite oxide. Carbon 2007;45:1558−65.

[24] Chen Y, Fu K, Zhu S, Luo W, Wang Y, Li Y, et al. Reduced graphene oxide films with ultrahigh conductivity as Li-ion battery current collectors. Nano Lett 2016;16:3616−23.

[25] Gusain R, Kumar N, Ray SS. Recent advances in carbon nanomaterial-based adsorbents for water purification. Coord Chem Rev 2020;405:213111.

[26] Verma S, Mungse HP, Kumar N, Choudhary S, Jain SL, Sain B, et al. Graphene oxide: an efficient and reusable carbocatalyst for aza-Michael addition of amines to activated alkenes. Chem Commun 2011;47:12673−5.

[27] Nanotechnology N. Ten years in two dimensions. Nat Nanotechnol 2014;9:725.

[28] Wan X, Huang Y, Chen Y. Focusing on energy and optoelectronic applications: a journey for graphene and graphene oxide at large scale. Acc Chem Res 2012;45:598−607.

[29] Novoselov KS, Geim AK, Morozov SV, Jiang D, Zhang Y, Dubonos SV, et al. Electric field effect in atomically thin carbon films. Science 2004;306:666−9.

[30] Somani PR, Somani SP, Umeno M. Planer nano-graphenes from camphor by CVD. Chem Phys Lett 2006;430:56−9.

[31] Wang X-Y, Narita A, Müllen K. Precision synthesis versus bulk-scale fabrication of graphenes. Nat Rev Chem 2017;2:0100.

[32] Staudenmaier L. Verfahren zur Darstellung der Graphitsäure. Berichte Deutsc. chemischen Ges 1898;31:1481−7.

[33] Brodie XIII BC. On the atomic weight of graphite. Philos Trans R Soc Lond 1859;149:249−59.

[34] Hummers WS, Offeman RE. Preparation of graphitic oxide. J Am Chem Soc 1958;80:1339.

[35] Chen J, Yao B, Li C, Shi G. An improved Hummers method for eco-friendly synthesis of graphene oxide. Carbon 2013;64:225−9.

[36] Ren W, Cheng H-M. The global growth of graphene. Nat Nanotechnol 2014;9:726−30.

[37] Wang X, Zhi L, Müllen K. Transparent, conductive graphene electrodes for dye-sensitized solar cells. Nano Lett 2008;8:323−7.

[38] Chua CK, Pumera M. Chemical reduction of graphene oxide: a synthetic chemistry viewpoint. Chem Soc Rev 2014;43:291−312.

[39] Pei S, Cheng H-M. The reduction of graphene oxide. Carbon 2012;50:3210−28.

[40] Zhou Y, Bao Q, Tang LAL, Zhong Y, Loh KP. Hydrothermal dehydration for the "green" reduction of exfoliated graphene oxide to graphene and demonstration of tunable optical limiting properties. Chem Mater 2009;21:2950−6.

[41] Voiry D, Yang J, Kupferberg J, Fullon R, Lee C, Jeong HY, et al. High-quality graphene via microwave reduction of solution-exfoliated graphene oxide. Science 2016;353:1413−16.

[42] Shin H-J, Kim KK, Benayad A, Yoon S-M, Park HK, Jung I-S, et al. Efficient reduction of graphite oxide by sodium borohydride and its effect on electrical conductance. Adv Funct Mater 2009;19:1987−92.

[43] Gao J, Liu F, Liu Y, Ma N, Wang Z, Zhang X. Environment-friendly method to produce graphene that employs vitamin c and amino acid. Chem Mater 2010;22:2213−18.

[44] Amarnath CA, Hong CE, Kim NH, Ku B-C, Kuila T, Lee JH. Efficient synthesis of graphene sheets using pyrrole as a reducing agent. Carbon 2011;49:3497−502.

[45] Zhang J, Yang H, Shen G, Cheng P, Zhang J, Guo S. Reduction of graphene oxide vial-ascorbic acid. Chem Commun 2010;46:1112−14.

[46] Lei Z, Lu L, Zhao XS. The electrocapacitive properties of graphene oxide reduced by urea. Energy Environ Sci 2012;5:6391−9.

[47] Kairi MI, Zuhan MKNM, Khavarian M, Vigolo B, Bakar SA, Mohamed AR. Co-synthesis of large-area graphene and syngas via CVD method from greenhouse gases. Mater Lett 2018;227:132−5.

[48] Di Gaspare L, Scaparro AM, Fanfoni M, Fazi L, Sgarlata A, Notargiacomo A, et al. Early stage of CVD graphene synthesis on Ge(001) substrate. Carbon 2018;134:183−8.

[49] Ge X, Zhang Y, Chen L, Zheng Y, Chen Z, Liang Y, et al. Mechanism of SiOx particles formation during CVD graphene growth on Cu substrates. Carbon 2018;139:989−98.

[50] Murdock AT, van Engers CD, Britton J, Babenko V, Meysami SS, Bishop H, et al. Targeted removal of copper foil surface impurities for improved synthesis of CVD graphene. Carbon 2017;122:207−16.

[51] Bayev VG, Fedotova JA, Kasiuk JV, Vorobyova SA, Sohor AA, Komissarov IV, et al. CVD graphene sheets electrochemically decorated with "core-shell" Co/CoO nanoparticles. Appl Surf Sci 2018;440:1252−60.

[52] Naghdi S, Rhee KY, Park SJ. A catalytic, catalyst-free, and roll-to-roll production of graphene via chemical vapor deposition: Low temperature growth. Carbon 2018;127:1−12.

[53] Britnell L, Gorbachev RV, Jalil R, Belle BD, Schedin F, Mishchenko A, et al. Field-effect tunneling transistor based on vertical graphene heterostructures. Science 2012;335:947−50.

[54] Chen X, Zhang L, Chen S. Large area CVD growth of graphene. Synth Met 2015;210:95—108.

[55] Li X, Cai W, Colombo L, Ruoff RS. Evolution of graphene growth on Ni and Cu by carbon isotope labeling. Nano Lett 2009;9:4268—72.

[56] Yan Z, Lin J, Peng Z, Sun Z, Zhu Y, Li L, et al. Toward the synthesis of wafer-scale single-crystal graphene on copper foils. ACS Nano 2012;6:9110—17.

[57] Li M, Liu D, Wei D, Song X, Wei D, Wee ATS. Controllable synthesis of graphene by plasma-enhanced chemical vapor deposition and its related applications. Adv Sci 2016;3:1600003.

[58] Woehrl N, Ochedowski O, Gottlieb S, Shibasaki K, Schulz S. Plasma-enhanced chemical vapor deposition of graphene on copper substrates. AIP Adv 2014;4:047128.

[59] Bo Z, Yang Y, Chen J, Yu K, Yan J, Cen K. Plasma-enhanced chemical vapor deposition synthesis of vertically oriented graphene nanosheets. Nanoscale 2013;5:5180—204.

[60] Chandrashekar BN, Deng B, Smitha AS, Chen Y, Tan C, Zhang H, et al. Roll-to-roll green transfer of CVD graphene onto plastic for a transparent and flexible triboelectric nanogenerator. Adv Mater 2015;27:5210—16.

[61] Sun J, Lindvall N, Cole MT, Angel KTT, Wang T, Teo KBK, et al. Low partial pressure chemical vapor deposition of graphene on copper. IEEE Trans Nanotechnol 2012;11:255—60.

[62] Hussain S, Iqbal MW, Park J, Ahmad M, Singh J, Eom J, et al. Physical and electrical properties of graphene grown under different hydrogen flow in low pressure chemical vapor deposition. Nanoscale Res Lett 2014;9:546.

[63] Choi J-H, Li Z, Cui P, Fan X, Zhang H, Zeng C, et al. Drastic reduction in the growth temperature of graphene on copper via enhanced London dispersion force. Sci Rep 2013;3:1925.

[64] Guermoune A, Chari T, Popescu F, Sabri SS, Guillemette J, Skulason HS, et al. Chemical vapor deposition synthesis of graphene on copper with methanol, ethanol, and propanol precursors. Carbon 2011;49:4204—10.

[65] Yazici MS, Azder MA, Salihoglu O. CVD grown graphene as catalyst for acid electrolytes. Int J Hydrogen Energy 2018;43:10710—16.

[66] Chen C-S, Hsieh C-K. Effects of acetylene flow rate and processing temperature on graphene films grown by thermal chemical vapor deposition. Thin Solid Films 2015;584:265—9.

[67] Dathbun A, Chaisitsak S. Effects of three parameters on graphene synthesis by chemical vapor deposition. In: The 8th annual IEEE international conference on nano/micro engineered and molecular systems; 2013. p. 1018—21.

[68] Zhang B, Lee WH, Piner R, Kholmanov I, Wu Y, Li H, et al. Low-temperature chemical vapor deposition growth of graphene from toluene on electropolished copper foils. ACS Nano 2012;6:2471—6.

[69] Dadkhah AA, Rabiee Faradonbeh M, Rashidi A, Tasharofi S, Mansourkhani F. One step synthesis of nitrogen-doped graphene from naphthalene and urea by atmospheric chemical vapor deposition. J Inorg Organomet Polym Mater 2018;28:1609—15.

[70] Weatherup RS, Dlubak B, Hofmann S. Kinetic control of catalytic CVD for high-quality graphene at low temperatures. ACS Nano 2012;6:9996—10003.

[71] Mukanova A, Tussupbayev R, Sabitov A, Bondarenko I, Nemkaeva R, Aldamzharov B, et al. CVD graphene growth on a surface of liquid gallium. Mater Today: Proc 2017;4:4548—54.

[72] Terasawa T-o, Saiki K. Growth of graphene on Cu by plasma enhanced chemical vapor deposition. Carbon 2012;50:869—74.

[73] Rao KS, Sentilnathan J, Cho H-W, Wu J-J, Yoshimura M. Soft processing of graphene nanosheets by glycine-bisulfate ionic-complex-assisted electrochemical exfoliation of graphite for reduction catalysis. Adv Funct Mater 2015;25:298—305.

[74] Niu T, Zhou M, Zhang J, Feng Y, Chen W. Growth intermediates for CVD graphene on Cu (111): carbon clusters and defective graphene. J Am Chem Soc 2013;135:8409—14.

[75] Cazzanelli E, Caruso T, Castriota M, Marino AR, Politano A, Chiarello G, et al. Spectroscopic characterization of graphene films grown on Pt(111) surface by chemical vapor deposition of ethylene. J Raman Spectrosc 2013;44:1393—7.

[76] Kim J, Ishihara M, Koga Y, Tsugawa K, Hasegawa M, Iijima S. Low-temperature synthesis of large-area graphene-based transparent conductive films using surface wave plasma chemical vapor deposition. Appl Phys Lett 2011;98:091502.

[77] Woo YS, Seo DH, Yeon D-H, Heo J, Chung H-J, Benayad A, et al. Low temperature growth of complete monolayer graphene films on Ni-doped copper and gold catalysts by a self-limiting surface reaction. Carbon 2013;64:315−23.

[78] Nang LV, Kim E-T. Low-temperature synthesis of graphene on Fe2O3 using inductively coupled plasma chemical vapor deposition. Mater Lett 2013;92:437−9.

[79] Uchida H, Aryal HR, Adhikari S, Umeno M. LOW temperature plasma CVD grown graphene by microwave surface-wave plasma CVD using camphor precursor. J Phys Sci Appl 2016;6:34−8.

[80] Yuan GD, Zhang WJ, Yang Y, Tang YB, Li YQ, Wang JX, et al. Graphene sheets via microwave chemical vapor deposition. Chem Phys Lett 2009;467:361−4.

[81] Qi JL, Zheng WT, Zheng XH, Wang X, Tian HW. Relatively low temperature synthesis of graphene by radio frequency plasma enhanced chemical vapor deposition. Appl Surf Sci 2011;257:6531−4.

[82] Hernandez Y, Nicolosi V, Lotya M, Blighe FM, Sun Z, De S, et al. High-yield production of graphene by liquid-phase exfoliation of graphite. Nat Nanotechnol 2008;3:563−8.

[83] Pu N-W, Wang C-A, Sung Y, Liu Y-M, Ger M-D. Production of few-layer graphene by supercritical CO2 exfoliation of graphite. Mater Lett 2009;63:1987−9.

[84] Xu Y, Cao H, Xue Y, Li B, Cai W. Liquid-phase exfoliation of graphene: an overview on exfoliation media, techniques, and challenges. Nanomaterials 2018;8:942.

[85] Choucair M, Thordarson P, Stride JA. Gram-scale production of graphene based on solvothermal synthesis and sonication. Nat Nanotechnol 2009;4:30.

[86] Karagiannidis PG, Hodge SA, Lombardi L, Tomarchio F, Decorde N, Milana S, et al. Microfluidization of graphite and formulation of graphene-based conductive inks. ACS Nano 2017;11:2742−55.

[87] Shen Z, Li J, Yi M, Zhang X, Ma S. Preparation of graphene by jet cavitation. Nanotechnology 2011;22:365306.

[88] Paton KR, Varrla E, Backes C, Smith RJ, Khan U, O'Neill A, et al. Scalable production of large quantities of defect-free few-layer graphene by shear exfoliation in liquids. Nat Mater 2014;13:624−30.

[89] Radushkevich LV, Lukyanovich VM. Structure of the carbon produced in the thermal decomposition of carbon monoxide on an iron catalyst. Russ J Phys Chem 1952;26:89−95.

[90] Iijima S. Helical microtubules of graphitic carbon. Nature 1991;354:56.

[91] Prasek J, Drbohlavova J, Chomoucka J, Hubalek J, Jasek O, Adam V, et al. Methods for carbon nanotubes synthesis—review. J Mater Chem 2011;21:15872−84.

[92] Eatemadi A, Daraee H, Karimkhanloo H, Kouhi M, Zarghami N, Akbarzadeh A, et al. Carbon nanotubes: properties, synthesis, purification, and medical applications. Nanoscale Res Lett 2014;9:393.

[93] Dai H, Rinzler AG, Nikolaev P, Thess A, Colbert DT, Smalley RE. Single-wall nanotubes produced by metal-catalyzed disproportionation of carbon monoxide. Chem Phys Lett 1996;260:471−5.

[94] Chrzanowska J, Hoffman J, Małolepszy A, Mazurkiewicz M, Kowalewski TA, Szymanski Z, et al. Synthesis of carbon nanotubes by the laser ablation method: Effect of laser wavelength. Phys Stat Solidi (b) 2015;252:1860−7.

[95] Bronikowski MJ, Willis PA, Colbert DT, Smith KA, Smalley RE. Gas-phase production of carbon single-walled nanotubes from carbon monoxide via the HiPco process: a parametric study. J Vac Sci Technol A 2001;19:1800−5.

[96] Kroto HW, Heath JR, O'Brien SC, Curl RF, Smalley RE. C60: Buckminsterfullerene. Nature 1985;318:162.

[97] Kratschmer W, Lamb LD, Fostiropoulos K, Huffman DR. Solid C60: A New Form of Carbon. Nature 1990;347:354−8.

[98] Ying ZC, Hettich RL, Compton RN, Haufler RE. Synthesis of nitrogen-doped fullerenes by laser ablation. J Phys B: Atomic Mol Opt Phys 1996;29:4935.

[99] Howard JB, McKinnon JT, Makarovsky Y, Lafleur AL, Johnson ME. Fullerenes C60 and C70 in flames. Nature 1991;352:139.

[100] Scott LT. Methods for the chemical synthesis of fullerenes. Angew Chem Int Ed 2004;43:4994—5007.

[101] Wang Z, Zhu F, Wang W, Ruan M. Synthesis of carbon nanostructures by ion sputtering. Phys Lett A 1998;242:261—5.

[102] Bunshah RF, Jou S, Prakash S, Doerr HJ, Isaacs L, Wehrsig A, et al. Fullerene formation in sputtering and electron beam evaporation processes. J Phys Chem 1992;96:6866—9.

[103] Enoki T, Takai K, Osipov V, Baidakova M, Vul' A. Nanographene and nanodiamond; new members in the nanocarbon family. Chem —Asian J 2009;4:796—804.

[104] Danilenko VV. On the history of the discovery of nanodiamond synthesis. Phys Solid State 2004;46:595—9.

[105] Tang CJ, Neves AJ, Grácio J, Fernandes AJS, Carmo MC. A new chemical path for fabrication of nanocrystalline diamond films. J Cryst Growth 2008;310:261—5.

[106] Yang Q, Yang S, Xiao C, Hirose A. Transformation of carbon nanotubes to diamond in microwave hydrogen plasma. Mater Lett 2007;61:2208—11.

[107] Kimura Y, Kaito C. Production of nanodiamond from carbon film containing silicon. J Cryst Growth 2003;255:282—5.

[108] Akaishi M, Kanda H, Yamaoka S. Synthesis of diamond from graphite-carbonate system under very high temperature and pressure. J Cryst Growth 1990;104:578—81.

[109] Kamali AR, Fray DJ. Preparation of nanodiamonds from carbon nanoparticles at atmospheric pressure. Chem Commun 2015;51:5594—7.

[110] Nee C-H, Yap S-L, Tou T-Y, Chang H-C, Yap S-S. Direct synthesis of nanodiamonds by femtosecond laser irradiation of ethanol. Sci Rep 2016;6:33966.

[111] Kharisov BI, Kharissova OV, Chávez-Guerrero L. Synthesis techniques, properties, and applications of nanodiamonds. Synth Reactivity Inorg Metal-Org Nano-Metal Chem 2010;40:84—101.

[112] Zhang W, Fan B, Zhang Y, Fan J. Hydrothermal synthesis of well crystallized C8 and diamond nanocrystals and pH-controlled C8 ↔ diamond phase transition. CrystEngComm 2017;19:1248—52.

[113] Gu CZ, Jiang X. Deposition and characterization of nanocrystalline diamond films prepared by ion bombardment-assisted method. J Appl Phys 2000;88:1788—93.

[114] Yang GW. Laser ablation in liquids: Applications in the synthesis of nanocrystals. Prog Mater Sci 2007;52:648—98.

[115] Chowdhury S, Borham J, Catledge SA, Eberhardt AW, Johnson PS, Vohra YK. Synthesis and mechanical wear studies of ultra smooth nanostructured diamond (USND) coatings deposited by microwave plasma chemical vapor deposition with He/H2/CH4/N2 mixtures. Diam Relat Mater 2008;17:419—27.

[116] Iakoubovskii K, Baidakova MV, Wouters BH, Stesmans A, Adriaenssens GJ, Vul AY, et al. Structure and defects of detonation synthesis nanodiamond. Diam Relat Mater 2000;9:861—5.

[117] Boudou J-P, Curmi PA, Jelezko F, Wrachtrup J, Aubert P, Sennour M, et al. High yield fabrication of fluorescent nanodiamonds. Nanotechnology. 2009;20:235602.

[118] Reina G, Zhao L, Bianco A, Komatsu N. Chemical functionalization of nanodiamonds: opportunities and challenges ahead. Angew Chem Int Ed 2019;58:17918—29.

[119] Khachatryan AK, Aloyan SG, May PW, Sargsyan R, Khachatryan VA, Baghdasaryan VS. Graphite-to-diamond transformation induced by ultrasound cavitation. Diam Relat Mater 2008;17:931—6.

[120] Chiang WH, Sankaran RM. Microplasma synthesis of metal nanoparticles for gas-phase studies of catalyzed carbon nanotube growth. Appl Phys Lett 2007;91:121503.

[121] Švrček V, Mariotti D, Kondo M. Microplasma-induced surface engineering of silicon nanocrystals in colloidal dispersion. Appl Phys Lett 2010;97:161502.

[122] Mariotti D, Švrček V, Hamilton JWJ, Schmidt M, Kondo M. Silicon nanocrystals in liquid media: optical properties and surface stabilization by microplasma-induced non-equilibrium liquid chemistry. Adv Funct Mater 2012;22:954—64.

[123] Bora B, Bhuyan H, Favre M, Chuaqui H, Wyndham E, Kakati M. Investigation on plasma parameters and step ionization from discharge characteristics of an atmospheric pressure Ar microplasma jet. Phys Plasmas 2012;19:064503.

[124] Barbosa D, Barreto P, Ribas V, Trava-Airoldi V, Corat E. Diamond nanostructures growth. Structure 2010;25:31.

[125] Iqbal S, Rafique Muhammad S, Zahid M, Bashir S, Ahmad Muhammad A, Ahmad R. Impact of carrier gas flow rate on the synthesis of nanodiamonds via microplasma technique. Mater Sci Semiconductor Process 2018;74:31–41.

[126] Dolmatov VY. On the mechanism of detonation nanodiamond synthesis. J Superhard Mater 2008;30:233–40.

[127] Krueger A. Diamond nanoparticles: jewels for chemistry and physics. Adv Mater 2008;20:2445–9.

[128] Lee J, Han S, Hyeon T. Synthesis of new nanoporous carbon materials using nanostructured silica materials as templates. J Mater Chem 2004;14:478–86.

[129] Armandi M, Bonelli B, Geobaldo F, Garrone E. Nanoporous carbon materials obtained by sucrose carbonization in the presence of KOH. Microporous Mesoporous Mater 2010;132:414–20.

[130] Chen Y, Gerald JF, Chadderton LT, Chaffron L. Nanoporous carbon produced by ball milling. Appl Phys Lett 1999;74:2782–4.

[131] Liang C, Hong K, Guiochon GA, Mays JW, Dai S. Synthesis of a large-scale highly ordered porous carbon film by self-assembly of block copolymers. Angew Chem Int Ed 2004;43:5785–9.

[132] Sun B, Li G, Wang X. Facile synthesis of microporous carbon through a soft-template pathway and its performance in desulfurization and denitrogenation. J Nat Gas Chem 2010;19:471–6.

[133] Kowalewski T, Tsarevsky NV, Matyjaszewski K. Nanostructured carbon arrays from block copolymers of polyacrylonitrile. J Am Chem Soc 2002;124:10632–3.

[134] Ozaki J, Endo N, Ohizumi W, Igarashi K, Nakahara M, Oya A, et al. Novel preparation method for the production of mesoporous carbon fiber from a polymer blend. Carbon. 1997;35:1031–3.

[135] Kyotani T. Control of pore structure in carbon. Carbon. 2000;38:269–86.

[136] Tamai H, Kakii T, Hirota Y, Kumamoto T, Yasuda H. Synthesis of extremely large mesoporous activated carbon and its unique adsorption for giant molecules. Chem Mater 1996;8:454–62.

[137] Pekala RW. Organic aerogels from the polycondensation of resorcinol with formaldehyde. J Mater Sci 1989;24:3221–7.

[138] Liang C, Li Z, Dai S. Mesoporous carbon materials: synthesis and modification. Angew Chem Int Ed 2008;47:3696–717.

[139] Knox JH, Kaur B, Millward GR. Structure and performance of porous graphitic carbon in liquid chromatography. J Chromatogr A 1986;352:3–25.

[140] Miao J, Miyauchi M, Simmons TS, Dordick J, Linhardt R. Electrospinning of nanomaterials and applications in electronic components and devices. J Nanosci Nanotech 2010;10:5507–5519.

[141] De Jong KP, Geus JW. Carbon nanofibers: catalytic synthesis and applications. Catal Rev 2000;42:481–510.

[142] Gugulothu D, Barhoum A, Nerella R, Ajmer R, Bechelany M. Fabrication of nanofibers: electrospinning and non-electrospinning techniques. In: Barhoum A, Bechelany M, Makhlouf ASH, editors. Handbook of nanofibers. Cham: Springer International Publishing; 2019. p. 45–77.

[143] Khalf A, Madihally SV. Recent advances in multiaxial electrospinning for drug delivery. Eur J Pharm Biopharm 2017;112:1–17.

[144] Xue J, Wu T, Dai Y, Xia Y. Electrospinning and electrospun nanofibers: methods, materials, and applications. Chem Rev 2019;119:5298–415.

[145] Guo Q, Zhou X, Li X, Chen S, Seema A, Greiner A, et al. Supercapacitors based on hybrid carbon nanofibers containing multiwalled carbon nanotubes. J Mater Chem 2009;19:2810–16.

[146] Ong W-J, Tan L-L, Ng YH, Yong S-T, Chai S-P. Graphitic carbon nitride (g-C3N4)-based photocatalysts for artificial photosynthesis and environmental remediation: are we a step closer to achieving sustainability? Chem Rev 2016;116:7159–329.

[147] Liu J, Li W, Duan L, Li X, Ji L, Geng Z, et al. A graphene-like oxygenated carbon nitride material for improved cycle-life lithium/sulfur batteries. Nano Lett 2015;15:5137–42.

[148] Song B, Zeng Z, Zeng G, Gong J, Xiao R, Ye S, et al. Powerful combination of g-C3N4 and LDHs for enhanced photocatalytic performance: a review of strategy, synthesis, and applications. Adv Colloid Interface Sci 2019;272:101999.

[149] Yuan S, Zhang Q, Xu B, Liu S, Wang J, Xie J, et al. A new precursor to synthesize g-C3N4 with superior visible light absorption for photocatalytic application. Catal Sci Technol 2017;7:1826−30.

[150] Kumar S, Karthikeyan S, Lee AF. g-C3N4-based nanomaterials for visible light-driven photocatalysis. Catalysts 2018;8:74.

[151] Wu P, Shi J, Chen J, Wang B, Guo L. Graphitic Carbon Nitride Modified by Silicon for Improved Visible-Light-Driven Photocatalytic Hydrogen Production, *Nanostructured Materials and Nanotechnology VI* Copyright © 2013 The American Ceramic Society, 137−148.

[152] Qin L, Huang D, Xu P, Zeng G, Lai C, Fu Y, et al. In-situ deposition of gold nanoparticles onto polydopamine-decorated g-C3N4 for highly efficient reduction of nitroaromatics in environmental water purification. J Colloid Interface Sci 2019;534:357−69.

[153] Thomas A, Fischer A, Goettmann F, Antonietti M, Müller J-O, Schlögl R, et al. Graphitic carbon nitride materials: variation of structure and morphology and their use as metal-free catalysts. J Mater Chem 2008;18:4893−908.

[154] Masteri-Farahani M, Mirshekar S. Covalent functionalization of graphene oxide with molybdenum-carboxylate complexes: New reusable catalysts for the epoxidation of olefins. Colloids Surf A: Physicochem Eng Asp 2018;538:387−92.

[155] Gusain R, Kumar P, Sharma OP, Jain SL, Khatri OP. Reduced graphene oxide−CuO nanocomposites for photocatalytic conversion of CO2 into methanol under visible light irradiation. Appl Catal B: Environ 2016;181:352−62.

[156] Mungse HP, Verma S, Kumar N, Sain B, Khatri OP. Grafting of oxo-vanadium Schiff base on graphene nanosheets and its catalytic activity for the oxidation of alcohols. J Mater Chem 2012;22:5427−33.

[157] Gusain R, Mungse HP, Kumar N, Ravindran TR, Pandian R, Sugimura H, et al. Covalently attached graphene−ionic liquid hybrid nanomaterials: synthesis, characterization and tribological application. J Mater Chem A 2016;4:926−37.

[158] Kumar N, Sinha Ray S. Synthesis and functionalization of nanomaterials. In: Sinha Ray S, editor. Processing of polymer-based nanocomposites: introduction. Cham: Springer International Publishing; 2018. p. 15−55.

[159] Hu X, Su E, Zhu B, Jia J, Yao P, Bai Y. Preparation of silanized graphene/poly(methyl methacrylate) nanocomposites in situ copolymerization and its mechanical properties. Compos Sci Technol 2014;97:6−11.

[160] Yang H, Shan C, Li F, Han D, Zhang Q, Niu L. Covalent functionalization of polydisperse chemically-converted graphene sheets with amine-terminated ionic liquid. Chem Commun 2009;3880−2.

[161] Zhang X, Huang Y, Wang Y, Ma Y, Liu Z, Chen Y. Synthesis and characterization of a graphene−C60 hybrid material. Carbon 2009;47:334−7.

[162] Englert JM, Dotzer C, Yang G, Schmid M, Papp C, Gottfried JM, et al. Covalent bulk functionalization of graphene. Nat Chem 2011;3:279−86.

[163] Gusain R, Kumar N, Fosso-Kankeu E, Ray SS. Efficient removal of Pb(II) and Cd(II) from industrial mine water by a hierarchical MoS2/SH-MWCNT nanocomposite. ACS Omega 2019;4:13922−35.

[164] Cheng HKF, Basu T, Sahoo NG, Li L, Chan SH. Current advances in the carbon nanotube/thermotropic main-chain liquid crystalline polymer nanocomposites and their blends. Polymers 2012;4:889−912.

[165] Karaulova N, Bagrii EI. Fullerenes: functionalisation and prospects for the use of derivatives. Russian Chem Rev 1999;68:889−907.

[166] Yan W, Seifermann SM, Pierrat P, Bräse S. Synthesis of highly functionalized C60 fullerene derivatives and their applications in material and life sciences. Org Biomol. Chem 2015;13:25−54.

[167] Britz DA, Khlobystov AN, Wang J, O'Neil AS, Poliakoff M, Ardavan A, et al. Selective host−guest interaction of single-walled carbon nanotubes with functionalised fullerenes. Chem Commun 2004;176−7.

[168] Hummelen JC, Knight BW, LePeq F, Wudl F, Yao J, Wilkins CL. Preparation and characterization of fulleroid and methanofullerene derivatives. J Org Chem 1995;60:532−8.

[169] Periya VK, Koike I, Kitamura Y, Iwamatsu S-i, Murata S. Hydrophilic [60]fullerene carboxylic acid derivatives retaining the original 60π electronic system. Tetrahedron Lett 2004;45:8311—13.

[170] Tinwala H, Wairkar S. Production, surface modification and biomedical applications of nanodiamonds: a sparkling tool for theranostics. Mater Sci Eng: C 2019;97:913—31.

[171] Georgakilas V, Tiwari JN, Kemp KC, Perman JA, Bourlinos AB, Kim KS, et al. Noncovalent functionalization of graphene and graphene oxide for energy materials, biosensing, catalytic, and biomedical applications. Chem Rev 2016;116:5464—519.

[172] Gupta P, Vermani K, Garg S. Hydrogels: from controlled release to pH-responsive drug delivery. Drug Discov Today 2002;7:569—79.

[173] Di Crescenzo A, Ettorre V, Fontana A. Non-covalent and reversible functionalization of carbon nanotubes. Beilstein J Nanotechnol 2014;5:1675—90.

[174] Jiang B-P, Hu L-F, Wang D-J, Ji S-C, Shen X-C, Liang H. Graphene loading water-soluble phthalocyanine for dual-modality photothermal/photodynamic therapy via a one-step method. J Mater Chem B 2014;2:7141—8.

[175] Hsiao S-T, Ma C-CM, Tien H-W, Liao W-H, Wang Y-S, Li S-M, et al. Using a non-covalent modification to prepare a high electromagnetic interference shielding performance graphene nanosheet/water-borne polyurethane composite. Carbon. 2013;60:57—66.

[176] Hirsch A. Functionalization of Single-Walled Carbon Nanotubes. Angew Chem Int Ed 2002;41:1853—9.

[177] Haddad R, Cosnier S, Maaref A, Holzinger M. Non-covalent biofunctionalization of single-walled carbon nanotubes via biotin attachment by π-stacking interactions and pyrrole polymerization. Analyst 2009;134:2412—18.

[178] Zhan K, Liu H, Zhang H, Chen Y, Ni H, Wu M, et al. A facile method for the immobilization of myoglobin on multi-walled carbon nanotubes: poly(methacrylic acid-co-acrylamide) nanocomposite and its application for direct bio-detection of H2O2. J Electroanalytical Chem 2014;724:80—6.

[179] Mukwevho N, Gusain R, Fosso-Kankeu E, Kumar N, Waanders F, Ray SS. Removal of naphthalene from simulated wastewater through adsorption-photodegradation by ZnO/Ag/GO nanocomposite. J Ind Eng Chem 2020;81:393—404.

[180] Ali GAM, Yusoff MM, Algarni H, Chong KF. One-step electrosynthesis of MnO2/rGO nanocomposite and its enhanced electrochemical performance. Ceram Int 2018;44:7799—807.

[181] Shi X, Gong H, Li Y, Wang C, Cheng L, Liu Z. Graphene-based magnetic plasmonic nanocomposite for dual bioimaging and photothermal therapy. Biomaterials 2013;34:4786—93.

[182] Bai S, Shen X. Graphene—inorganic nanocomposites. RSC Adv 2012;2:64—98.

[183] Kumar N, Fosso-Kankeu E, Ray SS. Achieving controllable MoS2 nanostructures with increased interlayer spacing for efficient removal of Pb(II) from aquatic systems. ACS Appl Mater Interfaces 2019;11:19141—55.

[184] Ntakadzeni M, Anku WW, Kumar N, Govender PP, Reddy L. PEGylated MoS2 nanosheets: a dual functional photocatalyst for photodegradation of organic dyes and photoreduction of chromium from aqueous solution. Bull Chem React Eng Catal 2019;14:142—152.

[185] Kumar N, Kumar V, Swart HC, Mishra AK, Catherine Ngila J, Parashar V. Controlled microstructural hydrothermal synthesis of strontium selenides host matrices for EuII and EuIII luminescence. Mater Lett 2015;146:51—4.

[186] Kumar N, George BPA, Abrahamse H, Parashar V, Ray SS, Ngila JC. A novel approach to low-temperature synthesis of cubic HfO2 nanostructures and their cytotoxicity. Sci Rep 2017;7:9351.

[187] Liu J, Fu S, Yuan B, Li Y, Deng Z. Toward a universal "adhesive nanosheet" for the assembly of multiple nanoparticles based on a protein-induced reduction/decoration of graphene oxide. J Am Chem Soc 2010;132:7279—81.

[188] Yang S, Feng X, Ivanovici S, Müllen K. Fabrication of graphene-encapsulated oxide nanoparticles: towards high-performance anode materials for lithium storage. Angew Chem Int Ed 2010;49:8408—11.

[189] Wen Y, Zhu H, Zhang L, Hao J, Wang C, Zhang S, et al. Beyond colloidal synthesis: nanofiber reactor to design self-supported core—shell Pd16S7/MoS2/CNFs electrode for efficient and durable hydrogen evolution catalysis. ACS Appl Energy Mater 2019;2:2013—21.

[190] Panahian Y, Arsalani N, Nasiri R. Enhanced photo and sono-photo degradation of crystal violet dye in aqueous solution by 3D flower like F-TiO2(B)/fullerene under visible light. J Photochem Photobiol A: Chem 2018;365:45−51.

[191] Kong X-K, Chen C-L, Chen Q-W. Doped graphene for metal-free catalysis. Chem Soc Rev 2014;43:2841−57.

[192] Lee H, Paeng K, Kim IS. A review of doping modulation in graphene. Synth Met 2018;244:36−47.

[193] Yamamoto T, Watanabe K, Hernández ER. Mechanical properties, thermal stability and heat transport in carbon nanotubes. In: Jorio A, Dresselhaus G, Dresselhaus MS, editors. Carbon nanotubes: advanced topics in the synthesis, structure, properties and applications. Berlin, Heidelberg: Springer Berlin Heidelberg; 2008. p. 165−95.

[194] Yang K, Wang J, Chen X, Zhao Q, Ghaffar A, Chen B. Application of graphene-based materials in water purification: from the nanoscale to specific devices. Environ Sci Nano 2018;5:1264−97.

[195] Jiang D-e, Sumpter BG, Dai S. Unique chemical reactivity of a graphene nanoribbon's zigzag edge. J Chem Phys 2007;126:134701.

[196] Nakada K, Fujita M, Dresselhaus G, Dresselhaus MS. Edge state in graphene ribbons: nanometer size effect and edge shape dependence. Phys Rev B 1996;54:17954−61.

[197] Hao F, Fang D, Xu Z. Mechanical and thermal transport properties of graphene with defects. Appl Phys Lett 2011;99:041901.

[198] Rossi A, Piccinin S, Pellegrini V, de Gironcoli S, Tozzini V. Nano-scale corrugations in graphene: a density functional theory study of structure, electronic properties and hydrogenation. J Phys Chem C 2015;119:7900−10.

[199] Chen X, Chen B. Macroscopic and spectroscopic investigations of the adsorption of nitroaromatic compounds on graphene oxide, reduced graphene oxide, and graphene nanosheets. Environ Sci Technol 2015;49:6181−9.

[200] Gupta K, Khatri OP. Reduced graphene oxide as an effective adsorbent for removal of malachite green dye: plausible adsorption pathways. J Colloid Interface Sci 2017;501:11−21.

[201] Ghosh A, Subrahmanyam KS, Krishna KS, Datta S, Govindaraj A, Pati SK, et al. Uptake of H2 and CO2 by graphene. J Phys Chem C 2008;112:15704−7.

[202] Peigney A, Laurent C, Flahaut E, Bacsa RR, Rousset A. Specific surface area of carbon nanotubes and bundles of carbon nanotubes. Carbon 2001;39:507−14.

[203] Monthioux M, Serp P, Flahaut E, Razafinimanana M, Laurent C, Peigney A, et al. Introduction to carbon nanotubes. In: Bhushan B, editor. Springer handbook of nanotechnology. Berlin, Heidelberg: Springer Berlin Heidelberg; 2007. p. 43−112.

[204] Su F, Lu C, Chen H-S. Adsorption, desorption, and thermodynamic studies of CO2 with high-amine-loaded multiwalled carbon nanotubes. Langmuir 2011;27:8090−8.

[205] Rodriguez NM. A review of catalytically grown carbon nanofibers. J Mater Res 2011;8:3233−50.

[206] Yoon S-H, Lim S, Song Y, Ota Y, Qiao W, Tanaka A, et al. KOH activation of carbon nanofibers. Carbon 2004;42:1723−9.

[207] Loutfy RO, Wexler EM, Li W. Unique fullerene-based highly microporous carbons for gas storage. In: Ōsawa E, editor. Perspectives of fullerene nanotechnology. Dordrecht: Springer Netherlands; 2002. p. 293−303.

[208] Mochalin VN, Shenderova O, Ho D, Gogotsi Y. The properties and applications of nanodiamonds. Nat Nanotechnol 2012;7:11−23.

[209] Chen W, Wang G, Qin S, Wang C, Fang J, Qi J, et al. The nonlinear optical properties of coupling and decoupling graphene layers. AIP Adv 2013;3:042123.

[210] Zhu Y, Murali S, Cai W, Li X, Suk JW, Potts JR, et al. Graphene and Graphene Oxide: Synthesis, Properties, and Applications. Adv Mater 2010;22:3906−24.

[211] Jaiswal MKK, Graphene M. A review of optical properties and photonic applications. Asian J Phys 2016;25:809−31.

[212] Boukhvalov DW, Katsnelson MI. Modeling of graphite oxide. J Am Chem Soc 2008;130:10697−701.

[213] Johari P, Shenoy VB. Modulating optical properties of graphene oxide: role of prominent functional groups. ACS Nano 2011;5:7640−7.

[214] Hashim U, Farehanim MA, Azizah N, Norhafiezah S, Fatin MF, Ruslinda AR, et al. Optical properties of MWCNTs dispersed in various solutions. In 2015 2nd International Conference on Biomedical Engineering (ICoBE); 2015. p. 1−3.

[215] Zhu W, Sun F, Goei R, Zhou Y. Construction of WO3−g-C3N4 composites as efficient photocatalysts for pharmaceutical degradation under visible light. Catal Sci Technol 2017;7:2591−600.

[216] Usoltseva LO, Volkov DS, Nedosekin DA, Korobov MV, Proskurnin MA, Zharov VP. Absorption spectra of nanodiamond aqueous dispersions by optical absorption and optoacoustic spectroscopies. Photoacoustics 2018;12:55−66.

[217] Inani H, Singhal R, Sharma P, Vishnoi R, Aggarwal S, Sharma GD. Effect of low fluence radiation on nanocomposite thin films of Cu nanoparticles embedded in fullerene C60. Vacuum 2017;142:5−12.

[218] Lee C, Wei X, Kysar JW, Hone J. Measurement of the elastic properties and intrinsic strength of monolayer graphene. Science 2008;321:385−8.

[219] Zhang P, Ma L, Fan F, Zeng Z, Peng C, Loya PE, et al. Fracture toughness of graphene. Nat Commun 2014;5:3782.

[220] Yu M-F, Lourie O, Dyer MJ, Moloni K, Kelly TF, Ruoff RS. Strength and breaking mechanism of multiwalled carbon nanotubes under tensile load. Science 2000;287:637−40.

[221] Sun X, Sun H, Li H, Peng H. Developing polymer composite materials: carbon nanotubes or graphene? Adv Mater 2013;25:5153−76.

[222] Li Y, Yang T, Yu T, Zheng L, Liao K. Synergistic effect of hybrid carbon nantube−graphene oxide as a nanofiller in enhancing the mechanical properties of PVA composites. J Mater Chem 2011;21:10844−51.

[223] Park S, Ruoff RS. Chemical methods for the production of graphenes. Nat Nanotechnol 2009;4:217−24.

[224] Novoselov KS, Fal'ko VI, Colombo L, Gellert PR, Schwab MG, Kim K. A roadmap for graphene. Nature 2012;490:192−200.

[225] Tang L, Li X, Ji R, Teng KS, Tai G, Ye J, et al. Bottom-up synthesis of large-scale graphene oxide nanosheets. J Mater Chem 2012;22:5676−83.

[226] Eda G, Fanchini G, Chhowalla M. Large-area ultrathin films of reduced graphene oxide as a transparent and flexible electronic material. Nat Nanotechnol 2008;3:270−4.

[227] Pei S, Zhao J, Du J, Ren W, Cheng H-M. Direct reduction of graphene oxide films into highly conductive and flexible graphene films by hydrohalic acids. Carbon 2010;48:4466−74.

[228] Wang Y, Weng GJ. Electrical conductivity of carbon nanotube- and graphene-based nanocomposites. In: Meguid SA, Weng GJ, editors. Micromechanics and nanomechanics of composite solids. Cham: Springer International Publishing; 2018. p. 123−56.

[229] Bronnikov S, Podshivalov A, Kostromin S, Asandulesa M, Cozan V. Electrical conductivity of polyazomethine/fullerene C60 nanocomposites. Phys Lett A 2017;381:796−800.

[230] Piña-Salazar E-Z, Sagisaka K, Hattori Y, Sakai T, Futamura R, Ōsawa E, et al. Electrical conductivity changes of water-adsorbed nanodiamonds with thermal treatment. Chem Phys Letters: X 2019;2:100018.

[231] Cardoso P, Silva J, Paleo AJ, van Hattum FWJ, Simoes R, Lanceros-Méndez S. The dominant role of tunneling in the conductivity of carbon nanofiber-epoxy composites. Phys Stat Solidi (a) 2010;207:407−10.

[232] Zhang S, Yin S, Rong C, Huo P, Jiang Z, Wang G. Synergistic effects of functionalized graphene and functionalized multi-walled carbon nanotubes on the electrical and mechanical properties of poly(ether sulfone) composites. Eur Polym J 2013;49:3125−34.

[233] Renteria JD, Ramirez S, Malekpour H, Alonso B, Centeno A, Zurutuza A, et al. Strongly anisotropic thermal conductivity of free-standing reduced graphene oxide films annealed at high temperature. Adv Funct Mater 2015;25:4664−72.

[234] Stankovich S, Dikin DA, Dommett GHB, Kohlhaas KM, Zimney EJ, Stach EA, et al. Graphene-based composite materials. Nature 2006;442:282−6.

[235] Deep N, Mishra P. Fabrication and characterization of thermally conductive PMMA/MWCNT nanocomposites. Mater Today: Proc 2018;5:28328−36.

[236] Han Z, Fina A. Thermal conductivity of carbon nanotubes and their polymer nanocomposites: a review. Prog Polym Sci 2011;36:914−44.

[237] Tea NH, Yu R-C, Salamon MB, Lorents DC, Malhotra R, Ruoff RS. Thermal conductivity of C60 and C70 crystals. Appl Phys A 1993;56:219−25.

[238] Chatterjee S, Nafezarefi F, Tai NH, Schlagenhauf L, Nüesch FA, Chu BTT. Size and synergy effects of nanofiller hybrids including graphene nanoplatelets and carbon nanotubes in mechanical properties of epoxy composites. Carbon 2012;50:5380−6.

[239] Yang S-Y, Ma C-CM, Teng C-C, Huang Y-W, Liao S-H, Huang Y-L, et al. Effect of functionalized carbon nanotubes on the thermal conductivity of epoxy composites. Carbon 2010;48:592−603.

[240] Yang S-Y, Lin W-N, Huang Y-L, Tien H-W, Wang J-Y, Ma C-CM, et al. Synergetic effects of graphene platelets and carbon nanotubes on the mechanical and thermal properties of epoxy composites. Carbon 2011;49:793−803.

CHAPTER EIGHT

Zero-dimensional carbon nanomaterials-based adsorbents

8.1 Introduction

Myriad kinds of carbon nanomaterials (CNMs) with varied functionalities, such as carbon nanotubes, graphene and derivatives, carbon nanofibers, nanodiamonds, fullerenes, and nanoporous carbons, have been produced and widely employed in various applications, including water treatments [1−7]. Zero-dimensional (0D) CNMs, namely nanodiamonds (NDs), fullerenes, and carbon dots (CDs), have garnered significant attention in a diverse array of applications, including printing ink, sensors, electronics, bioimaging, batteries, adsorption, and photocatalysis. Ease in surface modification, high surface area, functional groups available on the surface, and the electrochemical and photoelectric properties of 0D-CNMs make them suitable materials for environmental pollution remediation [4,8−12].

NDs are nanoscale diamonds (average particle size 3−5 nm) with a complex diamond core structure and an amorphous carbon shell with sp^3 hybridized carbons [13]. NDs can be synthesized by detonation [14,15], high-energy ball milling of diamond microcrystals under high pressure and temperature [16], plasma-assisted CVD [17], laser ablation [18], ultrasound cavitation [19], and ion irradiation of graphite [20]. Due to their high chemical stability, nontoxicity, inertness, high specific area (> 200 m^2/g), morphology, and high affinity for organic molecules, NDs are attractive for applications in water contaminant adsorption and water purification [21−25]. Adsorption potential of NDs has also been exploited in the medical field, by adsorbing drugs onto the surface of an ND carrier for release at a targeted location [26].

Fullerenes, also called C_{60} or Buckyballs, were first observed in 1985 and quickly became a topic of great interest to researchers [27]. C_{60} is a sphere composed of 60 carbon atoms, with a football-like structure. The C_{60} structure consists of 32 faces, of which 12 are pentagons and 20 are hexagons, has a diameter of approximately 7Å, and truncated icosahedral symmetry. Subsequently, other forms of fullerene with different numbers of carbons, including C_{20}, C_{70}, C_{80}, and C_{94}, have been reported [28,29]. Fullerenes can be synthesized by laser ablation [27], arc discharge methods with graphite electrodes [30], thermal plasma synthesis [31], combustion synthesis [32], and CVD [33]. Fullerenes have a high surface area and are cost-effective,

Carbon Nanomaterial-Based Adsorbents for Water Purification.
DOI: https://doi.org/10.1016/B978-0-12-821959-1.00008-8

environment-friendly, have a small size, ordered structures, excellent mechanical and electro/thermoconductive properties, and an electron–accepting nature, which makes them attractive for use in environmental applications [34]. C_{60} is the most studied material in the fullerene family and has been investigated as an adsorbent for environmental remediation [35–37]. The π electrons present in the inner and outer spheres of fullerene interact with contaminants via π–π stacking. The adsorptive characteristics of C_{60} can be tailored by controlling surface chemical functionalities and pore structures [38,39].

Carbon dots (CDs) or carbon quantum dots, a new class material of the CNMs family have found a wide array of applications (e.g., sensors, bioimaging, and photocatalysis) due to their extraordinary advantages like biocompatibility, chemical stability, tunable fluorescence, and high optical absorptivity [40]. They can easily be prepared to form both top-down and bottom-up synthesis strategies. Among others, owing to easy controls of reactions parameters, hydrothermal/solvothermal and microwave pyrolysis methods are mostly employed for CDs preparation at low temperature, considering carbon precursors such as ginger, glucose, orange juice, food waste, soy milk, waste paper, citric acid, frying oil, and various kinds of polymers as well as CNMs (e.g., graphene, CNTs, and fullerene) [40–48]. Depending on their starting materials, they are classified as polymer dots, graphene quantum dots (GQDs) and carbon nanodots. CDs are uniform spherical nanoparticles with a C-based skeleton and an abundant amount of oxygen-carrying functional groups on the surface [49]. Owing to nontoxicity, insolubility in water, high conductivity, and ample surface functionalities, there have been attempts to utilize them in water purification using adsorption and photocatalysis methods [8,50].

0D-CNMs have a very small size (<10 nm) and spherical morphologies (Fig. 8.1). They have a tendency to aggregate irreversibly that leads to a significant decrease in surface area and adsorption performance. They also have superb substrate-free water-dispersed properties, which leads to difficulties in their removal from the water after adsorption and reusability, and thus actively hinders their direct application in water purification. Numerous effective strategies to eliminate the aggregation of 0D-CNM

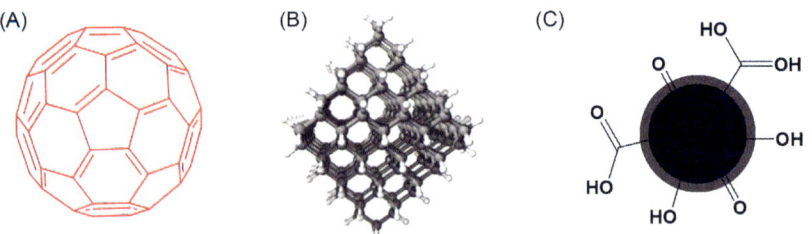

Figure 8.1 The structures of 0D-CNMs such as fullerene (A), octahedral NDs (B), and CDs (C).

in aqueous solutions and surface functionalization or immobilization of 0D-CNM on the proper substrate to enhance the adsorption characteristics, separability, and reusability have been investigated in detail. Therefore this chapter will discuss the recent advances in the adsorption performance of 0D-CNMs (CDs, fullerenes, and NDs) in the elimination of organic and inorganic pollutants from aqueous solutions.

8.2 Nanodiamonds as adsorbents

NDs are mostly produced by detonation, using high-pressure/high-temperature and CVDs approaches [51]. As NDs consist of the same core sp^3 structure, they exhibit optical and mechanical properties similar to bulk diamond. They have properties of high thermal conductivity, high stiffness, high Young's modulus, high refractive index, and stability in harsh environments [51,52]. The high-temperature treatments can induce surface graphitization, which leads to a loss in surface functionalities. NDs obtained by a detonation (DND) process offer the highest variability in functional groups, such as carboxylic, lactone, epoxide, and hydroxyl groups. Whereas NDs prepared by high-pressure/high-temperature (HPHT-ND) processes have largely surface functionalities of hydroxyl and carboxylic groups (Fig. 8.2A) [53]. The versatile surface chemistry of NDs allows the tuning of both their microscopic (viz., reactivity) and macroscopic properties (viz., colloidal stability and hydrophilicity) using various surface modification, enabling the preparation of NDs appropriate for water purification (Fig. 8.2B).

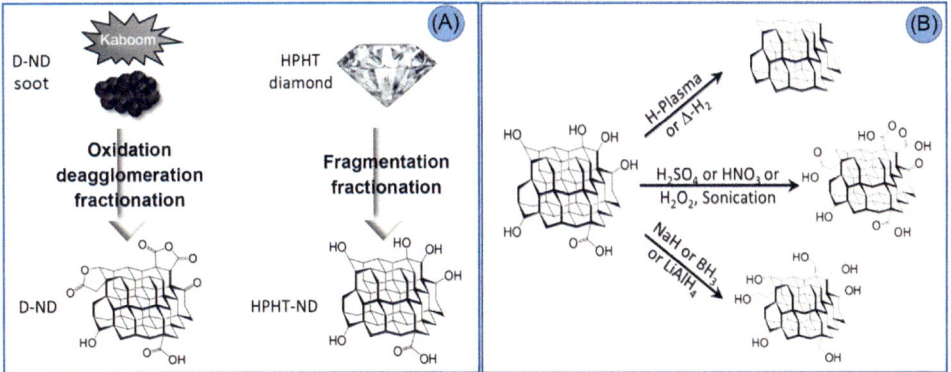

Figure 8.2 (A) The diagram shows the synthesis and postsynthetic treatments of D-NDs and HPHT-NDs. (B) Surface oxidation and reduction approach used for NDs modification. *Reproduced with permission from Reina G, Zhao L, Bianco A, Komatsu N. Chemical functionalization of nanodiamonds: opportunities and challenges ahead. Angew Chem Int Ed. 2019;58:17918–17929. Copyrights 2019, Wiley.*

Like other members of the carbon family, nanodiamonds (NDs) can be used as adsorbents to remove contaminants from wastewater. The excess availability of oxygen-carrying functional groups on the ND surface makes it a potential adsorbent for the removal of organic and inorganic contaminants through electrostatic interaction, hydrogen bonding, and complexation. Functionalized NDs have recently been employed as adsorbents to determine their adsorption potential for contaminants from wastewater. For example, Zhao et al. modified oxidized ND using single-armed (SA) and double-armed (DA) benzoylthiourea ligands for the adsorption of different metal ions. Due to moderate complexation tendency and high flexible spatial configuration, single–armed (ND-SA) nanostructures have exhibited better adsorption selectivity for UO_2 (II) than double-armed (ND-DA) nanostructures [25] (Fig. 8.3A). Badea et al. studied the

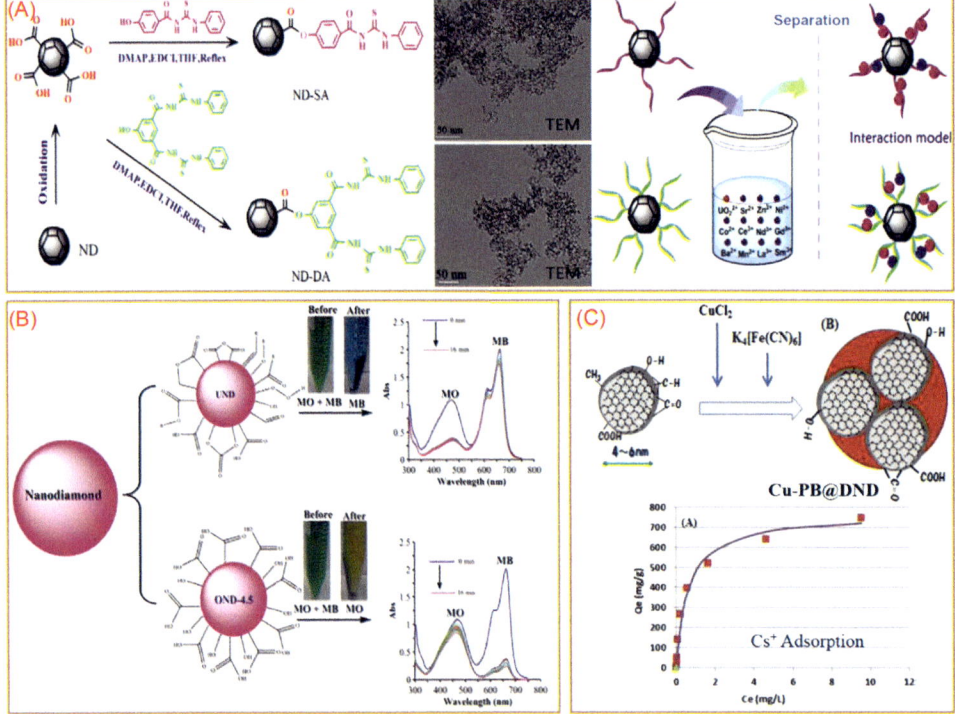

Figure 8.3 Modifications of ND using various strategies: (A) Functionalization of ND particles using single/double-armed benzoylthiourea ligands for removal of metal ions, TEM image of ND-SA and ND-DA, and interaction behavior of metals ions [25]. (B) Rapid and selective adsorption of dyes (MB and MO) using untreated nanodiamonds and thermally oxidized nanodiamonds [54]. (A and B) Reproduced with permission from Elsevier Science Ltd. (C) Detonated ND as support for Cu-Prussian Blue (Cu-PB@DND) for removal of Cs^+ from simulated contaminated water. *Reproduced with permission from Matsumoto K, Yamato H, Kakimoto S, Yamashita T, Wada R, Tanaka Y, et al. A highly efficient adsorbent Cu-Perusian blue@nanodiamond for cesium in diluted artificial seawater and soil-treated wastewater. Sci Rep. 2018;8:5807. Copyright 2018, Nature.*

adsorption of an azo dye (acid orange 7 (AO7)) onto the ND surface [23]. The selective adsorption of azo molecules (AO7 and sulfasalazine (SSZ)) in the presence of other molecules (5-aminosalicylic acid (5-ASA) and sulfapyridine (SPY)) was also examined. ND had adsorption capacities of 1288 and 925 mmol/kg for AO7 and SSZ in wastewater, respectively. In addition to electrostatic interactions between the ND and the azo molecule, hydrogen bonding also played a significant role in the adsorption. However, the positive surface charge of NDs renders it less effective as an adsorbent for cationic species. Nevertheless, surface modifications can alter the surface charges and selective adsorption capacities of NDs. Pourghaderi et al. reported the selective adsorption of organic dyes (MB, MO) on thermally oxidized NDs [54]. MB dye was successfully and preferentially adsorbed to the oxidized NDs (ONDs) over MO dye, due to the reduction in the zeta potential of NDs with thermal oxidation (Fig. 8.3B). A low zeta potential indicates that the ND surface is negative, which efficiently attracts the cationic dye (MB) via electrostatic interactions. In contrast, when untreated ND (UND) was used as an adsorbent for a mixture of MB and MO in contaminated water, MO was selectively adsorbed within 16 minutes, leaving a blue aqueous solution containing MB. In another report, Fujimura et al. used NDs as support for Cu-Prussian Blue (Cu-PB@DND) to study cesium adsorption from simulated contaminated water [55] (Fig. 8.3C). The use of air-treated detonated ND resulted in a negative zeta potential for Cu-PB@DND, whereas positive zeta values were observed for NaOH-treated and untreated detonated ND. Consequently, the adsorption capacity of Cu-PB@DND for positively charged Cs (I) was high (759 mg/g) due to the negative surface charge and high surface area. Similarly, Yang et al. modified ND using the ionic liquid 3-n-hexadecyl-1-vinylimidazolium bromide ([C$_{16}$VIm$^+$] [Br$^-$]) through a simple and effective Michael addition reaction [11]. The as-made ND@IL composite was employed for the removal of Congo red dye (CR) with high removal efficiency (92.78%) from a neutral solution at room temperature. The CR removal capacities of pristine ND and ND@IL are 61.1 and 226.4 mg/g, respectively. The high adsorption performance of composite was attributed to electrostatic interaction and hydrophobic interaction between alkyl chains and aromatic rings of CR [11].

8.3 Fullerene as adsorbents

Owing to their high surface area, specific morphology, small size, ordered structures, and good mechanical properties, fullerenes have been applied in water purification for the elimination of various pollutants from aqueous solution. The abundant π electrons in the inner and outer spheres are responsible for the high adsorption of pollutants through $\pi-\pi$ interactions. The adsorption performance of pristine fullerene can be improved by surface functionalization and forming nanocomposites with highly reactive inorganic compounds. However, studies on the adsorption capacity of fullerene for pollutants in water treatment have been limited. Tomson et al. studied the absorption potential of nanosized aggregates of C_{60} fullerene for naphthalene and 1,2-dichlorobenzene (1,2-DCB) in comparison with that of larger fullerene aggregates [37]. The adsorption and desorption of contaminants on the C_{60} exhibited hysteresis. Adsorption/desorption hysteresis was described by a two-compartment desorption model, which includes, first, the adsorption of contaminants onto the external surface of the adsorbent, which is in contact with water, and second, adsorption to the internal surfaces of the aggregates. The adsorption of polyaromatic compounds (pyrene, naphthalene, and phenanthrene) onto C_{60} can be fitted to the Polanyi-Manes model (PMM) [36]. C_{60} was found to adsorb phenols (phenol 10%, 3,4-dimethylphenol 48%, 2-tert-butylphenol 54%, 2-isopropoxyphenol 62%, and 4-chlorophenol 45%), polyaromatic compounds (naphthalene 22%, phenanthrene 37%, anthracene 42%, and perylene 62%), and amines (methylethylenediamine 10%, aniline 14%, and 1,2-phenylenediamine 28%) successfully [56].

Elessawy et al. proposed a new facile, one-step route for the preparation of functionalized magnetic fullerene nanocomposite (FMFN) through catalytic thermal decomposition of polyethylene terephthalate (PET) bottle wastes [57]. Here, the waste PET was used as a feedstock, while the ferrocene was considered to be a catalyst and precursor of Fe_3O_4 (Fig. 8.4A). The as-made nanocomposite has demonstrated a relatively high surface area (336.84 m^2/g) with micropores and mesopores. FMFN exhibited a substantial adsorption rate for ciprofloxacin (CIP) removal from water and adsorption followed a pseudo-second-order kinetic model and obeyed Langmuir and Freundlich isotherm models (Fig. 8.4A). The thermodynamic study suggested that the CIP removal process was exothermic and spontaneous. The adsorption of CIP onto FMFN is governed by $\pi-\pi$ electron donor–acceptor (EDA) interactions, hydrogen bonding, electrostatic interactions between CIP and the adsorbent surfaces, as well as high surface area and porosity (Fig. 8.4C). Later, the same research group studied FMFN for the removal of cationic methylene blue dye (MB), and anionic dye Acid Blue 25 dye (AB25) from aqueous water [58].

8.4 Carbon dots as adsorbents

Carbon dots (CDs), the latest member of the nanocarbon family, have garnered huge attention in a wide array of applications, including fluorescent probes, biosensing, drug delivery, bioimaging, and photovoltaic devices, due to their exceptional optical

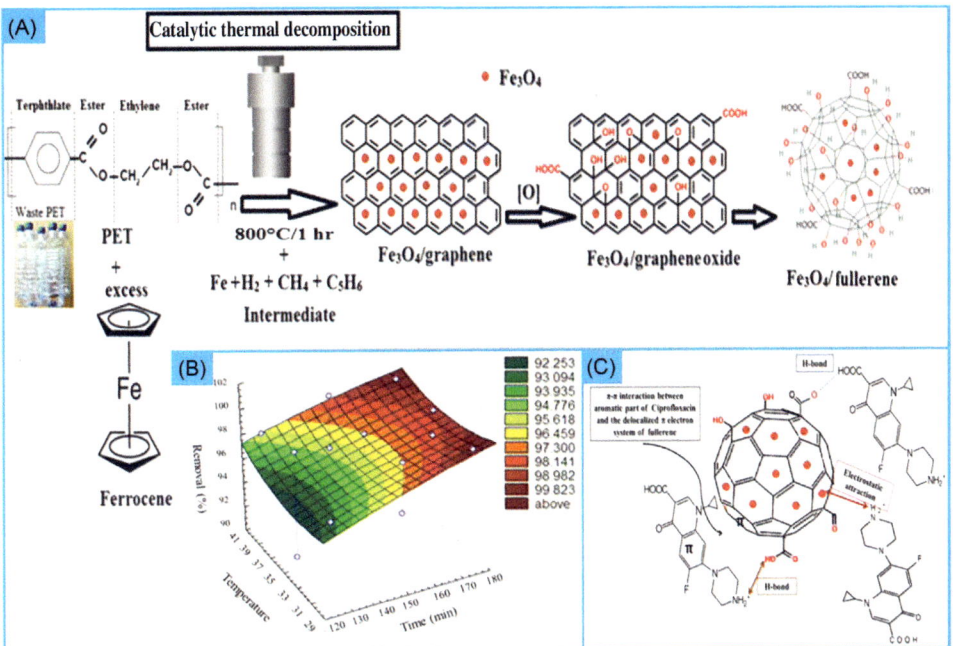

Figure 8.4 (A) Diagram represents the one-step preparation of functionalized magnetic fullerene nanocomposites (FMFN) through catalytic thermal dissociation of waste PET. (B) The three-dimensional plot depicts the CIP removal (%) versus process variables (temperature and time) onto as-prepared FMFN. (C) Various interaction mechanisms involved in ciprofloxacin removal using FMFN adsorbent. *Reproduced with permission from Elessawy NA, Elnouby M, Gouda MH, Hamad HA, Taha NA, Gouda M, et al. Ciprofloxacin removal using magnetic fullerene nanocomposite obtained from sustainable PET bottle wastes: adsorption process optimization, kinetics, isotherm, regeneration and recycling studies. Chemosphere. 2020;239:124728. Copyrights 2020, Elsevier.*

properties, high chemical stability, high dispersibility, and low toxicity [49]. Adsorption performance of CDs is attributed to the available functional groups such as hydroxyl, carbonyl, epoxide, and carboxylic groups on the surface (Fig. 8.5) [60]. Various strategies have been developed to modify CDs surface via functionalization and form nanocomposites to improve the adsorption characteristics for efficient removal of inorganic and organic contaminants.

Kang et al. demonstrated the hydrothermal synthesis of CDs decorated magnetic $ZnFe_2O_4$ nanoparticles (CDs/ZFO) composite for the removal of methyl orange (MO) dye from water [61]. The composite with 5 wt% CDs/ZFO exhibited the highest adsorption performance (181.2 mg/g) following the Langmuir isotherm model with monolayer adsorption. The significantly increased adsorption capacity of CDs/ZFO composites was credited to ample oxygen-carrying functional groups on the surface of CDs. In another study, heavy metal Ni^{2+} ions were effectively removed from waste-water using CDs, which were synthesized from frying oil using the hydrothermal

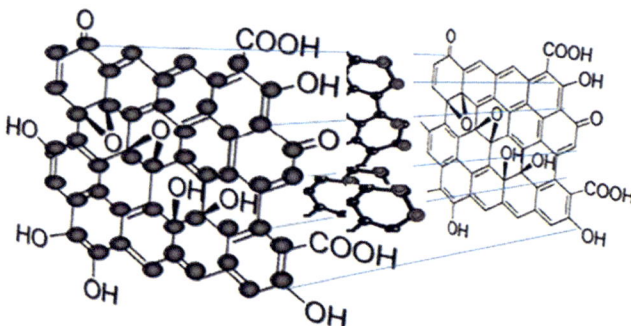

Figure 8.5 Chemical structure of CDs. *Reproduced with permission from Demchenko AP, Dekaliuk MO. Novel fluorescent carbonic nanomaterials for sensing and imaging. Methods Appl Fluorescence. 2013;1:042001 [59]. Copyright 2013, IOP science.*

method [8]. Basavaiah et al. fabricated N-doped carbon quantum dots-based magnetite nanocomposite (Fe$_3$O$_4$@NCQDs NCs) using cost-effective lemon juice as a starting material for the elimination of MB from aqueous solution [62]. Ethylenediamine was used for nitrogen doping of CQDs using a hydrothermal method. The average size of spherical Fe$_3$O$_4$@NCQDs NCs was determined to be 5 nm using a high-resolution transmission electron microscope. Batch adsorption experiments exhibited improved fast removal of MB dye with an adsorption efficiency of $\sim 90.84\%$ within 20 minutes at optimum conditions. The adsorption experiments demonstrated proper fitting with Freundlich isotherms and pseudo-second-order kinetics. Wang et al. investigated the efficient removal of U(VI) ions using layered double oxides (LDO) and LDO/CDs nanocomposite, which were synthesized by a simple calcination approach at 500°C for 6 hours under N$_2$ conditions [63]. CDs were successfully grown on flower-like morphology of LDO (Mg-Al oxides) (Fig. 8.6A and B) using chitosan as CDs source. The adsorption experiments suggested that U(VI) removal on LDO/CDs nanocomposites was independent of ionic strength at pH < 6, and considerably dependent on ionic strength and pH at pH > 6 (Fig. 8.6C). The LDO/CDs nanocomposites showed enhanced adsorption capacity (354.2 mg/g) compared with LDO (237.6 mg/g) for U(VI) uptake at pH = 5.0 and $T = 298$K due to the incorporation of CDs, the high number of oxygen functionalities, and improved surface area, which were confirmed by X-ray photoelectron spectroscopy (XPS), extended X-ray absorption fine structure (EXAFS), and Brunauer—Emmett—Teller (BET) analysis (Fig. 8.6D and E). Thus LDO/CDs nanocomposites are promising materials for the uptake of radionuclides through precipitation, ion-exchange, surface complexation, and electrostatic interactions (Fig. 8.6F). Moreover, CDs are mostly used in photocatalyst applications rather than as adsorbents. The richness in surface functional groups has the potential to enhance the adsorption characteristics of any support material for the elimination of numerous contaminants from water.

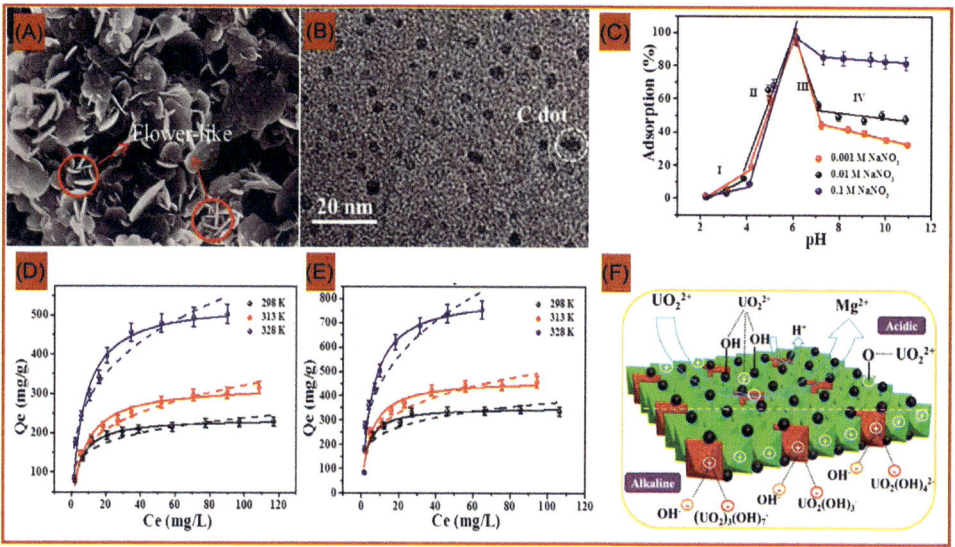

Figure 8.6 (A) SEM image, (B) HRTEM image, and (C) pH study of LDO/CDs nanocomposite. Adsorption isotherms of U(VI) uptake on LDO (D) and LDO/CDs nanocomposite (E). (F) Schematic shows the U(VI) adsorption mechanism. *Reproduced with permission from Yao W, Wang X, Liang Y, Yu S, Gu P, Sun Y, et al. Synthesis of novel flower-like layered double oxides/carbon dots nanocomposites for U(VI) and 241Am(III) efficient removal: Batch and EXAFS studies. Chem Eng J. 2018;332:775−786. Copyright 2018, Elsevier.*

8.5 Conclusion

The zero-dimensional carbon nanomaterials (0D-CNMs) include fullerenes, NDs, and CDs and have a small size (<10 nm) with spherical morphologies. 0D-CNMs can chemically amendable to improve and render some special functional characteristics. Various properties such as high surface area, availability of a large number of functional groups on the surface, ease in surface modification, and the excellent electrochemical and photoelectric features of 0D-CNMs make them suitable materials for environmental remediation, including wastewater treatments. NDs are mostly produced by detonation, using high–pressure/high–temperature, and CVDs approaches. Owing to their high chemical stability, nontoxicity, inertness, and high specific area, NDs have demonstrated good affinity toward water contaminants. Due to their high cost and complicated synthetic processes, literature has shown limited adsorption studies for water purification using NDs. Fullerenes can be synthesized by laser ablation, arc discharge methods with graphite electrodes, thermal plasma synthesis, combustion synthesis, and CVD method. They offer significant characteristics such as high surface area, small size, ordered structures, and excellent mechanical properties as well as an excess of π electrons in their inner and outer spheres for adsorptive removal of pollutants from water. Furthermore, CDs, the newest member of the carbon family, are generally prepared by hydrothermal/solvothermal and microwave pyrolysis methods using various carbon precursors. The exceptional optical properties, high chemical stability, high dispersibility, low toxicity, and excess number of functional groups (e.g., $-COOH$, $-OH$, and $-C = O$) on the surface of CDs, make them promising materials for water purification.

Mostly, 0D-CNMs have been utilized in water purification after postfunctionalization or formation of a composite with other reactive materials. It is noteworthy that they can impose serious health risks due to the small size of their particles and their high dispersibility. Magnetization of these materials is considered to be the best strategy because it promises fast and complete recovery of adsorbents as well as pollutants. Overall, 0D-CNMs have not been explored much, although they have strong potential to eliminate pollutants from wastewater. In future, research needs to be directed toward making effective three-dimensional superstructures by incorporating 0D-CNMs for enhanced adsorption performance. Additionally, an integrated system or device may be fabricated utilizing 0D-CNMs-based composites for water purification by considering the combined functions of detection, energy harvesting, and treatment.

References

[1] Gusain R, Kumar N, Ray SS. Recent advances in carbon nanomaterial-based adsorbents for water purification. Coord Chem Rev 2020;405:213111.

[2] Maiti D, Tong X, Mou X, Yang K. Carbon-based nanomaterials for biomedical applications: a recent study. Front Pharmacol. 2019;9.

[3] Bai L, Zhang Y, Tong W, Sun L, Huang H, An Q, et al. Graphene for energy storage and conversion: synthesis and interdisciplinary applications. Electrochem Energy Rev 2019.

[4] Speranza G. The role of functionalization in the applications of carbon materials: an overview. C. J Carbon Res 2019;5:84.

[5] Mukwevho N, Gusain R, Fosso-Kankeu E, Kumar N, Waanders F, Ray SS. Removal of naphthalene from simulated wastewater through adsorption-photodegradation by ZnO/Ag/GO nanocomposite. J Ind Eng Chem 2020;81:393—404.

[6] Gusain R, Kumar N, Fosso-Kankeu E, Ray SS. Efficient removal of Pb(II) and Cd(II) from industrial mine water by a hierarchical MoS_2/SH-MWCNT nanocomposite. ACS Omega 2019;4:13922—35.

[7] Verma S, Mungse HP, Kumar N, Choudhary S, Jain SL, Sain B, et al. Graphene oxide: an efficient and reusable carbocatalyst for aza-Michael addition of amines to activated alkenes. Chem Commun 2011;47:12673—5.

[8] Aji MP, Wiguna PA, Karunawan J, Wati AL, Sulhadi. Removal of heavy metal nickel-ions from wastewaters using carbon nanodots from frying oil. Procedia Eng 2017;170:36—40.

[9] Vorobiev AK, Gazizov RR, Borschevskii AY, Markov VY, Ioutsi VA, Brotsman VA, et al. Fullerene as photocatalyst: visible-light induced reaction of perfluorinated α,ω-diiodoalkanes with C_{60}. J Phys Chem A 2017;121:113—21.

[10] Cheng X, Kan AT, Tomson MB. Naphthalene adsorption and desorption from aqueous C_{60} fullerene. J Chem Eng Data 2004;49:675—83.

[11] Yang G, Huang H, Chen J, Gan D, Deng F, Huang Q, et al. Preparation of ionic liquids functionalized nanodiamonds-based composites through the Michael addition reaction for efficient removal of environmental pollutants. J Mol Liq 2019;296:111874.

[12] Coro J, Suárez M, Silva LSR, Eguiluz KIB, Salazar-Banda GR. Fullerene applications in fuel cells: a review. Int J Hydrog Energy 2016;41:17944−59.

[13] Zhang Y, Rhee KY, Hui D, Park S-J. A critical review of nanodiamond based nanocomposites: synthesis, properties and applications. Compos Part B: Eng 2018;143:19−27.

[14] Kovalenko I, Bucknall DG, Yushin G. Detonation nanodiamond and onion-like-carbon-embedded polyaniline for supercapacitors. Adv Funct Mater 2010;20:3979−86.

[15] Krueger A, Stegk J, Liang Y, Lu L, Jarre G. Biotinylated nanodiamond: simple and efficient functionalization of detonation diamond. Langmuir 2008;24:4200−4.

[16] Lai L, Barnard A. Functionalized nanodiamonds for biological and medical applications, J Nanosci Nanotechnol 2015;15:989−99.

[17] Shi Y, Tan M, Jiang X. Deposition of diamond/β−SiC gradient composite films by microwave plasma-assisted chemical vapor deposition. J Mater Res 2011;17:1241−3.

[18] Amans D, Chenus A-C, Ledoux G, Dujardin C, Reynaud C, Sublemontier O, et al. Nanodiamond synthesis by pulsed laser ablation in liquids. Diam Relat Mater 2009;18:177−80.

[19] Mochalin VN, Shenderova O, Ho D, Gogotsi Y. The properties and applications of nanodiamonds. Nat Nanotechnol 2011;7:11.

[20] Daulton TL, Kirk MA, Lewis RS, Rehn LE. Production of nanodiamonds by high-energy ion irradiation of graphite at room temperature. Nucl Instrum Meth Phys Res Sect B: Beam Interact Mater At 2001;175-177:12−20.

[21] Shalaginov MY, Naik GV, Ishii S, Slipchenko MN, Boltasseva A, Cheng JX, et al. Characterization of nanodiamonds for metamaterial applications. Appl Phys B 2011;105:191.

[22] Gibson NM, Luo TJM, Shenderova O, Choi YJ, Fitzgerald Z, Brenner DW. Fluorescent dye adsorption on nanocarbon substrates through electrostatic interactions. Diam Relat Mater 2010;19:234−7.

[23] Wang H-D, Yang Q, Hui Niu C, Badea I. Adsorption of azo dye onto nanodiamond surface. Diam Relat Mater 2012;26:1−6.

[24] Spitsyn BV, Denisov SA, Skorik NA, Chopurova AG, Parkaeva SA, Belyakova LD, et al. The physical−chemical study of detonation nanodiamond application in adsorption and chromatography. Diam Relat Mater 2010;19:123−7.

[25] Zhao X, Zhang S, Bai C, Li B, Li Y, Wang L, et al. Nano-diamond particles functionalized with single/double-arm amide−thiourea ligands for adsorption of metal ions. J Colloid Interface Sci 2016;469:109−19.

[26] Giammarco J, Mochalin VN, Haeckel J, Gogotsi Y. The adsorption of tetracycline and vancomycin onto nanodiamond with controlled release. J Colloid Interface Sci 2016;468:253−61.

[27] Kroto HW, Heath JR, O'Brien SC, Curl RF, Smalley RE. C_{60}: buckminsterfullerene. Nature. 1985;318:162.

[28] Chang TM, Naim A, Ahmed SN, Goodloe G, Shevlin PB. On the mechanism of fullerene formation. Trapping of some possible intermediates. J Am Chem Soc 1992;114:7603−4.

[29] Scott L. Methods for the chemical synthesis of fullerenes, Angew Chem Int Ed 2004;43:4994−5007.

[30] Krätschmer W, Lamb LD, Fostiropoulos K, Huffman DR. Solid C_{60}: a new form of carbon. Nature. 1990;347:354.

[31] Wang C, Imahori T, Tanaka Y, Sakuta T, Takikawa H, Matsuo H. Synthesis of fullerenes from carbon powder by using high power induction thermal plasma. Thin Solid Films 2001;390:31−6.

[32] Howard JB, Lafleur AL, Makarovsky Y, Mitra S, Pope CJ, Yadav TK. Fullerenes synthesis in combustion. Carbon. 1992;30:1183−201.

[33] Wang X, Xu B, Liu X, Guo J, Ichinose H. Synthesis of Fe-included onion-like Fullerenes by chemical vapor deposition. Diam Relat Mater 2006;15:147−50.

[34] Pan B, Lin D, Mashayekhi H, Xing B. Adsorption and hysteresis of bisphenol A and 17α-ethinyl estradiol on carbon nanomaterials. Environ Sci Technol 2008;42:5480−5.

[35] Gupta VK, Saleh TA. Sorption of pollutants by porous carbon, carbon nanotubes and fullerene- an overview. Environ Sci Pollut Res 2013;20:2828—43.

[36] Yang K, Zhu L, Xing B. Adsorption of polycyclic aromatic hydrocarbons by carbon nanomaterials. Environ Sci Technol 2006;40:1855—61.

[37] Cheng X, Kan AT, Tomson MB. Uptake and sequestration of naphthalene and 1,2-dichloroben-zene by C_{60}. J Nanopart Res 2005;7:555—67.

[38] Su DS, Centi G. A perspective on carbon materials for future energy application. J Energy Chem 2013;22:151—73.

[39] Chae S-R, Hotze EM, Wiesner MR. Chapter 21 - possible applications of fullerene nanomaterials in water treatment and reuse. In: Street A, Sustich R, Duncan J, Savage N, editors. Nanotechnology applications for clean water. 2nd ed. Oxford: William Andrew Publishing; 2014. p. 329—38.

[40] Wang X, Feng Y, Dong P, Huang J. A mini review on carbon quantum dots: preparation, proper-ties, and electrocatalytic application. Front Chem 2019;7.

[41] Liu ML, Chen BB, Li CM, Huang CZ. Carbon dots: synthesis, formation mechanism, fluorescence origin and sensing applications. Green Chem 2019;21:449—71.

[42] Aji M, Wiguna P, Susanto, Wicaksono R, Sulhadi S. Identification of carbon dots in waste cooking oil. Adv Mater Res 2015;1123:402—5.

[43] Wei J, Zhang X, Sheng Y, Shen J, Huang P, Guo S, et al. Simple one-step synthesis of water-soluble fluorescent carbon dots from waste paper. N J Chem 2014;38:906—9.

[44] Li C-L, Ou C-M, Huang C-C, Wu W-C, Chen Y-P, Lin T-E, et al. Carbon dots prepared from ginger exhibiting efficient inhibition of human hepatocellular carcinoma cells. J Mater Chem B 2014;2:4564—71.

[45] Tian R, Zhong S, Wu J, Jiang W, Wang T. Facile hydrothermal method to prepare graphene quan-tum dots from graphene oxide with different photoluminescences. RSC Adv 2016;6:40422—6.

[46] Kumar N, Kumar V, Swart HC, Mishra AK, Catherine Ngila J, Parashar V. Controlled microstruc-tural hydrothermal synthesis of strontium selenides host matrices for EuII and EuIII luminescence. Mater Lett 2015;146:51—4.

[47] Kumar N, Sinha Ray S, Ngila JC. Ionic liquid-assisted synthesis of Ag/Ag_2Te nanocrystals via a hydrothermal route for enhanced photocatalytic performance. N J Chem 2017;41:14618—26.

[48] Kumar N, George BPA, Abrahamse H, Parashar V, Ray SS, Ngila JC. A novel approach to low-temperature synthesis of cubic HfO_2 nanostructures and their cytotoxicity. Sci Rep 2017;7:9351.

[49] Lim SY, Shen W, Gao Z. Carbon quantum dots and their applications. Chem Soc Rev 2015;44:362—81.

[50] Han M, Zhu S, Lu S, Song Y, Feng T, Tao S, et al. Recent progress on the photocatalysis of carbon dots: classification, mechanism and applications. Nano Today 2018;19:201—18.

[51] Reina G, Zhao L, Bianco A, Komatsu N. Chemical functionalization of nanodiamonds: opportu-nities and challenges ahead. Angew Chem Int Ed 2019;58:17918—29.

[52] Krueger A. The structure and reactivity of nanoscale diamond. J Mater Chem 2008;18:1485—92.

[53] Neburkova J, Vavra J, Cigler P. Coating nanodiamonds with biocompatible shells for applications in biology and medicine. Curr Op Solid State Mater Sci 2017;21:43—53.

[54] Molavi H, Shojaei A, Pourghaderi A. Rapid and tunable selective adsorption of dyes using thermally oxidized nanodiamond. J Colloid Interface Sci 2018;524:52—64.

[55] Matsumoto K, Yamato H, Kakimoto S, Yamashita T, Wada R, Tanaka Y, et al. A Highly Efficient adsorbent Cu-Perusian blue@nanodiamond for cesium in diluted artificial seawater and soil-treated wastewater. Sci Rep 2018;8:5807.

[56] Ballesteros E, Gallego M, Valcárcel M. Analytical potential of fullerene as adsorbent for organic and organometallic compounds from aqueous solutions. J Chromatogr A 2000;869:101—10.

[57] Elessawy NA, Elnouby M, Gouda MH, Hamad HA, Taha NA, Gouda M, et al. Ciprofloxacin removal using magnetic fullerene nanocomposite obtained from sustainable PET bottle wastes: adsorption process optimization, kinetics, isotherm, regeneration and recycling studies. Chemosphere. 2020;239:124728.

[58] Elessawy NA, El-Sayed EM, Ali S, Elkady MF, Elnouby M, Hamad HA. One-pot green synthesis of magnetic fullerene nanocomposite for adsorption characteristics. J Water Process Eng 2019;101047.

[59] Demchenko AP, Dekaliuk MO. Novel fluorescent carbonic nanomaterials for sensing and imaging. Methods Appl Fluorescence 2013;1:042001.

[60] Liu R, Liu J, Kong W, Huang H, Han X, Zhang X, et al. Adsorption dominant catalytic activity of a carbon dots stabilized gold nanoparticles system. Dalton Trans 2014;43:10920−9.

[61] Shi W, Guo F, Wang H, Liu C, Fu Y, Yuan S, et al. Carbon dots decorated magnetic $ZnFe_2O_4$ nanoparticles with enhanced adsorption capacity for the removal of dye from aqueous solution. Appl Surf Sci 2018;433:790−7.

[62] Tadesse A, RamaDevi D, Hagos M, Battu G, Basavaiah K. Synthesis of nitrogen doped carbon quantum dots/magnetite nanocomposites for efficient removal of methyl blue dye pollutant from contaminated water. RSC Adv 2018;8:8528−36.

[63] Yao W, Wang X, Liang Y, Yu S, Gu P, Sun Y, et al. Synthesis of novel flower-like layered double oxides/carbon dots nanocomposites for U(VI) and 241Am(III) efficient removal: batch and EXAFS studies. Chem Eng J 2018;332:775−86.

CHAPTER NINE

One-dimensional carbon nanomaterials-based adsorbents

9.1 Introduction

The conventional adsorbents, such as activated carbon, iron oxide, silica, zero--valent iron, zeolite, titanium oxide, gums, fly ash, clays, granular biomass, and polymers, have been utilized for wastewater treatments, but all of these materials have limitations of low adsorption capacity, low reusability, and low levels of adsorption performance for heavy metals [1−5]. Thus the hunting of new efficient and effective low-cost adsorbents is continuing. A recent development in nanotechnology provides a varied opportunity for new emerging adsorption processes. Various cutting-edge nanomaterials are slowly finding applications in environmental remediation. Owing to their high surface area, high pore volume, and advantageous surface characteristics, carbon-based nanomaterials (CNMs) have been considered to be in the forefront of nanomaterials for water conservation and management [6]. In particular, one-dimensional (1D) CNMs, namely carbon nanotubes (CNTs), and carbon nanofibers (CNFs) are emerging as excellent adsorbents for water treatment because of their distinctive physical and chemical properties [7]. The 1D CNMs are quite stable compounds [8]. The pristine 1D CNMs have demonstrated reduced adsorption capacity for the elimination of various inorganic and organic pollutants, but introducing new functionalities, including nitrogen, oxygen, and sulfur-carrying functional groups, to their surface can hugely enhance their selectivity, and sensitivity toward heavy metals and other contaminants, resulting in high adsorption performance of 1D CNMs. Furthermore, the surface of 1D CNMs needs to be functionalized with active molecules and activated under high temperature to make them highly reactive toward various noxious water effluents. 1D CNMs are also modified with a wide range of nanoparticles (NPs) to make their nanocomposite, magnetic nanocomposite, and nanohybrids, which are employed for efficient removal of contaminants from water [1,9,10]. The magnetic nanocomposites of 1D CNMs offer advantages of easy recovery of complete adsorbent after adsorption and high reusability [6,11]. Overall, these modifications in 1D CNMs might improve the surface properties (pore size/volume and surface area), improve surface functionalities, and boost their structural stability. Herein, the main objective of this chapter is to deliver a comprehensive review of

Carbon Nanomaterial-Based Adsorbents for Water Purification.
DOI: https://doi.org/10.1016/B978-0-12-821959-1.00009-X

structures, adsorption properties, and surface modifications of 1D CNMs and related removal performances to inorganic and organic effluents in aqueous solutions.

9.2 One-dimensional carbon nanomaterials

The 1D CNMs comprise single-walled carbon nanotubes (SWCNTs) multi-walled carbon nanotubes (MWCNTs), nanohorns and CNFs. CNTs are hollow, one-dimensional (1D) tube-like structures, with thin graphitic sp^2 carbon walls, which have fascinating mechanical, electrical, and magnetic properties [12]. CNTs have nanosized diameters (<100 nm) with high aspect ratios. They can be thought of as graphene nanosheets rolled into cylindrical tubes, and can be categorized as SWCNTs and MWCNTs [13]. As depicted in Fig. 9.1, three different types of SWCNTs (namely, chiral, armchair, and zigzag) can be achieved by rolling up graphene sheets, based on the alignment of the tube axis with respect to the hexagonal lattice. As represented by the chiral indices (n, m), chiral ($n \neq m$) possesses two enantiomers with left- and right-handed helicity, while zigzag ($m = 0$) and armchair ($n = m$) both have mirror symmetry. In the case of MWCNTs, they consist of a cylindrical multilayer of graphene, and each layer can have a unique chirality.

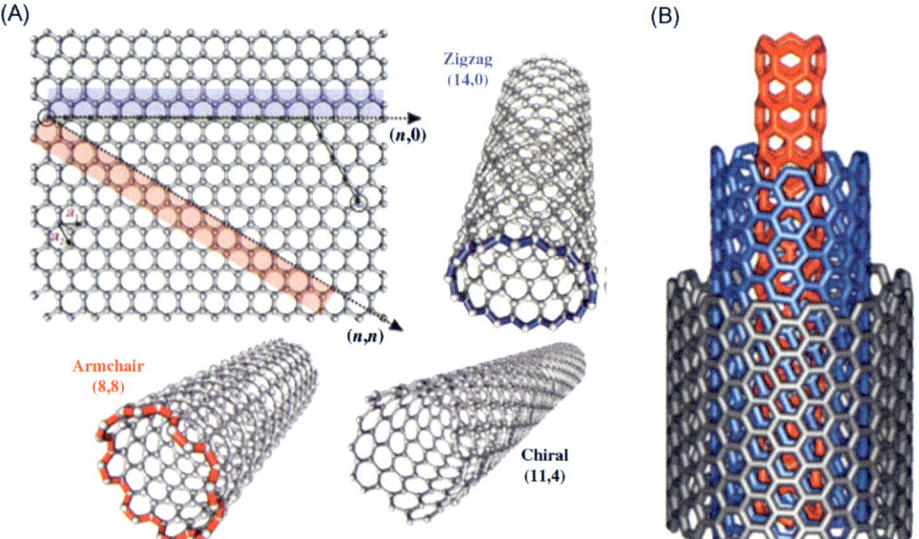

Figure 9.1 (A) Three different types of SWCNTs, namely, chiral, armchair, and zigzag can be achieved by rolling up graphene sheet; (B) structure of MWCNTs consists of three-layers of different chirality. *Reproduced with permission from Balasubramanian K, Burghard M. Chemically functionalized carbon nanotubes. Small. 2005;1:180–192. Copyright 2004, Wiley.*

Iijima discovered CNTs in 1991. The first synthesized CNTs were multiwalled, made up of several long graphitic shells with intershell separation gap of ~ 0.34 nm, diameters of 1 nm, and high aspect ratio [14]. However, Oberlin et al. reported the first image of CNTs, which was obtained by the pyrolysis of ferrocene and benzene (C_6H_6) at 1000°C [15]. Later in 1993, Iijima and Ichihashi [16], and Bethune et al. [17] independently reported SWCNTs using the arc-discharge approach with Fe/Co as catalysts. The three most common methods for the production of CNTs are laser ablation, arc-discharge, and chemical vapor deposition (CVD) (Fig. 9.2). In laser ablation/arc-discharge approaches, a solid carbon source is heated in the temperature range of 3000°C−4000°C, which produces CNTs upon condensation, whereas CVD uses pyrolysis of a carbon source on the catalytic surface at the elevated temperatures (600°C−1100°C). In the laser ablation technique, in a closed chamber with inert (Ar) gas environment, a high-energy laser beam is directed to the carbon source mixed with a metal catalyst to evaporate it. The catalytic particles attach to carbon atoms of the source to expedite the growth of CNTs. In the arc-discharge approach, the graphitic rod, considering as cathode and anode, are placed with a small gap apart in a

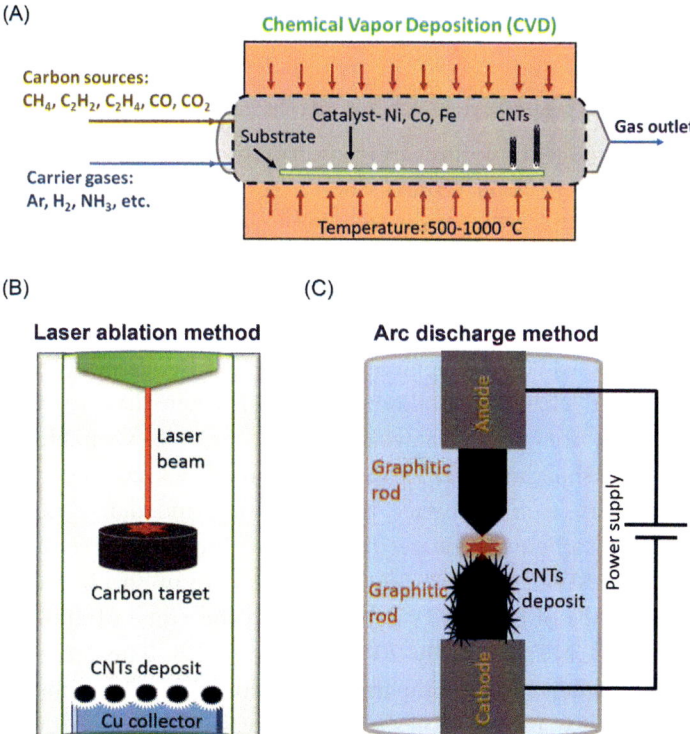

Figure 9.2 Schematic represents the commonly used methods for CNTs production: (A) CVD; (B) laser ablation method; and (C) arc-discharge method.

low-pressure inert gas atmosphere. On passing electric current between anode and cathode, high-speed electrons are emitted from the cathode directly to the anode and evaporate it by heating. Upon condensation, the evaporated carbon is converted into CNTs on the cathode surface. CVD is a vacuum deposition approach which involves the development of CNTs on the metal catalysts including Ni, Co, and Fe using hydrocarbons as carbon sources at a temperature range of $500°C-1100°C$ in an inert atmosphere [18]. The used catalysts are preheated in a reducing gas atmosphere (viz., hydrogen and ammonia) to activate them. Then, a carbonaceous gas (e.g., C_2H_2, C_2H_4, CO, CO_2) is allowed in the reaction chamber for the formation of CNTs. The oxide support layer, namely alumina and silica, is commonly applied under the catalyst, although the growth of CNTs on the conductive surface is comparatively challenging. CVD is the most favored CNTs growth method as it allows reasonable control over parameters (viz., diameter, length, chirality, and metallicity), CNTs quality, and can develop CNTs in predefined zones on a substrate [19]. This method can produce CNTs in large quantity, but attaining repeatability is a serious issue associated with it. Concurrently, a plasma-enhanced CVD process has been introduced to expedite decomposition of carbonaceous gas at comparatively lower temperatures for growth of CNTs [19].

The unique structure of CNTs offers excellent intrinsic properties, including a large active surface area, high chemical and mechanical stabilities, and high electrical conductivity, which provides immense potential for use in sensors for biomedical applications, electronics, composites, and adsorption applications [8,20−22]. Besides providing inbuilt graphene characteristics due to sp^2 carbons, CNTs have their distinctive properties owing to their curvature and chiral nature which controls their bandgap because of the periodic boundary surroundings in the circumferential path [23]. Dissimilar to graphene, CNTs can be semiconducting or metallic based on their chirality [24]. This allows the use of CNTs in a broad range of applications, particularly in electronics where they can act as semiconductors and as interconnectors due to their metallic property for charge mobility. Recently, there have been attempts to commercialize CNTs for paints, bicycles frames, flexible wearable strain sensors, electrostatic-discharge shielding, CNTs yarn, and superblack coating [19,25−27]. The exceptionally high surface area, surface functionalities, and defect-rich surface make 1D CNMs suitable adsorbents for water purification. From the literature, a theoretical surface area of closed SWCNTs was determined as 1315 m^2/g [28], while an experimental BET surface area of SWCNTs is achieved in the range of $400-900$ m^2/g, and a surface area of MWCNTs is usually $200-400$ m^2/g [29]. The surface area of CNFs ($50-350$ m^2/g) was much lower compared with activated carbon fibers and activated carbons [30]. The surface area of CNFs can be increased to $400-1200$ m^2/g on heating at $800°C-1000°C$ during activation using KOH [31].

The surface of pristine CNTs is highly hydrophobic. To get rid of this problem, various methods have been pursued for surface modification, namely covalent functionalization, noncovalent functionalization, and modification using nanomaterials [6,32]. Covalent functionalization of CNTs involves the reaction of carbon atoms with hydrophilic organic molecules on the surface. It can occur either as "ends and defects" functionalization or "sidewalls" functionalization. The ends and defects functionalization strategy is more reactive and specific than sidewalls functionalization [33,34]. In some cases, CNTs are first oxidized using strong oxidizing agents (concentrated $HNO_3:H_2SO_4:3:1$) to generate oxygen-carrying functionalities, namely hydroxyl and carboxylic groups at ends and defects sites of CNTs. This oxidation process results in oxygen functionalized CNTs as well as shortens CNTs compared with pristine CNTs. The sidewall modifications lead to a large number of molecules linked to CNTs, which in turn introduce significant perturbation in the π-electron-based electronic structure. Moreover, the oxygen functional groups ($-OH$, $-COOH$) of the CNTs surface can be covalently functionalized using hydrophilic, hydrophobic, and amphiphilic molecules to improve the surface properties. In noncovalent functionalization, the intrinsic properties of CNTs remain intact because it is not affecting the sp^2 hybridized carbon network. This can be obtained by considering $\pi-\pi$ interaction between graphitic sidewalls of CNTs and conjugated molecules. Other functional molecules can be noncovalently anchored on oxidized CNTs by taking advantage of electrostatic interaction, hydrogen bonding, and van der Waals forces [35]. The amphiphilic molecules have been anchored on CNTs in aqueous media using noncovalent hydrophobic interactions, which results in a reduction of the hydrophobic interface between polar molecules and CNTs [35]. The modification of CNTs using oxidizing agents and different organic molecules leads to a high adsorption property and separate surface area.

Additionally, functionalization can enhance the solubility of CNTs as well as increase the dispersion of CNTs in a uniform solution. CNTs can also be modified with NPs and activated at high temperature to improve the selective adsorption characteristics. Fig. 9.3 exhibits the different stage modifications of CNTs with acids, chemicals, and heat treatments to ensure the selective adsorption of polar and nonpolar chemicals or water pollutants. Overall, modified CNTs offer high affinity to a wide variety of organic and inorganic contaminants.

Similar to CNTs, CNFs also have a cylindrical carbon structure with high aspect ratio, with a diameter in nanosize and length in microsize. Owing to their excellent chemical and mechanical properties, CNFs have found their uses in numerous areas such as adsorption, catalysis, composites, nanodevices, and many more. Among several synthetic routes of CNFs, electrospinning is the efficient, facile, and low-cost approach. The electrospun CNFs with various structures can be achieved by employing different electrospinning approaches (Fig. 9.4). Generally, CNFs are produced by

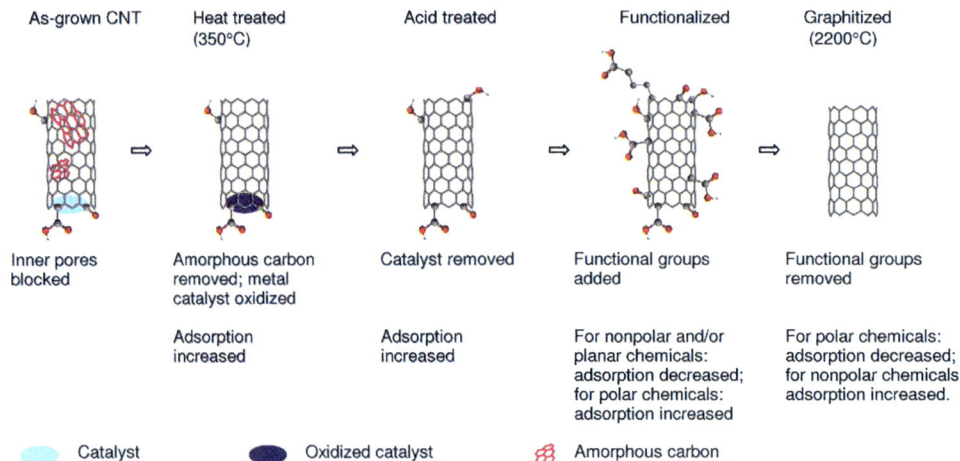

Figure 9.3 Adsorption performances of modified CNTs to polar and nonpolar chemicals. *Reproduced with permission from Pan B, Xing B. Adsorption mechanisms of organic chemicals on carbon nanotubes. Environ Sci Technol. 2008;42:9005—9013 [36]. Copyright 2008, American Chemical Society.*

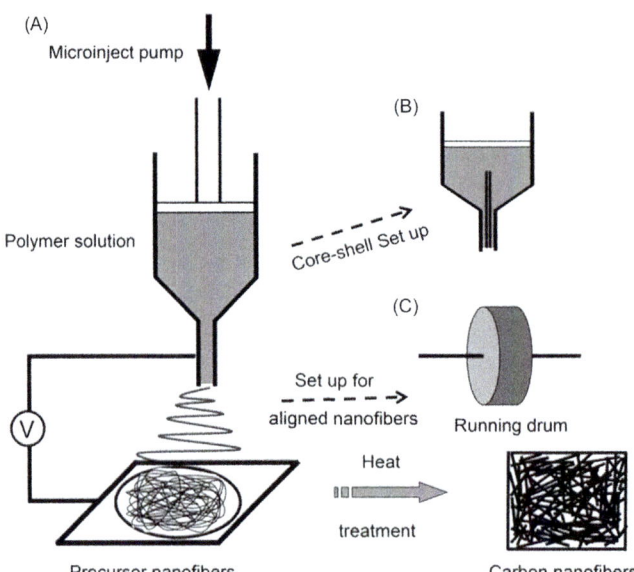

Figure 9.4 Electrospinning setup to produce CNFs of different morphologies: (A) radon fiber; (B) core—shell fiber; and (C) aligned fiber structures. *Reproduced with permission from Peng S, Li L, Kong Yoong Lee J, Tian L, Srinivasan M, Adams S, et al. Electrospun carbon nanofibers and their hybrid composites as advanced materials for energy conversion and storage. Nano Energy 2016;22:361—395. Copyright 2016, Elsevier.*

two steps: (1) precursor nanofibers are synthesized by an electrospinning method, and (2) then the formed precursor nanofibers undergo preoxidation treatment and carbonation process to develop electrospun CNFs. CNFs are mostly prepared using the carbonization of polymer precursors including polyacrylonitrile, polyimide, cellulose, lignin, polycarbosilane, poly(vinylidene fluoride), polyphenylsilane, poly(vinyl alcohol), polyvinylpyrrolidone, and phenolic resin [37].

Similar to CNTs, CNFs can be modified using oxidation, surface functionalization (covalent/noncovalent), and incorporation of NPs to enhance their adsorption properties.

9.3 CNTs-based adsorbents

The high surface area and porosity, the hollow layered architecture, small diameter, and high aspect ratio of CNTs make them effective for pollutant adsorption, and they are widely employed to adsorb water contaminants [38,39]. The basic driving forces for pollutant adsorption on CNTs are $\pi-\pi$ interactions, hydrophobic interactions, electrostatic interactions, and charge transfer [40]. CNTs offer four possible active adsorption sites: (1) the hollow tube structure, that is, internal sites; (2) the space between the walls, that is, interstitial channels; (3) the grooves between the peripheral nanobundles, that is, grooves; and (4) the external surface area, that is, exposed active surface area (Fig. 9.5A and B) [10,41].

Calvete et al. compared the adsorption capacities of MWCNTs to those of powdered activated carbons (PAC) for Direct Blue 53 (DB-53) dye in a contaminated aqueous system [43]. The maximum adsorption capacities (Q_m) of MWCNTs and PAC were found to be 409.4 and 135.2 mg/g, respectively, at pH 2. The higher adsorption capacity of MWCNTs compared with PAC is due to similarities between its textural properties and those of the dye molecules. The aggregate pores in MWCNTs (diagonal length 3.7 nm) allowed the insertion of the large DB-53 molecule (diagonal length 2.04 nm) into the tube, and approximately three molecules adhered easily onto the surface, whereas PAC could only accommodate one molecule due to its small diagonal length (1.7 nm). A similar trend was observed for the adsorption of dioxin on MWCNTs and ACs [44]. The compatible pore size and structural properties of the hexagonal array of carbon atoms in CNTs with the aromatic dioxin molecule allowed strong interactions favorable to adsorption. In contrast, Xia et al. found that the adsorption capacity of CNTs was less than those of GO and ACs for methylene blue (MB) dye [45]. The low absorbance of MB to the CNTs was due to its small particle size and surface area. Therefore the textural properties of CNTs play a significant role in adsorption and can be altered to accommodate the adsorbate.

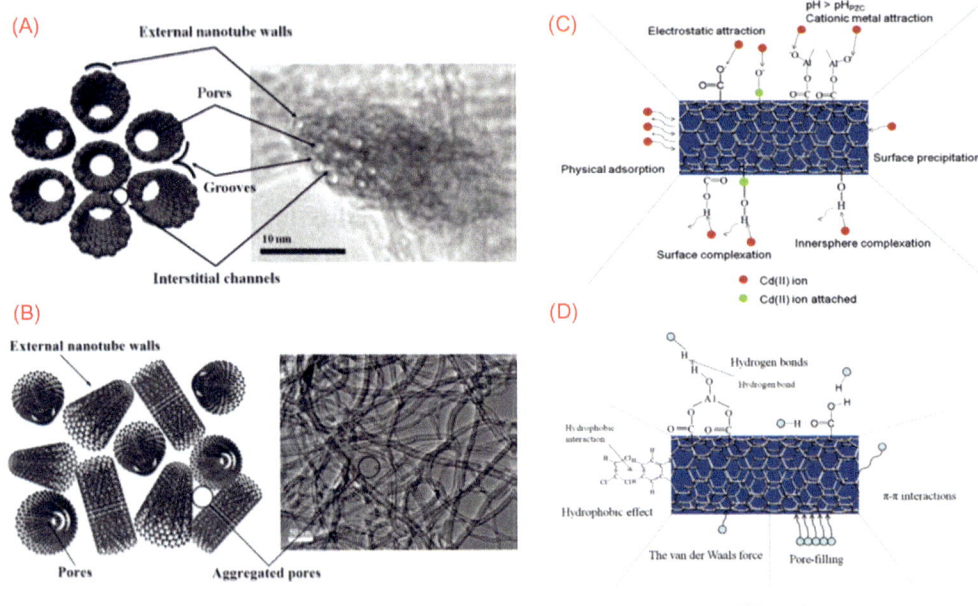

Figure 9.5 (A) Schematic model showing possible adsorption active sites of SWCNTs (left), and TEM image of SWCNTs (right). (B) Schematic model representing aggregated pores of MWCNTs (left) and TEM image of MWCNTs, which highlights the aggregated pores (right) [41]. Copyright, 2015, Springer Nature. (C) Possible interactions of Al$_2$O$_3$/MWCNTs with Cd (II) ions and TCE (D). *Reproduced with permission from Liang J, Liu J, Yuan X, Dong H, Zeng G, Wu H, et al. Facile synthesis of alumina-decorated multi-walled carbon nanotubes for simultaneous adsorption of cadmium ion and trichloroethylene. Chem Eng J. 2015;273:101–110. Copyright 2015, Elsevier Science Ltd.*

For example, Stacie et al. reported that the adsorption rate of heavy metals decreased in the order Cu (II) > Pb (II) > Co (II) > Zn (II) > Mn (II), following the Freundlich model, whereas Tofighy et al. observed the adsorption decreasing in the order Pb (II) > Cd (II) > Co (II) > Zn (II) > Cu (II), while the adsorption mechanism followed both the Freundlich and Langmuir adsorption models [46,47]. Therefore adsorption affinity and mechanism can be controlled by the radius and other structural and physical properties of CNTs. Contaminants can become trapped in the pores or adsorbed onto the surface via different interactions. The adsorption of dyes onto CNTs is also affected by the morphology of the adsorbate, for example, planar polynuclear organic dyes (rhodamine B (RhB), Alizarin Red (AR), Acridine orange (AO) and anthracene (AN)) easily adhered to the surface of CNTs, however, nonplanar, conjugated molecules (Orange G (OG), Xylenol Orange (XO), and 1-(2-pyridylazo)-2-naphthol (PAN)) adsorbed to the surface of CNTs less efficiently, and nonplanar, nonconjugated molecules (diidofluorescein (DIF) and bromothymol blue (BTB)) adsorbed onto the CNT the least [48]. The adsorption capacity of SWCNTs is higher than that of MWCNTs, possibly due to the higher surface area of SWCNTs. In contrast, MWCNTs are approximately 100 times more cost-effective than SWCNTs [49]. Therefore economic concerns have motivated investigation into enhancing the adsorption capacity of MWCNTs by surface functionalization. Surface functionalization plays a crucial role in the adsorption of contaminants [50]. Table 9.1 summarizes the use of CNTs as adsorbents for various organic pollutants.

Adsorption of heavy metal ions onto CNTs occurs via both chemisorption and physisorption. Yang et al. successfully removed Pb(II) from wastewater by adsorption onto CNTs and found that chemisorption processes govern 75.3 % of the adsorption, via the formation of complexes, and the remaining 24.7% is accounted for by physisorption processes, principally via electrostatic interactions [72]. In an interesting report, an Al_2O_3-decorated MWCNT hybrid was synthesized using a pyrolysis approach for simultaneous removal of Cd(II) ions and trichloroethylene (TCE) from groundwater (Fig. 9.5C and D) [42]. The high adsorption of Cd(II) ions was attributed to a variety of forces, including strong physical adsorption, surface precipitation, inner-sphere complexation, cationic metal attraction, electrostatic interactions, and surface complexation, between the hybrid MWCNTs and Cd(II). However, for TCE, the high removal rate was attributed to van der Waals forces, $\pi-\pi$ interactions, hydrogen bonding, pore-filling, and hydrophobic effects. Gusain et al. synthesized a nanocomposite (MoS_2/SH-MWCNT) based on molybdenum sulfide (MoS_2) and thiol-functionalized multiwalled carbon nanotube (SH-MWCNT) for effective removal of Pb(II) and Cd(II) ions from industrial mine water [9]. It was prepared by acid treatment and desulfurization of MWCNTs, then followed by a hydrothermal reaction using diethyldithiocarbamate and sodium molybdate as MoS_2 precursors (Fig. 9.6A). The morphological features of MoS_2/SH-MWCNT nanocomposite were characterized by field emission-scanning electron microscopy (FE-SEM) and transmission electron microscopy (TEM) (Fig. 9.6B−D). The electrostatic interactions, ion-exchange, and metal−sulfur complexation were attributed to the high adsorption capacity of the MoS_2/SH-MWCNT nanocomposite for Pb(II) (90.0 mg/g) and Cd(II) (66.6 mg/g) [9] (Fig. 9.6E). Isotherm and kinetics experiments exhibited that the adsorption process proceeds by Freundlich adsorption isotherm, and pseudo–second–order models, respectively. Similar to the previous report [73], the spent adsorbents in this study can be reused for photocatalytic environmental remediation.

The adsorption capacity and selectivity of CNTs can also be enhanced by surface carbon functionalization via grafting, oxidation, and physical adsorption of molecules [59,74,75]. Oxygen functionalities on the CNTs can be generated via oxidation [76] or removed by heating [77]. Treatment of MWCNTs with concentrated HNO_3 is a common way to generate oxygen functionalities and introduce hydrophilicity onto the surface [78]. Oxidized CNTs can successfully and selectively adsorb anionic dyes

Table 9.1 Adsorption of organic pollutants using CNT-based nanomaterials.

CNT-based nanomaterial	Organic pollutant	Adsorption conditions	Adsorption isotherm model	Adsorption kinetics	Maximum adsorption capacity (mg/g)	References
Ozone-treated MWCNT	Acetaminophen (ACE)	pH = 4, $T = 298K$, $t = 60$ min	Freundlich	Pseudo-second-order	250	[51]
Ni/PC-CNT	Malachite Green (MG), Congo Red (CR), RhB, MB, Methyl Orange (MO)	$T = RT$, $t = 60$ min	Langmuir	Pseudo-second-order	898, 818, 395, 312, 271	[52]
CNT-IL1	Ketoprofen (KET)	pH = >5, $T = 298K$, $t = 24$ h	Freundlich	Pseudo-second-order	207	[53]
CNT-IL2	Sulfamethoxazole (SMZ), KET, SMZ				198, 184, 100	
CNT-DFB	Carbamazepine (CBZ), Tetracycline (TC)	pH = 6, $T = 298K$, $t = 72$ h	Freundlich	Pseudo-second-order	403, 456	[54]
ZIF-CNT	MG	pH = 3.5, $T = 293K$, $t = 240$ min	Langmuir	Pseudo-second-order	2034	[55]
Al-doped CNTs	MO	pH = 4.5, $T = RT$	Langmuir	Pseudo-second-order	69.7	[56]
MWCNTs	CBZ, Dorzolamide (DA)	pH = 5.8–6, $T = 298K$, $t = 30$ h	Brouers-Sotolongo	—	224, 78	[57]

Adsorbent	Adsorbate	Conditions	Isotherm	Kinetics	q	Ref.
P-CNT5	p-Chlorophenol	pH = 7 T = 298K t = 24 h	Freundlich	Pseudo-second-order	86	[58]
CNT-NH$_2$	Acid Blue 45 (AB45) Acid Black 1 (AB1)	pH = 3.5 T = 293K t = 240 min	Langmuir	Pseudo-second-order	714 666	[59]
CNT-UV + H$_2$O$_2$	Naproxen (NAP)	pH = 7 T = 296K t = 144 h	Langmuir and Freundlich	Pseudo-second-order	763	[60]
MWCNTs	Chloramphenicol Thiamphenicol Florfenicol Sulfadiazine Sulfapyridine Sulfamethoxazole Sulfathiazole Sulfamerazine Sulfamethazine Sulfaquinoxaline Ibuprofen Diclofenac Carbamazepine	pH = neutral T = 298K t = 72 h	Freundlich	Pseudo-second-order	107 64 112 166 233 71 119 295 139 589 94 209 243	[61]
CNT/Mg(Al)O	CR	pH = 7 T = RT t = 24 h	Langmuir	–	1250	[62]
CNT	Basic Blue 41 (BB41) Basic Red 18 (BR18) Basic Violet 16 (BV16)	pH = 7.5 T = RT t = 60 min	Langmuir	Pseudo-second-order	123 80 64	[63]

(continued)

Table 9.1 (Continued)

CNT-based nanomaterial	Organic pollutant	Adsorption conditions	Adsorption isotherm model	Adsorption kinetics	Maximum adsorption capacity (mg/g)	References
MWCNTs	Phenol	pH = 7 T = RT t = 900 min	Langmuir	Pseudo-second-order and intraparticle diffusion	32	[64]
P-CSCNT	MB	pH = 6 T = RT t = 180 min	Langmuir	Pseudo-second-order	319	[65]
CNT@NiCo$_2$O$_4$	MO	pH = 3–11 T = RT t = 120 min	Langmuir and Freundlich	Pseudo-second-order	1188	[66]
CNT@MnCo$_2$O$_4$ CNT@CuCo$_2$O$_4$ CNT@ZnCo$_2$O$_4$					790 826 935	
CNT–MSN	MB	pH = 8 T = 303K	Langmuir	Pseudo First Order	524	[67]
SF-CNT	Direct Red 80 (DR80) Direct Red 23 (DR23)	pH = 7.5 T = 298K t = 60 min	Langmuir	Pseudo-second-order	120 188	[68]
mCS/CNT	Acid Red 18 (AR18)	pH = 3 T = 303K t = 60 min	Freundlich and Redlich – Peterson	Pseudo-second-order	771	[69]
Magnetic CNTs	MO	pH = 2–12 T = 298K t = 5 h	Freundlich	Pseudo-second-order	28	[70]
G–CNT	MB	pH = 2–10 T = RT t = 180 min	Freundlich	Pseudo-second-order	82	[71]

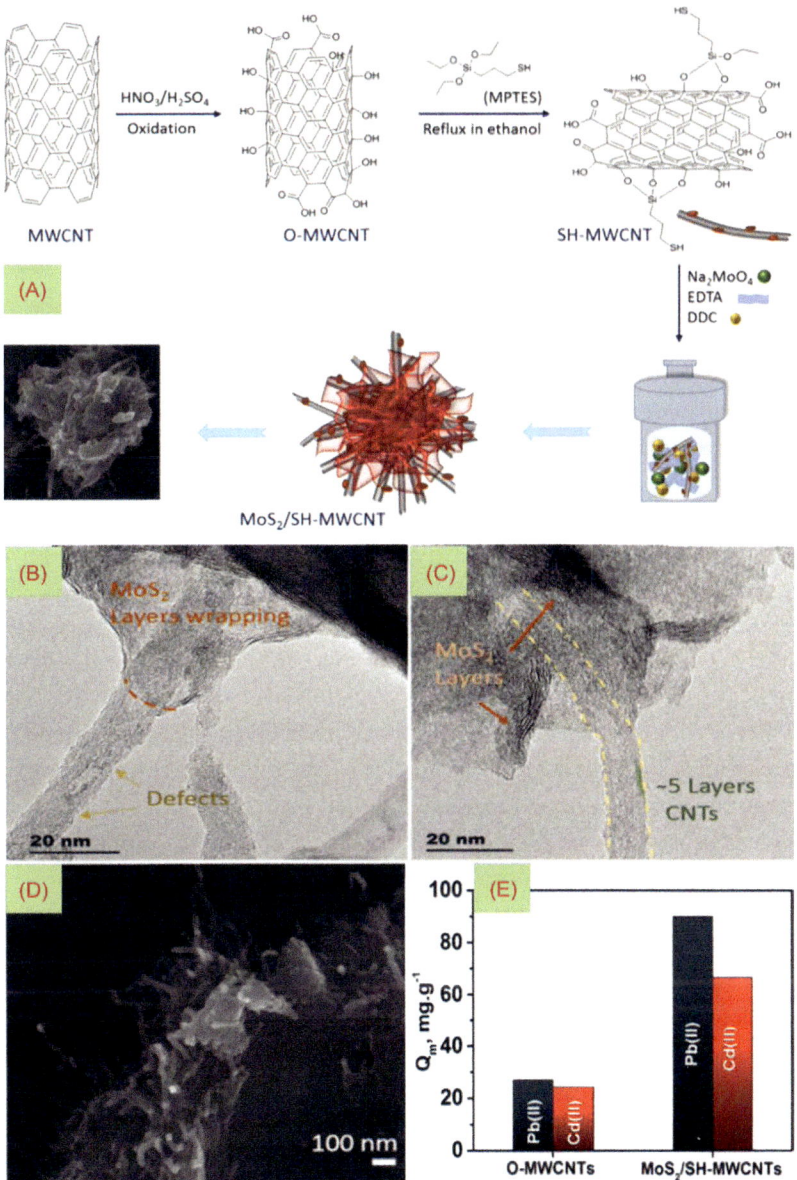

Figure 9.6 (A) Schematic illustration of the preparation route of the MoS_2/SH-MWCNT nanocomposite. TEM images (B, C) and FE-SEM image of MoS_2/SH-MWCNT nanocomposite. (D) Comparative adsorption performance of O-MWCNT and MoS_2/SH-MWCNT nanocomposite for Pb(II) and Cd(II) ions. *Reproduced with permission from Gusain R, Kumar N, Fosso-Kankeu E, Ray SS. Efficient removal of Pb(II) and Cd(II) from industrial mine water by a hierarchical MoS_2/SH-MWCNT nanocomposite. ACS Omega 2019;4:13922–13935. Copyright 2019, American Chemical Society.*

from wastewater through electrostatic and van der Waals interactions, following a rejection mechanism [79]. These oxygen functionalities can provide a partial negative charge on the surface of CNTs, and attract heavy metal ions from waste effluent by donating an electron pair [80]. Therefore oxidation of CNTs significantly improves the adsorption capacity for the selective removal of hydrophilic contaminants from wastewater. Schwab et al. compared the adsorption capacities of SWCNTs, MWCTs, and oxidized MWCNTs (O-MWCNTs) for the adsorption of hydrophobic pharmaceutical and personal care products (PPCPs) [81]. They concluded that PPCPs were most successfully adsorbed onto SWCNTs due to their large specific surface areas, and least successfully adsorbed onto O-MWCNTs due to the larger number of oxygen functionalities on the surface, which render it hydrophilic. Yu et al. modified oxidized MWCNTs using pentaerythritol (ox-MWCNT-PER) to remove Alizarin Yellow R (AYR) and Alizarin Red S (ARS) from the contaminated aqueous system via adsorption [82]. They suggested that aside from other interactions such as electrostatic and $\pi-\pi$ interactions, efficient hydrogen bonding between the hydroxyl groups on the surface of ox-MWCNT-PER and the dyes played a significant role in the excellent adsorption capacity observed for ox-MWCNT-PER. These hydroxyl functional groups were introduced during the oxidation treatment and grafting of pentaerythritol onto the MWCNTs. ARS is an anthraquinone compound and contains many phenolic (hydroxyl groups) and carbonyl groups, which is comparable to ox-MWCNT-PER, and hence it interacted strongly with ox-MWCNT-PER to achieve efficient removal. In contrast, AYR is an azo dye with a number of $-N=N-$ groups and fewer oxygen-containing functional groups, which restricts the hydrogen bonding and leads to lower adsorption onto ox-MWCNT-PER. Table 9.2 summarizes the use of CNTs as adsorbents for various inorganic pollutants.

Moodley et al. theoretically predicted and compared the adsorption potentials of ionic liquid functionalized CNTs (CNT-ILs) with those of CNTs, using DFT calculations [53]. They thoroughly studied the effect of pH (pH 2−10), at 298K, on the adsorption behavior of CNT-ILs and CNTs for sulfamethoxazole (SMZ) and ketoprofen (KET) from an aqueous system. CNT-ILs strongly adsorbed SMZ and KET at higher pH (pH ≥ 5), whereas CNT successfully adsorbed the pharmaceutical molecules at lower pH (pH < 5). This was attributed to the behavior of organic molecules in solution at different pH values. SMZ antibiotics and KET antiinflammatory drugs contain substituted benzene and heterocyclic rings, which are strong electron acceptors. At high pH, SMZ and KET dissociated into negatively charged, hydrophilic species [98,99]. CNT-IL (CNT-IL1 & CNT-IL2) surfaces are positively charged due to the cationic character of the ionic liquid at pH > 4.5, which significantly enhanced its adsorption capacity for SMZ and KET at higher pH (pH ≥ 5), via strong electrostatic interactions [53]. In contrast, bare CNTs demonstrated efficient adsorption at pH < 4, which is due to the electrostatic and hydrophobic interactions between CNTs and the

Table 9.2 Adsorption of inorganic pollutants using CNT-based nanomaterials.

CNT-based nanomaterial	Inorganic pollutant	Adsorption conditions	Adsorption isotherm model	Adsorption kinetics	Maximum adsorption capacity (mg/g)	References
NiO/CNT	Pb(II)	pH = 7 t = 10 min	Freundlich	Pseudo–second–order	24	[83]
PAMAM/CNT	Cu(II) Pb(II)	pH = 8 T = 298K	Langmuir	Pseudo–second–order	3333 4870	[84]
mHAP-oMWCNTs	Pb(II)	pH = 4.1 T = 298K t = 40 min	Langmuir and Freundlich	Pseudo–second–order	698	[85]
MNP/MWCNTs	Cr(VI)	pH = 2 T = 298K t = 360 min	Langmuir	Pseudo–second–order and intraparticle diffusion	17	[86]
KTEG-CNTs	Pb(II)	pH = 5 T = RT t = 30 min	Langmuir	Pseudo–second–order	288	[87]
MWCNT/SiO$_2$	Pb (II)	pH = 5–6 T = RT t = 100 min	Langmuir and Temkin	Pseudo–second–order	13	[88]
DTC-MWCNT	Cd(II) Cu(II) Zn(II)	pH = 6 T = 298K t = 150 min	Langmuir	Pseudo–second–order	167 98 11	[89]
O-MWCNTs	Pb (II) Cu(II) Zn(II) Cd(II)	pH = 5 T = 298K t = 10 min	Langmuir	Pseudo–second–order	76 15 13 32	[90]
MnO$_2$/CNT	Hg(II)	pH = 5–7 T = 323K t = 80 min	Freundlich	Pseudo–second–order	58	[91]

(continued)

Table 9.2 (Continued)

CNT-based nanomaterial	Inorganic pollutant	Adsorption conditions	Adsorption isotherm model	Adsorption kinetics	Maximum adsorption capacity (mg/g)	References
MWCNT-f	Pb (II)	pH = 5.5 T = 323K t = 60 min	Freundlich	Pseudo–second-order	14	[92]
Graphene–c–MWCNT hybrid aerogel	Pb (II)	T = 293K t = 5 days	Freundlich	Pseudo–second-order	105	[93]
	Hg(II)				93	
	Ag(I)				64	
	Cu(II)				33	
CNT/sand	Pb (II)	pH = 5.6 T = RT t = 12 h	Langmuir	Pseudo–second-order	92	[94]
	Cu(II)				67	
Oxidized CNT	Pb (II)	pH = 7 T = RT	Freundlich	Pseudo–second-order	101	[47]
	Cd(II)				75	
	Co(II)				69	
	Zn(II)				58	
	Cu(II)				50	
TA-MWCNT	La	pH = 5 T = 293K t = 120 min	Langmuir	Pseudo–second-order	5	[95]
	Tb				8	
	Lu				4	
CNT/TNT	Pb (II)	pH = 5 T = 298K t = 2 h	Langmuir	–	588	[96]
	Cu(II)				116	
MWCNT	Pb (II)	pH = 5.2 T = RT t = 4 h	Langmuir	–	97	[97]
	Cu(II)				24	
	Cd(II)				10	
MoS₂/SH-MWCNT	Pb (II)	pH = 6 T = 298K	Freundlich	Pseudo–second-order	90.0	[9]
	Cd(II)				66.6	

pharmaceutical molecules. However, the adsorption of SMZ and KET on CNT-ILs at low pH indicates that the adsorption process does not occur solely via electrostatic interactions. Other interactions, such as hydrophobic interactions, pore trapping, etc. are also involved. The adsorption of SMZ and KET on CNTs and CNT-ILs was best represented by the Freundlich isotherm and followed pseudo-second-order kinetics.

Aerogels of graphene and CNTs have a highly porous structure and high specific surface area in addition to their intrinsic properties, which increase their demand for adsorption applications [100,101]. However, the neat CNT aerogels and graphene aerogels exhibit weak elasticity, whereas the graphene−CNT aerogel composite has excellent mechanical properties [102,103]. The incorporation of CNTs into the graphene aerogels improves the adsorption potential of the nanohybrid more than 100-fold in comparison to a graphene aerogel [104]. Kim et al. employed a graphene−CNT nanohybrid aerogel to remove organic contaminants from wastewater [105]. In the nanohybrid, CNTs bound to the graphene aerogel, and the surface area increased by up to 57%. The graphene−CNT nanohybrid aerogel successfully adsorbed two cationic dyes, crystal violet (CV) and MB, and two anionic dyes, congo red (CR) and methyl orange (MO), with adsorption capacities of 575, 626, 560, and 532 mg/g, respectively, principally via $\pi-\pi$ and van der Waals interactions. Thus nanohybrid aerogels facilitate the simultaneous adsorption of positively and negatively charged water contaminants.

Several examples of functionalization of CNTs for removal of metal ions and organics from wastewater have been reported in the literature (Fig. 9.7). The abundant availability, the large number of possible functionalizations of CNTs to improve the selectivity, and the excellent adsorption capacity will lead to future demand for CNTs for water purification.

9.3.1 CNTs magnetic nanocomposites for adsorption

Similar to graphene, CNTs can be functionalized with magnetic nanoparticles to facilitate separation from the solution, for reuse [107]. Various strategies for magnetic functionalizing of MWCNTs has been described in Fig. 9.8. The adsorption capacity of CNT/Fe_3O_4 was enhanced in comparison to that of nonfunctionalized CNT, which was attributed to the active participation of the oxygen functionalities of Fe_3O_4 in the adsorption of heavy metal ions (Cr(III)), which provided additional active sites [108]. Vadahanambi et al. fabricated a three-dimensional (3D) graphene−CNT−Fe nanohybrid using a simple microwave approach for the removal of arsenic from water (Fig. 9.8A). Magnetic graphene−CNT−Fe nanohybrids exhibited higher adsorption capacity (6 mg/g) for As than graphene−Fe_2O_3 (2.5 mg/g) due to a combination of excess surface active sites of CNT, Fe_2O_3, and graphene in the nanohybrids [109]. Jin et al. synthesized high surface area (999 m^2/g) nanohybrids that comprised Ni NPs

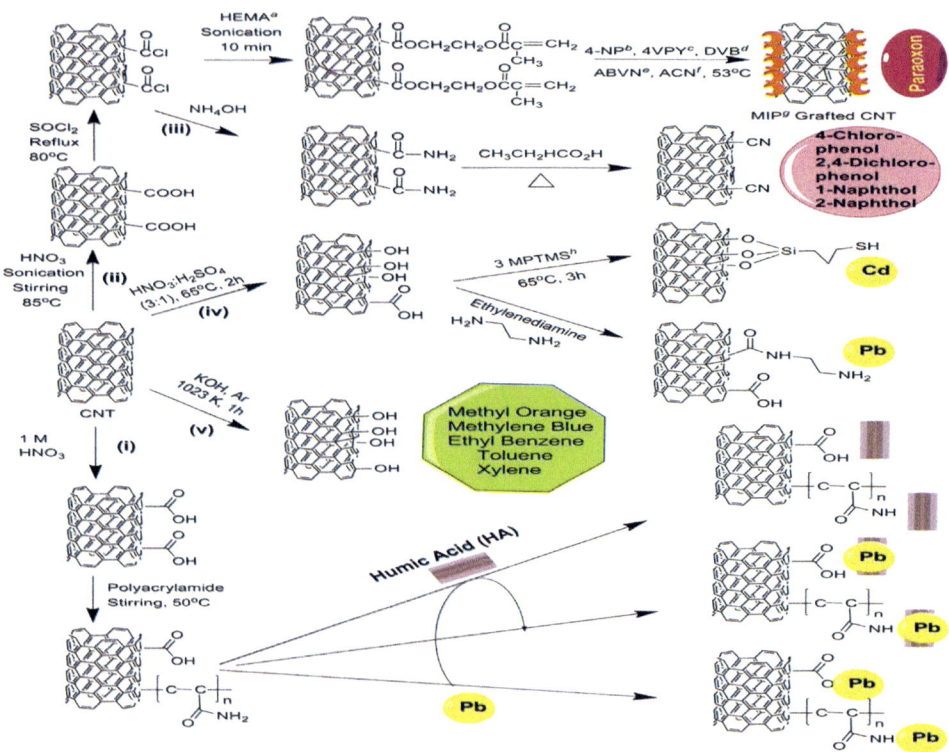

Figure 9.7 Functionalization of CNTs for adsorption of metal ions and organics from water. (A) 2-hydroxyethyl methacrylate; (B) 4-nitrophenol; (C) 4-vinyl pyridine; (D) divinylbenzene; (E) 2,2-azobis (2,4-dimethyl) valeronitrile; (F) acetonitrile; (G) molecularly imprinted polymers; and (H) 3-mercaptopropyltrimethoxysilane. *Reprinted with permission from Das R. Nanohybrid catalysts based on carbon nanotubes. Springer; 2017 [106]. Copyright 2017, Springer.*

and porous carbon/CNTs (Ni/PC-CNT) for the adsorption of organic dyes (malachite green (MG), MO, RhB, MB, and CR). This adsorbent demonstrated the highest adsorption capacity (898 mg/g) toward MG, among others (Fig. 9.8B). After adsorption, nanohybrids could be simply segregated from solution by employing a magnet [52]. Lu et al. modified MWCNTs via the reaction of 1,6-hexanediamine incorporated magnetic Fe_3O_4 NPs (MNP) with $-COOH$ groups of MWCNTs using simple solvothermal chemistry. The as-prepared MNP/MWCNTs were employed for the removal of Cr (VI) ions from contaminated water (Fig. 9.8C). The maximum adsorption performance of MNP/MWCNTs for Cr (VI) was achieved at a solution of pH 2.0 [86].

Cai et al. functionalized CNT/Fe_3O_4 nanomaterials with mercaptopropyltriethoxysilane (MPTS) to form a superparamagnetic $MPTS/CNT/Fe_3O_4$ nanomaterial with an enhanced adsorption capacity (Fig. 9.7D) [110]. The calculated specific active surface area for $MPTS/CNT/Fe_3O_4$ was observed to be higher than that of CNT/Fe_3O_4. Thiol groups in MPTS have strong affinities to heavy metal ions because of

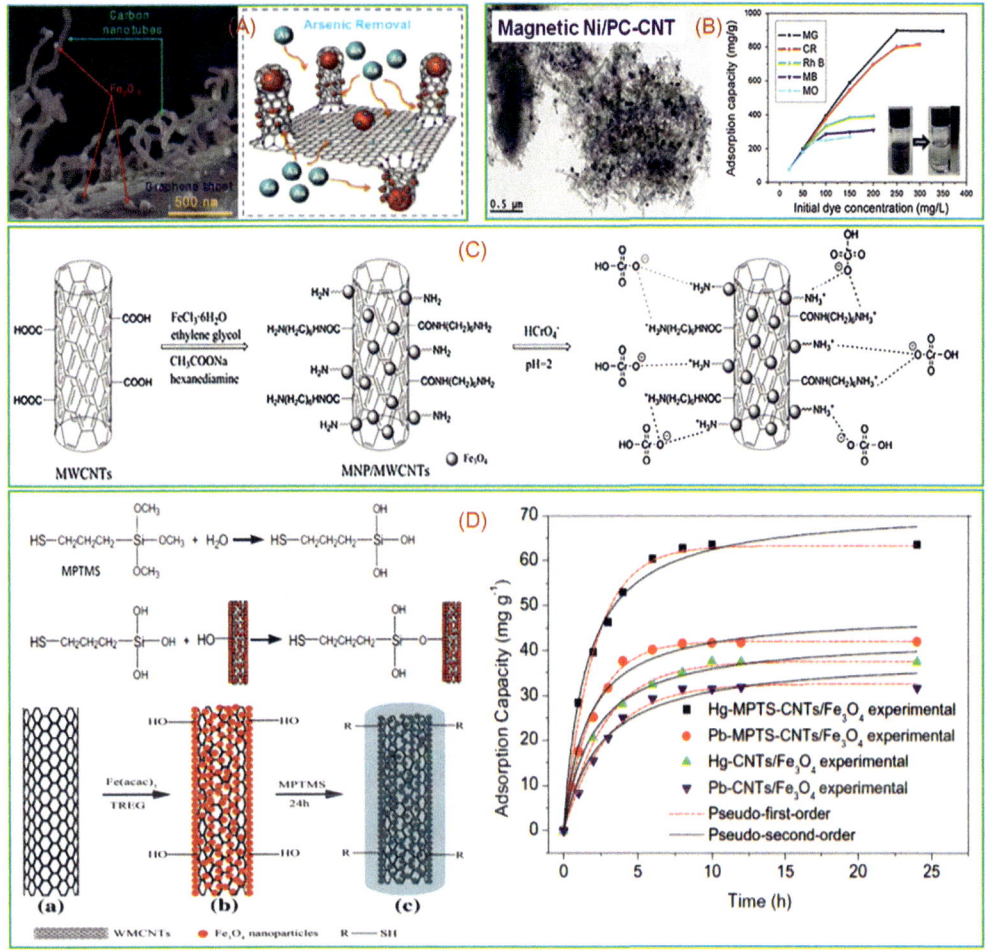

Figure 9.8 Magnetic CNTs for the removal of various organic and inorganic pollutants. (A) Fabrication of 3D graphene—CNT—Fe magnetic nanohybrids using microwave chemistry for removal of As. Reproduced with permission from [109]. Copyright 2013, American Chemical Society. (B) Ni NPs embedded nanoporous carbon/CNT (Ni/PC-CNT) for removal of organic dyes [52]. (C) Synthesis of MNP/MWCNTs for removal of Cr(VI) by employing external magnet [86]. (D) MPTS-CNTs/Fe$_3$O$_4$ nanocomposites for removal of Pb (II) and Hg (II) [110]. *For B, C, D reproduced with permission Elsevier Science Ltd.*

Lewis acid—base interactions [111]. Thus the MPTS/CNT/Fe$_3$O$_4$ nanomaterial successfully adsorbed heavy metal ions from wastewater. Azizi et al. synthesized MnFe$_2$O$_4$/MWCNTs via a hydrothermal route, and applied them as an adsorbent for the removal of cationic (Yellow 40, Y40) and anionic (Direct red 16, DR16) dyes [112]. DR16 was successfully adsorbed onto MnFe$_2$O$_4$/MWCNTs (Q$_m$ = 607.79 mg/g) at low pH values (pH = 2), and 328K, within 300 minutes. At lower pH, the H^+ ion concentration in the system increased, which intensified the electrostatic interactions between the negatively charged dye and the positively charged surface of $MnFe_2O_4/$ MWCNTs. In contrast, the adsorption of Y40 was observed to be minimal at pH 2, and increased with an increase in pH. The improvement in adsorption behavior of Y40 on $MnFe_2O_4/$MWCNTs (280 mg/g), at 328K, in 300 minutes, at high pH (pH 6), was attributed to electrostatic interactions between the negatively charged surface of $MnFe_2O_4/$MWCNTs and the Y40 zwitterion. The adsorption kinetics is pseudo-second-order and can be modeled with Sips (a combination of the Langmuir and Freundlich isotherms).

9.3.2 Alkali-activation of CNTs for enhanced adsorption

The specific surface area, pore volume, and excess oxygen functionalities of CNTs can also be enhanced via an activation method using alkali, namely K_2CO_3, Rb_2CO_3, Cs_2CO_3, KOH, Li_2CO_3, Na_2CO_3, and NaOH [113,114]. This promotes the possibility of trapping large water pollutants and the improved adsorption potential of CNTs [115]. The alkali-activation strategy produces materials with a large number of porous structures. The activation mechanism has basically involved alkali-based materials (viz., KOH) and redox processes [113,116]. The KOH activation of carbon-based materials can be explained using the following Eqs. (9.1−9.5):

$$6KOH + C \rightarrow 2K + 3H_2 + 2K_2CO_3 \tag{9.1}$$

At temperature $>700°C$, the chemical reaction occurs as follows.

$$K_2CO_3 + C \rightarrow K_2O + 2CO \tag{9.2}$$

$$K_2CO_3 \rightarrow K_2O + CO_2 \tag{9.3}$$

$$2K + CO_2 \rightarrow K_2O + CO \tag{9.4}$$

At temperature $>800°C$, the chemical reaction proceeds as follows.

$$K_2O + C \rightarrow 2K + CO \tag{9.5}$$

The entanglement of the long aspect ratio CNTs decreases the advantageous surface area. While the alkali-activation process opens the CNTs tips, reduces the entangled CNTs, and produces the large number of new mesopores and micropores with open apertures, which results in enhancement of the specific surface area of CNTs. Chen et al. reported alkali (KOH)-activated CNTs (A-CNTs) with a high specific surface area (534.6 m^2/g) and pore volume (1.61 cm^2/g) for the removal of anionic (MO) and cationic (MB) dye [115]. As-produced A-CNTs were shortened in length and carry a large number of defects, opened tips, destroyed hollow structures, and many flaky apertures (Fig. 9.9A). The high adsorption capacity for MB (400 mg/g) and MO (149 mg/g) onto A-CNTs was attributed to manifold adsorption

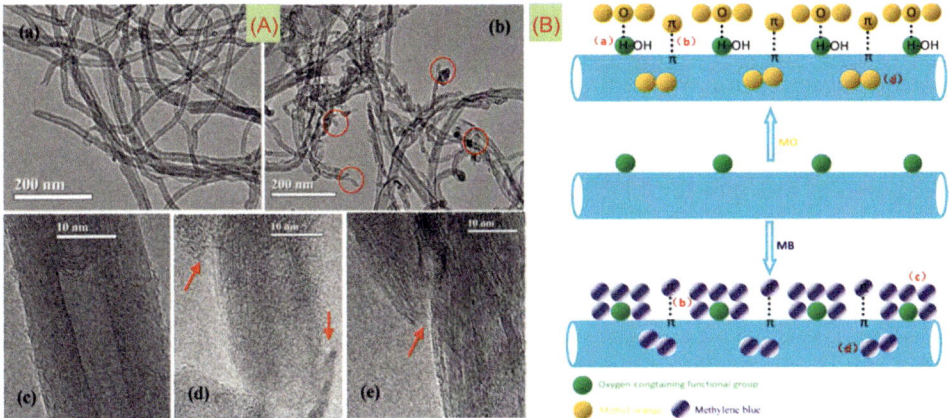

Figure 9.9 (A) TEM images of CNTs (a, c) and activated CNTs (A-CNTs) (b, d, e). (B) Schematic shows the adsorption mechanism of MB or MO dye on A-CNTs: (a) hydrogen linkage, (b) π-π EDA interactions, (c) electrostatic interactions, (d) mesopores filling. *Reproduced with permission from Ma J, Yu F, Zhou L, Jin L, Yang M, Luan J, et al. Enhanced adsorptive removal of methyl orange and methylene blue from aqueous solution by alkali-activated multiwalled carbon nanotubes. ACS Appl Mater Interfaces. 2012;4:5749–5760. Copyright 2012, American Chemical Society.*

mechanisms such as π—π electron donor—acceptor (EDA) interactions, electrostatic interactions, H-bonding, and pore-filling (Fig. 9.9B). In another study, KOH-activated CNTs were used to prepare magnetic iron oxide MI-CNTs hybrids for the removal of As(V) and As(III) with their excellent adsorption and magnetic separation properties [117]. The maximum monolayer adsorption capacities of ∼8.13 mg/g for As(III) and ∼9.74 mg/g for As(V) were attributed to the high surface area (662.1 m²/g), high dispersibility, and desired magnetic properties of MI-CNTs hybrids.

9.4 CNFs-based adsorbents

CNFs are an important class of carbon compound and have been investigated extensively, both for elementary scientific research and for practical applications [118,119]. 1D CNFs have diameters of 3—200 nm and micrometer lengths [120]. CNFs are readily prepared by electrospinning, which is powerful and facile and can impart the hierarchical porous structure crucial for multifunctional integration [120,121]. CNFs have different properties than conventional carbon fibers (CFs), principally due to the larger size of the latter, which are out of the nanorange. CNFs exhibit remarkable physicochemical characteristics, including high tensile strength, excellent thermal and chemical stabilities, low density, and high conductivity, which have led to their application in a variety of fields, for example, as sensors [122], batteries [123], for catalysis [124], and in aerospace technology [125]. The high surface area, porosity, and selective active surface sites of CNFs suggest applications in adsorption [119]. However, CNFs have not been as extensively studied as other carbon nanomaterials for the adsorption of water contaminants.

Bera et al. prepared mesoporous CNFs using a sol−gel/electrospinning process and examined their use in the adsorption of large dye molecules from wastewater [126]. Three different mesoporous CNFs (C-1, C-2, and C-3) were prepared using phenolic resins as carbon precursors and the triblock copolymer Pluronic F127 as a template, with varying amounts of tetraethyl orthosilicate (TEOS), that is, 1, 2, and 3 g, respectively. C-1, C-2, and C-3 exhibited very high active surface areas (1088, 1176, and 1642 m^2/g, respectively) and pore volumes (0.65, 0.78, and 1.02 cm^3/g, respectively) which suggest their application in adsorption (Fig. 9.10A−C). To determine the effect of pH on the adsorption capacity of CNFs, zero point charge (pH_{pzc}) was measured using the zeta potential and determined to be 2.9, 3.0, and 3.2, for C-1, C-2, and

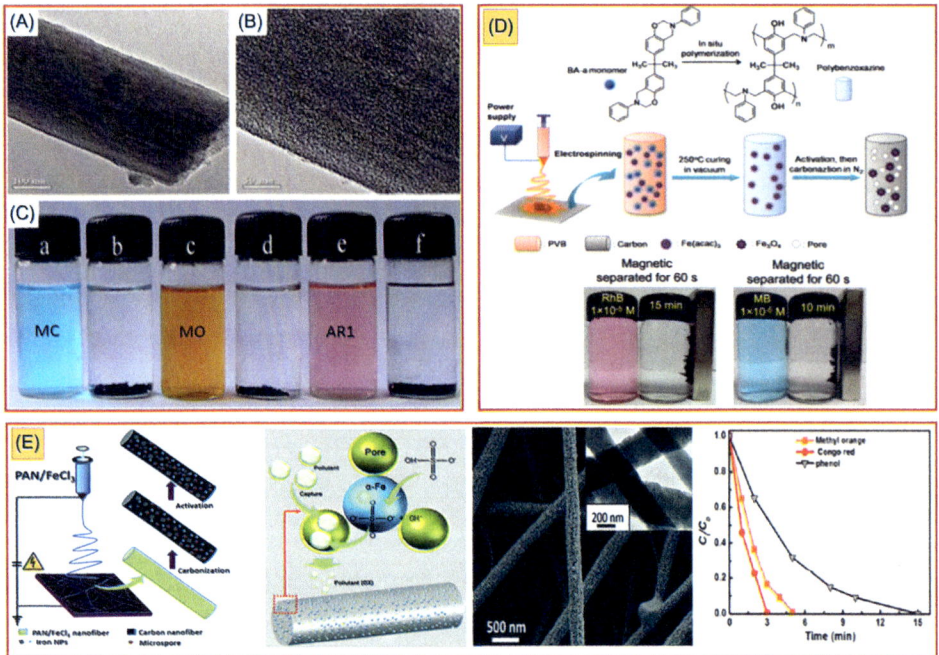

Figure 9.10 (A, B) TEM images of mesoporous carbon nanofibers (CNFs). (C) Photographs of dyes before and after adsorption using CNFs. [126] (D) Hierarchical porous magnetic Fe_3O_4@CNF synthesized via a combined approach of in situ polymerization and electrospinning using polybenzoxazine precursors for the adsorption of MB and RhB [127]. (A−D) Reproduced with permission from Elsevier Science Ltd. (E) Fabrication of CNFs with α-Fe nanoparticles (nano-Fe@CNFs) using electrospinning for adsorption-intensified degradation of organics pollutants (phenol, CR, and MO). *Reproduced with permission from Zhu Z, Xu Y, Qi B, Zeng G, Wu P, Liu G, et al. Adsorption-intensified degradation of organic pollutants over bifunctional α-Fe@carbon nanofibres. Environ Sci Nano. 2017;4:302−306. Copyright 2017, Royal Society of Chemistry.*

C-3, respectively. Therefore the adsorption of methylthionine chloride (MC, cationic dye) increased with increasing solution pH, while for anionic dyes (MO and AR1) the maximum adsorption was recorded at low pH. C-3 had the highest adsorption capacity due to its large pores, which can accommodate the large dye molecules. Therefore a comparable adsorbate and adsorbent size enhances the adsorption potential. AlSaadi et al. reported a theoretical study that used Design-Expert software to model the adsorption of Pb(II) from industrial wastewater [129]. The adsorption capacity of CNFs for Pb(II) was determined to be 166.6 mg/g, which was significantly higher than for the other adsorbents studied.

Similar to other CNMs used in adsorption, the introduction of magnetic groups into CNFs facilitates the separation process after the adsorption is complete [127,130]. Si et al. fabricated hierarchical porous magnetic $Fe_3O_4@CNF$ for the adsorption of MB and RhB [131,132]. This nanocomposite has an extremely high surface area ($1885\ m^2/g$) and pore volume ($2.083\ cm^3/g$), which facilitated the adsorption of organic dyes, and its susceptibility to magnetic separation encourages its use [132] (Fig. 9.10D). Sun et al. fabricated CNFs with α-Fe nanoparticles (nano-Fe@CNFs) which had high adsorption capacities for dyes and phenol and were easily separated from aqueous systems using an external magnet [128]. The driving force for the adsorption of contaminants was an electrostatic attraction, along with hydrogen bonding and $\pi-\pi$ interactions (Fig. 9.10E). Verma et al. found that the adsorption behavior of Fe-grown CNF for As(V) could be modeled with the Freundlich adsorption isotherm due to the heterogeneous active sites on the adsorbent [130].

In parallel to other carbon derivatives, chemical activation treatment of CNFs is done to enhance the specific surface area, pore sizes, and pore volume, which regulate its surface application characteristics. Yasumori et al. reported the activation of CNFs using several activating chemical agents (K_2CO_3, Rb_2CO_3, Cs_2CO_3, KOH, Li_2CO_3, Na_2CO_3, and NaOH) and the highest surface area was obtained in Cs_2CO_3-treated CNF (CNF-Cs_2CO_3), followed by CNF-Rb_2CO_3 > CNF-KOH > CNF-NaOH > CNF-Na_2CO_3 > CNF-Li_2CO_3 [114]. Such trend is obtained due to the sizes of intercalated cations (Cs^+ (0.338), Rb^+ (0.3), K^+ (0.26 nm), Na (0.19 nm), and Li (0.12 nm)) into the sheets of CNFs. Activation of CNFs could also efficiently open the closed pores in pristine material and also develop new pores. This two- to threefold increase in specific surface area and pore volume helps to endorse its application in adsorption [133].

9.5 Conclusion

In summary, among 1D CNMs, CNTs have garnered huge attention from industry and academicians because of their distinctive properties, including high thermal and electrical conductivity, high aspect ratio, high mechanical strength, chemical resistance, and large surface area. CNTs are produced by arc-discharge method, CVD, and laser ablation methods, but among these, CVD is a more facile and popular approach for large-scale production of CNTs. The CVD method permits the high production of excellent quality CNTs with high-level control of their numerous parameters, namely, diameter, length, metallicity, and chirality. The active surface characteristics, available defects, delocalized π-electrons network, curvaceous surface, and large surface area and porosity make CNTs appropriate adsorbents for the elimination of a wide variety of water pollutants. Modification of CNTs using organic molecules via covalent and noncovalent functionalization and inorganic NPs by forming composites enhances the adsorption performance, selectivity, and sensitivity for the removal of inorganic and organic pollutants from aqueous water. Furthermore, the alkali-activation process used to open the CNTs tips reduces the entangled CNTs, and produces a large number of new mesopores and micropores with open apertures. The activation process yields CNTs with very high surface area, high defects, and surface functionalities, which consequently lead to high adsorption characteristics. In addition, 1D CNFs have diameters in the nanometer range and lengths of micrometers. CNFs are easily prepared by a facile electrospinning method which can produce the hierarchical porous structure with a varied aspect ratio that is crucial for multifunctional integration. Owing to good porosity, high surface area, and selective surface sites, CNFs have been utilized for water treatment via adsorption. Moreover, a magnetic nanocomposite of 1D CNMs offers the advantages of high adsorption capacity and fast recovery of adsorbents for recyclability and reusability. The study of the adsorption capabilities of CNFs for water purification is still minimal, and further research, with respect to different heterostructures, porosities, and the adsorption of various contaminants, is required. Eco-friendly and cost-effective CNF materials will provide new directions and avenues to explore for the removal of water contaminants.

References

[1] Fiyadh SS, AlSaadi MA, Jaafar WZ, AlOmar MK, Fayaed SS, Mohd NS, et al. Review on heavy metal adsorption processes by carbon nanotubes. J Clean Prod 2019;230:783—93.

[2] Kumar N, Mittal H, Alhassan SM, Ray SS. Bionanocomposite hydrogel for the adsorption of dye and reusability of generated waste for the photodegradation of ciprofloxacin: a demonstration of the circularity concept for water purification. ACS Sustain Chem Eng 2018;6:17011—25.

[3] Crini G, Lichtfouse E, Wilson LD, Morin-Crini N. Conventional and non-conventional adsorbents for wastewater treatment. Environ Chem Lett 2019;17:195—213.

[4] Kumar N, Mittal H, Parashar V, Ray SS, Ngila JC. Efficient removal of rhodamine 6G dye from aqueous solution using nickel sulphide incorporated polyacrylamide grafted gum karaya bionanocomposite hydrogel. RSC Adv 2016;6:21929—39.

[5] Kumar N, Reddy L, Parashar V, Ngila JC. Controlled synthesis of microsheets of ZnAl layered double hydroxides hexagonal nanoplates for efficient removal of Cr(VI) ions and anionic dye from water. J Environ Chem Eng 2017;5:1718—31.

[6] Gusain R, Kumar N, Ray SS. Recent advances in carbon nanomaterial-based adsorbents for water purification. Coord Chem Rev 2020;405:213111.

[7] Thostenson ET, Ren Z, Chou T-W. Advances in the science and technology of carbon nanotubes and their composites: a review. Compos Sci Technol 2001;61:1899—912.

[8] Gao C, Guo Z, Liu J-H, Huang X-J. The new age of carbon nanotubes: an updated review of functionalized carbon nanotubes in electrochemical sensors. Nanoscale. 2012;4:1948—63.

[9] Gusain R, Kumar N, Fosso-Kankeu E, Ray SS. Efficient removal of Pb(II) and Cd(II) from industrial mine water by a hierarchical MoS_2/SH-MWCNT nanocomposite. ACS Omega 2019;4:13922—35.

[10] Ren X, Chen C, Nagatsu M, Wang X. Carbon nanotubes as adsorbents in environmental pollution management: a review. Chem Eng J 2011;170:395—410.

[11] Nas MS, Kuyuldar E, Demirkan B, Calimli MH, Demirbaş O, Sen F. Magnetic nanocomposites decorated on multiwalled carbon nanotube for removal of Maxilon Blue 5G using the sono-Fenton method. Sci Rep 2019;9:10850.

[12] Janas D. Towards monochiral carbon nanotubes: a review of progress in the sorting of single-walled carbon nanotubes. Mater Chem Front 2018;2:36—63.

[13] Balasubramanian K, Burghard M. Chemically functionalized carbon nanotubes. Small. 2005;1:180—92.

[14] Popov VN. Carbon nanotubes: properties and application. Mater Sci Eng R Rep 2004;43:61—102.

[15] Oberlin A, Endo M, Koyama T. Filamentous growth of carbon through benzene decomposition. J Cryst growth 1976;32:335—49.

[16] Iijima S, Ichihashi T. Single-shell carbon nanotubes of 1-nm diameter. Nature. 1993;363:603.

[17] Bethune D, Kiang CH, De Vries M, Gorman G, Savoy R, Vazquez J, et al. Cobalt-catalysed growth of carbon nanotubes with single-atomic-layer walls. Nature. 1993;363:605.

[18] Jourdain V, Bichara C. Current understanding of the growth of carbon nanotubes in catalytic chemical vapour deposition. Carbon. 2013;58:2—39.

[19] Ahmad M, Silva SRP. Low temperature growth of carbon nanotubes — a review. Carbon. 2020;158:24—44.

[20] Peng L-M, Zhang Z, Wang S. Carbon nanotube electronics: recent advances. Mater Today 2014;17:433—42.

[21] Alshehri R, Ilyas AM, Hasan A, Arnaout A, Ahmed F, Memic A. Carbon nanotubes in biomedical applications: factors, mechanisms, and remedies of toxicity: miniperspective. J Med Chem 2016;59:8149—67.

[22] Liu Y, Kumar S. Polymer/carbon nanotube nano composite fibers—a review. ACS Appl Mater Interfaces 2014;6:6069—87.

[23] Dresselhaus MS, Dresselhaus G, Eklund P, Rao A. Carbon nanotubes. The physics of fullerene-based and fullerene-related materials. Springer; 2000. p. 331—79.

[24] Hamada N, Sawada S-i, Oshiyama A. New one-dimensional conductors: graphitic microtubules. Phys Rev Lett 1992;68:1579.

[25] De Volder MF, Tawfick SH, Baughman RH, Hart AJ. Carbon nanotubes: present and future commercial applications. Science. 2013;339:535—9.

[26] Yamada T, Hayamizu Y, Yamamoto Y, Yomogida Y, Izadi-Najafabadi A, Futaba DN, et al. A stretchable carbon nanotube strain sensor for human-motion detection. Nat Nanotechnol 2011;6:296.

[27] Liu F, Wagterveld R, Gebben B, Otto M, Biesheuvel P, Hamelers H. Carbon nanotube yarns as strong flexible conductive capacitive electrodes. Colloid Interface Sci Commun 2014;3:9—12.

[28] Peigney A, Laurent C, Flahaut E, Bacsa RR, Rousset A. Specific surface area of carbon nanotubes and bundles of carbon nanotubes. Carbon. 2001;39:507—14.

[29] Monthioux M, Serp P, Flahaut E, Razafinimanana M, Laurent C, Peigney A, et al. Introduction to carbon nanotubes. In: Bhushan B, editor. Springer handbook of nanotechnology. Berlin, Heidelberg: Springer Berlin Heidelberg; 2007. p. 43—112.

[30] Rodriguez NM. A review of catalytically grown carbon nanofibers. J Mater Res 2011;8:3233—50.

[31] Yoon S-H, Lim S, Song Y, Ota Y, Qiao W, Tanaka A, et al. KOH activation of carbon nanofibers. Carbon. 2004;42:1723—9.

[32] Kumar N, Sinha Ray S. Synthesis and functionalization of nanomaterials. In: Sinha Ray S, editor. Processing of polymer-based nanocomposites: introduction. Cham: Springer International Publishing; 2018. p. 15—55.

[33] Moore VC, Strano MS, Haroz EH, Hauge RH, Smalley RE, Schmidt J, et al. Individually suspended single-walled carbon nanotubes in various surfactants. Nano Lett 2003;3:1379−82.

[34] Kam NWS, Dai H. Carbon nanotubes as intracellular protein transporters: generality and biological functionality. J Am Chem Soc 2005;127:6021−6.

[35] Zhou Y, Fang Y, Ramasamy RP. Non-covalent functionalization of carbon nanotubes for electrochemical biosensor development. Sensors. 2019;19:392.

[36] Pan B, Xing B. Adsorption mechanisms of organic chemicals on carbon nanotubes. Environ Sci Technol 2008;42:9005−13.

[37] Peng S, Li L, Kong Yoong Lee J, Tian L, Srinivasan M, Adams S, et al. Electrospun carbon nanofibers and their hybrid composites as advanced materials for energy conversion and storage. Nano Energy 2016;22:361−95.

[38] Ihsanullah, Abbas A, Al-Amer AM, Laoui T, Al-Marri MJ, Nasser MS, et al. Heavy metal removal from aqueous solution by advanced carbon nanotubes: critical review of adsorption applications. Sep Purif Technol 2016;157:141−61.

[39] Upadhyayula VKK, Deng S, Mitchell MC, Smith GB. Application of carbon nanotube technology for removal of contaminants in drinking water: a review. Sci Total Environ 2009;408:1−13.

[40] Yan Y, Zhang M, Gong K, Su L, Guo Z, Mao L. Adsorption of methylene blue dye onto carbon nanotubes: a route to an electrochemically functional nanostructure and its layer-by-layer assembled nanocomposite. Chem Mater 2005;17:3457−63.

[41] Machado FM, Fagan SB, da Silva IZ, de Andrade MJ. Carbon Nanoadsorbents. In: Bergmann CP, Machado FM, editors. Carbon nanomaterials as adsorbents for environmental and biological applications. Cham: Springer International Publishing; 2015. p. 11−32.

[42] Liang J, Liu J, Yuan X, Dong H, Zeng G, Wu H, et al. Facile synthesis of alumina-decorated multi-walled carbon nanotubes for simultaneous adsorption of cadmium ion and trichloroethylene. Chem Eng J 2015;273:101−10.

[43] Prola LDT, Machado FM, Bergmann CP, de Souza FE, Gally CR, Lima EC, et al. Adsorption of Direct Blue 53 dye from aqueous solutions by multi-walled carbon nanotubes and activated carbon. J Environ Manag 2013;130:166−75.

[44] Long RQ, Yang RT. Carbon nanotubes as superior sorbent for dioxin removal. J Am Chem Soc 2001;123:2058−9.

[45] Li Y, Du Q, Liu T, Peng X, Wang J, Sun J, et al. Comparative study of methylene blue dye adsorption onto activated carbon, graphene oxide, and carbon nanotubes. Chem Eng Res Des 2013;91:361−8.

[46] Stafiej A, Pyrzynska K. Adsorption of heavy metal ions with carbon nanotubes. Sep Purif Technol 2007;58:49−52.

[47] Tofighy MA, Mohammadi T. Adsorption of divalent heavy metal ions from water using carbon nanotube sheets. J Hazard Mater 2011;185:140−7.

[48] Liu C-H, Li J-J, Zhang H-L, Li B-R, Guo Y. Structure dependent interaction between organic dyes and carbon nanotubes. Colloids Surf A Physicochem Eng Asp 2008;313:9−12.

[49] Gupta VK, Kumar R, Nayak A, Saleh TA, Barakat M. Adsorptive removal of dyes from aqueous solution onto carbon nanotubes: a review. Adv Colloid Interface Sci 2013;193:24−34.

[50] Lu C, Chiu H, Liu C. Removal of zinc (II) from aqueous solution by purified carbon nanotubes: kinetics and equilibrium studies. Ind Eng Chem Res 2006;45:2850−5.

[51] Yanyan L, Kurniawan TA, Albadarin AB, Walker G. Enhanced removal of acetaminophen from synthetic wastewater using multi-walled carbon nanotubes (MWCNTs) chemically modified with NaOH, HNO_3/H_2SO_4, ozone, and/or chitosan. J Mol Liq 2018;251:369−77.

[52] Jin L, Zhao X, Qian X, Dong M. Nickel nanoparticles encapsulated in porous carbon and carbon nanotube hybrids from bimetallic metal-organic-frameworks for highly efficient adsorption of dyes. J Colloid Interface Sci 2018;509:245−53.

[53] Lawal IA, Lawal MM, Akpotu SO, Azeez MA, Ndungu P, Moodley B. Theoretical and experimental adsorption studies of sulfamethoxazole and ketoprofen on synthesized ionic liquids modified CNTs. Ecotoxicol Environ Saf 2018;161:542−52.

[54] Shan D, Deng S, He C, Li J, Wang H, Jiang C, et al. Intercalation of rigid molecules between carbon nanotubes for adsorption enhancement of typical pharmaceuticals. Chem Eng J 2018;332:102–8.

[55] Abdi J, Vossoughi M, Mahmoodi NM, Alemzadeh I. Synthesis of metal-organic framework hybrid nanocomposites based on GO and CNT with high adsorption capacity for dye removal. Chem Eng J 2017;326:1145–58.

[56] Kang D, Yu X, Ge M, Xiao F, Xu H. Novel Al-doped carbon nanotubes with adsorption and coagulation promotion for organic pollutant removal. J Environ Sci 2017;54:1–12.

[57] Ncibi MC, Sillanpää M. Optimizing the removal of pharmaceutical drugs Carbamazepine and Dorzolamide from aqueous solutions using mesoporous activated carbons and multi-walled carbon nanotubes. J Mol Liq 2017;238:379–88.

[58] Xu L, Wang Z, Ye S, Sui X. Removal of p-chlorophenol from aqueous solutions by carbon nanotube hybrid polymer adsorbents. Chem Eng Res Des 2017;123:76–83.

[59] Maleki A, Hamesadeghi U, Daraei H, Hayati B, Najafi F, McKay G, et al. Amine functionalized multi-walled carbon nanotubes: single and binary systems for high capacity dye removal. Chem Eng J 2017;313:826–35.

[60] Czech B, Oleszczuk P. Sorption of diclofenac and naproxen onto MWCNT in model wastewater treated by H_2O_2 and/or UV. Chemosphere. 2016;149:272–8.

[61] Zhao H, Liu X, Cao Z, Zhan Y, Shi X, Yang Y, et al. Adsorption behavior and mechanism of chloramphenicols, sulfonamides, and non-antibiotic pharmaceuticals on multi-walled carbon nanotubes. J Hazard Mater 2016;310:235–45.

[62] Yang S, Wang L, Zhang X, Yang W, Song G. Enhanced adsorption of Congo red dye by functionalized carbon nanotube/mixed metal oxides nanocomposites derived from layered double hydroxide precursor. Chem Eng J 2015;275:315–21.

[63] Mahmoodi NM, Ghobadi J. Extended isotherm and kinetics of binary system dye removal using carbon nanotube from wastewater. Desalination Water Treat 2015;54:2777–93.

[64] Abdel-Ghani NT, El-Chaghaby GA, Helal FS. Individual and competitive adsorption of phenol and nickel onto multiwalled carbon nanotubes. J Adv Res 2015;6:405–15.

[65] Gong J, Liu J, Jiang Z, Wen X, Mijowska E, Tang T, et al. A facile approach to prepare porous cup-stacked carbon nanotube with high performance in adsorption of methylene blue. J Colloid Interface Sci 2015;445:195–204.

[66] Li H, Sun Z, Tian Y, Cui G, Yan S. Facile and cost-effective synthesis of $CNT@MCo_2O_4$ (M = Ni, Mn, Cu, Zn) core–shell hybrid nanostructures for organic dye removal. RSC Adv 2015;5:79765–73.

[67] Karim AH, Jalil AA, Triwahyono S, Kamarudin NHN, Ripin A. Influence of multi-walled carbon nanotubes on textural and adsorption characteristics of in situ synthesized mesostructured silica. J Colloid Interface Sci 2014;421:93–102.

[68] Ghobadi J, Arami M, Bahrami H, Mahmoodi NM. Modification of carbon nanotubes with cationic surfactant and its application for removal of direct dyes. Desal Water Treat 2014;52:4356–68.

[69] Wang S, Zhai Y-Y, Gao Q, Luo W-J, Xia H, Zhou C-G. Highly efficient removal of acid Red 18 from aqueous solution by magnetically retrievable chitosan/carbon nanotube: batch study, isotherms, kinetics, and thermodynamics. J Chem Eng Data 2014;59:39–51.

[70] Yu F, Chen J, Chen L, Huai J, Gong W, Yuan Z, et al. Magnetic carbon nanotubes synthesis by Fenton's reagent method and their potential application for removal of azo dye from aqueous solution. J Colloid Interface Sci 2012;378:175–83.

[71] Ai L, Jiang J. Removal of methylene blue from aqueous solution with self-assembled cylindrical graphene–carbon nanotube hybrid. Chem Eng J 2012;192:156–63.

[72] Huang Z-n, Wang X-l, Yang D-s. Adsorption of Cr (VI) in wastewater using magnetic multi-wall carbon nanotubes. Water Sci Eng 2015;8:226–32.

[73] Kumar N, Fosso-Kankeu E, Ray SS. Achieving controllable MoS_2 nanostructures with increased interlayer spacing for efficient removal of Pb(II) from aquatic systems. ACS Appl Mater Interfaces 2019;11:19141–55.

[74] Ahmad A, Razali MH, Mamat M, Mehamod FSB, Anuar Mat Amin K. Adsorption of methyl orange by synthesized and functionalized-CNTs with 3-aminopropyltriethoxysilane loaded TiO_2 nanocomposites. Chemosphere. 2017;168:474−82.

[75] Saber-Samandari S, Saber-Samandari S, Joneidi-Yekta H, Mohseni M. Adsorption of anionic and cationic dyes from aqueous solution using gelatin-based magnetic nanocomposite beads comprising carboxylic acid functionalized carbon nanotube. Chem Eng J 2017;308:1133−44.

[76] Gotovac S, Song L, Kanoh H, Kaneko K. Assembly structure control of single wall carbon nanotubes with liquid phase naphthalene adsorption. Colloids Surf A Physicochem Eng Asp 2007;300:117−21.

[77] Peng X, Li Y, Luan Z, Di Z, Wang H, Tian B, et al. Adsorption of 1, 2-dichlorobenzene from water to carbon nanotubes. Chem Phys Lett 2003;376:154−8.

[78] Jiang X, Gu J, Bai X, Lin L, Zhang Y. The influence of acid treatment on multi-walled carbon nanotubes. Pigment Resin Technol 2009;38:165−73.

[79] Mishra AK, Arockiadoss T, Ramaprabhu S. Study of removal of azo dye by functionalized multi walled carbon nanotubes. Chem Eng J 2010;162:1026−34.

[80] Li Y-H, Wang S, Wei J, Zhang X, Xu C, Luan Z, et al. Lead adsorption on carbon nanotubes. Chem Phys Lett 2002;357:263−6.

[81] Cho H-H, Huang H, Schwab K. Effects of solution chemistry on the adsorption of ibuprofen and triclosan onto carbon nanotubes. Langmuir. 2011;27:12960−7.

[82] Yang J-Y, Jiang X-Y, Jiao F-P, Yu J-G. The oxygen-rich pentaerythritol modified multi-walled carbon nanotube as an efficient adsorbent for aqueous removal of alizarin yellow R and alizarin red S. Appl Surf Sci 2018;436:198−206.

[83] Navaei Diva T, Zare K, Taleshi F, Yousefi M. Synthesis, characterization, and application of nickel oxide/CNT nanocomposites to remove Pb^{2+} from aqueous solution. J Nanostruct Chem 2017;7:273−81.

[84] Hayati B, Maleki A, Najafi F, Daraei H, Gharibi F, McKay G. Super high removal capacities of heavy metals (Pb^{2+} and Cu^{2+}) using CNT dendrimer. J Hazard Mater 2017;336:146−57.

[85] Wang Y, Hu L, Zhang G, Yan T, Yan L, Wei Q, et al. Removal of Pb(II) and methylene blue from aqueous solution by magnetic hydroxyapatite-immobilized oxidized multi-walled carbon nanotubes. J Colloid Interface Sci 2017;494:380−8.

[86] Lu W, Li J, Sheng Y, Zhang X, You J, Chen L. One-pot synthesis of magnetic iron oxide nanoparticle-multiwalled carbon nanotube composites for enhanced removal of Cr(VI) from aqueous solution. J Colloid Interface Sci 2017;505:1134−46.

[87] AlOmar MK, Alsaadi MA, Hayyan M, Akib S, Ibrahim RK, Hashim MA. Lead removal from water by choline chloride based deep eutectic solvents functionalized carbon nanotubes. J Mol Liq 2016;222:883−94.

[88] Saleh TA. Nanocomposite of carbon nanotubes/silica nanoparticles and their use for adsorption of Pb(II): from surface properties to sorption mechanism. Desal Water Treat 2016;57:10730−44.

[89] Li Q, Yu J, Zhou F, Jiang X. Synthesis and characterization of dithiocarbamate carbon nanotubes for the removal of heavy metal ions from aqueous solutions. Colloids Surf A Physicochem Eng Asp 2015;482:306−14.

[90] Ma X, Yang S-T, Tang H, Liu Y, Wang H. Competitive adsorption of heavy metal ions on carbon nanotubes and the desorption in simulated biofluids. J Colloid Interface Sci 2015;448:347−55.

[91] Moghaddam HK, Pakizeh M. Experimental study on mercury ions removal from aqueous solution by MnO_2/CNTs nanocomposite adsorbent. J Ind Eng Chem 2015;21:221−9.

[92] Jahangiri M, Kiani F, Tahermansouri H, Rajabalinezhad A. The removal of lead ions from aqueous solutions by modified multi-walled carbon nanotubes with 1-isatin-3-thiosemicarbazone. J Mol Liq 2015;212:219−26.

[93] Sui Z, Meng Q, Zhang X, Ma R, Cao B. Green synthesis of carbon nanotube−graphene hybrid aerogels and their use as versatile agents for water purification. J Mater Chem 2012;22:8767−71.

[94] Tian Y, Gao B, Morales VL, Wu L, Wang Y, Muñoz-Carpena R, et al. Methods of using carbon nanotubes as filter media to remove aqueous heavy metals. Chem Eng J 2012;210:557−63.

[95] Tong S, Zhao S, Zhou W, Li R, Jia Q. Modification of multi-walled carbon nanotubes with tannic acid for the adsorption of La, Tb and Lu ions. Microchim Acta 2011;174:257—64.

[96] Doong R-A, Chiang L-F. Coupled removal of organic compounds and heavy metals by titanate/carbon nanotube composites. Water Sci Technol 2008;58:1985—92.

[97] Li Y-H, Ding J, Luan Z, Di Z, Zhu Y, Xu C, et al. Competitive adsorption of Pb^{2+}, Cu^{2+} and Cd^{2+} ions from aqueous solutions by multiwalled carbon nanotubes. Carbon. 2003;41:2787—92.

[98] Ji L, Chen W, Zheng S, Xu Z, Zhu D. Adsorption of sulfonamide antibiotics to multiwalled carbon nanotubes. Langmuir. 2009;25:11608—13.

[99] Meloun M, Bordovská S, Galla L. The thermodynamic dissociation constants of four non-steroidal anti-inflammatory drugs by the least-squares nonlinear regression of multiwavelength spectrophotometric pH-titration data. J Pharm Biomed Anal 2007;45:552—64.

[100] Li J, Li J, Meng H, Xie S, Zhang B, Li L, et al. Ultra-light, compressible and fire-resistant graphene aerogel as a highly efficient and recyclable absorbent for organic liquids. J Mater Chem A 2014;2:2934—41.

[101] Wu L, Qin Z, Zhang L, Meng T, Yu F, Ma J. CNT-enhanced amino-functionalized graphene aerogel adsorbent for highly efficient removal of formaldehyde. N J Chem 2017;41:2527—33.

[102] Qiu L, Liu JZ, Chang SLY, Wu Y, Li D. Biomimetic superelastic graphene-based cellular monoliths. Nat Commun 2012;3:1241.

[103] Nardecchia S, Carriazo D, Ferrer ML, Gutiérrez MC, del Monte F. Three dimensional macroporous architectures and aerogels built of carbon nanotubes and/or graphene: synthesis and applications. Chem Soc Rev 2013;42:794—830.

[104] Wan W, Zhang R, Li W, Liu H, Lin Y, Li L, et al. Graphene—carbon nanotube aerogel as an ultra-light, compressible and recyclable highly efficient absorbent for oil and dyes. Environ Sci Nano 2016;3:107—13.

[105] Lee B, Lee S, Lee M, Jeong DH, Baek Y, Yoon J, et al. Carbon nanotube-bonded graphene hybrid aerogels and their application to water purification. Nanoscale. 2015;7:6782—9.

[106] Das R. Nanohybrid catalysts based on carbon nanotubes. Springer; 2017.

[107] Chen C, Hu J, Shao D, Li J, Wang X. Adsorption behavior of multiwall carbon nanotube/iron oxide magnetic composites for Ni(II) and Sr(II). J Hazard Mater 2009;164:923—8.

[108] Gupta V, Agarwal S, Saleh TA. Chromium removal by combining the magnetic properties of iron oxide with adsorption properties of carbon nanotubes. Water Res 2011;45:2207—12.

[109] Vadahanambi S, Lee S-H, Kim W-J, Oh I-K. Arsenic removal from contaminated water using three-dimensional graphene-carbon nanotube-iron oxide nanostructures. Environ Sci Technol 2013;47:10510—17.

[110] Zhang C, Sui J, Li J, Tang Y, Cai W. Efficient removal of heavy metal ions by thiol-functionalized superparamagnetic carbon nanotubes. Chem Eng J 2012;210:45—52.

[111] Vieira EF, Simoni JA, Airoldi C. Interaction of cations with SH-modified silica gel: thermochemical study through calorimetric titration and direct extent of reaction determination. J Mater Chem 1997;7:2249—52.

[112] Kafshgari LA, Ghorbani M, Azizi A. Fabrication and investigation of $MnFe_2O_4$/MWCNTs nanocomposite by hydrothermal technique and adsorption of cationic and anionic dyes. Appl Surf Sci 2017;419:70—83.

[113] Raymundo-Piñero E, Azaïs P, Cacciaguerra T, Cazorla-Amorós D, Linares-Solano A, Béguin F. KOH and NaOH activation mechanisms of multiwalled carbon nanotubes with different structural organisation. Carbon. 2005;43:786—95.

[114] Okada K, Yamamoto N, Kameshima Y, Yasumori A. Porous properties of activated carbons from waste newspaper prepared by chemical and physical activation. J Colloid Interface Sci 2003;262:179—93.

[115] Ma J, Yu F, Zhou L, Jin L, Yang M, Luan J, et al. Enhanced adsorptive removal of methyl orange and methylene blue from aqueous solution by alkali-activated multiwalled carbon nanotubes. ACS Appl Mater Interfaces 2012;4:5749—60.

[116] Lillo-Ródenas MA, Cazorla-Amorós D, Linares-Solano A. Understanding chemical reactions between carbons and NaOH and KOH: an insight into the chemical activation mechanism. Carbon. 2003;41:267−75.

[117] Ma J, Zhu Z, Chen B, Yang M, Zhou H, Li C, et al. One-pot, large-scale synthesis of magnetic activated carbon nanotubes and their applications for arsenic removal. J Mater Chem A 2013;1:4662−6.

[118] Huang X. Fabrication and properties of carbon fibers. Materials. 2009;2:2369−403.

[119] De Jong KP, Geus JW. Carbon nanofibers: catalytic synthesis and applications. Catal Rev 2000;42:481−510.

[120] Feng L, Xie N, Zhong J. Carbon nanofibers and their composites: a review of synthesizing, properties and applications. Materials. 2014;7:3919.

[121] Roberts AD, Li X, Zhang H. Porous carbon spheres and monoliths: morphology control, pore size tuning and their applications as Li-ion battery anode materials. Chem Soc Rev 2014;43:4341−56.

[122] Zhu J, Wei S, Ryu J, Guo Z. Strain-sensing elastomer/carbon nanofiber "metacomposites". J Phys Chem C 2011;115:13215−22.

[123] Qie L, Chen W-M, Wang Z-H, Shao Q-G, Li X, Yuan L-X, et al. Nitrogen-doped porous carbon nanofiber webs as anodes for lithium ion batteries with a superhigh capacity and rate capability. Adv Mater 2012;24:2047−50.

[124] Serp P, Corrias M, Kalck P. Carbon nanotubes and nanofibers in catalysis. Appl Catal A Gen 2003;253:337−58.

[125] Barcena J, Coleto J, Zhang SC, Hilmas GE, Fahrenholtz WG. Processing of carbon nanofiber reinforced ZrB2 matrix composites for aerospace applications. Adv Eng Mater 2010;12:623−6.

[126] Teng M, Qiao J, Li F, Bera PK. Electrospun mesoporous carbon nanofibers produced from phenolic resin and their use in the adsorption of large dye molecules. Carbon. 2012;50:2877−86.

[127] Si Y, Ren T, Li Y, Ding B, Yu J. Fabrication of magnetic polybenzoxazine-based carbon nanofibers with Fe_3O_4 inclusions with a hierarchical porous structure for water treatment. Carbon. 2012;50:5176−85.

[128] Zhu Z, Xu Y, Qi B, Zeng G, Wu P, Liu G, et al. Adsorption-intensified degradation of organic pollutants over bifunctional α-Fe@carbon nanofibres. Environ Sci Nano 2017;4:302−6.

[129] Ahmed YM, Al-Mamun A, Al Khatib MaFR, Jameel AT, AlSaadi M. Efficient lead sorption from wastewater by carbon nanofibers. Environ Chem Lett 2015;13:341−6.

[130] Gupta AK, Deva D, Sharma A, Verma N. Fe-Grown carbon nanofibers for removal of arsenic(V) in wastewater. Ind Eng Chem Res 2010;49:7074−84.

[131] Ren T, Si Y, Yang J, Ding B, Yang X, Hong F, et al. Polyacrylonitrile/polybenzoxazine-based Fe_3O_4@carbon nanofibers: hierarchical porous structure and magnetic adsorption property. J Mater Chem 2012;22:15919−27.

[132] Si Y, Ren T, Ding B, Yu J, Sun G. Synthesis of mesoporous magnetic Fe_3O_4@carbon nanofibers utilizing in situ polymerized polybenzoxazine for water purification. J Mater Chem 2012;22:4619−22.

[133] Li X, Chen S, Fan X, Quan X, Tan F, Zhang Y, et al. Adsorption of ciprofloxacin, bisphenol and 2-chlorophenol on electrospun carbon nanofibers: in comparison with powder activated carbon. J Colloid Interface Sci 2015;447:120−7.

CHAPTER TEN

Two-dimensional carbon nanomaterials-based adsorbents

10.1 Introduction

Water is an essential constituent for living beings on the earth and this valuable resource requires preservation and protection. A massive amount of hazardous pollutants are released into water bodies due to substantial urbanization and growing industrialization. Various pollutants, such as dyes, heavy metals, antibiotics, polycyclic hydrocarbons (PAHs), polychlorinated biphenyls (PCBs), endocrine-disrupting chemicals, and personal care products, are regularly released into wastewater and industrial streams, without any treatment. These contaminants pose a serious threat to all living beings because of their nonbiodegradability, toxicity, and tendency to amass in living organisms [1]. Therefore a wide range of methodologies have been studied to avail the affordable technologies and effective materials that can control the number of contaminants to permissible limits before allowing them to enter water streams. Previously developed techniques, including coagulation/flocculation [2], ion exchange [3], precipitation [4], photocatalysis [5−7], oxidation [8], and membrane separation [9], are employed for wastewater treatments, but these techniques suffer from serious shortcomings in areas such as operation efficiency, reusability, high operation cost, and applicability for broad-spectrum pollutants. However, adsorption has been considered as an economically achievable, versatile, and extremely efficient technology for water purification [10−12]. Owing to their high surface area and low-cost, conventionally carbon-rich materials such as biochar, soot, carbon black, and charcoal have been used as adsorbents for the elimination of various pollutants from water [13]. These carbon-rich materials have limited adsorption active sites and internal pores for the removal of various effluents from polluted water [14]. Thus in order to fulfill the increasing demand for pure water, water sustainability, and to achieve rigorous environmental guidelines, there is the need for dedicated research to develop efficient carbon nanomaterials (CNMs)-based adsorbents with high performance, partition coefficients, excellent removal capacities, higher surface area, and porosity.

Due to the fast growth of nanotechnology in the last decade, CNMs have been extensively studied for water treatments with hope that they might offer highly effective adsorbents and new advanced strategies to solve the conundrums of water. Two-dimensional (2D) CNMs cover graphene, graphene oxide (GO), reduced

Carbon Nanomaterial-Based Adsorbents for Water Purification.
DOI: https://doi.org/10.1016/B978-0-12-821959-1.00010-6

graphene oxide (rGO), graphene nanoplatelets (GNPs), graphitic-carbon nitride (g-C$_3$N$_4$), and sheet-like nanoporous carbons (NPCs). Among the 2D CNMs, graphene and analogues have been at the forefront of cutting edge research in multidisciplinary fields due to their exceptional chemical and physical properties [15]. They have been transcended into multipurpose and supporting platforms to develop a disruptive myriad of technologies in sensing, energy, environment, printable electronics, optoelectronics, and biomedical areas. Due to their high surface area, sp^2 hybridized carbons, high active sites, decent chemical stability, and large pi-(π) electron systems, graphene has been used as a versatile adsorbent for water contaminants management [16]. The very hydrophobic carbon surfaces of graphene and reduced graphene are greatly exploited for the removal of organic pollutants using $\pi-\pi$ interactions.

Moreover, the step-edges, groove/wrinkle sites, and voids/defects on the graphene act as high-energy, active sites for adsorption of foreign molecules. The GO is a highly oxidized form of graphene, which is obtained by oxidation of graphite, followed by exfoliation. Like graphene, GO has also been employed as an adsorbent because of its hydrophilicity, high surface area, and various oxygen functionalities ($-$OH, $-$COOH, and $-$C$=$O) [17]. The adsorption of various pollutants on GO occur through the electrostatic attraction, complexation/coordination, and $\pi-\pi$ stacking [18].

Nevertheless, the serious problem associated with graphene that hinders its practical applicability in water treatments is posttreatment recovery as it has a tendency to aggregate or rolls to form graphite in water because of van der Waals forces and $\pi-\pi$ interactions between graphene layers [19,20]. The aggregation of graphene sheets results in a decline in surface area, a slowdown of mass transport, and an overall decrease in adsorption performance [16]. In the case of GO, it exhibits very low adsorption characteristics toward anionic molecules as it carries a negative charged surface, resulting in vigorous electrostatic repulsions. These shortcomings of graphene/GO can be resolved by functionalization with appropriate molecules, and anchoring of economical and reactive inorganic compounds to make hybrids or composites, which can enhance adsorption efficiency, sensitivity, selectivity, and versatility for the removal of emerging pollutants from waste and industrial water. This chapter covers the fundamental characteristics of 2D CNMs and their recent advancements in water remediation application as adsorbents.

10.2 Two-dimensional carbon nanomaterials

2D CNMs include graphene, GO, rGO, GNPs, g-C$_3$N$_4$, and sheet-like NPCs. Among these 2D CNMs, graphene and its derivatives are most popular for a wide range of applications due to their exceptional physicochemical properties. Graphene is a single-atom-thick, two-dimensional (2D) nanosheet of sp^2 hybridized carbon atoms, arranged in a honeycomb structure. Generally, it exists as a stack of 2D layers in the 3D graphite structure and as the basic building block of other carbon allotropes. Due to its extraordinary characteristics, including excellent mechanical strength, high thermal and electrical conductivities, chemical inertness, high current density, optical transmittance, etc., graphene has captured intense attention from researchers worldwide for use in different applications, including as sensors, photovoltaic cells, energy storage, catalysis, lubrication, and separation membrane technology [17,21−25].

The graphene derivative, GO, can be obtained by strong oxidation of graphite using commonly used modified/improved Hummers methods in the presence of acids (H_2SO_4 and/or H_3PO_4) and oxidants ($KMnO_4$ or $NaNO_3$) [1]. Some researchers pretreated graphite powders using compounds ($K_2S_2O_8$ and P_2O_5) before Hummer's oxidation [26,27] to decrease the C/O ratio in GO. In many cases, expanded graphite is formed first through thermal treatment or using different oxidizers to increase interlayer distances, then it is easily exfoliated into GO [28]. The "improved Hummer's method" is the most popular modification that excludes $NaNO_3$ with the inclusion of H_3PO_4 and the extra addition of $KMnO_4$. This method produces no toxic gases (NOx), allowing easy control of the reaction temperature, and produces GO with a high number of oxygen functionalities [29]. All these methods yield highly oxidized hydrophilic GO. In GO, carboxyl and hydroxyl groups are present on the edges, while hydroxyl, epoxy, and carbonyl groups are present on basal planes. The oxygen contents or C/O ratio in GO is dependent on the selection of pretreatment methods, oxidizing agent, and protonated solvent. Later, the prepared GO is reduced to rGO using different routes including chemical, thermal, hydrothermal, microwave, and laser reduction (Fig. 10.1A). In thermal reduction, decomposition of oxygen moieties into CO_2/CO gases takes place at high temperatures, resulting in delaminated individual graphene sheets [30]. While in chemical reduction GO is reduced by reducing agents such as sodium borohydride [31], hydrazine [32], and metal hydrides or hydrohalic acids [30]. Concurrently, a wide range of green reducing agents such as sugars, ascorbic acid, and microorganisms, have been effectively utilized for the synthesis of rGO [33]. The different reduction approaches come with various advantages of scalability, maintaining or improving characteristics, energy intensiveness, C/O ratio control, selective removal of oxygen functional groups, decrease in level of defects, and chemical waste generation. The structure of GO and rGO is slightly different from pristine graphene and they can be distinguished by their surface carrying oxygen functional moieties. Pristine graphene has no oxygen with interlayer spacing of 0.34 nm, while GO and rGO have oxygen functionalities with interlayer spacing of 0.57−0.90 and 0.37 nm, respectively. The C/O ratios of GO and rGO are in the range of 2−4 and 8−246, respectively [34]. Table 10.1 shows the structures and properties of graphene, GO, and rGO.

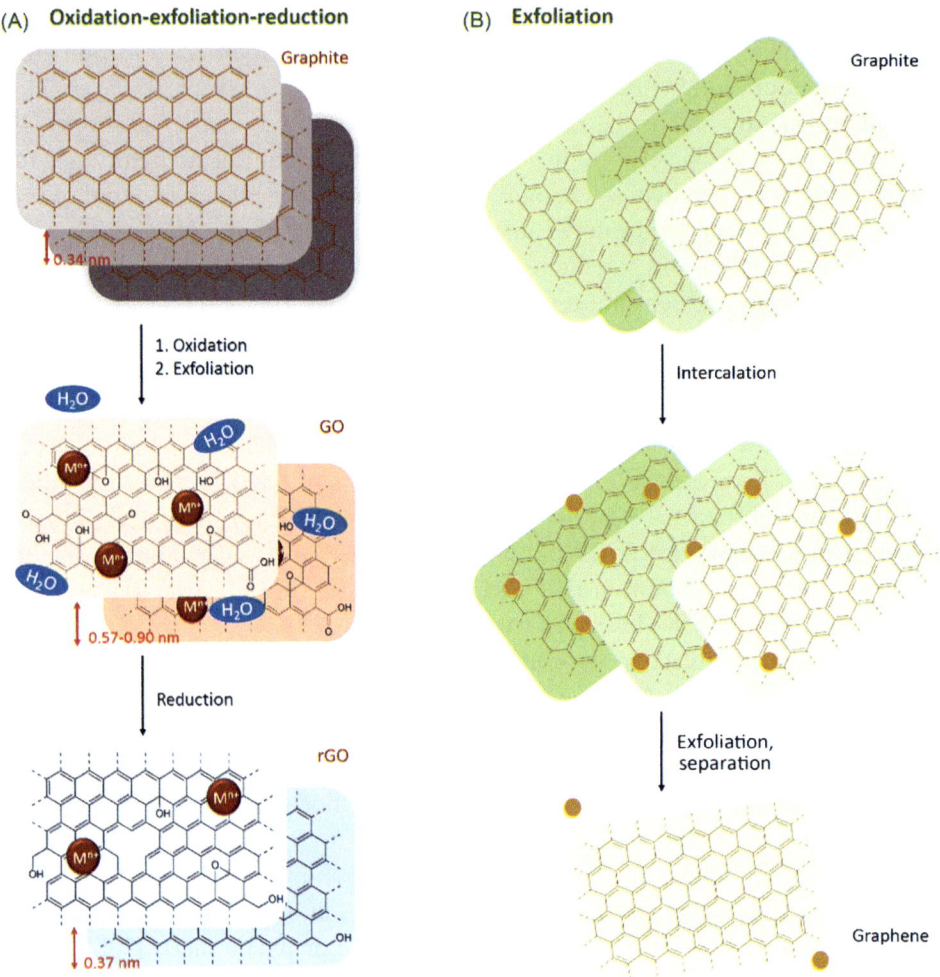

Figure 10.1 Schematic illustration of a synthesis route for (A) preparation of GO and rGO using oxidation-exfoliation-reduction method, and (B) preparation of graphene using interaction/exfoliation approach.

The CVD is the most appropriate technique to develop single-layer graphene [26]. However, GNPs, few-layer graphene, and graphene quantum dots (GQDs) are prepared by liquid-phase exfoliations (LPE) (Fig. 10.1B). In LPE numerous dispersants such as water, ionic liquids, supercritical fluids, organic solvents, surfactants, polymers, and other green solvents have been used for the delamination of graphite into graphene. A number of techniques, namely high-shear mixing, sonication, jet cavitation, and microfluidization, are used in the LPE process [44,45]. Among the LPE techniques, the high-shear mixing or microfluidization methods are considered as emerging technologies, which can successfully yield defect-rich few-layer graphene/multilayer

Table 10.1 Structure, synthesis methods, oxygen contents, the surface area of graphene, GO and rGO.

Graphene and derivatives	Synthesis methods	Oxygen contents (%)	Surface area (m²/g)	References
Graphene	Graphite exfoliation, chemical vapor deposition, etc.	0.8—5	120—1556	[35,36]
Graphene oxide (GO)	Graphite oxidation and exfoliation	10.4—54.4	4—771	[17,37,38]
Reduced graphene oxide (rGO)	Reduction of GO	17.5—31.3	128—931	[39,40]
Graphene nanoplatelets (GNPs)	Graphite liquid/solid exfoliation, graphite intercalation—exfoliation	<1	20—750	[37,41—43]

graphene in bulk using high-shear mixer-based fluid dynamics [45]. For instance, Wang et al. developed a novel and effective method to produce high-grade graphene from graphite using an orderly sonication and microfluidization in N-methyl-2-pyrrolidone (NMP) and sodium hydroxide (NaOH) [46]. The concentration of obtained few-layer graphene was 0.47 mg/mL and the lateral size was in the range of 0.5−2.0 μm. During microfluidization exfoliation, graphite undergoes intersheets collision, shear stress, and a cavitation mechanism to produce few-layer graphene [46]. Moreover, the occurrence of oxygen-carrying functional groups on the GO/rGO surface makes it an appropriate material for the adsorbent. The organic species easily adsorb on the GO/rGO through electrostatic attractions and π−π interactions, while they can adsorb on pristine graphene/GNPs via only π−π interactions. The adsorption performance of GO/rGO can be enhanced by covalent and noncovalent functionalization. Among various functional groups on GO, the carboxylic acid is highly reactive as, being present at the edges, it can be activated initially using the addition of thionyl chloride, followed by nucleophile attack to create covalent bonds [47]. Likewise, the epoxy group present on the basal plane of GO can be opened using nucleophilic addition [48]. The present hydroxyl group is also reactive and acts as a nucleophile that can attack reactive ketones giving functionalized ketone carrying functionality on rGO/GO [49]. Moreover, silanization is one of the most popular strategies to modify GO surface with amine, amide, and mercapto groups [50]. Besides organic functionalization, graphene and its derivatives are also functionalized with inorganic nanomaterials and polymer, with different dimensionalities to improve their adsorption properties [18,34].

In the literature, five types of carbon nitride with different phases (α-C_3N_4, β-C_3N_4, c-C_3N_4, g-C_3N_4, and p-C_3N_4) are reported [51]. Except for graphitic-carbon nitride (g-C_3N_4), all carbon nitrides are superhard materials [52]. Therefore the structures and morphology of g-C_3N_4 can easily be modified for their use in various applications. Owing to their nitrogen-rich functionalities, high surface area, and porosity, g-C_3N_4 and its composites are successfully employed for water purification application. They also have a network of sp^2 hybridized carbon, −C−N bonds, and a greatly delocalized conjugated π-system, which offer versatile functionalities for adsorption applications. The g-C_3N_4 is applied as an adsorbent and photocatalyst for water treatments. The g-C_3N_4 can be prepared using various methods, namely solid-state reaction, solvothermal reaction, electrochemical deposition, and thermal decomposition (Fig. 10.2). Among these methods, the thermal decomposition approach is more commonly used for the preparation of g-C_3N_4 using various precursors, such as urea, cyanamide, guanidine hydrochloride, melamine, dicyanamide, guanidinium thiocyanate, and sulfocarbamide, by heating at elevated temperatures [53]. Furthermore, carbon nitrides can be functionalized through organic and inorganic compounds to enhance their intrinsic properties.

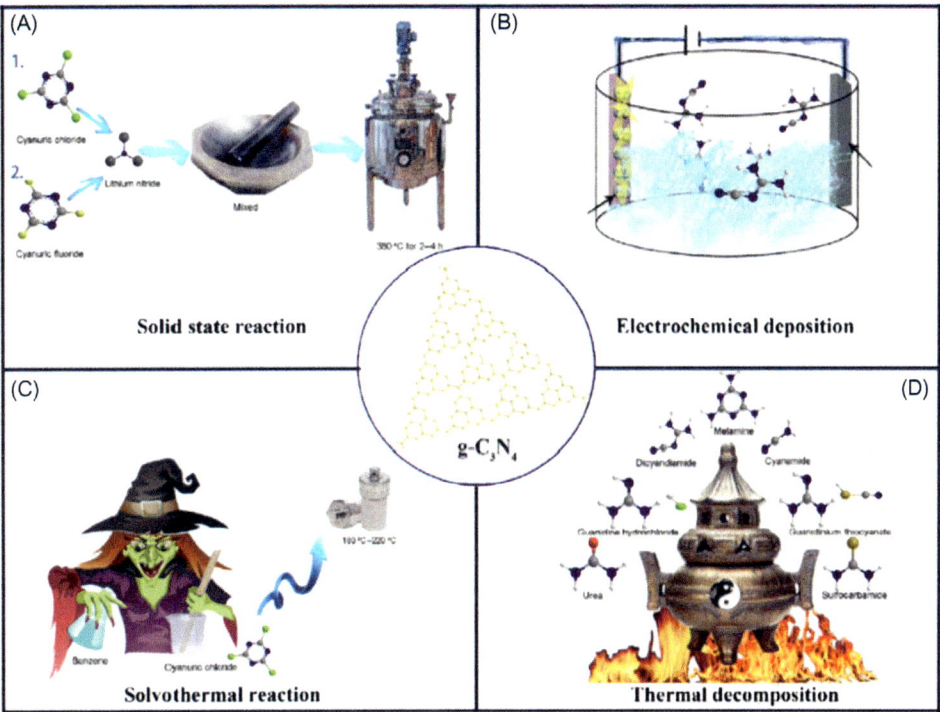

Figure 10.2 Various synthetic methods for the preparation of g-C$_3$N$_4$: (A) solid-state reaction; (B) electrochemical deposition; (C) solvothermal reaction; and (D) thermal decomposition. *Reproduced with permission from Hao Q, Jia G, Wei W, Vinu A, Wang Y, Arandiyan H, et al. Graphitic carbon nitride with different dimensionalities for energy and environmental applications. Nano Res 2019 [53]. Copyright 2019, Springer.*

Large surface area and high porosity are the fundamental characteristics of NPCs, which make them strong candidates for use in the adsorption of organic and inorganic pollutants from aqueous media [54]. They are low–cost, nontoxic, abundant, originate from renewable sources, and have excellent chemical resistivity and high thermal stability, which has increased the demand for NPCs as economically feasible adsorbents [55]. Furthermore, the synthesis of NPCs using waste material relieves an environmental burden, which encourages NPC use in a variety of applications [56,57].

10.3 Graphene-based adsorbents

A high surface to weight ratio (theoretical surface area ~ 2600 m^2/g) and chemical stability make graphene a potent candidate for the adsorption of organic and inorganic effluents from water [58]. Recently, graphene oxide (GO), reduced graphene oxide (rGO), GO sponges, chemically functionalized GO, and GO nanocomposites have been proven to be successful adsorbents in the removal of pollutants from industrial, domestic, and agricultural effluents [59,60]. Several studies report on the role of graphene and its derivatives for the dynamic adsorption of organic dyes [e.g., rhodamine (RG), malachite green (MG), methyl orange (MO), methylene blue (MB), Congo red (CR)], heavy metal ions [e.g., Cr(VI), Zn(II), Cd(II), Fe(III), Co(II), Pb(II)], phenols, and pharmaceutical compounds [60−62]. Adsorption to graphene and its derivatives is principally via physisorption and is governed by electrostatic interactions. GO is more suitable for adsorption of positively charged water contaminants due to the negatively charged oxygen functionalities on the graphene oxide surface [40,63]. Moreover, the structural transformation and stability behavior of graphene and its derivatives during surface adsorption have been demonstrated in Fig. 10.3A and B. Besides the amphipathic nature of GO, it is a soft material with a large aspect ratio, and its morphology and structure are simply distorted by applying external influences. Aggregation of GO could be noticed in the presence of metal ion adsorption and on varying the pH of the solution. Chowdhury et al. highlighted that GO aggregation follows the DLVO theory and the Schulze−Hardy rule, in which aggregation is highly reliant on the ionic strength and cation type but feebly depends on pH in the range of 4−10 [67]. It was further revealed that GO aggregation occurs via face-to-face and edge-to-edge interactions (Fig. 10.3A). In the presence of M(II), GO is linked to creating an edge-to-edge aggregation due to surface complexation. The face-to-face aggregation occurs in low pH solution due to excessive protonation of GO [64]. During the heavy metal removal process using GO, it is noted that heavy metal cations Ag(I), Cu(II), Cd(II), Pb(II), and Cr(III) could destabilize GO more effectively than normal cations such as K(I), Na(I), Ca(II), and Mg(II) via surface adsorption. Heavy metal ions might easily surpass electric double-layer suppression and anchor to the GO surface and then disturb the surface potential which eventually results in the structural transformation of 2D GO into 1D tube-like, 2D multioverlapped sheets, and 3D sphere-like structures via π-cation interactions (Fig. 10.3B). Kinetics observation revealed that morphological conversion of GO during metal adsorption can be governed by the metal's electronegativity and thickness of the hydration shell [65].

In general, graphene or graphene-based nanocomposites follow Langmuir adsorption isotherms and demonstrate monolayer formation of the contaminant onto the graphene nanocomposite surface. GO, and rGO provide oxygen functionalities, such as −OH, −COO−, −C−O−C−, and −COOH, on the surface in different quantities, which are available to bond with water contaminants. Electrostatic interactions between the electron-rich graphene oxide and cationic dyes, and π−π interactions between the π electron cloud of the GO rings and aromatic rings of the dye are two major interactions between the dye (adsorbate) and GO (adsorbent) [68]. Many studies have shown excellent adsorption properties of intrinsic GO/rGO for metal ions removal as well as organic pollutants removal. Gupta et al. synthesized high surface area (931 m^2/g) rGO with a

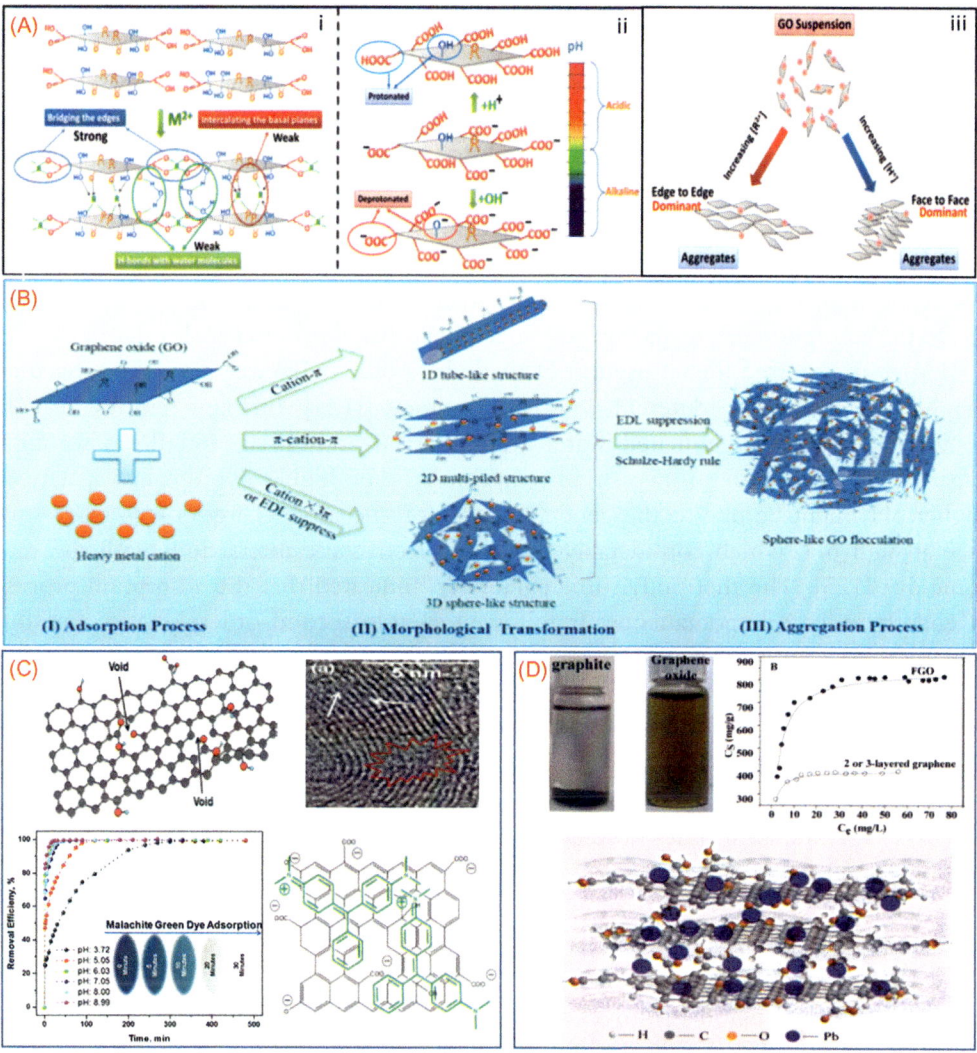

Figure 10.3 (A) Exhibit the aggregation mechanisms of GO on interactions with metal cations and protons. (B) Morphological transformation of GO upon the influence of heavy metal cations and aggregation. (C) Reduced GO with a high number of voids for efficient removal of malachite green (MG) dye using π—π interaction and electrostatic interactions. (D) High dispersion of FGO even after 5 months, and adsorption capacities of Pb(II) on graphene and FGO. *(A) Reproduced with permission from Wu L, Liu L, Gao B, Muñoz-Carpena R, Zhang M, Chen H, et al. Aggregation kinetics of graphene oxides in aqueous solutions: experiments, mechanisms, and modeling. Langmuir 2013;29:15174—81 [64]. Copyright 2013, American Chemical Society. (B) Reproduced with permission from Yang K, Chen B, Zhu X, Xing B. Aggregation, adsorption, and morphological transformation of graphene oxide in aqueous solutions containing different metal cations. Environ Sci Technol 2016;50:11066—75 [65]. Copyright 2016, American Chemical Society. (C) Reproduced with permission (Continued)*

high number of voids for efficient removal of MG dye from simulated wastewater (Fig. 10.3C). The high adsorption capacity (476.2 mg/g) of rGO toward MG dye was associated with effective $\pi-\pi$ interaction between graphene skeleton and aromatic moieties of dye and electrostatic interactions of residual oxygen functional groups and $\pi-e^-$ clouds with the positive center of MG dye [40]. The oxidation of graphene into GO not only increases the solubility by introducing hydrophilic oxygen functional groups but also enhanced adsorption performance for heavy metal ions removal (Fig. 10.3D). The maximum removal capacity of Pb(II) ions on few-layered GO (FGO) was found to be 842 mg/g at 20°C, which was much higher than the original graphene [66]. Table 10.2 shows the adsorption of organic dyes using graphene-based nanomaterials.

Liu et al. reported the adsorption of methylene blue (MB) dye onto the surface of graphene nanosheets in different working conditions [124]. Efficient removal of MB ($\sim$99.68%) from contaminated water was observed at pH 10, and the adsorption capacity increased with an increase in temperature. The higher dye uptake capacity at higher pH might be due to the generation of more negative charges on the GO surface at high pH, which considerably enhances electrostatic interactions with the cationic dye [125]. The thermodynamic parameters indicated that the adsorption process is endothermic and spontaneous. It followed pseudo-second-order kinetics and the Langmuir adsorption isotherm model, indicating the formation of a monolayer of dye on the adsorbent. Similar kinetics, adsorption isotherms, and thermodynamic parameters were observed by other researchers for the adsorption of MB onto graphene/graphene oxide nanosheets [125–130]. Electrostatic interactions dominated the MB dye uptake onto GO, and the adsorption capacity was measured to be 2255.35 mg/g [126]. Oxygen functionalities on the graphene oxide nanosheets strongly influence the adsorption potential for contaminants [92,131]. With an increase in oxygen density on the graphene surface, the uptake of cationic pollutant increases exponentially [132,133]. Reduced graphene oxide is suitable for the efficient adsorption of anionic dyes [63]. Minitha et al. comparatively studied the adsorption of cationic (MB) and anionic [methyl orange (MO)] dyes on GO and reduced graphene oxide (rGO) under identical conditions [63]. The adsorption of anionic dye was found to be faster and more efficient on rGO due to noncovalent and charge interactions.

Similarly, the surface oxygen functionalities of GO are involved in the adsorption of inorganic pollutants (heavy metal ions) from wastewater. Wu et al. demonstrated that adsorption of Cu(II) onto the GO nanosheets followed the Freundlich model, and the

◀ *from Gupta K, Khatri OP. Reduced graphene oxide as an effective adsorbent for removal of malachite green dye: plausible adsorption pathways. J Colloid Interface Sci 2017;501:11–21 [40]. Copyright 2017, Elsevier Science Ltd. (D) Reproduced with permission from Zhao G, Ren X, Gao X, Tan X, Li J, Chen C, et al. Removal of Pb(ii) ions from aqueous solutions on few-layered graphene oxide nanosheets. Dalton Trans 2011;40:10945–52 [66]. Copyright 2011, Royal Society of Chemistry.*

Table 10.2 Adsorption of organic dyes using graphene-based nanomaterials.

Graphene-based nanomaterial	Dye	Adsorption conditions	Adsorption isotherm model	Adsorption kinetics	Maximum adsorption capacity (mg/g)	References
CTAB-GO	CR	pH = 3 T = 298K t = 60 min	Langmuir	Pseudo-second-order	2767	[69]
rGO/ZIF-67 aerogel	CV, MO	pH = 3−9 T = 298K t = 0−16 h		Pseudo-first-order	1714 426	[41]
GO/Fauasite SGO rSGO	MB MB	t = 5 min t = 2−300 min T = 298K	Freundlich Langmuir	Pseudo-second-order	92.4 2530 2187	[70] [71]
G-Fe	CR CV MO MB	pH = 6.5, t = 30−180 min	Langmuir	Pseudo-second-order	970 909 664 402	[72]
A-rGO/Co$_3$O$_4$	RhB	t = 30−480 min	Freundlich and Temkin	Pseudo-first-order	434	[73]
GO0123 GO0125	MB	pH = 5.5 T = 303K t = 10 min	Langmuir and Temkin	Pseudo-second-order	365 504	[74]
SIS/RGO-2.5%	RhB	pH = 3 t = 60 min	Langmuir	Pseudo-second-order	174	[75]
GO-MNP	MB	pH = 9 T = ambient t = 10−30 min	Langmuir and Freundlich	Pseudo-second-order	1428	[76]
GO-calcium alginate Gel	MB	t = 60 min T = 298K 306K 313K	Freundlich	Pseudo-first-order	122	[77]

(continued)

Table 10.2 (Continued)

Graphene-based nanomaterial	Dye	Adsorption conditions	Adsorption isotherm model	Adsorption kinetics	Maximum adsorption capacity (mg/g)	References
GO–FeCl$_3$	MB			Pseudo-second-order	200	[78]
MgO–GO	MB	pH = 6–7 T = 303K	Langmuir	Pseudo-second-order	237	[79]
PDA/RGO/HNTs	MB CR	T = 313.15K	Langmuir	Pseudo-second-order	12 11	[80]
GO/MOF	MB	T = 298K t = 12 h	Langmuir	Pseudo-second-order	274	[81]
MGO	Eriochrome Black T (EBT)	pH = 2 T = 303K t = 5 h	Langmuir	Pseudo-second-order	210	[82]
SDS exfoliated graphene	MB	pH = 7 T = 298K = 308K t = 2–48 h	Freundlich	Pseudo-second-order	782	[83]
rGO/Cu(4)	MB CV	pH = 6 t = 20 min	Langmuir	Pseudo-second-order	1200 238	[84]
5% GO/MIL–100(Fe)	MB MO	pH = 4 T = 298K t = 48 h	Langmuir	Pseudo-second-order	1231 1189	[85]
Fe@G	Basic Yellow 28 (BY28) Basic Red 46 (BR46)	pH = 7 T = 303K t = 90 min	Langmuir	Pseudo-second-order	52 46	[86]
MnFe$_2$O$_4$–GO nanocomposite	MB	pH = 7 T = RT t = 20 min	Langmuir	Pseudo-second-order	177	[87]

Adsorbent	Dye	Conditions	Isotherm	Kinetics	Capacity	Ref.
rGO−MMT	MB	pH = 8 $T = 313K$ $t = 4\,h$	Langmuir	Pseudo-second-order	227	[88]
GO−Au	MG Ethyl Violet (EV)	pH = 2 $T = 308K$ $t = 50\,min$	Freundlich	Pseudo-second-order	71 39	[89]
Bi$_2$O$_3$@GO	RhB	pH = 4 $T = 308K$ $t = 65\,min$	Langmuir and Temkin	pseudo-first-order and intraparticle diffusion	320	[90]
NG-2 NG-6	MB MO	$T = RT$ $t = 24\,h$	Langmuir	Pseudo-first-order	156 232	[91]
Fe$_3$O$_4$@GNs	MB	$T = 318K$ $t = 5\,h$	Langmuir	Pseudo-second-order	211	[92]
RGO/NMA	CR	$T = 298K$ $t = 24\,h$	Langmuir	Pseudo-second-order	474	[93]
Fe$_3$O$_4$/porous graphene nanocomposite	CV	pH = 7 $T = RT$ $t = 0–300\,min$	Langmuir	Pseudo-second-order	460	[94]
GS	RhB	pH = 3 $T = 295K$ $t = 60\,min$	Langmuir	Pseudo-second-order	56	[95]
Graphene	MB Acid Blue 25 (AB25)	pH = 12 $T = 308K$ $t = 30–50\,min$	Langmuir and Freundlich	Pseudo-second-order	481 460	[96]
γ-Fe$_2$O$_3$/N-rGO	R6G	pH = 7 $T = 303K$ $t = 120\,min$	Langmuir	Pseudo-second-order	44	[97]
N-doped graphene hydrogels	Acridine orange (AO)	pH = 5 $T = 298K$ $t = 60–400\,min$	Langmuir	Pseudo-first-order and Pseudo-second-order	124	[98]

(continued)

Table 10.2 (Continued)

Graphene-based nanomaterial	Dye	Adsorption conditions	Adsorption isotherm model	Adsorption kinetics	Maximum adsorption capacity (mg/g)	References
Fe_3O_4@SiO_2@CS-TETA-GO	MB	pH = 10 T = 303K t = 20 min	Langmuir	Pseudo-second-order	529	[99]
PmPD/rGO/NFO	MB MO CR	pH = 7 T = RT t = 5−90 min	Langmuir	Pseudo-second-order	103 136 285	[100]
rGO/CTAB	DR80 DR23		Langmuir	Pseudo-second-order	213 79	[101]
rGO	MG	pH = 3.7 T = RT t = 75−600 min	Langmuir	Pseudo-second-order	476	[40]
Amino functionalized SiO_2@$CoFe_2O_4$-GO	AB 1	pH = 2 T = RT t = 120 min	Langmuir	Pseudo-second-order	130	[102]
mimGO sponge	DR80	pH = 2 T = 296K t = 20 min	Langmuir	Pseudo-second-order	501	[103]
rGO	MB MO	pH = 7 T = RT t = 60 min	Langmuir	Pseudo-second-order	145 244	[63]
pTSA-Pani@GO-CNT	CR	pH = 5 T = 303K t = 0−640 min	Langmuir	Pseudo-second-order	66	[104]
GN-CTAB	Acid Red 265 (AR265) Acid Orange 7 (AO7)	pH = 2 T = 298 K t = 0−210 min	Langmuir and Freundlich	Pseudo-second-order	510 355	[105]

Material	Dye	Conditions	Isotherm	Kinetics	Capacity	Ref.
GO	Proflavine	pH = 8.5 T = 303K t = 20−180 min	Freundlich	Pseudo-second-order	240	[106]
MoS_2-rGO	CR	pH = 3 T = 298K t = 120 min	Langmuir	Pseudo-first-order	441	[107]
GO-SBA-16	MG MV	pH = 8 and 9 T = 308K t = 40 min	Langmuir	Pseudo-second-order	358 536	[108]
LI-MGO	Glenn Black R (GR) Orange IV (OIV) AO CV	pH = 4 T = 298K t = 300 min	Langmuir	Pseudo-second-order	588 57 132 69	[109]
N/S-GHs	MG	pH = 7 T = 293K t = 600 min	Langmuir	Pseudo-second-order	738	[110]
CA-mGO5CS	MB	pH = 12 T = 298K t = 20 min	Freundlich	Pseudo-second-order	315	[111]
PPD-GO	CR	pH = 3 T = 298K t = 60 min	Langmuir	Pseudo-second-order	892	[112]
RGO-Cys	Indigo Carmine (IC) Neutral Red (NR)	pH = 2 and 7 T = 303K t = 12 h	Langmuir	Pseudo-second-order	1005 1301	[113]
GO-SH GO-N	MB	pH = 6 T = 298K t = 48 h	Langmuir	Pseudo-second-order	763 636	[114]

(continued)

Table 10.2 (Continued)

Graphene-based nanomaterial	Dye	Adsorption conditions	Adsorption isotherm model	Adsorption kinetics	Maximum adsorption capacity (mg/g)	References
HPA–GO	MB	pH = 6 $T = 298K$ $t = 20-180$ min	Langmuir	Pseudo-second-order	740	[115]
GOKOH	OIV	pH = 7 $T = RT$ $t = 3-24$ h	Langmuir	Pseudo-second-order	606	[116]
MrGO/TiO$_2$	MB	pH = 10.8 $T = 298K$ $t = 220$ min	Langmuir	Pseudo-second-order	845	[117]
Fe$_3$O$_4$–GS	MB	pH = 6 $T = 308K$ $t = 24$ h	Temkin	Pseudo-second-order	526	[118]
TA–G	RhB	pH = 11 $T = 298K$ $t = 150$ min	Langmuir	Pseudo-second-order	201	[119]
mGO/PVA-50%	MB	pH = 7 $T = 298K$ $t = 3$ h	Langmuir	Pseudo-second-order	271	[120]
MgSi/RGO	MB	pH = 7 $T = RT$ $t = 600$ min	Langmuir	Pseudo-second-order	433	[121]
Ce–Fe/RGO	CR	$T = RT$ $t = 4$ h	Langmuir	Pseudo-second-order	179	[122]
3D RGO-based hydrogels	MB RhB	pH = 6.4 $T = 298K$ $t = 2$ h	Freundlich Freundlich	Pseudo-second-order	8	[123]

adsorption mechanism is attributed to ion exchange, electrostatic interactions, and complexation [134]. The adsorption of toxic metal ions onto GO nanosheets is mainly chemisorption and is accompanied by the formation of metal complexes with the oxygen functionalities on the surface of GO [135]. The mechanism of the adsorption of pharmaceutical compounds from industrial effluent, using GO and rGO, is usually attributed to hydrogen bonding between the O/H of the adsorbent and the H/O and nitrogen/sulfur-containing groups (electron-rich elements) of the adsorbate [135]. Other interactions, such as $\pi-\pi$ and electrostatic interactions, equally contributed to the adsorption of drug precursors and phenol. Drug precursors are rich in amino acids and aromatic rings, which make the interaction stronger and more efficient.

Despite several advantages of graphene as an adsorbent, it tends to agglomerate to form graphite, due to strong interplanar interactions and stacking [136]. Therefore surface modifications of graphene, GO, or rGO nanosheets, with surfactants or organic compounds, are performed to restrict aggregation. Mahmoodi et al. modified GO and rGO nanosheet surfaces with cetyltrimethylammonium bromide (CTAB), to form GO/CTAB and rGO/CTAB, which successfully adsorbed the anionic dyes DR80 and DR23 from contaminated water [101]. The cationic surfactant on the rGO surface facilitated the electrostatic interactions between the anionic dye and adsorbent. The adsorption model fitted well to the Langmuir adsorption isotherms and followed pseudo-second-order kinetics. Li et al. functionalized GO with a long alkyl chain CTAB (CTAB-GO) via ionic interactions to enhance the hydrophobicity of GO and facilitate its removal from aqueous solution after adsorption [69]. The dye uptake capacity of CTAB-GO for CR dye was found to be 2767 mg/g at pH 3, and 298K, in a period of 1 hour. CR dye contains a negatively charged sulfonated group, which enhanced the adsorption at low pH through electrostatic interactions of the dye with the positively charged head group of the CTAB-GO composite. CTAB-GO demonstrated excellent reusability, with an adsorption capacity of 2000 mg/g up to the 10th cycle.

The noncovalent functionalization of the two-component organic gelator (1-OA), that is, tetrazolyl derivative (1) [1, 3-di(1H-tetrazol-5-yl) benzene] and octadecylamine (OA) on GO (GO/1-OA), allowed simultaneous adsorption of primary, binary, or mixed pollutant mixtures in water [137]. 1-OA restricted the agglomeration of GO nanosheets and simultaneously contributed to a variety of synergistic noncovalent interactions, including electrostatic interactions, hydrogen bonding, $\pi-\pi$ stacking, van der Waal forces, and hydrophobic interactions, to bind the contaminants (organic dyes, pharmaceuticals, and metal ions) onto the adsorbate surface. A higher pH promoted the adsorption of cationic pollutants, whereas a lower pH encouraged the adsorption of anionic pollutants due to protonation of the substrate at amino groups of OA, tetrazolyl groups and carbonic units of GO. GO/1-OA also displayed excellent recyclability up to the fifth cycle of regeneration. Priestley et al. compared the adsorption capacities of EDTA-GO, amine-GO, GO, and activated carbon (AC) for Ni(II) and found that functionalized GO, that is, EDTA-GO and amine-GO, performed better than the others [138]. They thoroughly studied the effects of pH and contact time and adsorption selectivity. The impact of the presence of competition for metal ions [Cu(II), Cd(II), and Fe(II)] on the selectivity and adsorption capacity for Ni (II) was also studied. The selectivity of EDTA-GO for Ni(II) adsorption was found to be higher, whereas amine-GO was more selective for Cu(II) in the mixed solution. Zanella et al. performed computational and experimental studies on the adsorption of the pharmaceutical drug sodium diclofenac (s-DCF) on pristine graphene (Grap pristine), graphene with a single vacancy (Grap vacancy), and graphene nanoribbons functionalized with different groups [139]. The introduction of functional groups enhanced the adsorption of s-DCF on functionalized graphene due to an increase in binding energies. The binding energies increase with functional groups in the following order: carboxyl > hydroxyl > carbonyl > epoxy. Some examples of surface modifications of GO for removal of different water pollutants are described in Fig. 10.4A—H. Table 10.3 shows the adsorption of heavy metal ions using graphene-based nanomaterials.

The recovery of GO from wastewater, for recycling after adsorption, is a challenging task. Integration of magnetic (Fe_3O_4) nanoparticles into GO or rGO nanosheets allows separation using an external magnet and expedites the recovery process for adsorbent recycling [189]. Conversely, functionalization with Fe_3O_4 reduces the active surface area of GO/rGO available for adsorption. The magnetic recovery of Fe_3O_4-rGO nanocomposites has also been reported in conjunction with the removal of triazine pesticides (e.g., prometryn, ametryn, simazine, simeton, and atrazine) from water (Fig. 10.5A) [190]. Electrostatic and $\pi-\pi$ interactions were principally responsible for the high adsorptions of a variety of contaminants onto the Fe_3O_4-rGO nanocomposite surface. Due to the magnetic characteristics, the adsorbent was easily recovered for recycling experiments, using an external magnet. Kumar et al. demonstrated that a magnetic nanohybrid of GO with $MnFe_2O_4$ NPs (Fig. 10.5B—D) had excellent adsorption capacity for effective removal of heavy metal ions As(III), As(IV), and Pb (II). It was attributed to the combined nature of the layered characteristic of GO with the high surface area and easy separation due to the magnetic nature of $MnFe_2O_4$ NPs (Fig. 10.5C), and the high adsorption capacities of GO and magnetic NPs [191]. Jiao et al. tuned the geometry and dimensions of Fe_3O_4 nanoparticles to reduce the contact area between GO and Fe_3O_4 [193]. They employed the nanohybrid material for the removal of cationic dyes [e.g., MB and Rhodamine B (RhB)] from wastewater and observed very high adsorption efficiencies and exceptionally high recyclability. Essential strategies to apply for heavy metal removal using graphene-related materials as adsorbents are depicted in Fig. 10.5E. Moreover, Yoon et al. have demonstrated

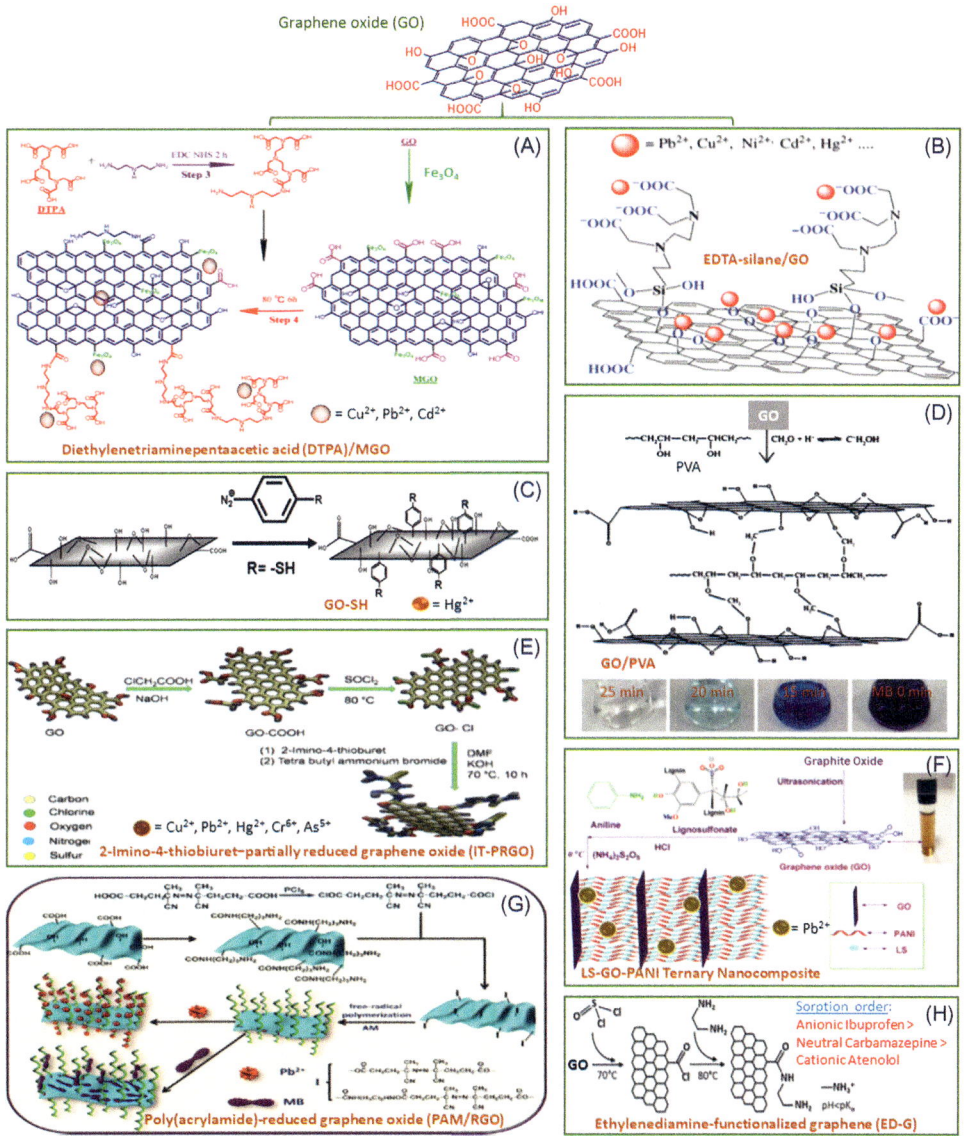

Figure 10.4 Surface modifications of GO using various strategies for removal of different water pollutants: (A) diethylenetriaminepentaacetic acid-modified magnetic GO (DTPA/MGO) [140]; (B) *N*-(trimethoxy-silylpropyl) ethylenediamine triacetic acid (EDTA-silane) functionalized GO [141]; (C) thiol-modified GO using diazonium chemistry (SH-GO) [142]; (D) poly(vinyl alcohol) (PVA) cross-linked GO (GO/PVA) [143]; (E) 2-imino-4-thiobiuret-partially reduced graphene oxide (IT-PRGO) [144]; (F) lignosulfonate−graphene oxide−polyaniline (LS-GO-PANI) nanocomposite [145]; (G) poly(acrylamide) (PAM) polymer brushes attached rGO (RGO/PAM) [146]; and (H) ethylenediamine-functionalized graphene (ED-G) [147]. *(A−G) Reproduced with permission from Refs. [140−146]. Copyright American Chemical Society. (H) Reproduced with permission from Ref. [147]. Copyright 2016, Elsevier Science Limited.*

Table 10.3 Adsorption of heavy metal ions using graphene-based nanomaterials.

Graphene-based nanomaterial	Metal ions	Adsorption conditions	Adsorption isotherm model	Adsorption kinetics	Maximum adsorption capacity (mg/g)	References
SA/PVA/GO (SPG)	Cu^{2+} UO_2^{2+}	pH = 4 $T = 298K$ $t = 270$ and 180 min	Langmuir	Pseudo-second-order	247 403	[156]
GO/SA	Mn^{2+}	pH = 6 $T = 318K$ $t = 210$ min	Freundlich	Pseudo-second-order	56	[157]
GO@SnS$_2$	Hg^{2+}	pH = 6.5 $T = $ RT $t = 24$ h	Langmuir	Pseudo-second-order	342	[158]
GO-BPEI	Cu^{2+} Cd^{2+} Pb^{2+}	pH = 5 $T = 298K$ $t = 300-400$ min	Langmuir	Pseudo-second-order	1096 2051 3390	[159]
TPGA	Cr(VI)	pH = 2 $T = $ RT $t = 40$ min	Langmuir	Pseudo-second-order	408	[160]
Bio-GM nanocomposite	Cr(VI)	pH = 4 $T = $ RT $t = 4$ h	Langmuir	Pseudo-second-order	189	[161]
Fe@MgO	Pb(II)	pH = 2–6 $T = $ RT $t = 360$ min	Langmuir	Pseudo-second-order	1476	[162]
GO-SO$_x$R@TiO$_2$	Cd(II) Pb(II) Zn(II) Ni(II)	$t = 0-300$ min	Redlich–Peterson	Pseudo-second-order	217 285 196 175	[163]
MnFe$_2$O$_4$–GO	As(V)	$T = $ RT $t = 20$ min	Langmuir	Pseudo-second-order	240	[87]

Adsorbent	Metal ion	Conditions	Isotherm	Kinetics	Capacity	Ref.
GO-NH$_2$	Mn(II)	pH = 1–6 T = 303K t = 5–80 min	Freundlich	Pseudo–second-order	161	[164]
IT-PRGO	Hg(II) Pb(II) Cr(VI) Cu(II) As(V)	pH = 2.5–5.5 T = 273K t = 5–420 min	Langmuir	Pseudo–second-order	624 10,163 3737 19	[144]
GO-CTPy	Cu(II)	pH = 6 T = 303K t = 24 h	Langmuir	Pseudo–second-order	119	[165]
GO-W-MC	Pb(II)	pH = 6 T = RT t = 180 min	Langmuir	Pseudo–second-order	253	[166]
CS/GO/Fe$_3$O$_4$-IIP	Cu(II)	pH = 6 T = RT t = 300 min	Freundlich	Pseudo–second-order	132	[167]
M-nOG	As(III) As(V)	pH = 7 T = 298K t = 24 h	Sips	Pseudo–second-order	38 14	[168]
GI-RGO	Pb(II)	pH = 3 T = RT t = 48 h	Freundlich	Elovich	101	[169]
PPy–GO	Hg(II)	pH = 7 T = 300K t = 20–720 min	Langmuir	Pseudo–second-order	400	[170]
GNP/Fe–Mg	As(V)	pH = 7 T = RT t = 24 h	Freundlich	Pseudo–second-order	104	[171]
MoS$_2$–rGO	Pb(II) Ni(II)	pH = 6.8 T = RT t = 15 min	Langmuir	Pseudo–second-order	322 294	[172]

(continued)

Table 10.3 (Continued)

Graphene-based nanomaterial	Metal ions	Adsorption conditions	Adsorption isotherm model	Adsorption kinetics	Maximum adsorption capacity (mg/g)	References
AMGO	Cr(VI)	pH = 2 T = 298K t = 12 h	Langmuir	Pseudo–second-order	123	[173]
GO/Fe–Mn	Hg(II)	pH = 7 T = RT t = 24 h	Sips	Pseudo–second-order	33	[174]
GN–SDS	Mn(II) Cu(II)	pH = 6 T = 298K t = 120 min	Langmuir	Pseudo–second-order	223 369	[175]
GO–DPA	Pb(II) Cd(II) Ni(II) Cu(II)	pH = ∼5 T = RT t = 4 min	Langmuir	Pseudo–second-order	360 253 169 347	[176]
MBT-functionalized graphene oxide	Hg(II)	pH = 5.4−6.9 T = RT t = 30 min	Langmuir	Pseudo–second-order	107	[177]
Xanthated Fe_3O_4–CS–GO	Cu(II)	pH = 5 T = 298K t = 45 min	Langmuir	Pseudo–second-order	426	[178]
HPA–GO	Pb(II)	pH = 5 T = 298K t = 20−120 min	Langmuir	Pseudo–second-order	819	[115]
GO–HPEI1.8K gel	Pb(II)	pH = 5.8 T = 288K t = 20−120 min	Langmuir	Pseudo–second-order	438	[179]
LS–GH	Pb(II)	pH = 5 T = 298K t = 40 min	Langmuir	Pseudo–second-order	1308	[180]

Material	Metal ion	Conditions	Isotherm	Kinetic model	Capacity	Reference
GO membranes	Cu(II) Cd(II) Ni(II)	pH = 5.7 T = 303K t = 60 min	Langmuir	Pseudo-second-order	72 83 62	[181]
3D-SRGO	Cd(II)	pH = 6 T = 298K t = 40 min	Langmuir	Pseudo-second-order	234	[182]
GO/CdS	Cu(II)	pH = 6 T = 298K t = 40 min	Langmuir	–	137	[183]
PAS-GO	Pb(II)	pH = 4–5 T = 303K t = 7 h	Langmuir	Pseudo-second-order	312	[184]
Amination graphene oxide	Co(II)	pH = 6 T = 298K t = 180 min	Langmuir	Pseudo-second-order	116	[185]
GO	Zn(II)	pH = 7 T = 293K t = 180 min	Langmuir	Pseudo-second-order	246	[186]
MGO	Cd(II)	pH = 6 T = 298K t = 24 h	Langmuir	Pseudo-second-order	91	[187]
GO	Cu(II) Zn(II) Cd(II) Pb(II)	pH = 5 t = 60 min	Langmuir	Pseudo-second-order	294 345 530 1119	[135]
Few-layered graphene oxide	Cd(II) Co(II)	pH = 6 T = 303K t = 24 h	Langmuir	–	106 68	[188]

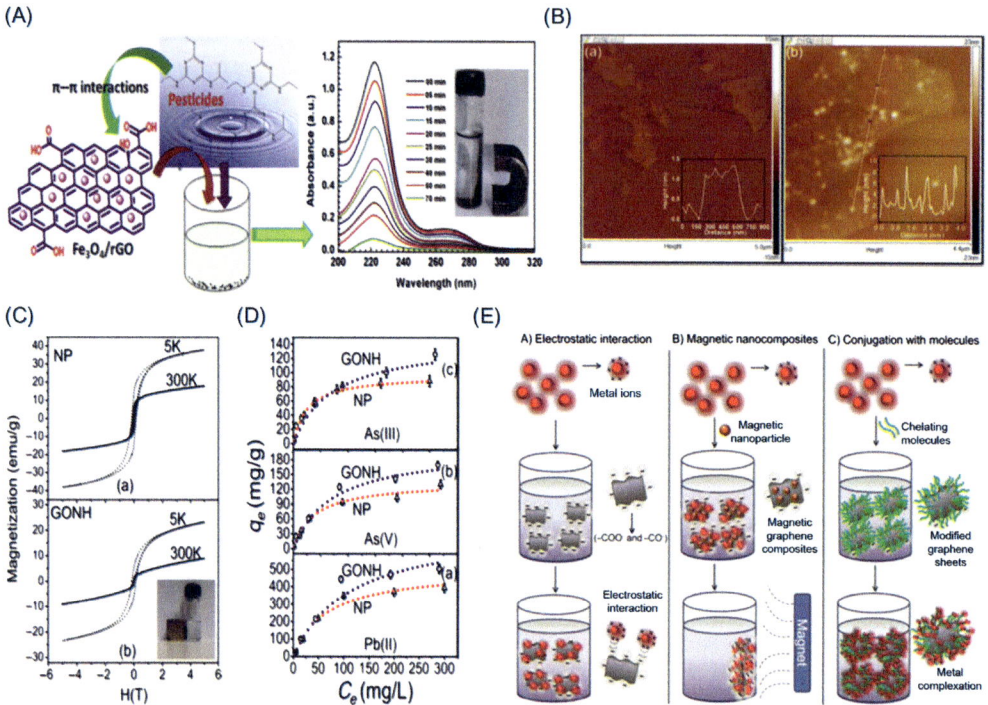

Figure 10.5 (A) Illustration of a possible mechanism for the interaction between a triazine pesticide molecule and Fe$_3$O$_4$/rGO. (B) Atomic force microscopic (AFM) image of GO and GO-MnFe$_2$O$_4$ nanohybrid (GONH). (C) Hysteresis loops, and (D) adsorption performance of MnFe$_2$O$_4$ nanoparticles (NPs) and GONH toward As (V), As(III) and Pb(II) ions. (E) Important strategies to apply for heavy metal removal using graphene-related materials as adsorbents. *(A) Reproduced with permission from Boruah PK, Sharma B, Hussain N, Das MR. Magnetically recoverable Fe$_3$O$_4$/graphene nanocomposite towards efficient removal of triazine pesticides from aqueous solution: investigation of the adsorption phenomenon and specific ion effect. Chemosphere 2017;168:1058−67 [190]. Copyright 2017, Elsevier Science Ltd. (B−D) Reproduced with permission from Kumar S, Nair RR, Pillai PB, Gupta SN, Iyengar MAR, Sood AK. Graphene oxide−MnFe$_2$O$_4$ magnetic nanohybrids for efficient removal of lead and arsenic from water. ACS Appl Mater Interfaces 2014;6:17426−36 [191]. Copyright 2014, American Chemical Society. (E) Reproduced with permission from Perreault F, Fonseca de Faria A, Elimelech M. Environmental applications of graphene-based nanomaterials. Chem Soc Rev 2015;44:5861−96 [192]. Copyright 2015, Royal Society of Chemistry.*

the synthesis of a Fe$_3$O$_4$/nonoxidative graphene composite for the efficient removal of As(III) and As(IV) from water. The active sites were regenerated using NaOH [168].

The efficiency and selectively of adsorption to GO/rGO can be enhanced by functionalizing it with metals, metal chalcogenides, or biopolymers, or by doping (N, S, etc.) [194−197]. Graphene sheets provide a surface to support the dopant, which enhances the adsorption of wastewater contaminants to the nanocomposite. Luo et al. have reported that doping of sulfur onto graphene sponges (GS) improved their adsorption capacity for heavy metal ions to nearly 40 times that of active carbon and five times that of GO [197]. The major interactions are between the metal ions and sulfur and are pH-independent. The GS exhibited excellent recyclability. In comparisons of the adsorption capacity of sulfonated graphene oxide (SGO), reduced sulfonated graphene oxide (rSGO), rGO, and GO for MB and Pb(II), SGO performed the best, for both the organic dye (2530 mg/g for MB) and the heavy metal ion [415 mg/g for Pb(II)] [71]. SGO has a variety of oxygen functionalities, including $-COOH$, $-OH$ and $-SO_3H$, which enhance the adsorption of MB, in combination with electrostatic and $\pi-\pi$ interactions. Furthermore, the presence of active functional groups at the edges of the sp^3 hybridized graphene of SGO and rGO influences the adsorption capacity for Pb(II). These functional groups are exposed to interact with the pollutant via electrostatic interactions, and the extent of exposure determines the adsorption capacity. The adsorption capacities decreased in the order SGO > rSGO > GO > rGO. Saha et al. observed that the removal of fulvic acid from the water via adsorption was faster using Fe-functionalized rGO (frGO) than powdered activated carbon, and more efficient in low pH conditions [198]. At higher pH, the surface of frGO became negative, and inter- and intramolecular electronic repulsive forces in fulvic acid increased, which hindered the adsorption process. Mamani et al. functionalized rGO with amine groups (rGO-NH$_2$) and immobilized them with horseradish peroxidase (HRP) onto the resulting nanoparticles, for the removal of a high concentration (2500 mg/L) of phenol from contaminated wastewater [199]. The adsorption capacity of free HRP was also evaluated to remove phenol, which was found to be very low in comparison to HRP-rGO-NH$_2$ nanocomposite. rGO-NH$_2$ provides an excellent matrix for the immobilization of HRP enzyme following physical adsorption and $\pi-\pi$ interactions. In a nanocomposite, rGO enhanced the stability of HRP and actively removed the phenol from the wastewater. The formation of the HRP$-$phenol complex is one of the driving forces for the adsorption of phenol onto the immobilized HRP-rGO-NH$_2$. Also, the rGO-NH$_2$ could adsorb the phenol on the matrix through $\pi-\pi$ interactions, electrostatic interactions, and hydrogen bonding. Thus the positive synergetic effect of rGO-NH$_2$ and stable immobilized HRP could successfully remove the high concentration of phenol from wastewater completely in 40 minutes. The nanocomposite behaved efficiently for 10 regeneration cycles.

Nimade et al. functionalized GO sponge with methylimidazolium ionic liquid (mimGO) via an amidation reaction between the carboxylate groups of GO and imidazolium-based ionic liquids [103]. The mechanism (Fig. 10.6) of the ultrafast adsorption of an anionic dye [Direct Red 80 (DR80)], approximately 588.2 mg/(g min) from water, using mimGO, is attributed to charge-induced adsorption. DR80 is exceptionally rich in aromatic rings and sulfonic groups. The electrostatic interactions between the sulfonic groups of the dye and the cationic imidazolium rings of the mimGO sponge enhanced the rate of adsorption. Further, the hydrogen bonding

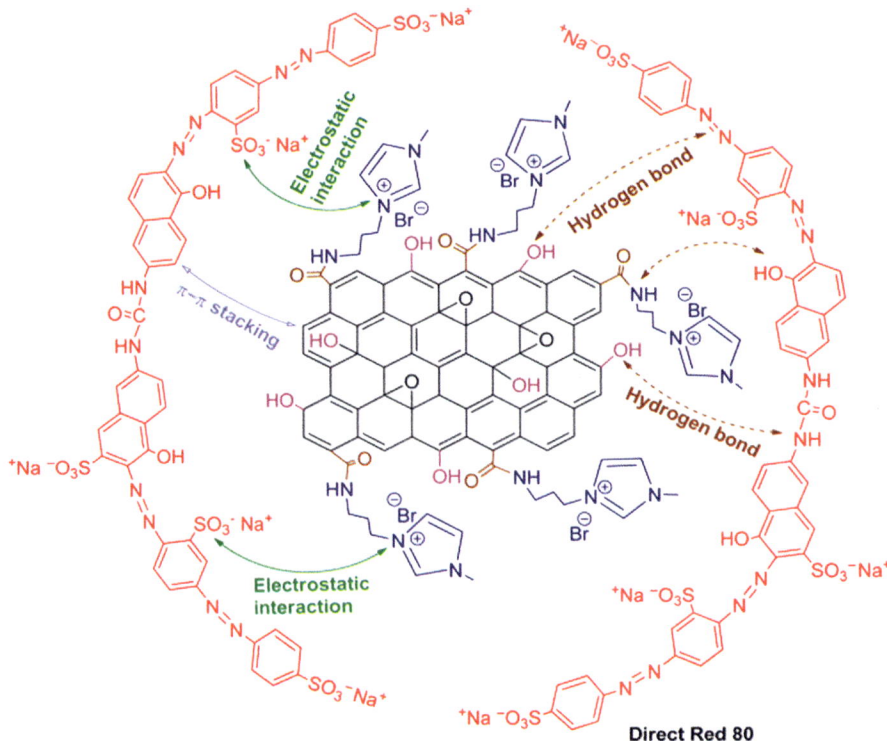

Figure 10.6 Plausible mechanism for the ultrafast adsorption of DR80 dye onto ionic liquid-functionalized graphene oxide (mimGO) sponge. *Reproduced with permission from Zambare R, Song X, Bhuvana S, Antony Prince JS, Nemade P. Ultrafast dye removal using ionic liquid–graphene oxide sponge. ACS Sustain Chem Eng 2017;5:6026–35 [103]. Copyright 2017, American Chemical Society.*

between the dye and adsorbent, and π–π interactions between the aromatic rings of the dye and mimGO also participated in the adsorption. Therefore combined electrostatic interactions, hydrogen bonding, and π–π interactions contribute to the ultrafast adsorption of the dye. After adsorption, mimGO was easily regenerated and performed with 99.2% dye removal efficiency until the fourth adsorption cycle. Ionic liquid-functionalized GOs were also employed for the removal of different types of phthalates (PAEs), used in eraser manufacturing, from aqueous solution [200]. Yakimova et al. performed DFT calculations to probe the interactions of neutral and charged heavy metals with GQDs [201]. Due to the different ionization energies of the neutral atoms Cd, Hg, and Pb, Pb bound more firmly to GQD than the others and acted as an electron donor. However, for charged heavy metals, due to chemisorption behavior, cationic heavy metals behaved like an electron acceptor in the GQD-HM complex. Armchair GQDs displayed better adsorption characteristics than zigzag GQDs. Table 10.4 lists the adsorption of pharmaceutical compounds using graphene and graphene derivatives.

Table 10.4 Adsorption of aromatic and pharmaceutical compounds using graphene-based nanomaterials.

Graphene-based nanomaterial	Pesticide/ aromatic compounds	Adsorption conditions	Adsorption isotherm model	Adsorption kinetics	Maximum adsorption capacity (mg/g)	References
RGO-based hydrogels	Naproxen (NPX)	pH = Natural T = RT t = 1 h	Langmuir	Pseudo-second-order	360	[202]
	Ibuprofen (IBP) Diclofenac (DFC)				466 489	
MCM-48 encapsulated RGO/GO	Caffeine (CAF) Phenacetin (PHE)	pH = 2–10 T = 298K t = 180 min	Langmuir	Pseudo-second-order	154 217	[203]
MCM-41-G	Acetaminophen (Acet) Aspirin (Asp)	pH = 2 T = 298K t = 1440 min	Freundlich	Pseudo-second-order	555 769	[204]
GOS	Ciprofloxacin (CPN)	pH = 6 T = 298K t = 200 min	Hill and Toth	Elovich	173	[205]
GO	Phenol 4-Chlorophenol 2,4-Dichlorophenol 2,4,6-Trichlorophenol 4-Nonylphenol (4-NP) Bisphenol A (BPA) Tetrabromobisphenol A (TBBPA)	pH = 4, 5, 9 T = RT t = 5–1440 min	Langmuir and Freundlich	Pseudo-second-order	29 19 18 18 10 21 10	[206]
SA/GO 1.0 wt.%	BPA	pH = 7 T = 298K t = 8 h	Langmuir	Pseudo-second-order	342	[207]

(continued)

Table 10.4 (Continued)

Graphene-based nanomaterial	Pesticide/ aromatic compounds	Adsorption conditions	Adsorption isotherm model	Adsorption kinetics	Maximum adsorption capacity (mg/g)	References
Cu(II)-binded GO	Aniline	pH = 7 T = RT t = 30 min	Langmuir	Pseudo-second-order	79	[208]
GAD	Tetracycline Ciprofloxacin	pH = 8 T = 298K t = 24 h	Langmuir	Pseudo-second-order	290	[209]
Fe_3O_4/rGO	Ametryn	pH = 5 T = 298K t = 90 min	Langmuir	Pseudo-second-order	55	[190]
GO-Fe_3O_4	2,4-Dichlorophenoxy acetic acid (2,4-D)	pH = 3 T = 298K t = 160–470 min	Langmuir	Pseudo-second-order	67	[210]
GO-Fe_3O_4 particle	Chlorpheniramine	pH = 10 t = 24 h	Freundlich	Pseudo-second-order	470	[211]
Graphene	Phenol	pH = 9.3 T = 295K t = 24 h	Langmuir	Pseudo-second-order	233	[212]
(GQDs + SDS)–LDH	2,4,6-Trichlorophenol	pH = 3 T = 308K t = 5–90 min	Langmuir	Pseudo-second-order	119	[213]
GO-SA gel	Ciprofloxacin (CPX)	pH = 2–12 T = 298K t = 48 h	Langmuir	Pseudo-second-order	86	[214]
MRGO-1 MRGO-2	BPA	pH = 6 T = 288.15K t = 4 h	Langmuir	Pseudo-second-order	93 71	[215]

Material	Pollutant	Conditions	Isotherm	Kinetics	Capacity	Reference
DGO	Acetaminophen (ACT)	pH = 8 $T = 298K$ $t = 300$ min	Langmuir	Pseudo-second-order	704	[216]
MIL-68(Al)/RGO	p-Nitrophenol	pH = 3–11 $T = 303K$ $t = 5$ h	Langmuir	Pseudo-second-order	332	[217]
T-rGO	BPA Phenol	pH = 6.4 $T = RT$	Langmuir	Pseudo-second-order	96 23	[218]
H-rGO	BPA Phenol	$t = 12$ h			81 38	
Gas-MS	Phenol Catechol Resorcinol Hydroquinone	$T = 298K$ $t = 24$ h	Langmuir and Freundlich	Pseudo-second-order	90 66 22 67	[219]
GO	Clofibric acid (CA)	pH = 4.5 $T = 293K$	Langmuir	Pseudo-second-order	994	[220]
Graphene	Phenol	pH = 6.3 $T = 285K$ $t = 400$ min	Langmuir and Freundlich	Pseudo-second-order	28	[221]
ZnO/Ag/GO	Naphthalene	pH = Natural $T = 298K$ $t = 20$ min	Freundlich	Pseudo-second-order	500	[222]

Several physical and chemical methods have been adopted for the activation of graphene and graphene derivatives [223]. Physical activation is performed in an inert atmosphere by carbonization of carbonaceous material at high temperature (generally $600°C-1200°C$) and subsequently, suitable oxidizing gaseous agents are passed through the carbonized material in a furnace or reactor for porosity generation. Chemical activation is done by carbonization of the carbon precursor with activating agents (e.g., KOH, $ZnCl_2$, H_3PO_4, CO_2) at high temperatures ($400°C-900°C$). The chemical activation process is preferable and consists of several advantages over the physical activation process including comparatively low-temperature requirements, less time, higher SSA, high pore volume, and higher yields. However, it also exhibits several disadvantages, such as corrosive behavior of activating chemical agents and washing of material thoroughly to remove all the chemical agents. The activation of graphene and graphene derivatives is generally done by chemical methods using KOH. Several methods have been attempted to activate the graphene material using KOH. GO or GO derivatives are immersed into KOH aqueous solution for treatment, and the resulting mixture is dried under vacuum oven at a high temperature ($100°C-150°C$) [224]. The dried product is again placed in the furnace at high temperature ($500°C-1000°C$) under inert atmosphere and the residue collected is activated graphene which is washed several times to remove the impurities. Liu et al. washed the obtained material with 15 wt.% HCl, followed by deionized water [224]. Again the washed activated graphene is dried in the oven. Liu et al. synthesized activated rGO using $ZnCl_2/CO_2$ in order to enhance the surface area, porosity, and oxygen content for the high adsorption of MB. Due to the presence of excess negative functional groups, electrostatic interactions were the favored mechanism over van der Waals interactions for the adsorption of MB onto activated rGO [223].

In summary, graphene nanosheets can be modified into GO, rGO, and other derivatives to adsorb contaminants from waste effluent. Graphene is low-cost, environmentally benign, abundantly available, and exhibits excellent physicochemical characteristics, which encourage its use in adsorption applications.

10.4 Graphitic-carbon nitride-based adsorbents

Among the many forms of carbon nitride, $g-C_3N_4$ is one of the hardest and most stable materials [225,226]. The basic structure of $g-C_3N_4$ is graphite-like planes of sp^2 hybridized C and N atoms [227]. Integration of nitrogen atoms (such as s-triazine and tri-s-triazine) into $g-C_3N_4$ makes it more attractive for applications including photocatalysis [228,229], energy storage [230], solar cells [231], and sensors [232]. Owing to the small bandgap suitable for visible light adsorption, it has

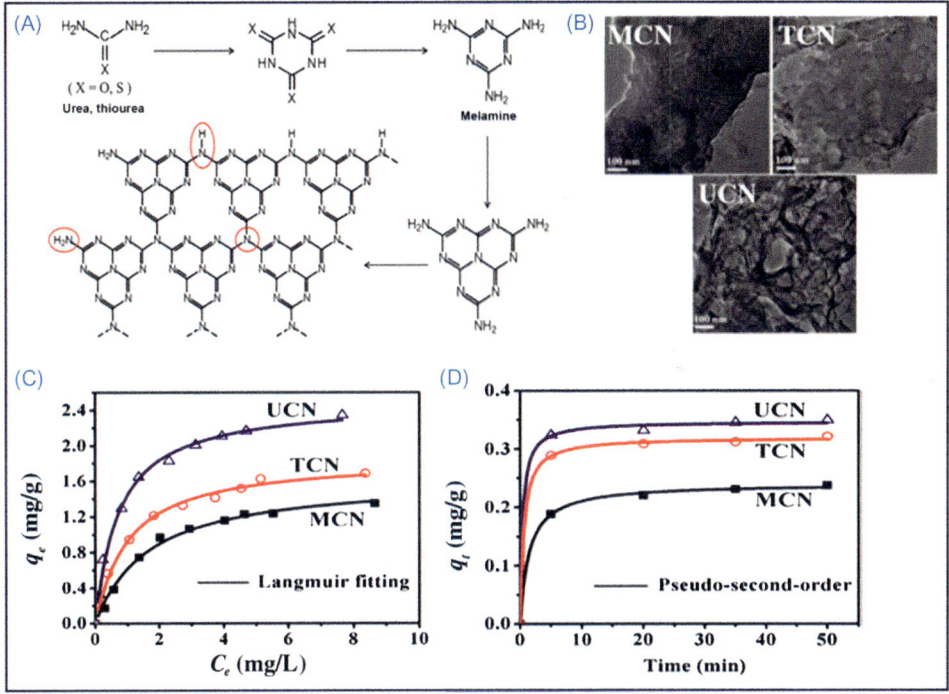

Figure 10.7 (A) Steps for the preparation of g-C$_3$N$_4$ using urea, melamine, and thiourea; (B) TEM images; (C) Langmuir fitting; and (D) pseudo-second-order fit for the adsorption of methylene blue onto UCN, TCN, and MCN nanoadsorbents. *Reproduced with permission from Zhu B, Xia P, Ho W, Yu J. Isoelectric point and adsorption activity of porous g-C3N4. Appl Surf Sci 2015;344:188–95 [235]. Copyright 2015, Elsevier Science Ltd.*

frequently been employed as a photocatalyst and successfully degraded toxic substances under solar radiation [233]. However, for efficient photocatalysis, adsorption of the substrate onto the g-C$_3$N$_4$ surface is the first, and crucial, step. g-C$_3$N$_4$ with a defect-rich surface and electron-rich nitrogen atoms and functional groups facilitates nucleophilic interactions, which enhance the adsorption, meaning it performs better than defect-free g-C$_3$N$_4$ [233]. The functional groups on the surface ($-NH_2$, $>N-N<$, $=N-$, $-NH-$, $=C-N<$, etc.) provide active sites for the adsorption of toxic substances from effluent [234]. Physical interactions (including electrostatic interactions, $\pi-\pi$ conjugated interactions, and hydrophobic interactions) and chemical interactions (including complex formation or acid–base formation) on the g-C$_3$N$_4$ surface enhance the adsorption of organic and inorganic pollutants from water. The electron-rich surface of g-C$_3$N$_4$ selectively adsorbed cationic dyes (methylene dyes) effectively from a water suspension, in the presence of anionic dyes (Fig. 10.7A–D) [235]. The use of different precursors during the synthesis of g-C$_3$N$_4$ results in products with different morphologies [235]. The pore sizes of g-C$_3$N$_4$ synthesized using thiourea (TCN) and urea (UCN) are larger than those of g-C_3N_4 synthesized using melamine (MCN). However, the surface area of UCN (40 m^2/g) is much larger than those of TCN (12 m^2/g) and MCN (37 m^2/g). g-C_3N_4 behaved as a negatively charged surface in the aqueous system and successfully adsorbed a cationic dye in the presence of an anionic dye. The adsorption mechanism for the dyes fitted better to the Langmuir adsorption isotherm model than to the Freundlich adsorption isotherm, which suggests the formation of a homogeneous monolayer of dye molecules on the g-C_3N_4 surface.

However, Zhu et al. reported that the adsorption of heavy metal ions [e.g., Pb(II)] to g-C_3N_4 fitted better to the Freundlich adsorption model than the Langmuir adsorption model [236]. The higher adsorption capacity of g-C_3N_4 at higher pH can be attributed to changes in surface charge with pH. At low pH, due to its high concentration, H^+ competed with the heavy metal ions to adsorb to the surface of g-C_3N_4. Additionally, the surface of g-C_3N_4 became positive as the electron lone pairs act as proton acceptors, and thus repelled the heavy metal ions and had a decreased affinity to metal ion adsorption. However, at higher pH, the surface of g-C_3N_4 became negatively charged and effectively attracted the positively charged metal ions, which were strongly adsorbed via electrostatic interactions. The adsorption followed pseudo-second-order kinetics. Zhang et al. synthesized a g-C_3N_4 hydrogel (h-CN) and used it to adsorb a variety of cationic (MB, Azure B, Acriflavine, and Safranin O) and anionic (RhB, eosin Y, and MO) dyes [237]. h-CN selectively adsorbed the cationic dyes with a remarkable 99% adsorption efficiency (Fig. 10.8A–D). The selective adsorption of the cationic dyes is due to the surface charge of h-CN, which was analyzed using zeta potential. The zeta potential of h-CN was recorded as being −63.1 mV, which is far more negative than that of pristine g-C_3N_4 (30–40 mV).

Furthermore, the active surface area was observed to be significantly higher for h-CN (190 m^2/g) than for pristine g-C_3N_4 ($\sim$50 m^2/g). Thus the higher surface area and highly negative surface charge of h-CN facilitated the efficient adsorption of positively charged species via electrostatic and π–π interactions. The selectivity of h-CN for cationic dyes could facilitate the removal of cationic dyes from mixtures of cationic and anionic pollutants in water.

Hydrophilic interactions on the surface of g-C_3N_4 can be generated by the oxidation of the material, which introduces oxygen functionalities, including −OH, −COOH, and −NO_3, to the surface [238−241]. The oxidized g-C_3N_4 displayed better dispersion in an aqueous medium, which increases the reaction efficiency. Reports of the use of g-C_3N_4 and modified g-C_3N_4 for the removal of toxic substances from aqueous suspensions have been limited. The potential of g-C_3N_4 for water pollutant removal could be enhanced by architectural modifications, for example, by introducing porosity for increased surface area [235], incorporating metal chalcogenides [242], or ions/atoms, for example, Fe(III) [243], B [244], P [245,246],

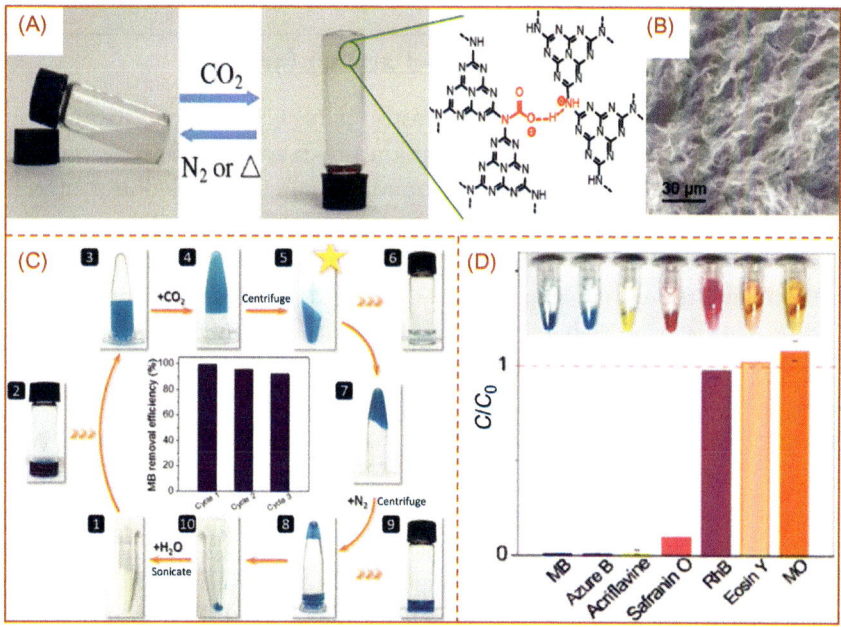

Figure 10.8 (A) Phase (sol—gel) transition in h-CN using CO_2 and N_2, and followed by CN fibers network after freeze-drying. (B) SEM image of freeze-dried h-CN. (C) Adsorption of MB, recyclability, and reusability of h-CN. (D) Selective adsorption characteristics of h-CN. *Reproduced with permission from Zhang Y, Zhou Z, Shen Y, Zhou Q, Wang J, Liu A, et al. Reversible assembly of graphitic carbon nitride 3D network for highly selective dyes absorption and regeneration. ACS Nano 2016;10:9036—43 [237]. Copyright 2016, American Chemical Society.*

doping with noble metals, for example, Pt [247], Ag [248,249], or Au [250], or by including semiconductor heterojunctions. The majority of remediation processes involving $g-C_3N_4$ occur via photodegradation of the toxic substances. Guan et al. compared the adsorptive capacity and photocatalytic remediation properties of $g-C_3N_4$, TiO_2-graphene aerogel (TiO_2-GA), and $g-C_3N_4$-TiO_2-GA nanohybrid [251]. $g-C_3N_4$-TiO_2-GA adsorbed 96.5% of MB dye from an aqueous medium, which was superior to the performances of TiO_2-GA and $g-C_3N_4$. The high adsorption potential of $g-C_3N_4$-TiO_2-GA was attributed to the impregnation of $g-C_3N_4$ between TiO_2 and GA, which enhanced the porosity and therefore the adsorption of the material. Modification of $g-C_3N_4$ with highly electronegative elements led to the adsorption of anionic dyes or electron-rich pollutants, which further contributed to the decomposition of toxic organic compounds [250].

Owing to its graphitic characteristics, low–cost, and high mechanical, chemical, and thermal stabilities, $g-C_3N_4$ is a genuinely environment-friendly adsorbent for water pollutants. Further investigation of the use of $g-C_3N_4$ for the adsorption of different kinds of toxic substances is an essential avenue for future development.

10.5 Nanoporous carbon-based adsorbents

The surface texture and surface chemistry of NPCs influence the adsorption, as the driving forces which facilitate the adsorption process are electrostatic and dispersive interactions ($\pi-\pi$ interactions, hydrogen bonding, and electron donor and acceptor complex chemistry) [252]. The excellent adsorption capacities of NPCs are due to their exceptional nanoporous constructions. Mukherji et al. synthesized NPCs following two distinctive routes: (1) using silica as a catalyst, under a hydrogen atmosphere; and (2) using cobalt as a catalyst, under a nitrogen atmosphere [253]. NPCs synthesized with a silica catalyst exhibited unique homogenous porosity and removed heavy metal ions efficiently. In contrast, NPCs synthesized with a cobalt catalyst had granular morphology and did not effectively remove contaminants via adsorption. Moreover, the extremely inert nature and hydrophobic characteristics of NPCs complicate its use as an adsorbent. This problem can be overcome by surface modification or functionalization of the NPCs, resulting in selective removal of pollutants from wastewater [148]. Han et al. used silica nanoparticles as a template to prepare NPCs and employed them to remove Direct Blue 78 dye from wastewater [149]. The adsorption capacity of the material was found to be 10 times higher than that of activated carbon. NPCs are dissimilar from activated carbon, as the former consists of interconnecting pores and channels, and the uniform porosity and honeycomb-like structure enhances adsorption.

Thomas et al. incorporated N and O into NPCs and observed significant improvement in the adsorption capacity for heavy metal ions, which could be modeled using the Langmuir isotherm [150]. The efficiency of heavy metal adsorption in aqueous media was found to decrease in the order Hg(II) > Cu(II) > Pb(II) > Cd(II) > Ca(II). The effect of pH on the adsorption potentials of oxidized N-functionalized NPCs [oxidized carbon from coconut shell (GN) and polyacrylnitrile (PANC)] was also investigated. At low pH, the adsorption of heavy metal ions to GN and PANC decreased due to competition between the metal ion [Pb(II)] and H^+ to adhere to the active sites. The concentration of the hydrolysis product of Pb $(OH)^+$ also decreased. Heavy metal ion-adsorbent complex formation was more efficient on PANC due to a high concentration of O and N donor sites. The oxygen functional groups of the adsorbent mainly participated in an ion-exchange mechanism, with proton displacement, whereas nitrogen adsorbs the heavy metal ions by coordination. Therefore the acidic character of the oxygen group and the basic character of the nitrogen group facilitate the adsorption of heavy metal ions on NPCs. Zhang et al. prepared an NPC complex using $ZnCl_2$ (APC) and calcined it at a very high temperature (700°C−900°C) for 2 hours under an N_2 atmosphere [54]. The adsorption properties of APC were compared with those of an NPC prepared without $ZnCl_2$ (PC). APC exhibited 3.5 times more active surface area than PC, with a high pore volume, which promotes the surface applications of APC. Thus the large active surface area provides the abundant active sites for dye to adhere to, the large pore size offers transport channels allowing the dye to easily adsorb onto the inner pores, and the functional groups on the surface of the adsorbate improve the dispersibility of APC in an aqueous system. The adsorption was fitted to the Langmuir model, with an observed adsorption capacity of 555.56 mg/g for MB dye, which is much higher than those of many commercially available adsorbents. In another study, graphene-like porous carbon nanostructures (BGBH-C-K) were obtained by carbonation of Bengal gram bean husk (BGBH) as agrowaste biomass and then alkali-mediated activation [151]. The BGBH-C-K exhibited extremely high surface area (1710 m^2/g) with high pore volume (0.834 cm^3/g) and abundance of micropores. The high adsorption performance of BGBH-C-K toward MB (469 mg/g) and MO (418 mg/g) dyes was attributed to a large number of active sites for hydrogen linkage, $\pi-\pi$ interaction, electrostatic interaction, and adsorption via surface defects (Fig. 10.9) [151].

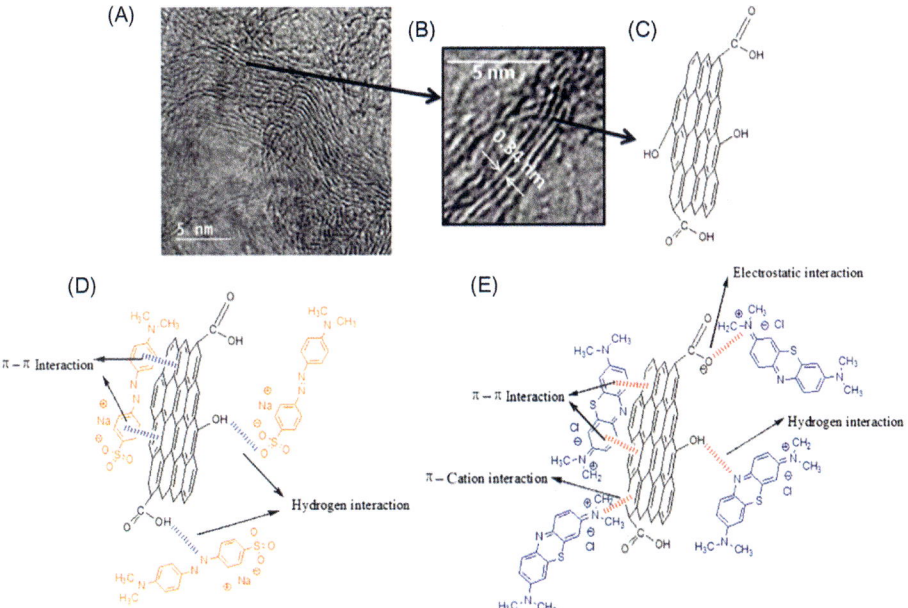

Figure 10.9 (A and B) TEM images of BGBH-C-K indicating the lamellar structure like graphene and amorphous areas. (C) The structural features of BGBH-C-K. (D and E) Schematic diagrams showing the probable mechanism for the adsorption of MO and MB dye, respectively. *Reproduced with permission from Gupta K, Gupta D, Khatri OP. Graphene-like porous carbon nanostructure from Bengal gram bean husk and its application for fast and efficient adsorption of organic dyes. Appl Surf Sci 2019;476:647–57 [151]. Copyright 2019, Elsevier.*

Table 10.5 Maximum adsorption capacities of NPC, SWCNT, and PAC for SMX, BPA, and MO [152].

Adsorbent	Antibiotic	Maximum adsorption capacity Q_m (mg/g)	Phenolic compound	Maximum adsorption capacity Q_m (mg/g)	Organic dye	Maximum adsorption capacity Q_m (mg/g)
NPCs	SMX	625	BPA	757	MO	872
SWCNTs		484		242		394
PAC		344		234		850

You et al. successfully removed an antibiotic (sulfamethoxazole (SMX)), a phenolic compound (Bisphenol A (BPA)), and a dye (methyl orange (MO)) using metal–organic framework (MOF)-derived NPCs [152]. These NPCs have a large surface area (1731 m^2/g) and high pore volume (1.68 cm^3/g). The adsorption potential of NPCs was compared with those of SWCNTs and powdered activated carbon (PAC) and was observed to be 1.0–3.2 times higher (Table 10.5), under similar conditions. During the adsorption experiment, the pH of the aqueous solution was unchanged (SMX: pH = 5.8, MO: pH = 6.0, and BPA: pH = 6.3) and higher than the pH_{iep} (pH = 5.6; pH_{iep} = where zeta potential = 0) value for NPC, which signifies that electrostatic interactions do not play a role in the adsorption. Hydrogen bonding, π-electron polarization, and pore-filling mechanisms contribute equally to the adsorption of the contaminants. The adsorption capacity (efficiency) of the NPC for the substrates decreased in the order MO > BPS > SMX. The adsorption of the organic contaminants onto the NPC followed the Langmuir model, which suggests the formation of a monolayer of adsorbate on the adsorbent.

Thomas et al. studied the competitive adsorption of heavy metal ions on the surface of oxidized NPCs and found the adsorption decreased in the following order: Hg(II) > Pb(II) > Cd(II) > Ca(II) [153]. The effective adsorption of heavy metal ions on oxidized NPCs was mainly due to ion exchange with the oxygen functionalities present on the NPC surface. Aside from ion exchange, electrostatic interactions and the ability of metal ions to get access to the porous structure contributed to the adsorption process. However, the adsorption capacity and selectivity for an adsorbate is highly dependent on the interactions, including electrostatic and ion-exchange interactions, and is influenced by the compatibility of the structure and size of the adsorbate and the pores of NPCs. Vega et al. reported a theoretical study of the adsorption of pharmaceuticals onto the surface of microporous activated carbon using the grand canonical Monte Carlo (GCMC) method, and found that absorption efficiency decreased in the following order: paracetamol (PRM) > diclofenac (DCF) > naproxen (NPX) > ibuprofen (IBF) > amoxicillin (AMX) [154]. PRM and IBF both had strong electrostatic interactions with the carbon surface and achieved similar levels of surface coverage. However, IBF was too large for the NPC pores and displayed less affinity for the surface. Thus the size compatibility of the adsorbate and the pores of the adsorbent are important for the efficient removal of adsorbate. The design and synthesis of NPCs with controlled pore sizes are crucial for effective adsorption [155].

10.6 Conclusion

In summary, 2D CNMs, including graphene, GO, rGO, GNPs, g-C_3N_4, and NPCs, are effectively used as adsorbents in wastewater purification due to their extraordinary properties, like high surface area and porosity, their abundance of functional groups, and their delocalized π-electron systems. 2D CNMs are environment-friendly, economical, nontoxic, and selective, and can contribute to the removal of organic (dyes, PAHs, PCBs, antibiotics) and inorganic (heavy metals and radioactive substances) water contaminants via adsorption. Adsorption characteristics of pristine 2D CNMs can be improved by functionalization (covalent, noncovalent, and NPs modification), and activation at elevated temperature. Among 2D CNMs, graphene and its modified compounds have been explored extensively as an adsorbent in wastewater treatments due to excellent adsorption performance toward a wide range of pollutants. Owing to its graphitic characteristics, low-cost, and high mechanical, chemical, and thermal stabilities, g-C_3N_4 is a genuinely environment-friendly adsorbent for water pollutants. Further investigation of the use of g-C_3N_4 for the adsorption of different kinds of toxic substances is an important field for future development. 1D CNMs might be a cheap replacement for activated carbon-based filters as they have the potential to eliminate all types of water pollutants effectively from water. To attain the full impact on society, these materials must be produced at a large scale for an affordable price, be better than current products, and be eco-friendly.

References

[1] Gusain R, Kumar N, Ray SS. Recent advances in carbon nanomaterial-based adsorbents for water purification. Coord Chem Rev 2020;405:213111.
[2] Neoh CH, Noor ZZ, Mutamim NSA, Lim CK. Green technology in wastewater treatment technologies: integration of membrane bioreactor with various wastewater treatment systems. Chem Eng J 2016;283:582–94.
[3] Dabrowski A, Hubicki Z, Podkościelny P, Robens E. Selective removal of the heavy metal ions from waters and industrial wastewaters by ion-exchange method. Chemosphere. 2004;56:91–106.
[4] Kazadi Mbamba C, Batstone DJ, Flores-Alsina X, Tait S. A generalised chemical precipitation modelling approach in wastewater treatment applied to calcite. Water Res 2015;68:342–53.
[5] Mukwevho N, Fosso-Kankeu E, Waanders F, Kumar N, Ray SS, Yangkou Mbianda X. Photocatalytic activity of Gd2O2CO3·ZnO·CuO nanocomposite used for the degradation of phenanthrene. SN Appl Sci 2018;1:10.

[6] Ntakadzeni M, Anku WW, Kumar N, Govender PP, Reddy L. PEGylated MoS_2 nanosheets: a dual functional photocatalyst for photodegradation of organic dyes and photoreduction of chromium from aqueous solution. Bull. Chem. React. Eng 2019;14:142–152.

[7] Kumar N, Sinha Ray S, Ngila JC. Ionic liquid-assisted synthesis of Ag/Ag2Te nanocrystals via a hydrothermal route for enhanced photocatalytic performance. N J Chem 2017;41:14618–26.

[8] Buthiyappan A, Abdul Raman AA, Daud WMAW. Development of an advanced chemical oxidation wastewater treatment system for the batik industry in Malaysia. RSC Adv 2016;6:25222–41.

[9] Alzahrani S, Mohammad AW. Challenges and trends in membrane technology implementation for produced water treatment: a review. J Water Process Eng 2014;4:107–33.

[10] Kumar N, Fosso-Kankeu E, Ray SS. Achieving controllable MoS_2 nanostructures with increased interlayer spacing for efficient removal of Pb(II) from aquatic systems. ACS Appl Mater Interfaces 2019;11:19141–55.

[11] Kumar N, Mittal H, Alhassan SM, Ray SS. Bionanocomposite hydrogel for the adsorption of dye and reusability of generated waste for the photodegradation of ciprofloxacin: a demonstration of the circularity concept for water purification. ACS Sustain Chem Eng 2018;6:17011–25.

[12] Kumar N, Mittal H, Parashar V, Ray SS, Ngila JC. Efficient removal of rhodamine 6G dye from aqueous solution using nickel sulphide incorporated polyacrylamide grafted gum karaya bionano-composite hydrogel. RSC Adv 2016;6:21929–39.

[13] Hagemann N, Spokas K, Schmidt H-P, Kägi R, Böhler MA, Bucheli TD. Activated carbon, biochar and charcoal: linkages and synergies across pyrogenic carbon's ABCs. Water. 2018;10:182.

[14] De Gisi S, Lofrano G, Grassi M, Notarnicola M. Characteristics and adsorption capacities of low-cost sorbents for wastewater treatment: a review. Sustain Mater Technol 2016;9:10–40.

[15] Zhu Y, Murali S, Cai W, Li X, Suk JW, Potts JR, et al. Graphene and graphene oxide: synthesis, properties, and applications. Adv Mater 2010;22:3906–24.

[16] Ersan G, Apul OG, Perreault F, Karanfil T. Adsorption of organic contaminants by graphene nanosheets: a review. Water Res 2017;126:385–98.

[17] Verma S, Mungse HP, Kumar N, Choudhary S, Jain SL, Sain B, et al. Graphene oxide: an efficient and reusable carbocatalyst for aza-Michael addition of amines to activated alkenes. Chem Commun 2011;47:12673–5.

[18] Mukwevho N, Gusain R, Fosso-Kankeu E, Kumar N, Waanders F, Ray SS. Removal of naphthalene from simulated wastewater through adsorption-photodegradation by ZnO/Ag/GO nanocomposite. J Ind Eng Chem 2020;81:393–404.

[19] Yang K, Chen B, Zhu L. Graphene-coated materials using silica particles as a framework for highly efficient removal of aromatic pollutants in water. Sci Rep 2015;5:11641.

[20] Wang H, Yuan X, Wu Y, Huang H, Peng X, Zeng G, et al. Graphene-based materials: fabrication, characterization and application for the decontamination of wastewater and wastegas and hydrogen storage/generation. Adv Colloid Interface Sci 2013;195-196:19–40.

[21] Allen MJ, Tung VC, Kaner RB. Honeycomb carbon: a review of graphene. Chem Rev 2009;110:132–45.

[22] Gusain R, Kumar P, Sharma OP, Jain SL, Khatri OP. Reduced graphene oxide–CuO nanocomposites for photocatalytic conversion of CO_2 into methanol under visible light irradiation. Appl Catal B: Environ 2016;181:352–62.

[23] Berman D, Erdemir A, Sumant AV. Graphene: a new emerging lubricant. Mater Today 2014;17:31–42.

[24] Song N, Gao X, Ma Z, Wang X, Wei Y, Gao C. A review of graphene-based separation membrane: materials, characteristics, preparation and applications. Desalination 2018;437:59–72.

[25] Mungse HP, Verma S, Kumar N, Sain B, Khatri OP. Grafting of oxo-vanadium Schiff base on graphene nanosheets and its catalytic activity for the oxidation of alcohols. J Mater Chem 2012;22:5427–33.

[26] Kuilla T, Bhadra S, Yao D, Kim NH, Bose S, Lee JH. Recent advances in graphene based polymer composites. Prog Polym Sci 2010;35:1350–75.

[27] Kovtyukhova NI, Ollivier PJ, Martin BR, Mallouk TE, Chizhik SA, Buzaneva EV, et al. Layer-by-layer assembly of ultrathin composite films from micron-sized graphite oxide sheets and polycations. Chem Mater 1999;11:771−8.

[28] Sun L, Fugetsu B. Mass production of graphene oxide from expanded graphite. Mater Lett 2013;109:207−10.

[29] Marcano DC, Kosynkin DV, Berlin JM, Sinitskii A, Sun Z, Slesarev A, et al. Improved synthesis of graphene oxide. ACS Nano 2010;4:4806−14.

[30] Pei S, Cheng H-M. The reduction of graphene oxide. Carbon. 2012;50:3210−28.

[31] Yang Z-z, Zheng Q-b, Qiu H-x, Li J, Yang J-h. A simple method for the reduction of graphene oxide by sodium borohydride with $CaCl_2$ as a catalyst. N Carbon Mater 2015;30:41−7.

[32] Stankovich S, Dikin DA, Piner RD, Kohlhaas KA, Kleinhammes A, Jia Y, et al. Synthesis of graphene-based nanosheets via chemical reduction of exfoliated graphite oxide. Carbon 2007;45:1558−65.

[33] De Silva K, Huang H-H, Joshi R, Yoshimura M. Chemical reduction of graphene oxide using green reductants. Carbon. 2017;119:190−9.

[34] Baig N, Ihsanullah, Sajid M, Saleh TA. Graphene-based adsorbents for the removal of toxic organic pollutants: a review. J Environ Manag 2019;244:370−82.

[35] Liu F-f, Zhao J, Wang S, Du P, Xing B. Effects of solution chemistry on adsorption of selected pharmaceuticals and personal care products (PPCPs) by graphenes and carbon nanotubes. Environ Sci Technol 2014;48:13197−206.

[36] Smith AT, LaChance AM, Zeng S, Liu B, Sun L. Synthesis, properties, and applications of graphene oxide/reduced graphene oxide and their nanocomposites. Nano Mater Sci 2019;1:31−47.

[37] Chen L, Batchelor-McAuley C, Rasche B, Johnston C, Hindle N, Compton RG. Surface area measurements of graphene and graphene oxide samples: dopamine adsorption as a complement or alternative to methylene blue? Appl Mater Today 2020;18:100506.

[38] Montes-Navajas P, Asenjo NG, Santamaría R, Menéndez R, Corma A, García H. Surface area measurement of graphene oxide in aqueous solutions. Langmuir 2013;29:13443−8.

[39] Jin Z, Wang X, Sun Y, Ai Y, Wang X. Adsorption of 4-n-nonylphenol and bisphenol-A on magnetic reduced graphene oxides: a combined experimental and theoretical studies. Environ Sci Technol 2015;49:9168−75.

[40] Gupta K, Khatri OP. Reduced graphene oxide as an effective adsorbent for removal of malachite green dye: plausible adsorption pathways. J Colloid Interface Sci 2017;501:11−21.

[41] Ravindran AR, Feng C, Huang S, Wang Y, Zhao Z, Yang J. Effects of graphene nanoplatelet size and surface area on the AC electrical conductivity and dielectric constant of epoxy nanocomposites. Polymers 2018;10:477.

[42] Lee MS, Choi H-J, Baek J-B, Chang DW. Simple solution-based synthesis of pyridinic-rich nitrogen-doped graphene nanoplatelets for supercapacitors. Appl Energy 2017;195:1071−8.

[43] Ren W, Cheng H-M. The global growth of graphene. Nat Nanotechnol 2014;9:726−30.

[44] Yi M, Shen Z. Fluid dynamics: an emerging route for the scalable production of graphene in the last five years. RSC Adv 2016;6:72525−36.

[45] Xu Y, Cao H, Xue Y, Li B, Cai W. Liquid-phase exfoliation of graphene: an overview on exfoliation media, techniques, and challenges. Nanomaterials 2018;8:942.

[46] Wang Y-Z, Chen T, Gao X-F, Liu H-H, Zhang X. Liquid phase exfoliation of graphite into few-layer graphene by sonication and microfluidization. Mater Express 2017;7:491−9.

[47] Tang X-Z, Li W, Yu Z-Z, Rafiee MA, Rafiee J, Yavari F, et al. Enhanced thermal stability in graphene oxide covalently functionalized with 2-amino-4,6-didodecylamino-1,3,5-triazine. Carbon 2011;49:1258−65.

[48] Thomas HR, Marsden AJ, Walker M, Wilson NR, Rourke JP. Sulfur-functionalized graphene oxide by epoxide ring-opening. Angew Chem Int Ed 2014;53:7613−18.

[49] McGrail BT, Rodier BJ, Pentzer E. Rapid functionalization of graphene oxide in water. Chem Mater 2014;26:5806−11.

[50] Abbas SS, Rees GJ, Kelly NL, Dancer CEJ, Hanna JV, McNally T. Facile silane functionalization of graphene oxide. Nanoscale 2018;10:16231−42.

[51] Teter DM, Hemley RJ. Low-compressibility carbon nitrides. Science 1996;271:53—5.

[52] Zhang J-S, Wang B, Wang X-C. Chemical synthesis and applications of graphitic carbon nitride. Acta Phys Chim Sin 2013;29.

[53] Hao Q, Jia G, Wei W, Vinu A, Wang Y, Arandiyan H, et al. Graphitic carbon nitride with different dimensionalities for energy and environmental applications. Nano Res **13**, 2019, 18-37;.

[54] Han X, Wang H, Zhang L. Efficient removal of methyl blue using nanoporous carbon from the waste biomass. Water Air Soil Pollut 2018;229:26.

[55] Han S, Kim S, Lim H, Choi W, Park H, Yoon J, et al. New nanoporous carbon materials with high adsorption capacity and rapid adsorption kinetics for removing humic acids. Microporous Mesoporous Mater 2003;58:131—5.

[56] Jedsada S, Taweechai A, Pasit P. Nanoporous carbon derived from agro-waste pineapple leaves for supercapacitor electrode. Adv Nat Sci: Nanosci Nanotechnol 2017;8:035017.

[57] Yuan S-J, Dai X-H. Facile synthesis of sewage sludge-derived in-situ multi-doped nanoporous carbon material for electrocatalytic oxygen reduction. Sci Rep 2016;6:27570.

[58] Zhao J, Ren W, Cheng H-M. Graphene sponge for efficient and repeatable adsorption and desorption of water contaminations. J Mater Chem 2012;22:20197—202.

[59] Chowdhury S, Balasubramanian R. Recent advances in the use of graphene-family nanoadsorbents for removal of toxic pollutants from wastewater. Adv Colloid Interface Sci 2014;204:35—56.

[60] Yusuf M, Elfghi F, Zaidi SA, Abdullah E, Khan MA. Applications of graphene and its derivatives as an adsorbent for heavy metal and dye removal: a systematic and comprehensive overview. RSC Adv 2015;5:50392—420.

[61] Carmalin Sophia A, Lima EC, Allaudeen N, Rajan S. Application of graphene based materials for adsorption of pharmaceutical traces from water and wastewater-a review. Desalination Water Treat 2016;57:27573—86.

[62] Cortés-Arriagada D, Toro-Labbé A. Improving As (iii) adsorption on graphene based surfaces: impact of chemical doping. Phys Chem Chem Phys 2015;17:12056—64.

[63] Minitha C, Lalitha M, Jeyachandran Y, Senthilkumar L. Adsorption behaviour of reduced graphene oxide towards cationic and anionic dyes: co-action of electrostatic and π—π interactions. Mater Chem Phys 2017;194:243—52.

[64] Wu L, Liu L, Gao B, Muñoz-Carpena R, Zhang M, Chen H, et al. Aggregation kinetics of graphene oxides in aqueous solutions: experiments, mechanisms, and modeling. Langmuir 2013;29:15174—81.

[65] Yang K, Chen B, Zhu X, Xing B. Aggregation, adsorption, and morphological transformation of graphene oxide in aqueous solutions containing different metal cations. Environ Sci Technol 2016;50:11066—75.

[66] Zhao G, Ren X, Gao X, Tan X, Li J, Chen C, et al. Removal of Pb(ii) ions from aqueous solutions on few-layered graphene oxide nanosheets. Dalton Trans 2011;40:10945—52.

[67] Chowdhury I, Duch MC, Mansukhani ND, Hersam MC, Bouchard D. Colloidal properties and stability of graphene oxide nanomaterials in the aquatic environment. Environ Sci Technol 2013;47:6288—96.

[68] Lee D-W, Kim T, Lee M. An amphiphilic pyrene sheet for selective functionalization of graphene. Chem Commun 2011;47:8259—61.

[69] Su J, He S, Zhao Z, Liu X, Li H. Efficient preparation of cetyltrimethylammonium bromide-graphene oxide composite and its adsorption of Congo red from aqueous solutions. Colloids Surf A: Physicochem Eng Asp 2018;.

[70] Kim DW, Han H, Kim H, Guo X, Tsapatsis M. Preparation of a graphene oxide/faujasite composite adsorbent. Microporous Mesoporous Mater 2018;268:243—50.

[71] Wei M-p, Chai H, Cao Y-l, Jia D-z. Sulfonated graphene oxide as an adsorbent for removal of Pb2+ and methylene blue. J Colloid Interface Sci 2018;524:297—305.

[72] Mahto A, Kumar A, Chaudhary JP, Bhatt M, Sharma AK, Paul P, et al. Solvent-free production of nano-FeS anchored graphene from Ulva fasciata: a scalable synthesis of super-adsorbent for lead, chromium and dyes. J Hazard Mater 2018;353:190—203.

[73] Alwan SH, Alshamsi HAH, Jasim LS. Rhodamine B removal on A-rGO/cobalt oxide nanoparticles composite by adsorption from contaminated water. J Mol Struct 2018;1161:356–65.

[74] Caroline Maria Bezerra de A, Romero Barbosa de Assis F, Ana Maria Salgueiro B, Gabriel Filipe Oliveira do N, Gabriel Rodrigues Bezerra da C, Marilda Nascimento C, et al. Systematic study of graphene oxide production using factorial design techniques and its application to the adsorptive removal of methylene blue dye in aqueous medium. Mater Res Express 2018;5:065042.

[75] Wu Z, Yuan X, Zhong H, Wang H, Jiang L, Leng L, et al. Effective removal of high-chroma rhodamine B over $Sn_{0.215}In_{0.38}S$/reduced graphene oxide composite: synergistic factors and mechanism of adsorption enrichment and visible photocatalytic degradation. Powder Technol 2018;329:217–31.

[76] Othman NH, Alias NH, Shahruddin MZ, Abu Bakar NF, Nik Him NR, Lau WJ. Adsorption kinetics of methylene blue dyes onto magnetic graphene oxide. J Environ Chem Eng 2018;6:2803–11.

[77] Balkız G, Pingo E, Kahya N, Kaygusuz H, Bedia Erim F. Graphene oxide/alginate quasi-cryogels for removal of methylene blue. Water Air Soil Pollut 2018;229:131.

[78] Ojha A, Thareja P. Electrolyte induced rheological modulation of graphene oxide suspensions and its applications in adsorption. Appl Surf Sci 2018;435:786–98.

[79] Xu J, Xu D, Zhu B, Cheng B, Jiang C. Adsorptive removal of an anionic dye Congo red by flower-like hierarchical magnesium oxide (MgO)-graphene oxide composite microspheres. Appl Surf Sci 2018;435:1136–42.

[80] Ge X, Zhang Y, Chen L, Zheng Y, Chen Z, Liang Y, et al. Mechanism of SiOx particles formation during CVD graphene growth on Cu substrates. Carbon 2018;139:989–98.

[81] Zhao S, Chen D, Wei F, Chen N, Liang Z, Luo Y. Synthesis of graphene oxide/metal−organic frameworks hybrid materials for enhanced removal of Methylene blue in acidic and alkaline solutions. J Chem Technol Biotechnol 2018;93:698–709.

[82] Khurana I, Shaw AK, Bharti, Khurana JM, Rai PK. Batch and dynamic adsorption of Eriochrome Black T from water on magnetic graphene oxide: experimental and theoretical studies. J Environ Chem Eng 2018;6:468–77.

[83] Sham AY, Notley SM. Adsorption of organic dyes from aqueous solutions using surfactant exfoliated graphene. J Environ Chem Eng 2018;6:495–504.

[84] Aboelfetoh EF, Elhelaly AA, Gemeay AH. Synergistic effect of Cu(II) in the one-pot synthesis of reduced graphene oxide (rGO/CuxO) nanohybrids as adsorbents for cationic and anionic dyes. J Environ Chem Eng 2018;6:623–34.

[85] Luo S, Wang J. MOF/graphene oxide composite as an efficient adsorbent for the removal of organic dyes from aqueous solution. Environ Sci Pollut Res 2018;25:5521–8.

[86] Konicki W, Hełminiak A, Arabczyk W, Mijowska E. Adsorption of cationic dyes onto Fe@ graphite core−shell magnetic nanocomposite: equilibrium, kinetics and thermodynamics. Chem Eng Res Des 2018;129:259–70.

[87] Huong PTL, Tu N, Lan H, Van Quy N, Tuan PA, Dinh NX, et al. Functional manganese ferrite/graphene oxide nanocomposites: effects of graphene oxide on the adsorption mechanisms of organic MB dye and inorganic As (V) ions from aqueous solution. RSC Adv 2018;8:12376–89.

[88] Zhang Y, Yan X, Yan Y, Chen D, Huang L, Zhang J, et al. The utilization of a three-dimensional reduced graphene oxide and montmorillonite composite aerogel as a multifunctional agent for wastewater treatment. RSC Adv 2018;8:4239–48.

[89] Naeem H, Ajmal M, Muntha S, Ambreen J, Siddiq M. Synthesis and characterization of graphene oxide sheets integrated with gold nanoparticles and their applications to adsorptive removal and catalytic reduction of water contaminants. RSC Adv 2018;8:3599–610.

[90] Das TR, Patra S, Madhuri R, Sharma PK. Bismuth oxide decorated graphene oxide nanocomposites synthesized via sonochemical assisted hydrothermal method for adsorption of cationic organic dyes. J Colloid Interface Sci 2018;509:82–93.

[91] Geng J, Si L, Guo H, Lin C, Xi Y, Li Y, et al. 3D nitrogen-doped graphene gels as robust and sustainable adsorbents for dyes. N J Chem 2017;41:15447–57.

[92] Wang F. Effect of oxygen-containing functional groups on the adsorption of cationic dye by magnetic graphene nanosheets. Chem Eng Res Des 2017;128:155−61.

[93] Wu Z, Yuan X, Zhong H, Wang H, Jiang L, Zeng G, et al. Highly efficient adsorption of Congo red in single and binary water with cationic dyes by reduced graphene oxide decorated NH2-MIL-68(Al). J Mol Liq 2017;247:215−29.

[94] Bharath G, Alhseinat E, Ponpandian N, Khan MA, Siddiqui MR, Ahmed F, et al. Development of adsorption and electrosorption techniques for removal of organic and inorganic pollutants from wastewater using novel magnetite/porous graphene-based nanocomposites. Sep Purif Technol 2017;188:206−18.

[95] Cheng Z-L, Li Y-x, Liu Z. Fabrication of graphene oxide/silicalite-1 composites with hierarchical porous structure and investigation on their adsorption performance for rhodamine B. J Ind Eng Chem 2017;55:234−43.

[96] El Essawy NA, Ali SM, Farag HA, Konsowa AH, Elnouby M, Hamad HA. Green synthesis of graphene from recycled PET bottle wastes for use in the adsorption of dyes in aqueous solution. Ecotoxicol Environ Saf 2017;145:57−68.

[97] Yang L, Hu J, He L, Tang J, Zhou Y, Li J, et al. One-pot synthesis of multifunctional magnetic N-doped graphene composite for SERS detection, adsorption separation and photocatalytic degradation of Rhodamine 6G. Chem Eng J 2017;327:694−704.

[98] Jiang Y, Chowdhury S, Balasubramanian R. Nitrogen-doped graphene hydrogels as potential adsorbents and photocatalysts for environmental remediation. Chem Eng J 2017;327:751−63.

[99] Wang F, Zhang L, Wang Y, Liu X, Rohani S, Lu J. Fe$_3$O$_4$@SiO$_2$@CS-TETA functionalized graphene oxide for the adsorption of methylene blue (MB) and Cu(II). Appl Surf Sci 2017;420:970−81.

[100] Wang W, Cai K, Wu X, Shao X, Yang X. A novel poly(m-phenylenediamine)/reduced graphene oxide/nickel ferrite magnetic adsorbent with excellent removal ability of dyes and Cr(VI). J Alloy Compd 2017;722:532−43.

[101] Mahmoodi NM, Maroofi SM, Mazarji M, Nabi-Bidhendi G. Preparation of modified reduced graphene oxide nanosheet with cationic surfactant and its dye adsorption ability from colored wastewater. J Surfactants Detergents 2017;20:1085−93.

[102] Santhosh C, Daneshvar E, Kollu P, Peräniemi S, Grace AN, Bhatnagar A. Magnetic SiO$_2$@CoFe$_2$O$_4$ nanoparticles decorated on graphene oxide as efficient adsorbents for the removal of anionic pollutants from water. Chem Eng J 2017;322:472−87.

[103] Zambare R, Song X, Bhuvana S, Antony Prince JS, Nemade P. Ultrafast dye removal using ionic liquid−graphene oxide sponge. ACS Sustain Chem Eng 2017;5:6026−35.

[104] Ansari MO, Kumar R, Ansari SA, Ansari SP, Barakat MA, Alshahrie A, et al. Anion selective pTSA doped polyaniline@graphene oxide-multiwalled carbon nanotube composite for Cr(VI) and Congo red adsorption. J Colloid Interface Sci 2017;496:407−15.

[105] Yusuf M, Khan MA, Otero M, Abdullah EC, Hosomi M, Terada A, et al. Synthesis of CTAB intercalated graphene and its application for the adsorption of AR265 and AO7 dyes from water. J Colloid Interface Sci 2017;493:51−61.

[106] Bhattacharyya A, Mondal D, Roy I, Sarkar G, Saha NR, Rana D, et al. Studies of the kinetics and mechanism of the removal process of proflavine dye through adsorption by graphene oxide. J Mol Liq 2017;230:696−704.

[107] Xie H, Xiong X. A porous molybdenum disulfide and reduced graphene oxide nanocomposite (MoS$_2$- rGO) with high adsorption capacity for fast and preferential adsorption towards Congo red. J Environ Chem Eng 2017;5:1150−8.

[108] Chaudhuri H, Dash S, Gupta R, Pathak DD, Sarkar A. Room-temperature in-situ design and use of graphene oxide-SBA-16 composite for water remediation and reusable heterogeneous catalysis. ChemistrySelect. 2017;2:1835−42.

[109] Wang H, Wei Y. Magnetic graphene oxide modified by chloride imidazole ionic liquid for the high-efficiency adsorption of anionic dyes. RSC Adv 2017;7:9079−89.

[110] Shi Y-C, Wang A-J, Wu X-L, Chen J-R, Feng J-J. Green-assembly of three-dimensional porous graphene hydrogels for efficient removal of organic dyes. J Colloid Interface Sci 2016;484:254−62.

[111] Ge H, Wang C, Liu S, Huang Z. Synthesis of citric acid functionalized magnetic graphene oxide coated corn straw for methylene blue adsorption. Bioresour Technol 2016;221:419—29.

[112] Wu Z-L, Liu F, Li C-K, Chen X-Q, Yu J-G. A sandwich-structured graphene-based composite: preparation, characterization, and its adsorption behaviors for Congo red. Colloids Surf A: Physicochem Eng Asp 2016;509:65—72.

[113] Xiao J, Lv W, Xie Z, Tan Y, Song Y, Zheng Q. Environmentally friendly reduced graphene oxide as a broad-spectrum adsorbent for anionic and cationic dyes via π—π interactions. J Mater Chem A 2016;4:12126—35.

[114] Chen D, Zhang H, Yang K, Wang H. Functionalization of 4-aminothiophenol and 3-aminopropyltriethoxysilane with graphene oxide for potential dye and copper removal. *J. Hazard. Mater.* **310**, 2016, 179—187.

[115] Hu L, Yang Z, Cui L, Li Y, Ngo HH, Wang Y, et al. Fabrication of hyperbranched polyamine functionalized graphene for high-efficiency removal of Pb(II) and methylene blue. Chem Eng J 2016;287:545—56.

[116] Guo Y, Deng J, Zhu J, Zhou C, Zhou C, Zhou X, et al. Removal of anionic azo dye from water with activated graphene oxide: kinetic, equilibrium and thermodynamic modeling. RSC Adv 2016;6:39762—73.

[117] Wang H, Gao H, Chen M, Xu X, Wang X, Pan C, et al. Microwave-assisted synthesis of reduced graphene oxide/titania nanocomposites as an adsorbent for methylene blue adsorption. Appl Surf Sci 2016;360:840—8.

[118] Yu B, Zhang X, Xie J, Wu R, Liu X, Li H, et al. Magnetic graphene sponge for the removal of methylene blue. Appl Surf Sci 2015;351:765—71.

[119] Liu K, Li H, Wang Y, Gou X, Duan Y. Adsorption and removal of rhodamine B from aqueous solution by tannic acid functionalized graphene. Colloids Surf A: Physicochem Eng Asp 2015;477:35—41.

[120] Cheng Z, Liao J, He B, Zhang F, Zhang F, Huang X, et al. One-step fabrication of graphene oxide enhanced magnetic composite gel for highly efficient dye adsorption and catalysis. ACS Sustain Chem Eng 2015;3:1677—85.

[121] Gui C-X, Wang Q-Q, Hao S-M, Qu J, Huang P-P, Cao C-Y, et al. Sandwichlike magnesium silicate/reduced graphene oxide nanocomposite for enhanced Pb2+ and methylene blue adsorption. ACS Appl Mater Interfaces 2014;6:14653—9.

[122] Zhan K, Liu H, Zhang H, Chen Y, Ni H, Wu M, et al. A facile method for the immobilization of myoglobin on multi-walled carbon nanotubes: poly(methacrylic acid-co-acrylamide) nanocomposite and its application for direct bio-detection of H_2O_2. J Electroanal Chem 2014;724:80—6.

[123] Tiwari JN, Mahesh K, Le NH, Kemp KC, Timilsina R, Tiwari RN, et al. Reduced graphene oxide-based hydrogels for the efficient capture of dye pollutants from aqueous solutions. Carbon 2013;56:173—82.

[124] Liu T, Li Y, Du Q, Sun J, Jiao Y, Yang G, et al. Adsorption of methylene blue from aqueous solution by graphene. Colloids Surf B: Biointerfaces 2012;90:197—203.

[125] Yang S-T, Chen S, Chang Y, Cao A, Liu Y, Wang H. Removal of methylene blue from aqueous solution by graphene oxide. J Colloid Interface Sci 2011;359:24—9.

[126] Peng W, Li H, Liu Y, Song S. Adsorption of methylene blue on graphene oxide prepared from amorphous graphite: effects of pH and foreign ions. J Mol Liq 2016;221:82—7.

[127] Ramesha G, Kumara AV, Muralidhara H, Sampath S. Graphene and graphene oxide as effective adsorbents toward anionic and cationic dyes. J Colloid Interface Sci 2011;361:270—7.

[128] Zamani S, Tabrizi NS. Removal of methylene blue from water by graphene oxide aerogel: thermodynamic, kinetic, and equilibrium modeling. Res Chem Intermed 2015;41:7945—63.

[129] Zhang W, Zhou C, Zhou W, Lei A, Zhang Q, Wan Q, et al. Fast and considerable adsorption of methylene blue dye onto graphene oxide. Bull Environ Contam Toxicol 2011;87:86.

[130] Li Y, Du Q, Liu T, Peng X, Wang J, Sun J, et al. Comparative study of methylene blue dye adsorption onto activated carbon, graphene oxide, and carbon nanotubes. Chem Eng Res Des 2013;91:361—8.

[131] Tan P, Bi Q, Hu Y, Fang Z, Chen Y, Cheng J. Effect of the degree of oxidation and defects of graphene oxide on adsorption of Cu2+ from aqueous solution. Appl Surf Sci 2017;423:1141−51.

[132] Yan H, Tao X, Yang Z, Li K, Yang H, Li A, et al. Effects of the oxidation degree of graphene oxide on the adsorption of methylene blue. J Hazard Mater 2014;268:191−8.

[133] Pandey A, Deb M, Tiwari S, Pawar PB, Saxena S, Shukla S. 3D oxidized graphene frameworks: an efficient adsorbent for methylene blue. JOM. 2018;70:469−72.

[134] Wu W, Yang Y, Zhou H, Ye T, Huang Z, Liu R, et al. Highly efficient removal of Cu (II) from aqueous solution by using graphene oxide. Water Air Soil Pollut 2013;224:1372.

[135] Sitko R, Turek E, Zawisza B, Malicka E, Talik E, Heimann J, et al. Adsorption of divalent metal ions from aqueous solutions using graphene oxide. Dalton Trans 2013;42:5682−9.

[136] Hernandez Y, Nicolosi V, Lotya M, Blighe FM, Sun Z, De S, et al. High-yield production of graphene by liquid-phase exfoliation of graphite. Nat Nanotechnol 2008;3:563.

[137] Lv M, Yan L, Liu C, Su C, Zhou Q, Zhang X, et al. Non-covalent functionalized graphene oxide (GO) adsorbent with an organic gelator for co-adsorption of dye, endocrine-disruptor, pharmaceutical and metal ion. Chem Eng J 2018;349:791−9.

[138] Zhao C, Ma L, You J, Qu F, Priestley RD. EDTA-and amine-functionalized graphene oxide as sorbents for Ni (II) removal. Desalination Water Treat 2016;57:8942−51.

[139] Jauris I, Matos C, Saucier C, Lima E, Zarbin A, Fagan S, et al. Adsorption of sodium diclofenac on graphene: a combined experimental and theoretical study. Phys Chem Chem Phys 2016;18:1526−36.

[140] Li X, Wang S, Liu Y, Jiang L, Song B, Li M, et al. Adsorption of Cu(II), Pb(II), and Cd(II) ions from acidic aqueous solutions by diethylenetriaminepentaacetic acid-modified magnetic graphene oxide. J Chem Eng Data 2017;62:407−16.

[141] Madadrang CJ, Kim HY, Gao G, Wang N, Zhu J, Feng H, et al. Adsorption behavior of EDTA-graphene oxide for Pb (II) removal. ACS Appl Mater Interfaces 2012;4:1186−93.

[142] Gao W, Majumder M, Alemany LB, Narayanan TN, Ibarra MA, Pradhan BK, et al. Engineered graphite oxide materials for application in water purification. ACS Appl Mater Interfaces 2011;3:1821−6.

[143] Li X, Liu T, Wang D, Li Q, Liu Z, Li N, et al. Superlight adsorbent sponges based on graphene oxide cross-linked with poly(vinyl alcohol) for continuous flow adsorption. ACS Appl Mater Interfaces 2018;10:21672−80.

[144] Awad FS, AbouZeid KM, El-Maaty WMA, El-Wakil AM, El-Shall MS. Efficient removal of heavy metals from polluted water with high selectivity for mercury(II) by 2-imino-4-thiobiuret−partially reduced graphene oxide (IT-PRGO). ACS Appl Mater Interfaces 2017;9:34230−42.

[145] Yang J, Wu J-X, Lü Q-F, Lin T-T. Facile preparation of lignosulfonate−graphene oxide−polyaniline ternary nanocomposite as an effective adsorbent for Pb(II) ions. ACS Sustain Chem Eng 2014;2:1203−11.

[146] Yang Y, Xie Y, Pang L, Li M, Song X, Wen J, et al. Preparation of reduced graphene oxide/poly (acrylamide) nanocomposite and its adsorption of Pb(II) and methylene blue. Langmuir 2013;29:10727−36.

[147] Cai N, Larese-Casanova P. Application of positively-charged ethylenediamine-functionalized graphene for the sorption of anionic organic contaminants from water. J Environ Chem Eng 2016;4:2941−51.

[148] Vinu A, Hossian KZ, Srinivasu P, Miyahara M, Anandan S, Gokulakrishnan N, et al. Carboxy-mesoporous carbon and its excellent adsorption capability for proteins. J Mater Chem 2007;17:1819−25.

[149] Han S, Sohn K, Hyeon T. Fabrication of new nanoporous carbons through silica templates and their application to the adsorption of bulky dyes. Chem Mater 2000;12:3337−41.

[150] Xiao B, Thomas KM. Adsorption of aqueous metal ions on oxygen and nitrogen functionalized nanoporous activated carbons. Langmuir 2005;21:3892−902.

[151] Gupta K, Gupta D, Khatri OP. Graphene-like porous carbon nanostructure from Bengal gram bean husk and its application for fast and efficient adsorption of organic dyes. Appl Surf Sci 2019;476:647−57.

[152] Li X, Yuan H, Quan X, Chen S, You S. Effective adsorption of sulfamethoxazole, bisphenol A and methyl orange on nanoporous carbon derived from metal-organic frameworks. J Environ Sci 2018;63:250—9.

[153] Xiao B, Thomas KM. Competitive adsorption of aqueous metal ions on an oxidized nanoporous activated carbon. Langmuir 2004;20:4566—78.

[154] Bahamon D, Vega LF. Pharmaceutical removal from water effluents by adsorption on activated carbons: a Monte Carlo simulation study. Langmuir 2017;33:11146—55.

[155] Pelekani C, Snoeyink VL. Competitive adsorption in natural water: role of activated carbon pore size. Water Res 1999;33:1209—19.

[156] Yi X, Sun F, Han Z, Han F, He J, Ou M, et al. Graphene oxide encapsulated polyvinyl alcohol/sodium alginate hydrogel microspheres for Cu (II) and U (VI) removal. Ecotoxicol Environ Saf 2018;158:309—18.

[157] Yang X, Zhou T, Ren B, Hursthouse A, Zhang Y. Removal of Mn (II) by sodium alginate/graphene oxide composite double-network hydrogel beads from aqueous solutions. Sci Rep 2018;8:10717.

[158] Rathore E, Biswas K. Selective and ppb level removal of Hg(ii) from water: synergistic role of graphene oxide and SnS2. J Mater Chem A 2018;6:13142—52.

[159] Pakulski D, Czepa W, Witomska S, Aliprandi A, Pawluć P, Patroniak V, et al. Graphene oxide-branched polyethylenimine foams for efficient removal of toxic cations from water. J Mater Chem A 2018;6:9384—90.

[160] Liang Q, Luo H, Geng J, Chen J. Facile one-pot preparation of nitrogen-doped ultra-light graphene oxide aerogel and its prominent adsorption performance of Cr(VI). Chem Eng J 2018;338:62—71.

[161] Ramalingam B, Parandhaman T, Choudhary P, Das SK. Biomaterial functionalized graphene-magnetite nanocomposite: a novel approach for simultaneous removal of anionic dyes and heavy-metal ions. ACS Sustain Chem Eng 2018;6:6328—41.

[162] Ge L, Wang W, Peng Z, Tan F, Wang X, Chen J, et al. Facile fabrication of Fe@MgO magnetic nanocomposites for efficient removal of heavy metal ion and dye from water. Powder Technol 2018;326:393—401.

[163] Pirveysian M, Ghiaci M. Synthesis and characterization of sulfur functionalized graphene oxide nanosheets as efficient sorbent for removal of Pb2+, Cd2+, Ni2+ and Zn2+ ions from aqueous solution: a combined thermodynamic and kinetic studies. Appl Surf Sci 2018;428:98—109.

[164] Suddai A, Nuengmatcha P, Sricharoen P, Limchoowong N, Chanthai S. Feasibility of hard acid—base affinity for the pronounced adsorption capacity of manganese(ii) using amino-functionalized graphene oxide. RSC Adv 2018;8:4162—71.

[165] Guo Y, Jia Z, Cao M. Surface modification of graphene oxide by pyridine derivatives for copper (II) adsorption from aqueous solutions. J Ind Eng Chem 2017;53:325—32.

[166] Hu Z, Qin S, Huang Z, Zhu Y, Xi L, Li Z. Recyclable graphene oxide-covalently encapsulated magnetic composite for highly efficient Pb(II) removal. J Environ Chem Eng 2017;5:4630—8.

[167] Kong D, Wang N, Qiao N, Wang Q, Wang Z, Zhou Z, et al. Facile preparation of ion-imprinted chitosan microspheres enwrapping Fe3O4 and graphene oxide by inverse suspension cross-linking for highly selective removal of copper(II). ACS Sustain Chem Eng 2017;5:7401—9.

[168] Yoon Y, Zheng M, Ahn Y-T, Park WK, Yang WS, Kang J-W. Synthesis of magnetite/non-oxidative graphene composites and their application for arsenic removal. Sep Purif Technol 2017;178:40—8.

[169] Fang Q, Zhou X, Deng W, Liu Z. Hydroxyl-containing organic molecule induced self-assembly of porous graphene monoliths with high structural stability and recycle performance for heavy metal removal. Chem Eng J 2017;308:1001—9.

[170] Zhou C, Zhu H, Wang Q, Wang J, Cheng J, Guo Y, et al. Adsorption of mercury(ii) with an Fe3O4 magnetic polypyrrole—graphene oxide nanocomposite. RSC Adv 2017;7:18466—79.

[171] La DD, Patwari JM, Jones LA, Antolasic F, Bhosale SV. Fabrication of a GNP/Fe—Mg binary oxide composite for effective removal of arsenic from aqueous solution. ACS Omega 2017;2:218—26.

[172] Aghagoli MJ, Shemirani F. Hybrid nanosheets composed of molybdenum disulfide and reduced graphene oxide for enhanced solid phase extraction of Pb(II) and Ni(II). Microchim Acta 2017;184:237−44.

[173] Zhao D, Gao X, Wu C, Xie R, Feng S, Chen C. Facile preparation of amino functionalized graphene oxide decorated with Fe3O4 nanoparticles for the adsorption of Cr(VI). Appl Surf Sci 2016;384:1−9.

[174] Chen X, Li Y, Pan X, Cortie D, Huang X, Yi Z. Photocatalytic oxidation of methane over silver decorated zinc oxide nanocatalysts. Nat Commun 2016;7:12273.

[175] Yusuf M, Khan MA, Abdullah EC, Elfghi M, Hosomi M, Terada A, et al. Dodecyl sulfate chain anchored mesoporous graphene: synthesis and application to sequester heavy metal ions from aqueous phase. Chem Eng J 2016;304:431−9.

[176] Zare-Dorabei R, Ferdowsi SM, Barzin A, Tadjarodi A. Highly efficient simultaneous ultrasonic-assisted adsorption of Pb(II), Cd(II), Ni(II) and Cu(II) ions from aqueous solutions by graphene oxide modified with 2,2′-dipyridylamine: central composite design optimization. Ultrason Sonochem 2016;32:265−76.

[177] Krishna Kumar AS, Jiang S-J, Tseng W-L. Facile synthesis and characterization of thiol-functionalized graphene oxide as effective adsorbent for Hg(II). J Environ Chem Eng 2016;4:2052−65.

[178] Liu J, Liu W, Wang Y, Xu M, Wang B. A novel reusable nanocomposite adsorbent, xanthated Fe$_3$O$_4$-chitosan grafted onto graphene oxide, for removing Cu(II) from aqueous solutions. Appl Surf Sci 2016;367:327−34.

[179] Liu Y, Xu L, Liu J, Liu X, Chen C, Li G, et al. Graphene oxides cross-linked with hyperbranched polyethylenimines: preparation, characterization and their potential as recyclable and highly efficient adsorption materials for lead(II) ions. Chem Eng J 2016;285:698−708.

[180] Li F, Wang X, Yuan T, Sun R. A lignosulfonate-modified graphene hydrogel with ultrahigh adsorption capacity for Pb(II) removal. J Mater Chem A 2016;4:11888−96.

[181] Tan P, Sun J, Hu Y, Fang Z, Bi Q, Chen Y, et al. Adsorption of Cu2+, Cd2+ and Ni2+ from aqueous single metal solutions on graphene oxide membranes. J Hazard Mater 2015;297:251−60.

[182] Wu S, Zhang K, Wang X, Jia Y, Sun B, Luo T, et al. Enhanced adsorption of cadmium ions by 3D sulfonated reduced graphene oxide. Chem Eng J 2015;262:1292−302.

[183] Jiang T, Liu W, Mao Y, Zhang L, Cheng J, Gong M, et al. Adsorption behavior of copper ions from aqueous solution onto graphene oxide−CdS composite. Chem Eng J 2015;259:603−10.

[184] Luo S, Xu X, Zhou G, Liu C, Tang Y, Liu Y. Amino siloxane oligomer-linked graphene oxide as an efficient adsorbent for removal of Pb(II) from wastewater. J Hazard Mater 2014;274:145−55.

[185] Fang F, Kong L, Huang J, Wu S, Zhang K, Wang X, et al. Removal of cobalt ions from aqueous solution by an amination graphene oxide nanocomposite. J Hazard Mater 2014;270:1−10.

[186] Wang H, Yuan X, Wu Y, Huang H, Zeng G, Liu Y, et al. Adsorption characteristics and behaviors of graphene oxide for Zn(II) removal from aqueous solution. Appl Surf Sci 2013;279:432−40.

[187] Deng J-H, Zhang X-R, Zeng G-M, Gong J-L, Niu Q-Y, Liang J. Simultaneous removal of Cd(II) and ionic dyes from aqueous solution using magnetic graphene oxide nanocomposite as an adsorbent. Chem Eng J 2013;226:189−200.

[188] Zhao G, Li J, Ren X, Chen C, Wang X. Few-layered graphene oxide nanosheets as superior sorbents for heavy metal ion pollution management. Environ Sci Technol 2011;45:10454−62.

[189] Yao Y, Miao S, Liu S, Ma LP, Sun H, Wang S. Synthesis, characterization, and adsorption properties of magnetic Fe$_3$O$_4$@ graphene nanocomposite. Chem Eng J 2012;184:326−32.

[190] Boruah PK, Sharma B, Hussain N, Das MR. Magnetically recoverable Fe$_3$O$_4$/graphene nanocomposite towards efficient removal of triazine pesticides from aqueous solution: investigation of the adsorption phenomenon and specific ion effect. Chemosphere 2017;168:1058−67.

[191] Kumar S, Nair RR, Pillai PB, Gupta SN, Iyengar MAR, Sood AK. Graphene oxide−MnFe$_2$O$_4$ magnetic nanohybrids for efficient removal of lead and arsenic from water. ACS Appl Mater Interfaces 2014;6:17426−36.

[192] Perreault F, Fonseca de Faria A, Elimelech M. Environmental applications of graphene-based nano-materials. Chem Soc Rev 2015;44:5861−96.

[193] Jiao T, Liu Y, Wu Y, Zhang Q, Yan X, Gao F, et al. Facile and scalable preparation of graphene oxide-based magnetic hybrids for fast and highly efficient removal of organic dyes. Sci Rep 2015;5:12451.

[194] Sreeprasad T, Maliyekkal SM, Lisha K, Pradeep T. Reduced graphene oxide−metal/metal oxide composites: facile synthesis and application in water purification. J Hazard Mater 2011;186:921−31.

[195] Kumar ASK, Jiang S-J. Chitosan-functionalized graphene oxide: a novel adsorbent an efficient adsorption of arsenic from aqueous solution. J Environ Chem Eng 2016;4:1698−713.

[196] Fan L, Luo C, Sun M, Li X, Qiu H. Highly selective adsorption of lead ions by water-dispersible magnetic chitosan/graphene oxide composites. Colloids Surf B: Biointerfaces 2013;103:523−9.

[197] Zhao L, Yu B, Xue F, Xie J, Zhang X, Wu R, et al. Facile hydrothermal preparation of recyclable S-doped graphene sponge for Cu2+ adsorption. J Hazard Mater 2015;286:449−56.

[198] Ray SK, Majumder C, Saha P. Functionalized reduced graphene oxide (fRGO) for removal of fulvic acid contaminant. RSC Adv 2017;7:21768−79.

[199] Vineh MB, Saboury AA, Poostchi AA, Mamani L. Physical adsorption of horseradish peroxidase on reduced graphene oxide nanosheets functionalized by amine: a good system for biodegradation of high phenol concentration in wastewater. Int J Environ Res 2018;12:45−57.

[200] Zhou X, Zhang Y, Huang Z, Lu D, Zhu A, Shi G. Ionic liquids modified graphene oxide composites: a high efficient adsorbent for phthalates from aqueous solution. Sci Rep 2016;6:38417.

[201] Shtepliuk I, Caffrey NM, Iakimov T, Khranovskyy V, Abrikosov IA, Yakimova R. On the interaction of toxic Heavy Metals (Cd, Hg, Pb) with graphene quantum dots and infinite graphene. Sci Rep 2017;7:3934.

[202] Umbreen N, Sohni S, Ahmad I, Khattak NU, Gul K. Self-assembled three-dimensional reduced graphene oxide-based hydrogel for highly efficient and facile removal of pharmaceutical compounds from aqueous solution. J Colloid Interface Sci 2018;527:356−67.

[203] Akpotu SO, Moodley B. MCM-48 encapsulated with reduced graphene oxide/graphene oxide and as-synthesised MCM-48 application in remediation of pharmaceuticals from aqueous system. J Mol Liq 2018;261:540−9.

[204] Akpotu SO, Moodley B. Application of as-synthesised MCM-41 and MCM-41 wrapped with reduced graphene oxide/graphene oxide in the remediation of acetaminophen and aspirin from aqueous system. J Environ Manag 2018;209:205−15.

[205] Rostamian R, Behnejad H. A comprehensive adsorption study and modeling of antibiotics as a pharmaceutical waste by graphene oxide nanosheets. Ecotoxicol Environ Saf 2018;147:117−23.

[206] Catherine HN, Ou M-H, Manu B, Shih Y-h. Adsorption mechanism of emerging and conventional phenolic compounds on graphene oxide nanoflakes in water. Sci Total Environ 2018;635:629−38.

[207] Gan L, Li H, Chen L, Xu L, Liu J, Geng A, et al. Graphene oxide incorporated alginate hydrogel beads for the removal of various organic dyes and bisphenol A in water. Colloid Polym Sci 2018;296:607−15.

[208] Zavareh S, Norouzi E. Impregnation of GO with Cu2+ for enhancement of aniline adsorption and antibacterial activity. J Water Process Eng 2017;20:160−7.

[209] Zhuang Y, Yu F, Ma J, Chen J. Enhanced adsorption removal of antibiotics from aqueous solutions by modified alginate/graphene double network porous hydrogel. J Colloid Interface Sci 2017;507:250−9.

[210] Nethaji S, Sivasamy A. Graphene oxide coated with porous iron oxide ribbons for 2, 4-Dichlorophenoxyacetic acid (2,4-D) removal. Ecotoxicol Environ Saf 2017;138:292−7.

[211] Li C-M, Chen C-H, Chen W-H. Different influences of nanopore dimension and pH between chlorpheniramine adsorptions on graphene oxide-iron oxide suspension and particle. Chem Eng J 2017;307:447−55.

[212] Gaber D, Abu Haija M, Eskhan A, Banat F. Graphene as an efficient and reusable adsorbent compared to activated carbons for the removal of phenol from aqueous solutions. Water Air Soil Pollut 2017;228:320.

[213] Yao Q, Wang S, Shi W, Lu C, Liu G. Graphene quantum dots in two-dimensional confined and hydrophobic space for enhanced adsorption of nonionic organic adsorbates. Ind Eng Chem Res 2017;56:583—90.

[214] Fei Y, Li Y, Han S, Ma J. Adsorptive removal of ciprofloxacin by sodium alginate/graphene oxide composite beads from aqueous solution. J Colloid Interface Sci 2016;484:196—204.

[215] Wang P, Zhou X, Zhang Y, Wang L, Zhi K, Jiang Y. Synthesis and application of magnetic reduced graphene oxide composites for the removal of bisphenol A in aqueous solution—a mechanistic study. RSC Adv 2016;6:102348—58.

[216] Moussavi G, Hossaini Z, Pourakbar M. High-rate adsorption of acetaminophen from the contaminated water onto double-oxidized graphene oxide. Chem Eng J 2016;287:665—73.

[217] Wu Z, Yuan X, Zhong H, Wang H, Zeng G, Chen X, et al. Enhanced adsorptive removal of p-nitrophenol from water by aluminum metal—organic framework/reduced graphene oxide composite. Sci Rep 2016;6:25638.

[218] Kwon J, Lee B. Bisphenol A adsorption using reduced graphene oxide prepared by physical and chemical reduction methods. Chem Eng Res Des 2015;104:519—29.

[219] Wang X, Lu M, Wang H, Pei Y, Rao H, Du X. Three-dimensional graphene aerogels—mesoporous silica frameworks for superior adsorption capability of phenols. Sep Purif Technol 2015;153:7—13.

[220] Zhang Y-L, Liu Y-J, Dai C-M, Zhou X-F, Liu S-G. Adsorption of clofibric acid from aqueous solution by graphene oxide and the effect of environmental factors. Water Air Soil Pollut 2014;225:2064.

[221] Li Y, Du Q, Liu T, Sun J, Jiao Y, Xia Y, et al. Equilibrium, kinetic and thermodynamic studies on the adsorption of phenol onto graphene. Mater Res Bull 2012;47:1898—904.

[222] Mukwevho N, Gusain R, Fosso-Kankeu E, Kumar N, Waanders F, Ray SS. Removal of naphthalene from simulated wastewater through adsorption-photodegradation by ZnO/Ag/GO nanocomposite. J Ind Eng Chem 2019;81:393—404.

[223] Liu S, Peng W, Sun H, Wang S. Physical and chemical activation of reduced graphene oxide for enhanced adsorption and catalytic oxidation. Nanoscale 2014;6:766—71.

[224] Kim T, Jung G, Yoo S, Suh KS, Ruoff RS. Activated graphene-based carbons as supercapacitor electrodes with macro- and mesopores. ACS Nano 2013;7:6899—905.

[225] Matthews M, Pimenta M, Dresselhaus G, Dresselhaus M, Endo M. Origin of dispersive effects of the Raman D band in carbon materials. Phys Rev B 1999;59:R6585.

[226] Liu AY, Wentzcovitch RM. Stability of carbon nitride solids. Phys Rev B 1994;50:10362.

[227] Xu H-Y, Wu L-C, Zhao H, Jin L-G, Qi S-Y. Synergic effect between adsorption and photocatalysis of metal-free g-C3N4 derived from different precursors. PLoS One 2015;10:e0142616.

[228] Wang X, Maeda K, Thomas A, Takanabe K, Xin G, Carlsson JM, et al. A metal-free polymeric photocatalyst for hydrogen production from water under visible light. Nat Mater 2009;8:76.

[229] Groenewolt M, Antonietti M. Synthesis of g-C3N4 nanoparticles in mesoporous silica host matrices. Adv Mater 2005;17:1789—92.

[230] Yang SJ, Cho JH, Oh GH, Nahm KS, Park CR. Easy synthesis of highly nitrogen-enriched graphitic carbon with a high hydrogen storage capacity at room temperature. Carbon. 2009;47:1585—91.

[231] Di Noto V, Negro E. Development of nano-electrocatalysts based on carbon nitride supports for the ORR processes in PEM fuel cells. Electrochim Acta 2010;55:7564—74.

[232] Xu Y, Lei W, Su J, Hu J, Yu X, Zhou T, et al. A high-performance electrochemical sensor based on g-C3N4-E-PEDOT for the determination of acetaminophen. Electrochim Acta 2018;259:994—1003.

[233] Xu J, Wang Z, Zhu Y. Enhanced visible-light-driven photocatalytic disinfection performance and organic pollutant degradation activity of porous g-C3N4 nanosheets. ACS Appl Mater Interfaces 2017;9:27727—35.

[234] Haque E, Jun JW, Talapaneni SN, Vinu A, Jhung SH. Superior adsorption capacity of mesoporous carbon nitride with basic CN framework for phenol. J Mater Chem 2010;20:10801—3.

[235] Zhu B, Xia P, Ho W, Yu J. Isoelectric point and adsorption activity of porous g-C3N4. Appl Surf Sci 2015;344:188—95.

[236] Zhu L, You L, Wang Y, Shi Z. The application of graphitic carbon nitride for the adsorption of Pb2 + ion from aqueous solution. Mater Res Express 2017;4:075606.

[237] Zhang Y, Zhou Z, Shen Y, Zhou Q, Wang J, Liu A, et al. Reversible assembly of graphitic carbon nitride 3D network for highly selective dyes absorption and regeneration. ACS Nano 2016;10:9036−43.

[238] Kim SY, Oh J, Park S, Shim Y, Park S. Production of metal-free composites composed of graphite oxide and oxidized carbon nitride nanodots and their enhanced photocatalytic performances. Chem-A Eur J 2016;22:5142−5.

[239] Liu J, Li W, Duan L, Li X, Ji L, Geng Z, et al. A graphene-like oxygenated carbon nitride material for improved cycle-life lithium/sulfur batteries. Nano Lett 2015;15:5137−42.

[240] Ming L, Yue H, Xu L, Chen F. Hydrothermal synthesis of oxidized gC 3 N 4 and its regulation of photocatalytic activity. J Mater Chem A 2014;2:19145−9.

[241] Dong G, Ai Z, Zhang L. Efficient anoxic pollutant removal with oxygen functionalized graphitic carbon nitride under visible light. RSC Adv 2014;4:5553−60.

[242] Mirzaei A, Chen Z, Haghighat F, Yerushalmi L. Hierarchical magnetic petal-like Fe_3O_4-$ZnO@gC_3N_4$ for removal of sulfamethoxazole, suppression of photocorrosion, by-products identification and toxicity assessment. Chemosphere 2018;205:463−74.

[243] Hu S, Jin R, Lu G, Liu D, Gui J. The properties and photocatalytic performance comparison of Fe^{3+}-doped gC_3N_4 and Fe_2O_3/gC_3N_4 composite catalysts. RSC Adv 2014;4:24863−9.

[244] Yan S, Li Z, Zou Z. Photodegradation of rhodamine B and methyl orange over boron-doped g-C_3N_4 under visible light irradiation. Langmuir 2010;26:3894−901.

[245] Jiang L, Yuan X, Zeng G, Chen X, Wu Z, Liang J, et al. Phosphorus-and sulfur-codoped g-C_3N_4: facile preparation, mechanism insight, and application as efficient photocatalyst for tetracycline and methyl orange degradation under visible light irradiation. ACS Sustain Chem Eng 2017;5:5831−41.

[246] Deng Y, Tang L, Zeng G, Zhu Z, Yan M, Zhou Y, et al. Insight into highly efficient simultaneous photocatalytic removal of Cr(VI) and 2, 4-diclorophenol under visible light irradiation by phosphorus doped porous ultrathin g-C_3N_4 nanosheets from aqueous media: performance and reaction mechanism. Appl Catal B: Environ 2017;203:343−54.

[247] Maeda K, Wang X, Nishihara Y, Lu D, Antonietti M, Domen K. Photocatalytic activities of graphitic carbon nitride powder for water reduction and oxidation under visible light. J Phys Chem C 2009;113:4940−7.

[248] Ge L, Han C, Liu J, Li Y. Enhanced visible light photocatalytic activity of novel polymeric g-C_3N_4 loaded with Ag nanoparticles. Appl Catal A: Gen 2011;409:215−22.

[249] Wang S, Li D, Sun C, Yang S, Guan Y, He H. Synthesis and characterization of g-C_3N_4/Ag_3VO_4 composites with significantly enhanced visible-light photocatalytic activity for triphenylmethane dye degradation. Appl Catal B: Environ 2014;144:885−92.

[250] Cheng N, Tian J, Liu Q, Ge C, Qusti AH, Asiri AM, et al. Au-nanoparticle-loaded graphitic carbon nitride nanosheets: green photocatalytic synthesis and application toward the degradation of organic pollutants. ACS Appl Mater Interfaces 2013;5:6815−19.

[251] Zhang J-J, Fang S-S, Mei J-Y, Zheng G-P, Zheng X-C, Guan X-X. High-efficiency removal of rhodamine B dye in water using g-C_3N_4 and TiO_2 co-hybridized 3D graphene aerogel composites. Sep Purif Technol 2018;194:96−103.

[252] Radovic LR, Silva IF, Ume JI, Menéndez JA, Leon CALY, Scaroni AW. An experimental and theoretical study of the adsorption of aromatics possessing electron-withdrawing and electron-donating functional groups by chemically modified activated carbons. Carbon 1997;35:1339−48.

[253] Ruparelia JP, Duttagupta SP, Chatterjee AK, Mukherji S. Potential of carbon nanomaterials for removal of heavy metals from water. Desalination 2008;232:145−56.

CHAPTER ELEVEN

Multifunctional three-dimensional carbon nanomaterials-based adsorbents

11.1 Introduction

In last 15 years, carbon-based nanomaterials (CNMs) with high surface and porosity, high stability, surface functionalities, unsaturated electrons, and high electrical conductivities have garnered huge attention in environment remediation, particularly in the area of water purification. Although they have outstanding adsorption characteristics, various disadvantages of CNMs, such as the difficulty in segregating the nanoadsorbent from the aqueous phase, particle agglomeration, compulsory extra process to treat spent adsorbent, and expensive production cost, have hindered their utilization in the potential practical application of water treatments. Moreover, the direct applications of CNMs would unavoidably discharge CNMs into the environment, which might pose severe health risks to living beings and the ecosystem [1]. In order to have safe applications and reduce the environmental transfer and related risks, pristine CNMs can be integrated into devices or amalgamated with other materials to form three-dimensional (3D) multifunctional superstructures. For instance, CNMs can be incorporated into 3D hydrogels or aerogels or sponges via structural modification and interfacial interactions. The integration of CNMs into multifunctional superstructures would avoid the associated safety risks during application and also bestow new functionalities.

Constructing carbon-based materials into 3D configurations is the subject of extensive research for a variety of applications, including adsorption, as the resulting materials retain the intrinsic properties of the one-dimensional (1D) or two-dimensional (2D) precursor while overcoming its disadvantages [2–6]. For example, in 3D graphene-based materials, the graphene nanosheets are prevented from restacking into macroscale aggregates during the adsorption process [7]. The majority of studies on 3D carbon-based materials for adsorption have concerned graphene and carbon nanotubes (CNTs) nanomaterials. CNMs can be converted into multifunctional 3D materials using template-assisted methods, self-assembly methods, and 3D printing methods [8]. A variety of 3D CNMs-based nanoadsorbents, for example, sponges,

Carbon Nanomaterial-Based Adsorbents for Water Purification.
DOI: https://doi.org/10.1016/B978-0-12-821959-1.00011-8

foams, aerogels, and hydrogels, have been designed and applied to adsorption processes, and exhibit superior adsorption capacities [9−13].

Basically, 3D CNMs-based porous superhydrophilic or hydrophobic materials (sponges, foams, aerogels, and hydrogels) can be fabricated using pyrolysis of organic precursors, by self-assembly of CNTs/graphene, carbon nanofibers/graphene, and forming polymer composites of graphene or CNTs and CNTs/graphene [5,14]. The multifunctional 3D materials formed directly from organic precursors are frail and have a high density with comparatively low adsorption capacity and poor recyclability [15,16]. Whereas CNTs or graphene-based multifunctional 3D materials demonstrate superb adsorption uptake with high reusability due to extremely high porosity, high elasticity, and ultralight weight.

The oil spills and industrial oil leakage during oil production and oil shipping processes cause severe and long-term damage to marine ecosystem and the natural environment, and also result in huge economic loss. Thus there is the need to develop low-cost, eco-friendly, effective approaches for cleaning the wide range of oil contaminants from water bodies. In order to develop ideal oil spill removal or oil/water separation technology, the proposed method should follow some important conditions, such as (1) easy operation and practical applicability; (2) high elimination performance; (3) rapid removal to evade oil spreading; and (4) no additional secondary environmental pollution. In this context, CNMs such as graphene and CNTs are extensively studied for oil spill removal due to characteristics of superhydrophobicity, and ecological benignity [17]. 3D CNMs-based porous sponges, foams, and aerogels are considered as appropriate candidates for oil spill cleanup or oil/water separation due to having superhydrophobic and superoleophilic surface with high oil wettability [18,19]. Whereas hydrogels are versatile in nature and can remove all kinds of pollutants, namely dyes, heavy metals, oil spills, and other organic molecules from aqueous water [18,20−22]. Electrostatic interactions, π−π interactions, surface complexation, ion-exchange, hydrogen bonding, and hydrophobic interactions are the major driving forces for the adsorption of water contaminants (dyes, heavy metal ions, pharmaceuticals, and pesticides) to CNMs-based 3D superstructures [5,8,23]. In contrast, hydrocarbons such as vegetable oils, paraffin, and gasoline are adsorbed on CNMs-based 3D superstructures through interaction between π electrons of adsorbent and sigma electrons of adsorbate [24]. Almost all dyes and phenolic compounds have an aromatic ring or π-electron network in their molecular structure, leading to π−π interactions with the CNMs-based multifunctional 3D materials, which acts as an environmental scavenger for these contaminants. Therefore, this chapter emphasizes the fabrication of carbon-based 3D superstructures including aerogels, hydrogels, sponges, and foams, and their recent advances in the elimination of hydrocarbons, oil spill, dyes, heavy metals, pharmaceuticals, and other organic substances from water.

11.2 Carbon aerogels

In 1931 aerogel was first synthesized by Samuel Stephens Kistler [25], and after a considerable amount of research, carbon aerogel was first created in 1980 [26]. Aerogels are synthetic extremely porous materials. They are ultralight solids with very low density. They are obtained from a gel, in which liquid is replaced with gas/air using supercritical drying. Early on, gels were made up of inorganic materials, such as silica, alumina, tin dioxide, and chromia, but later research work produced aerogels based on polymer and carbon [26]. Owing to their ultralow density, large surface area, and high porosity, carbon aerogels have been extensively studied in environment remediation including water treatments [27,28]. Aerogels with superb mechanical properties, superhydrophobicity, and superoleophilicity are required for water applications. Previously, some techniques such as atomic layer deposition and chemical vapor deposition have been used to enhance the oleophilicity and hydrophobicity of aerogels, but these processes are complicated and expensive, which hinder their industrial applications. Recently, directional freeze-casting has been considered as a promising method for the preparation of directional gels with excellent mechanical property. In this method, the starting material solution is frozen in a unidirectional way, followed by anisotropic growth of ice crystals. The obtained porous gel after freeze-drying reproduces the shapes of the anisotropic developed ice flakes. The high mechanical anisotropy was attributed to the alignment of directional growth of the microstructure of the aerogel as it can easily sustain a high load in anisotropic directions [19,29].

Zhang et al. demonstrated the directional freeze-casting approach for the preparation of superhydrophobic and superoleophilic and anisotropic fluorinated aerogels using an aqueous suspension of agarose and graphene oxide (GO) in the presence of fluorine functionalized organosilane sol [19]. The prepared fluorinated GO aerogels exhibited excellent recoverability and compression properties because of the strong hydrogen-bonding affinity between oxygen-functional groups of GO and agarose, and anisotropically aligned porous structure. The fluorinated GO aerogels could remove the various oils and organic solvents with exceptional absorption capacities (80.7−187.4 g/g), which were ascribed to excellent superoleophilic and superhydrophobic properties, and porosity of aerogel. Due to superoleophilic and superhydrophobic characteristics, fluorinated GO aerogels showed high selectivity for floating hexane (mixed with dye to impart color) and kept floating on the water surface whilst holding hexane, suggesting strong selectivity of aerogels for oil absorption (Fig. 11.1A−F) [19].

Chen et al. constructed a 3D carbon aerogel material, using 2D GO nanosheets and 1D oxidized CNTs, via a self-assembly route, for use in the adsorption of oxytetracycline (OTC), diethyl phthalate (DEP), MB, Cd(II), and diesel from water

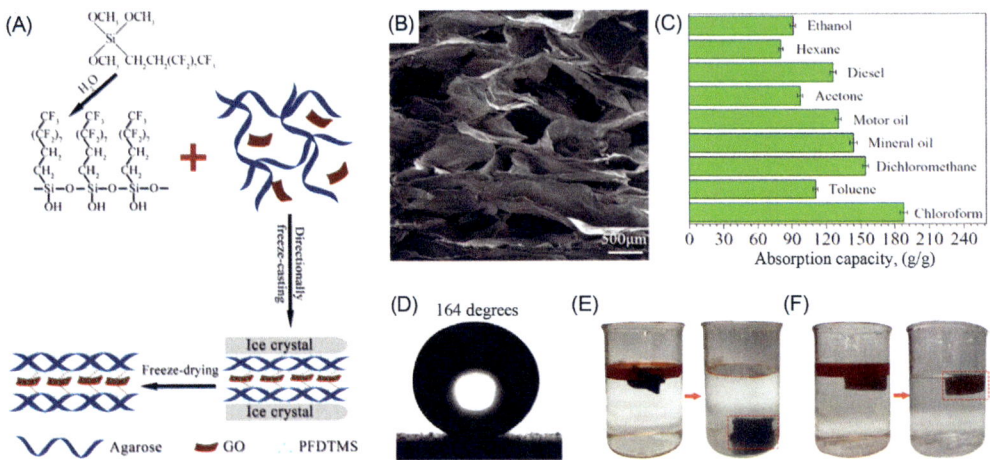

Figure 11.1 Schematic diagram shows the preparation (A), SEM image (B), absorption capacity for different oils and chemicals (C), contact angle (D) of fluorinated aerogel. Photographs exhibit the elimination of hexane (dyed with Sudan III) from water using the pristine aerogel (E) and fluorinated aerogel (F). *Reproduced with permission from Wang J, Zhang W, Zhang C. Versatile fabrication of anisotropic and superhydrophobic aerogels for highly selective oil absorption. Carbon 2019;155:16−24. Copyright 2019, Elsevier.*

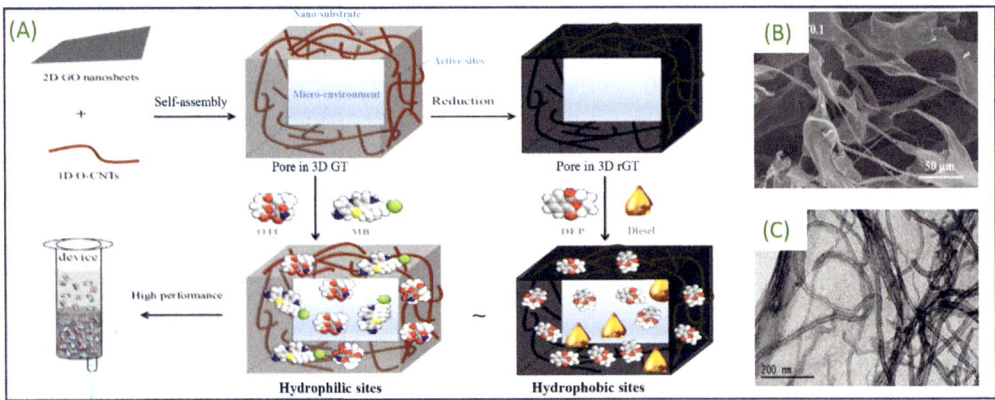

Figure 11.2 (A) 3D macrostructures of GO and CNTs, for enhanced adsorption performance via the synergistic effect. (B) SEM image, (C) TEM image of 3D macrostructures. *Reproduced with permission from Shen Y, Zhu X, Zhu L, Chen B. Synergistic effects of 2D graphene oxide nanosheets and 1D carbon nanotubes in the constructed 3D carbon aerogel for high performance pollutant removal. Chem Eng J 2017;314:336−46. Copyright 2017, Elsevier Science Ltd.*

[30] (Fig. 11.2A−C). They observed that synergistic effects between GO and CNT in the 3D aerogel structure resulted in more adsorption sites, higher porosity, larger interspaces, and better performance than those reported for graphene aerogels and nanomaterial suspensions of adsorbents. High porosity, large active surface area, rich negative

surface area, and efficient interactions between sulfur groups and heavy metal ions facilitate the adsorption of heavy metal ions onto sulfonated 3D graphene aerogels (3D-SRGO) [11]. The adsorption capacity of 3D-SRGO for Cd(II) was determined to be 235 mg/g at pH 6.0 and fitted to the Langmuir isotherm model. Cation exchange and electrostatic interactions were the key mechanisms in the removal of Cd(II) using the 3D-SRGO aerogel. Efficient regeneration of the adsorbent was carried out at pH 0.3 using HNO_3 solution, and the aerogel was reused as an adsorbent for further water purification. In another study, a graphene-based aerogel was synthesized using an in situ reducing assembly approach with the support of gelatin for the removal of organic dye such as rhodamine B (RhB), methylene blue (MB), and crystal violet (CV) from water [31]. The adsorption of dyes onto graphene aerogels decreases in the order: RhB > MB > CV > neutral red (NR). The adsorption of dyes onto the 3D graphene material was dominated by electrostatic interactions, $\pi-\pi$ interactions, and hydrogen bonding. Ding et al. fabricated MWCNTs-silica aerogels with 3D intersected networks using an acid—base sol—gel approach and hydrophobic treatment for oily water treatment [32]. The prepared aerogels demonstrated high adsorption capacity (24.42 g/g still water and 15.13 g/g for continuous water) toward diesel due to the presence of abundant mesopores and interspace between aerogel wreckages.

3D graphene aerogels are efficient adsorbents for the removal of pharmaceutical compounds from wastewater. A 3D network of 2D GO and cellulose nanofibrils (CNF) effectively adsorbed 21 antibiotics from six categories from the water via electrostatic attractions, $\pi-\pi$ interactions, p-π interactions, and hydrogen bonding [33]. The adsorption efficiency of 3D CNF/GO for the classes of antibiotics decreases in the following order: tetracycline (82%—90%, 450—500 mg/g) > quinolones (81%—82%, 128—135 mg/g) > sulfonamides (80%—81%, 219—328 mg/g) > chloramphenicol (77%—78%, 418—421 mg/g) > β-lactams (70%—78%, 230—475 mg/g) > macrolides (74%—75%, 291—307 mg/g), under similar adsorption conditions. The more highly conjugated system, and hydroxyl groups of tetracycline, resulted in the highest adsorption onto the CNF/GO aerogel. The efficient adsorption was attributed to the interconnected 3D network structure of the aerogel, which facilitated the diffusion of antibiotics through the material, increasing access to active sites, and thus the adsorption via electrostatic interactions. Furthermore, $\pi-\pi$ interactions between the aerogel and unsaturated double bonds or conjugated structures in antibiotics enhanced the adsorption process. Hydrogen bonding between the antibiotic molecules and the hydroxyl groups of CNF/GO aerogel also contributed to efficient adsorption of antibiotics onto the aerogel. Therefore all four interactions, that is, electrostatic attractions, $\pi-\pi$ interactions, p—π interactions, and hydrogen bonding, are driving forces for the adsorption of antibiotics onto the aerogel (Fig. 11.3).

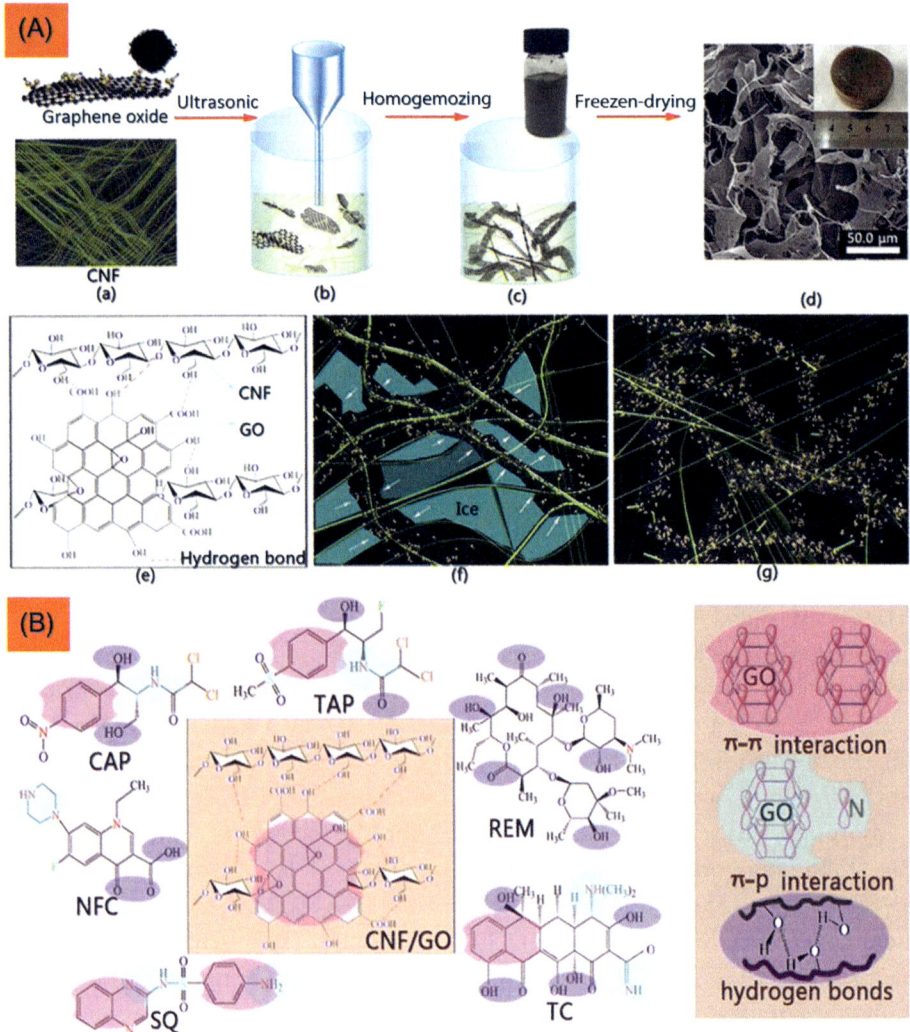

Figure 11.3 (A) The fabrication steps of cellulose nanofibril/graphene oxide (CNF/GO) aerogel. Interaction occurs between GO and cellulose nanofibril via H-bonding. (B) Antibiotic adsorption mechanism on the surface of CNF/GO aerogel. *Reproduced with permission from Yao Q, Fan B, Xiong Y, Jin C, Sun Q, Sheng C. 3D assembly based on 2D structure of cellulose nanofibril/graphene oxide hybrid aerogel for adsorptive removal of antibiotics in water. Sci Rep 2017;7:45914. Copyright 2017, Nature.*

11.3 Carbon sponges and foams

Sponges and foams are 3D porous polymer materials, which have high absorption capacity for both water and oil. Mostly, they are made of commercially existing polyurethane (PU) or melamine polymer, and have characteristics of good elasticity, low cost, advantage of large-scale production, and good absorption ability. PU is a favorable material for the preparation of a superhydrophobic and superoleophilic sponge/foam for oil/water separation [34−36]. The hydrophilicity and hydrophobicity of PU can be modified by choosing an appropriate chain extender or polyol components during synthesis. The porosity of sponge and foam should be controlled to improve the well-regulated characteristics such as elasticity, adsorption performance, and durability. The challenges with sponges/foams can be presented as: (1) they do not have good superhydrophobic stability over a particular range of temperatures and in corrosive solutions; and (2) they have poor mechanical strength. The development of sponges and foams with long-term superhydrophobic and superoleophilic stability, as well as with a high mechanical strength and excellent elasticity characteristics, is greatly needed for oil/water separation and oil spillage cleanup applications. Generally, sponges and foams are modified with CNMs, especially graphene, CNTs, and carbon nanofibers, in order to improve durable recyclability, retrievability, selectivity, hydrophobicity, mechanical strength, and absorption/adsorption properties [37]. Nanomaterials modified sponges/foams are fabricated by dip-coating, direct-deposition, hydrothermal, coprecipitation, sonication, template directed, and polymerization methods [37].

The incorporation of nanomaterials can change the wettability of sponges/foams from superhydrophilic to superhydrophobic by decreasing the surface energy and improving the surface roughness. For example, CNTs-modified sponges have been broadly studied as reusable oil adsorbents with high reusability and excellent oil absorption capacity [35,37,38]. In a particular study, the coating of hydrophobic CNTs and poly(dimethylsiloxane) (PDMS) was performed on PU using a facile dip-coating technique to adjust its nature to absorb only oily substance and resist H_2O completely [35]. The as-made CNT/PDMS-coated PU sponge has practical utility in the continuous absorption of oils or organic solvents from water (Fig. 11.4A, B, and F). When used in a vacuum system, in a one-step process, it could separate oils up to 35,000 times its own weight from water. CNT/PDMS-coated PU sponge had superhydrophobicity (water contact angle: 162 ± 2 degrees) and superoleophilicity (contact angles of n-hexane, n-hexadecane, and gasoline: all close to 0 degree) (Fig. 11.4C). A unique PU sponge showed hierarchical structures that occurred in the form of craterlike nanostructures. The coating provides a micro/nanotextured surface which is responsible for superhydrophobicity (Fig. 11.4D and E). The presence of CNTs also significantly enhances their mechanical and thermal properties. After absorption, oils can be reclaimed by vacuum sucking and mechanical squeezing.

Seo et al. investigated the adsorption potential of 3D graphene sponges (GS) for cationic dyes (MB and MV) [39]. Approximately 99% of the dye (MB and MV) was adsorbed onto the 3D GS nanomaterial within 2 minutes, via $\pi-\pi$ interactions and

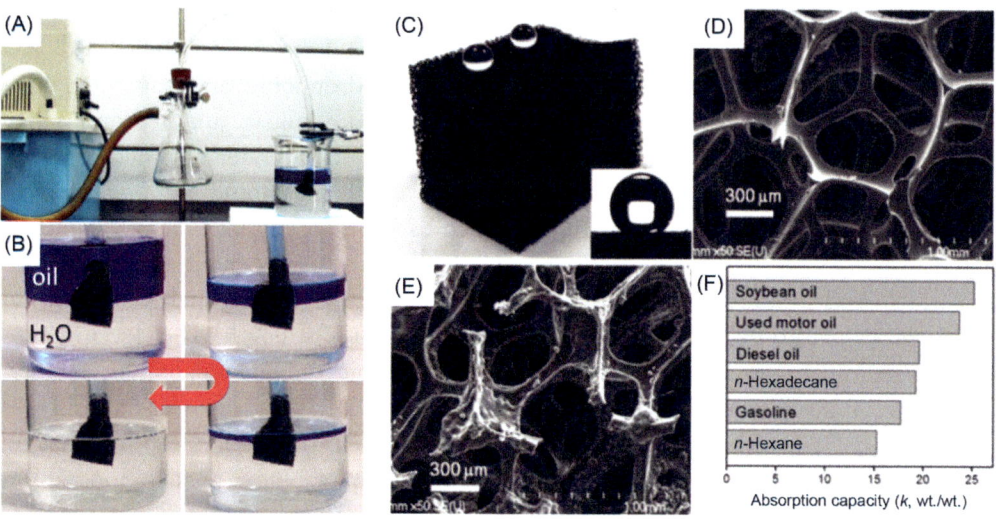

Figure 11.4 (A) Pictures of the continuous oil—water separation system; (B) the evolvement of the continuous absorption and elimination of an organic solvent from the water surface. (C) Picture of the CNT/PDMS-coated PU sponge; inset displays the contact angle of the as-prepared sponge. SEM images of unmodified PU sponge (D); CNTs-modified PU sponge (E); absorption capacity of CNT/PDMS-coated PU sponge for removal of various oils and organic solvents (F). *Reproduced with permission from Wang C-F, Lin S-J. Robust superhydrophobic/superoleophilic sponge for effective continuous absorption and expulsion of oil pollutants from water. ACS Appl Mater Interfaces 2013;5:8861—4. Copyright 2013, American Chemical Society.*

ionic interactions, due to the aromatic structure and cationic center of the dye molecules. The adsorption capacities of 3D GS reached 397 and 467 mg/g for MB and MV dyes from contaminated water, respectively. The recovery of the adsorbent was much easier for the 3D GS structure compared with 2D graphene. 3D graphene materials can be modified in order to target specific pollutants, for example, to adsorb cationic pollutants, oxygen–containing groups could be added by oxidation, whereas oxygen groups act as a barrier for anionic adsorbates. Chen et al. modified a 3D GS with chitosan (CS) to provide both positively and negatively charged surfaces on the adsorbent, which successfully adsorbed dyes (cationic MB and anionic eosin Y) and heavy metal ions (Cu(II) and Pb(II)) via electrostatic interactions and ion complex formation, respectively, with adsorption capacities of more than 300 mg/g for the dyes, and 70 and 90 mg/g for Cu(II) and Pb(II), respectively [40].

Concurrently, integrative strategies have been used for the fast removal of oil from water. For instance, a CNTs-functionalized PU sponge was designed as a photothermal sorbent with superhydrophobicity and superoleophilicity and excellent absorption ability for heavy oil [41]. This approach takes advantage of the photothermal effect on heavy oil, using sunlight as the energy source to decrease its viscosity, which consequently leads to the fast removal of heavy oil. The CNT-modified PU sponge obtained approximatively full sunlight absorption (99%). Similarly, Liu et al fabricated an effective conductive nanosponge filtration device using CNTs as conducting materials [42]. The incorporation of electrochemistry has greatly elevated the efficiency of this device for the adsorption and oxidation of organic species in aqueous solution.

Liu et al. fabricated a cheese-like foam of 3D carbon—boron—nitride (3D-C-BN) with small pore size (2—100 nm) using simple heating without any template support [43]. It has hydrophobic (contact angle ∼112.1 degrees) and lipophilic characteristics, making it a suitable floatable adsorbent for oil pollutants' removal. Due to the presence of abundant active sites of C, B-N, and small pore size, the 3D-C-BN acted as a versatile adsorbent for the adsorption of organic dyes (CR and MB), metal ions (Ni(II), Cd (II), and Cr(III)) and oils (gasoline, pump, and salad oil) from aqueous solution. The adsorbent can be re-employed by regeneration using heating or burning in air. The adsorption potentials of 3D graphene foams (GFs) for dyes in polluted water have been also evaluated [44]. The adsorption of dyes onto GFs decreases in the order: RhB > MG > acriflavine (AF). The adsorption of dyes onto the 3D graphene material is dominated by electrostatic interactions, $\pi-\pi$ interactions, and hydrogen bonding.

Chen et al. reported a modified foam ODA-rGO@MF via the coating of octadecylamine-grafted reduced graphene oxide (ODA-rGO) on the cross-linked structure of commercial melamine foam (MF) [45] (Fig. 11.5A). The as-made ODA-rGO@MF has properties such as hierarchical pore structure, mechanical stability, compressibility of original melamine foam, and also demonstrates superhydrophobicity, high water resistance, and possesses stability under harsh conditions such as salty and strong alkaline and acidic solutions. The superhydrophobicity and superoleophilicity characteristics of ODA-rGO@MF were accredited to the microarchitecture of the melamine foam, the formation of the rough rGO nanosheet wrinkles on the foam cross-linked structure, and the auxiliary reduction of the surface energy by octadecylamine (ODA) (Fig. 11.5B). Here, the superhydrophobicity of ODA-rGO@MF depends on the concentration of rGO and ODA. On increasing the ODA concentration, superhydrophobicity of the functionalized foam is increased, while by increasing the rGO concentration, the superhydrophobicity of the functionalized foam is initially increased and then decreased after the optimum value. These outstanding properties of functionalized foam make it superb absorbent for the selective absorption of numerous organic solvents and oils from water with excellent absorption performance (44—111 times its own weight) and high recyclability (Fig. 11.5C and D). The high absorption of organic solvents and oils on functionalized foam (ODA-rGO@MF) is controlled by capillary effect, van der Waals forces, and hydrophobic interaction (Fig. 11.5E). When functionalized foam was placed into a container for oil—water separation, oil filled the pore space by removing the air caught in the foam due to the hydrophobic interaction, capillary effect, and van der Waals forces. Here, the water was repelled by the

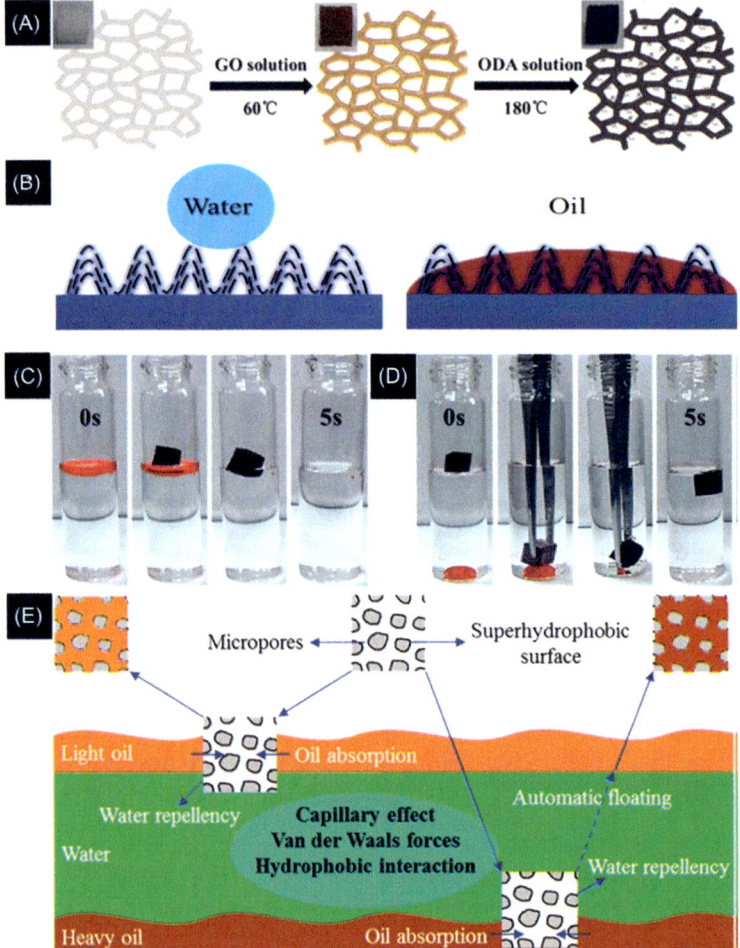

Figure 11.5 (A) Schematic of the synthesis of ODA-rGO@MF through coating; (B) depicts the super-hydrophobicity and superoleophilicity characteristics of the ODA-rGO@MF. (C) Photographs presenting the absorption process of cyclohexane (dyed with Sudan IV) on water. (D) Photographs presenting the absorption process of dichloromethane (dyed with Sudan IV) on water by the ODA-rGO@MF. (E) Diagram depicting the selective oil absorption mechanism for light oil and heavy oil. *Reproduced with permission from Chen C, Zhu X, Chen B. Durable superhydrophobic/superoleophilic graphene-based foam for high-efficiency oil spill cleanups and recovery. Environ Sci Technol 2019;53:1509—17. Copyright 2019, American Chemical Society.*

pores because of the superhydrophobicity of the functionalized foam. Therefore during the oil—water separation process, the air–occupied foam was converted into oil-occupied foam carrying a negligible quantity of water on its surface [45].

The separation of 3D nanomaterials from wastewater after the adsorption process can be facilitated by the introduction of magnetic groups to the adsorbents [46,47].

Yang et al. functionalized 3D GS with Fe_3O_4 and employed it for the adsorption of tetracycline from contaminated water [46]. Highly oxidized graphene sheets and the small size and negative surface charge of iron oxide enhanced the adsorption of tetracycline. The absorption capacity was observed to be 473 mg/g, which is 50% higher than that of GO. The $\pi-\pi$ interactions and hydrogen bonding are two major driving forces, along with electrostatic interactions, which facilitated the adsorption process of tetracycline onto magnetic GS. Table 11.1 shows some reported examples of CNMs-based 3D foams/sponges for the removal of oil and organic substances from aqueous solution.

11.4 Carbon hydrogel

Hydrogels are two- or multicomponent systems comprising 3D networks of hydrophilic polymer chains in which a large amount of water occupies the space between the pores [54]. Since the 1950s, many hydrogels have been explored for numerous applications, including emulsifiers, stabilizers, drug delivery devices, and thickeners in textiles, pharmaceuticals, lithography, and cosmetics [55]. Natural polymers, namely, polysaccharides such as chitin, alginate, gum, glycogen, cellulose, starch, and agarose, and proteins such as gelatin and collagen are used in the formation of a wide variety of hydrogels. Owing to cross-linked hydrophilic 3D polymer networks, hydrogels are proficient at swelling or deswelling reversibly in water and have the ability to keep huge volumes of liquid in a swollen condition.

They can be categorized in two groups. (1) Physical/reversible hydrogels, in which polymeric chains are held together through molecular secondary forces, including ionic and hydrogen bonding, or hydrophobic interactions, or self-assembly or entanglements [56]. These hydrogels are reversible in nature and can be disordered via changes in physical conditions or applying stress. (2) Chemical/permanent hydrogels, in which polymeric chains are held together through nonreversible covalent bonds. The swelling capacity of these hydrogel depends on cross-link density and water—polymerinteraction [57]. Additionally, over physical hydrogels, chemical hydrogels have the advantages of easy control of pore size, gelation time, and chemical functionalization. Hydrogels can be fabricated with controllable responses (as shrink or expand) with changes in surrounding environment via physical changes (including electric or magnetic field, pressure, light, temperature, and sound) [58] and chemical changes (pH, ionic strength, electrolyte, solvent composition, and molecular species) [55,59].

The cross-linked chemical hydrogels have an improved property profile, which makes them potential candidates for various industrial applications including water treatments. The cross-link networks between the different polymer chains provide a

Table 11.1 Brief details of various sorbents based on sponges/foams for removal of oil and organic solvents.

Sorbent	Synthesis methods	Water contact angle (degrees)	Removal capacity (g/g)	Reusability	Reference
rGO Sponge	Dip-coating Functionalization Reduction	—	32.0—46.0	Poor	[48]
Reduced magnetic GO melamine foam (r-MGMF)	Dip-coating Annealing	164.4	35.23—54.13	Good	[17]
Magnetic graphene foam	Gas reduction, coprecipitation	—	12.0—27.0	Good	[49]
Magnetic polymer-based graphene foam (MPG)	Deposition Dip-coating	158.0	9.0—27.0	Good	[50]
α-Fe_2O_3/CNT-N sponge	Oxidation, amination Deposition	11.8	0.04135	Good	[51]
CNT/PDMS-coated PU sponge	Dip-coating	162	15—25	Good	[35]
ODA-rGO@ Melamine foam	Sonication Dip-coating	153.5	44 — 111	Good	[45]
Carbon nanofibers/ carbon foam	Template synthesis of carbon foam and CVD deposition	140	15—28	—	[52]
CNTs coated sponges	Self-polymerization	158	22—35	Excellent	[38]
SiO_2/GO nanohybrids-modified PU sponge	Dip-casting	145	80—180	Good	[53]

gel viscoelastic or elastic behavior structure (hardness) and stickiness characteristics. The water holding tendency, biocompatibility, permeability, and abundant surface functionalities are the most vital properties of a hydrogel. Chemical modifications with inorganic nanomaterials (e.g., oxides, sulfides, metallic nanoparticles, and carbon nanostructures) and organic species (e.g., drugs, enzymes) have been employed to change the surface chemistry, polarity, and to improve the swelling behavior and optical and mechanical properties [61]. For instance, magnetic hydrogel composite can be developed through the addition of magnetic nanoparticles (NPs), which facilitate easy separation and recyclability [62]; NPs can be added to hydrogels to enhance their swelling characteristics [21]; stimuli-sensitive hydrogels incorporated with NPs containing drugs or enzymes have been synthesized for biomedical applications [63]. Fig. 11.6 demonstrates the concept for the formation of a few functional materials using NPs and hydrogels. The major preparation methods implemented for the synthesis of hydrogels are physical cross-linking (complex coacervation, H-bonding, heating/cooling, ionic interaction, freeze—thawing, heat induced aggregation), chemical cross-linking (using cross-linkers), grafting polymerization, and radiation cross-linking (using aqueous/paste/solid state radiation) [64].

The amalgamation of hydrogel and nanomaterials results in hybrid materials with multifunctional network characteristics and synergic effects that enhance swellability and hydrophilicity for water purification and various other applications. Various NPs, including CNMs, can be incorporated into hydrogels using five important strategies to achieve hydrogel—nanoparticle composites or hydrogel nanocomposites: (1) hydrogel development in an NP's suspension; (2) physically addition of the NPs into hydrogel matrix after gelation; (3) in situ NPs formation within a hydrogel; (4) cross-linking

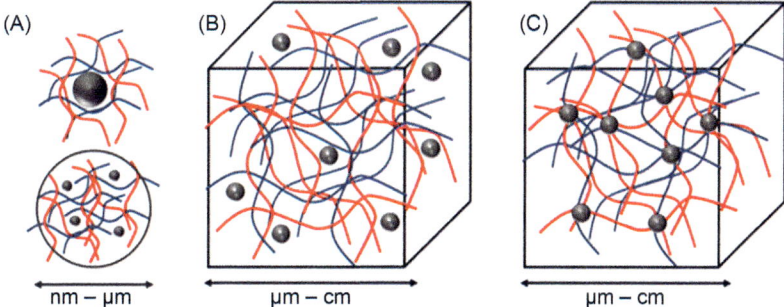

Figure 11.6 Schematic explaining the concept for the formation of a few functional materials using NPs and hydrogels. (A) Micro- or nanosized particles stabilizing hydrogel; (B) NPs noncovalently incorporated in a hydrogel matrix; (C) NPs covalently incorporated in a hydrogel matrix. *Reproduced with permission from Dannert C, Stokke BT, Dias RS. Nanoparticle-hydrogel composites: from molecular interactions to macroscopic behavior. Polymers 2019;11:275 [60]. Copyright 2015, Wiley.*

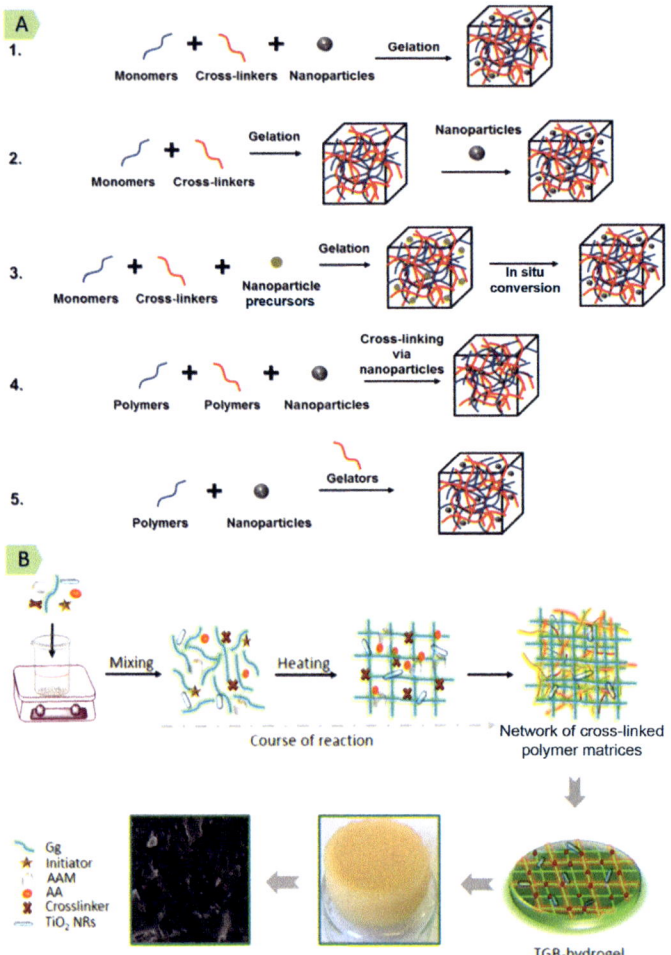

Figure 11.7 (A) Five important methods used to achieve hydrogel−nanoparticle composite: (1) hydrogel development in an NP's suspension; (2) physically addition of the NPs into hydrogel matrix after gelation; (3) in situ NPs formation within a hydrogel; (4) cross-linking using NPs to form hydrogels; and (5) hydrogel preparation using NPs, distinct gelator molecules, and polymers. (B) Schematic demonstrating the construction of the TGB-hydrogel using free-radical graft polymer-ization. *(A) Reproduced with permission from Dannert C, Stokke BT, Dias RS. Nanoparticle-hydrogel composites: from molecular interactions to macroscopic behavior. Polymers 2019;11:275. Copyright 2015, Wiley. (B) Reproduced with permission from Kumar N, Mittal H, Alhassan SM, Ray SS. Bionanocomposite hydrogel for the adsorption of dye and reusability of generated waste for the photo-degradation of ciprofloxacin: a demonstration of the circularity concept for water purification. ACS Sustain Chem Eng 2018;6:17011−25. Copyright 2018, American Chemical Society.*

using NPs to form hydrogels; and (5) hydrogel preparation using NPs, distinct gelator molecules, and polymers (Fig. 11.7A). For example, Kumar et al. synthesized TiO_2 NPs and a functionalized gum ghatti (Gg)-based bionanocomposite hydrogel

(TGB-hydrogel) for the adsorption of brilliant green dye from aqueous solution [20] (Fig. 11.7B). Here, hydrogel is developed in an NP's suspension using a free-radical graft copolymerization method, in which acrylamide (AAM) and acrylic acid (AA) is grafted on a Gg polymer backbone in the presence of N,N'-methylenebis(acrylamide) (MBA) (cross-linking agent) and the ascorbic acid (ASC)/potassium persulfate (KPS) redox pair (initiator) [20]. Using the aforementioned synthesis methods, CNMs such as graphene, GO, CNTs, and carbon nanofibres have been incorporated into the 3D polymeric network of hydrogels to explore their application in wastewater treatments to remove various pollutants such as dyes (MB, MG, RhB, Congo red (CR), etc.) and heavy metal ions (Pb^{2+}, Cd^{2+}, Cu^{2+}, and Hg^{2+}) [40,65–68].

Recently, 3D GO hydrogel membranes (GO-HMs) were prepared by gelation of GO with Fe^{2+} ions through vacuum filtration [65]. Fe^{2+} ions acts as cross-linkers to enhance the interactions between the GO nanosheets and to initiate microstructure conversion of GO through cation–π interactions to create a 3D lamellar porous network. GO-HMs showed high stability in water and also demonstrated considerably enhanced retention efficiency ($>99\%$) and water permeability ($111.5/Lm^2/h/bar$) for MB due to the interconnected nanopores present in hydrogel and high adsorption characteristics. Liang et al. fabricated magnetic GO–oxidized carbon nanotubes (OCNTs)-based hydrogels for the effective removal of As(V) and Se(VI) using combined strategies of sorption and catalysis [69] (Fig. 11.8A). The synergic influence of catalysis and sorption resulted in the high removal of As(V) (258.2 mg/g) and Se(VI) (46.2 mg/g) within a few minutes using as-prepared 3D magnetic GOs-OCNTs

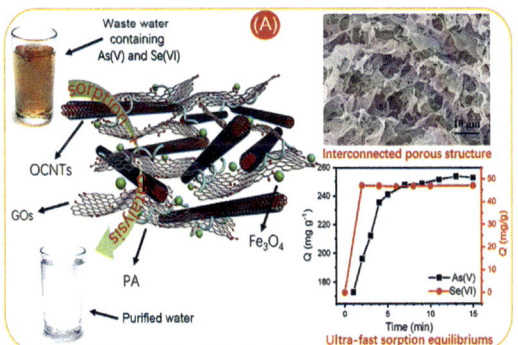 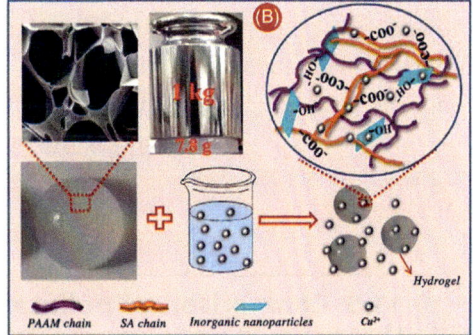

Figure 11.8 (A) 3D magnetic GOs-OCNTs hydrogel for effective removal of As(V) and Se(VI) using combined strategies of sorption and catalysis. (B) NPs (CNTs, NS, and NC)-incorporated PAAM-SA hydrogel for removal of Cu^{2+} ions from polluted water. *(A) Reproduced with permission from Liang J, Ding Z, Qin H, Li J, Wang W, Luo D, et al. Ultra-fast enrichment and reduction of As(V)/Se(VI) on three dimensional graphene oxide sheets-oxidized carbon nanotubes hydrogels. Environ Pollut 2019;251:945–51. Copyright 2019, Elsevier Science Ltd. (B) Reproduced with permission from Yue Y, Wang X, Wu Q, Han J, Jiang J. Assembly of polyacrylamide-sodium alginate-based organic-inorganic hydrogel with mechanical and adsorption properties. Polymers 2019;11:1239. Copyright 2019, MDPI.*

hydrogel. In another study, polyacrylamide (PAAM)–sodium alginate (SA) interconnected polymer hydrogels were synthesized via in situ polymerization reactions. CNTs, nanosilica (NS), and nanoclay (NC) were further integrated into a PAAM-SA matrix of hydrogel via H-bonding (Fig. 11.8B). In comparison to neat hydrogel, PAAM-SA hydrogel with inorganic NPs showed excellent elasticity and mechanical strength. The compression strength of PAAM-SA-NS hydrogel was found to be 1.3 MPa at $\varepsilon = 60\%$ on adding 0.036 g of NS in a 30 g polymer matrix that was 4.8 times higher than pure hydrogel. The achieved hydrogels demonstrated macroporous structural features with high water content ($\approx 83\%$) and low density (≈ 1.4 g/cm^3). However, CNTs were reported as the most appropriate filler to increase the adsorption capability due to their hollow multiwalled structures. The adsorption performance of PAAM-SA-CNT for Cu^{2+} ions was 1.28-fold higher than that of PAAM-SA hydrogel. Owing to their porous features, large carboxyl functionalities, and interpenetrating network structure, as-made hydrogels exhibited high adsorption capacity toward Cu^{2+} ions' removal from water. Additionally, strong coordination complex formation between Cu^{2+} ions and electron-rich groups such as hydroxyl ($-$OH) and carbonyl ($-$COOH) groups resulted in high adsorption [70].

Furthermore, a chitosan$-$GO hydrogel was developed by the self-assembly of GO sheets and CS chains. The 3D network with cross-linked GO-CS (hydrogel) exhibited high adsorption capacity for the removal of different contaminants, such as cationic (MB) and anionic (Eosin Y) dyes, and heavy metal (Cu^{2+} and Pb^{2+}) ions [40]. Tuning of the adsorbent behavior to preferentially adsorb cationic or anionic contaminants was possible by altering the concentration of GO or CS in the 3D hydrogel. Increasing the GO concentration in 3D hydrogel favors the adsorption of cationic contaminants, whereas an increased concentration of CS enhances the adsorption of anionic contaminants. Another report described a reusable polydopamine-modified graphene hydrogel for the elimination of a broad spectrum of pollutants, such as heavy metals, aromatic pollutants, and synthetic dyes [67].

11.5 Comparison of pollutants removal performance of three-dimensional macrostructures

Overall, the self-assembly of GO nanosheets, CNTs, or carbon nanofibres into 3D macrostructures (3DMs) solves the problem of high-colloidal stability of these nanomaterials in aqueous media. The highly porous 3DMs geometries, such as aerogels, foams, hydrogels, sponges, membranes, and fibers, can be prepared for water treatments. Among CNMs, graphene/GO-based 3DMs are the most investigated materials for water purification applications as the GO offers versatile chemistry with

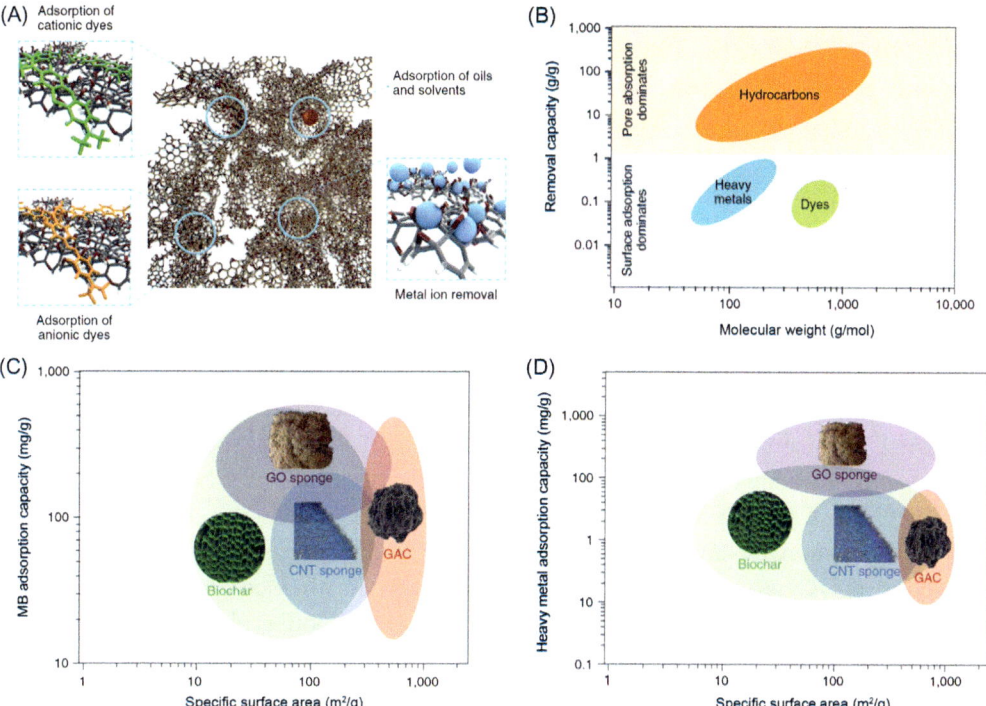

Figure 11.9 3DMs for pollutants removal form aqueous water. (A) 3DMs offer versatile chemistry with an abundance of active sites and have the ability to react by numerous mechanisms with pollutant molecules such as anionic and cationic dyes, metal ions, oils, and organic solvents. (B) Investigation of literature data displays that GO-based 3DMs have higher removal performance for hydrocarbons than for metals and dyes. (C) Dye removal capacity of GO sponges is compared with other CNMs-based 3DMs including granular activated carbon (GAC), CNTs sponges, and biochars with the function of their surface area. (D) Heavy metal removal capacity of GO sponges is also equated with other kinds of carbon-based 3DMs with the function of their surface area. *Reproduced with permission from* N. Yousefi, X. Lu, M. Elimelech and N. Tufenkji, Environmental performance of graphene-based 3D macrostructures. Nat Nanotechnol **14**, 2019, 107−119. *Copyright 2019, Nature.*

an abundance of active sites and has ability to react by numerous mechanisms with the pollutant molecules [65]. Contaminants such as dyes and metal ions are adsorbed onto surface pores and the interfacial surface, whereas contaminants such as organic solvents and oils are absorbed in the pores of 3DMs (Fig. 11.9A). Owing to their high surface area, tailorable surface chemistry, and wide range of pore structures, graphene/GO sponges demonstrate high removal performance for various oils, including diesel, crude oil, gasoline, vacuum oils, and motor, as well as organic solvents, namely, chloroform, methanol, toluene, xylene, and ethanol [6,71]. The versatile pores and high pore volume of GO sponges can uptake a larger number of pollutants than its surface, resulting in larger elimination performance for hydrocarbons than other kinds of pollutants (Fig. 11.9B). Therefore most studies are focused on the bulk absorption of organic solvents and oils, and a few investigations also demonstrated their removal from seawater and freshwater [68,72–74]. Moreover, the dye removal capacity of GO sponges is compared with other CNMs-based 3DMs including granular activated carbon (GAC), CNTs sponges, and biochars, based on their surface area (Fig. 11.9C) [72]. The dye uptake by GO sponges is competitive with other varieties of 3DMs, and comparatively better than that of GAC (an industry grade) at a lower surface area. The high dye removal performance of GO sorbents is attributed to excessive oxygen functionalities for electrostatic interactions, complexation, and H-bonding, and plentiful sp^2 carbon sites for $\pi-\pi$ bonding. Similarly, the heavy metal removal capacity of GO sponges is also equated with other kinds of carbon-based 3DMs, such as CNT sponges, biochar, and GAC, based on their surface area (Fig. 11.9D) [71,72]. GO sponges have higher adsorption performance for heavy metal ions than other types of CNMs-based 3DMs (such as GAC, CNTs sponges, and biochar). Although CNMs-based 3DMs (such as GAC, CNTs sponges, and biochar) have a slightly higher specific surface area than GO sponges, GO sponges exhibit higher adsorption performance due to the presence of innumerable oxygen-containing functionalities, which offer plentiful sites for ion adsorption.

11.6 Conclusion

Self-assembly of CNMs such as CNTs, graphene/GO, and carbon nanofibres into 3D superstructures/macrostructures is a robust approach for conserving the high surface area and versatile surface characteristics of these nanomaterials while eliminating or decreasing the disadvantages (i.e., associated potential risks to living beings and ecosystem) of their airborne and colloidal forms, and performance loss due to aggregation. CNMs-based 3D superstructures/macrostructures with diverse geometries such as aerogels, hydrogels, sponges, foams, and beads have been effectively used to eliminate a variety of oils, organic solvents, dyes, and metals from the water. The pollutant removal performance of sorbents could be enhanced by the regulation of materials' properties. Despite the considerable progress made in recent years, various hurdles must be resolved in order to realize the full potential of CNMs-based 3D superstructures/macrostructures in water treatment and other related applications.

Surface area is a very critical factor for monitoring the contaminants adsorption/absorption on the sorbents surface. While very few investigations show CNMs-based 3DMs with exceptional high surface area (> 1000 m^2/g), still more dedicated research is required to capitalize on the full potential of CNMs. The logical fabrication of versatile pore construction is needed in order to obtain higher specific surface area. Excessive porous structures lead to mechanically weak 3D superstructures, and poor interconnectivity poses low performance. Moreover, CNMs-based 3DMs have been used for small-scale or laboratory batch-scale removal of single pollutant systems. In order to check their full removal performance, these sorbents should be investigated for multicomponent pollutants systems. In regard to CNMs-based 3DMs commercialization, these sorbents must be examined for emerging pollutants, including pharmaceuticals, personal care products, hormones, and toxins, and their regeneration and recyclability must be systematically studied and discussed in future studies. Moreover, the production cost of CNMs is a major bottleneck in the commercialization of CNMs-based 3DMs in environmental applications.

To date, mainly CNMs (CNTs, graphene/GO) have been processed into 3DMs and employed for the removal of various contaminants from water. 3DMs materials prepared using other forms of nanocarbons and with other metal oxides, metal chalcogenides (i.e., MoS_2, WS_2), and 2D materials (i.e., g-C_3N_4, hexagonal boron nitride, phosphorene, MXenes) remain to be investigated for their utility in water treatments and other environmental applications.

References

[1] Zhao J, Wang Z, White JC, Xing B. Graphene in the aquatic environment: adsorption, dispersion, toxicity and transformation. Environ Sci Technol 2014;48:9995−10009.

[2] Cao X, Yin Z, Zhang H. Three-dimensional graphene materials: preparation, structures and application in supercapacitors. Energy Environ Sci 2014;7:1850−65.

[3] Li W, Gao S, Wu L, Qiu S, Guo Y, Geng X, et al. High-density three-dimension graphene macroscopic objects for high-capacity removal of heavy metal ions. Sci Rep 2013;3:2125.

[4] Mao S, Wen Z, Kim H, Lu G, Hurley P, Chen J. A general approach to one-pot fabrication of crumpled graphene-based nanohybrids for energy applications. ACS Nano 2012;6:7505−13.

[5] Hiew BYZ, Lee LY, Lee XJ, Thangalazhy-Gopakumar S, Gan S, Lim SS, et al. Review on synthesis of 3D graphene-based configurations and their adsorption performance for hazardous water pollutants. Process Saf Environ Prot 2018;116:262−86.

[6] Shen Y, Fang Q, Chen B. Environmental applications of three-dimensional graphene-based macrostructures: adsorption, transformation, and detection. Environ Sci Technol 2015;49:67−84.

[7] He Y, Liu Y, Wu T, Ma J, Wang X, Gong Q, et al. An environmentally friendly method for the fabrication of reduced graphene oxide foam with a super oil absorption capacity. J Hazard Mater 2013;260:796−805.

[8] Yang K, Wang J, Chen X, Zhao Q, Ghaffar A, Chen B. Application of graphene-based materials in water purification: from the nanoscale to specific devices. Environ Sci Nano 2018;5:1264−97.

[9] Wu R, Yu B, Liu X, Li H, Wang W, Chen L, et al. One-pot hydrothermal preparation of graphene sponge for the removal of oils and organic solvents. Appl Surf Sci 2016;362:56−62.

[10] Speyer L, Fontana S, Ploneis S, Hérold C. Influence of the precursor alcohol on the adsorptive properties of graphene foams elaborated by a solvothermal-based process. Microporous Mesoporous Mater 2017;243:254−62.

[11] Wu S, Zhang K, Wang X, Jia Y, Sun B, Luo T, et al. Enhanced adsorption of cadmium ions by 3D sulfonated reduced graphene oxide. Chem Eng J 2015;262:1292−302.

[12] Gan L, Shang S, Hu E, Yuen CWM, Jiang S-x. Konjac glucomannan/graphene oxide hydrogel with enhanced dyes adsorption capability for methyl blue and methyl orange. Appl Surf Sci 2015;357:866—72.

[13] He Y, Li J, Luo K, Li L, Chen J, Li J. Engineering reduced graphene oxide aerogel produced by effective γ-ray radiation-induced self-assembly and its application for continuous oil—water separation. Ind Eng Chem Res 2016;55:3775—81.

[14] Meng LY, Park S-J. Superhydrophobic carbon-based materials: a review of synthesis, structure, and applications. Carbon Lett 2014;15:89—104.

[15] Yang Y, Tong Z, Ngai T, Wang C. Nitrogen-rich and fire-resistant carbon aerogels for the removal of oil contaminants from water. ACS Appl Mater Interfaces 2014;6:6351—60.

[16] Bi H, Huang X, Wu X, Cao X, Tan C, Yin Z, et al. Carbon microbelt aerogel prepared by waste paper: an efficient and recyclable sorbent for oils and organic solvents. Small. 2014;10:3544—50.

[17] Lv X, Tian D, Peng Y, Li J, Jiang G. Superhydrophobic magnetic reduced graphene oxide-decorated foam for efficient and repeatable oil-water separation. Appl Surf Sci 2019;466:937—45.

[18] Tran VT, Xu X, Mredha MTI, Cui J, Vlassak JJ, Jeon I. Hydrogel bowls for cleaning oil spills on water. Water Res 2018;145:640—9.

[19] Wang J, Zhang W, Zhang C. Versatile fabrication of anisotropic and superhydrophobic aerogels for highly selective oil absorption. Carbon. 2019;155:16—24.

[20] Kumar N, Mittal H, Alhassan SM, Ray SS. Bionanocomposite hydrogel for the adsorption of dye and reusability of generated waste for the photodegradation of ciprofloxacin: a demonstration of the circularity concept for water purification. ACS Sustain Chem Eng 2018;6:17011—25.

[21] Kumar N, Mittal H, Parashar V, Ray SS, Ngila JC. Efficient removal of rhodamine 6G dye from aqueous solution using nickel sulphide incorporated polyacrylamide grafted gum karaya bionanocomposite hydrogel. RSC Adv 2016;6:21929—39.

[22] Mittal H, Maity A, Sinha Ray S. The adsorption of Pb^{2+} and Cu^{2+} onto gum ghatti-grafted poly (acrylamide-co-acrylonitrile) biodegradable hydrogel: isotherms and kinetic models. J Phys Chem B 2015;119:2026—39.

[23] Sahraei R, Ghaemy M. Synthesis of modified gum tragacanth/graphene oxide composite hydrogel for heavy metal ions removal and preparation of silver nanocomposite for antibacterial activity. Carbohydr Polym 2017;157:823—33.

[24] Kabiri S, Tran DNH, Altalhi T, Losic D. Outstanding adsorption performance of graphene—carbon nanotube aerogels for continuous oil removal. Carbon. 2014;80:523—33.

[25] Kistler SS. Coherent expanded aerogels and jellies. Nature. 1931;127:741.

[26] Pekala RW. Organic aerogels from the polycondensation of resorcinol with formaldehyde. J Mater Sci 1989;24:3221—7.

[27] Li J, Li J, Meng H, Xie S, Zhang B, Li L, et al. Ultra-light, compressible and fire-resistant graphene aerogel as a highly efficient and recyclable absorbent for organic liquids. J Mater Chem A 2014;2:2934—41.

[28] Bidgoli H, Mortazavi Y, Khodadadi AA. A functionalized nano-structured cellulosic sorbent aerogel for oil spill cleanup: synthesis and characterization. J Hazard Mater 2019;366:229—39.

[29] Nelson I, Naleway SE. Intrinsic and extrinsic control of freeze casting. J Mater Res Technol 2019;8:2372—85.

[30] Shen Y, Zhu X, Zhu L, Chen B. Synergistic effects of 2D graphene oxide nanosheets and 1D carbon nanotubes in the constructed 3D carbon aerogel for high performance pollutant removal. Chem Eng J 2017;314:336—46.

[31] Liu C, Liu H, Xu A, Tang K, Huang Y, Lu C. In situ reduced and assembled three-dimensional graphene aerogel for efficient dye removal. J Alloy Compd 2017;714:522—9.

[32] Huang J, Liu H, Chen S, Ding C. Hierarchical porous MWCNTs-silica aerogel synthesis for high-efficiency oily water treatment. J Environ Chem Eng 2016;4:3274—82.

[33] Yao Q, Fan B, Xiong Y, Jin C, Sun Q, Sheng C. 3D assembly based on 2D structure of cellulose nanofibril/graphene oxide hybrid aerogel for adsorptive removal of antibiotics in water. Sci Rep 2017;7:45914.

[34] Jiang G, Hu R, Xi X, Wang X, Wang R. Facile preparation of superhydrophobic and superoleophilic sponge for fast removal of oils from water surface. J Mater Res 2013;28:651−6.

[35] Wang C-F, Lin S-J. Robust superhydrophobic/superoleophilic sponge for effective continuous absorption and expulsion of oil pollutants from water. ACS Appl Mater Interfaces 2013;5:8861−4.

[36] Zhu Q, Pan Q, Liu F. Facile removal and collection of oils from water surfaces through superhydrophobic and superoleophilic sponges. J Phys Chem C 2011;115:17464−70.

[37] Gupta S, Tai N-H. Carbon materials as oil sorbent: a review on synthesis and performance. J Mater Chem A 2015;4.

[38] Wang H, Wang E, Liu Z, Gao D, Yuan R, Sun L, et al. A novel carbon nanotubes reinforced superhydrophobic and superoleophilic polyurethane sponge for selective oil−water separation through a chemical fabrication. J Mater Chem A 2015;3:266−73.

[39] Qiu L, Liu JZ, Chang SLY, Wu Y, Li D. Biomimetic superelastic graphene-based cellular monoliths. Nat Commun 2012;3:1241.

[40] Chen Y, Chen L, Bai H, Li L. Graphene oxide−chitosan composite hydrogels as broad-spectrum adsorbents for water purification. J Mater Chem A 2013;1:1992−2001.

[41] Chang J, Shi Y, Wu M, Li R, Shi L, Jin Y, et al. Solar-assisted fast cleanup of heavy oil spills using a photothermal sponge. J Mater Chem A 2018;6:9192−9.

[42] Liu Y, Li F, Xia Q, Wu J, Liu J, Huang M, et al. Conductive 3D sponges for affordable and highly-efficient water purification. Nanoscale. 2018;10:4771−8.

[43] Liu Z, Fang Y, Jia H, Wang C, Song Q, Li L, et al. Novel multifunctional cheese-like 3D carbon-BN as a highly efficient adsorbent for water purification. Sci Rep 2018;8:1104.

[44] Jayanthi S, KrishnaRao Eswar N, Singh SA, Chatterjee K, Madras G, Sood AK. Macroporous three-dimensional graphene oxide foams for dye adsorption and antibacterial applications. RSC Adv 2016;6:1231−42.

[45] Chen C, Zhu X, Chen B. Durable superhydrophobic/superoleophilic graphene-based foam for high-efficiency oil spill cleanups and recovery. Environ Sci Technol 2019;53:1509−17.

[46] Yu B, Bai Y, Ming Z, Yang H, Chen L, Hu X, et al. Adsorption behaviors of tetracycline on magnetic graphene oxide sponge. Mater Chem Phys 2017;198:283−90.

[47] Yu B, Zhang X, Xie J, Wu R, Liu X, Li H, et al. Magnetic graphene sponge for the removal of methylene blue. Appl Surf Sci 2015;351:765−71.

[48] Tjandra R, Lui G, Veilleux A, Broughton J, Chiu G, Yu A. Introduction of an enhanced binding of reduced graphene oxide to polyurethane sponge for oil absorption. Ind Eng Chem Res 2015;54:3657−63.

[49] Yang S, Chen L, Mu L, Ma P-C. Magnetic graphene foam for efficient adsorption of oil and organic solvents. J Colloid Interface Sci 2014;430:337−44.

[50] Liu C, Yang J, Tang Y, Yin L, Tang H, Li C. Versatile fabrication of the magnetic polymer-based graphene foam and applications for oil−water separation. Colloids Surf A Physicochem Eng Asp 2015;468:10−16.

[51] Wu Y, Wang Y, Lin Z, Wang Y, Li Y, Liu S, et al. Three-dimensional α-Fe2O3/amino-functionalization carbon nanotube sponge for adsorption and oxidative removal of tetrabromobisphenol A. Sep Purif Technol 2019;211:359−67.

[52] Xiao N, Zhou Y, Ling Z, Qiu J. Synthesis of a carbon nanofiber/carbon foam composite from coal liquefaction residue for the separation of oil and water. Carbon. 2013;59:530−6.

[53] Lü X, Cui Z, Wei W, Xie J, Jiang L, Huang J, et al. Constructing polyurethane sponge modified with silica/graphene oxide nanohybrids as a ternary sorbent. Chem Eng J 2016;284:478−86.

[54] Ahmed EM. Hydrogel: preparation, characterization, and applications: a review. J Adv Res 2015;6:105−21.

[55] Mittal H, Ray SS, Okamoto M. Recent progress on the design and applications of polysaccharide-based graft copolymer hydrogels as adsorbents for wastewater purification. Macromol Mater Eng 2016;301:496−522.

[56] Rosales AM, Anseth KS. The design of reversible hydrogels to capture extracellular matrix dynamics. Nat Rev Mater 2016;1:15012.

[57] Caló E, Khutoryanskiy VV. Biomedical applications of hydrogels: a review of patents and commercial products. Eur Polym J 2015;65:252—67.

[58] Weber C, Hoogenboom R, Schubert US. Temperature responsive bio-compatible polymers based on poly (ethylene oxide) and poly (2-oxazoline) s. Prog Polym Sci 2012;37:686—714.

[59] Sullad AG, Manjeshwar LS, Aminabhavi TM. Novel pH-sensitive hydrogels prepared from the blends of poly(vinyl alcohol) with acrylic acid-graft-guar gum matrixes for isoniazid delivery. Ind Eng Chem Res 2010;49:7323—9.

[60] Dannert C, Stokke BT, Dias RS. Nanoparticle-hydrogel composites: from molecular interactions to macroscopic behavior. Polymers. 2019;11:275.

[61] Zhu H-Y, Fu Y-Q, Jiang R, Yao J, Xiao L, Zeng G-M. Novel magnetic chitosan/poly (vinyl alcohol) hydrogel beads: preparation, characterization and application for adsorption of dye from aqueous solution. Bioresour Technol 2012;105:24—30.

[62] Gaharwar AK, Peppas NA, Khademhosseini A. Nanocomposite hydrogels for biomedical applications. Biotechnol Bioeng 2014;111:441—53.

[63] Thoniyot P, Tan MJ, Karim AA, Young DJ, Loh XJ. Nanoparticle—hydrogel composites: concept, design, and applications of these promising, multi-functional materials. Adv Sci 2015;2:1400010.

[64] Gulrez S, Al-Assaf S, Phillips G. Hydrogels: methods of preparation, characterisation and applications, in: Angelo Carpi (Eds.), Progress in Molecular and Environmental Bioengineering: From Analysis and Modeling to Technology Applications, 2011, Intechopen Publisher, London.

[65] Chen Z, Wang J, Duan X, Chu Y, Tan X, Liu S, et al. Facile fabrication of 3D ferrous ion cross-linked graphene oxide hydrogel membranes for excellent water purification. Environ Sci Nano 2019;6:3060—71.

[66] Thakur VK, Voicu SI. Recent advances in cellulose and chitosan based membranes for water purification: a concise review. Carbohydr Polym 2016;146:148—65.

[67] Gao H, Sun Y, Zhou J, Xu R, Duan H. Mussel-inspired synthesis of polydopamine-functionalized graphene hydrogel as reusable adsorbents for water purification. ACS Appl Mater Interfaces 2013;5:425—32.

[68] Chen B, Ma Q, Tan C, Lim T-T, Huang L, Zhang H. Carbon-based sorbents with three-dimensional architectures for water remediation. Small. 2015;11:3319—36.

[69] Liang J, Ding Z, Qin H, Li J, Wang W, Luo D, et al. Ultra-fast enrichment and reduction of As (V)/Se(VI) on three dimensional graphene oxide sheets-oxidized carbon nanotubes hydrogels. Environ Pollut 2019;251:945—51.

[70] Yue Y, Wang X, Wu Q, Han J, Jiang J. Assembly of polyacrylamide-sodium alginate-based organic-inorganic hydrogel with mechanical and adsorption properties. Polymers. 2019;11:1239.

[71] Yousefi N, Lu X, Elimelech M, Tufenkji N. Environmental performance of graphene-based 3D macrostructures. Nat Nanotechnol 2019;14:107—19.

[72] Wan S, Bi H, Sun L. Graphene and carbon-based nanomaterials as highly efficient adsorbents for oils and organic solvents. Nanotechnol Rev 2016;5.

[73] Noamani S, Niroomand S, Rastgar M, Sadrzadeh M. Carbon-based polymer nanocomposite membranes for oily wastewater treatment. npj Clean Water 2019;2:20.

[74] Cao N, Lyu Q, Li J, Wang Y, Yang B, Szunerits S, et al. Facile synthesis of fluorinated polydopamine/chitosan/reduced graphene oxide composite aerogel for efficient oil/water separation. Chem Eng J 2017;326:17—28.

CHAPTER TWELVE

Biopolymer-functionalized carbon nanomaterials—based adsorbents

12.1 Introduction

Biopolymers, or natural polymers, are polymers that occur from natural sources, for example, animals, plants, fungi, and bacteria, during their growth cycles [1]. Biopolymers can also be synthesized from biologically sourced materials, including sugars, resins, proteins, amino acids, and vegetable oils [2], and can be classified either based on their monomeric unit or their origin [3]. Based on the monomeric unit, biopolymers are subdivided into three categories: (1) polysaccharides, (2) polynucleotides, and (3) polypeptides. Biopolymers are renewable, nontoxic, economical, and are therefore suitable for applications in environmental problems and in the biomedical, packaging, pharmaceutical, energy, catalysis, and food industries [4−12]. They have been employed extensively for wastewater treatment via adsorption and photocatalytic degradation [13−17]. The adsorption efficacy of biopolymers varies according to their chemical structure and origin, and with the composition of the organic or inorganic adsorbates [18−22].

Polysaccharides are biodegradable, renewable, nontoxic, biocompatible, cost-effective, and abundant and are most frequently employed as nanoadsorbents to remove toxic substances [5,18,23−28]. The monomer units, that is, monosaccharides (sugars), are linked through glycosidic bonds to form polysaccharides. The unique adsorption characteristics of polysaccharides are mainly attributed to the (1) presence of hydroxyl groups from monosaccharides that provide hydrophilic interactions, (2) number of functional groups on the polymeric chain which provide chemical interactions, and (3) flexibility of the polymeric chain. Further, the selectivity and adsorption capacity of polysaccharides can be enhanced by functionalization or modification [29,30]. The most commonly used biopolymers in the adsorption of water contaminants are starch, cellulose, gum, chitin, chitosan, and lignin [24,28,31−36]. However, the adsorption of contaminants onto biopolymers is not commercially viable, as in an aqueous environment water molecules reduce the affinity between dyes and the biopolymer adsorbents, which impedes the adsorption process [37,38]. Further, the recovery of biopolymers after the adsorption of the contaminant is difficult [39]. Polysaccharides have poor mechanical stabilities, which further restricts the use of

Carbon Nanomaterial-Based Adsorbents for Water Purification.
DOI: https://doi.org/10.1016/B978-0-12-821959-1.00012-X

biopolymers as unaided adsorbents in water [39,40]. Therefore biopolymers are impregnated with carbon-based nanomaterials (CNMs) to improve their mechanical properties, chemical stabilities, and hydrophobicities, while, in return, the biopolymer acts as stabilizer for the carbon-based material, to avoid aggregation [41,42]. Several reports have been published concerning the potential of biopolymer/carbon-based nanocomposites for contaminant removal via adsorption [43–46].

12.2 Chitosan-functionalized carbon nanomaterials

Chitosan is a widely available and nontoxic mucopolysaccharide biopolymer, which has been exploited in a variety of applications, including adsorption [47,48]. Chitosan is a derivative of chitin and consists of 2-acetamido-2-deoxy-β-D-glucopyranose from chitin, and 2-amino-2-deoxy-β-D-glucopyranose residues [34,47]. Chitosan is a suitable nanoadsorbent for the removal of heavy metal ions, dyes, organochloride pesticides, and other organic pollutants from water due to its low cost in comparison with activated carbon and with other adsorbents containing amino and hydroxyl functional groups, thus enhancing the adsorption capacity [34]. Additionally, chitosan has excellent chemical stability, outstanding chelation behavior, and is biodegradable, nontoxic, and selective. Chitosan is rich in hydroxyl and amine groups, which actively participate in the adsorption of pollutants; however, at low pH values these functional groups are protonated and do not contribute to the removal of water contaminants [39]. Impregnation of graphene into chitosan or vice versa is the most common way to achieve both mutual stabilization and improvement in the adsorption process [49–52].

Recently Zou et al. fabricated graphene nanoplates cross-linked with chitosan spheres (CS/GNPs) using a syringe dropping method and employed them for the removal of MO (anionic) and AR1 (cationic) organic dyes from contaminated water [53]. The main driving force for the adsorption of dyes onto the surface of CS/GNP was the electrostatic attraction between the dye molecules and the protonated amino groups in the chitosan spheres (CS), along with π–π interactions between the aromatic rings of GNPs and the π electrons of the dye molecule. Adsorption of dyes was found to be more effective at lower pH (pH <4), as the amino groups of the CS are protonated and promote the process. The adsorption process follows pseudo–second-order kinetics and the Freundlich isotherm. A similar isotherm and kinetics were measured for the adsorption of picric acid on a graphene–chitosan nanocomposite (GO-Chi) [50]. The functionalization of chitosan on graphene oxide nanosheets was performed at alkaline medium (pH 11), followed by the addition of glutaraldehyde

(GLA) for the cross-linking of chitosan (Fig. 12.1A). The introduction of chitosan to GO nanosheets increased the surface area, total pore volume, and pore size, by 2.4, 2.87, and 7.68 times, respectively, which supports surface-based processes, including adsorption. Electrostatic interactions between the −OH group of picric acid and the amine group of chitosan, and hydrogen bonding facilitated efficient adsorption on the nanocomposite (Fig. 12.1A). Jhang et al. also briefly explained the functionalization of chitosan on the surface of GO nanosheets (chitosan−GO) via physical interactions such as peptide and hydrogen bonding (Fig. 12.1B), which was confirmed by XRD analysis [54]. The adsorption of As(V) as $H_2AsO_4^-$ and $HAsO_4^{2-}$ on chitosan−GO was supported by the electrostatic interaction. At lower pH value, due to protonation of adsorbent surface, the electrostatic interaction between positively charged adsorbent and negatively charged arsenate species was enhanced and resulted in increased removal efficiency. Furthermore, the adsorbent was regenerated using 1.0 mol/L

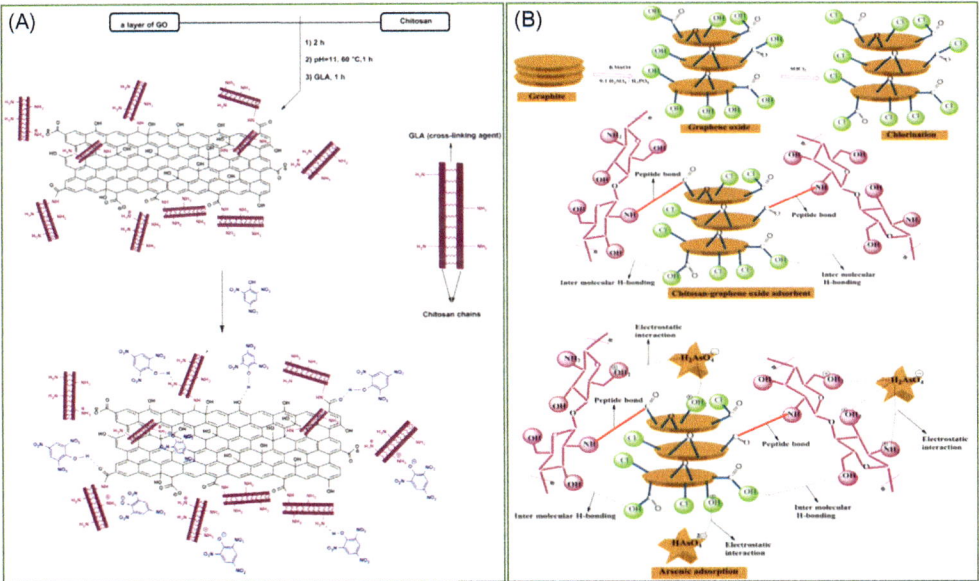

Figure 12.1 (A) Preparation route of graphene oxide/chitosan (GO-Chi) nanocomposite using glutaraldehyde (GLA) as cross-linker and adsorption of picric acid. (B) Synthesis route of chitosan-modified graphene oxide (chitosan−GO), adsorption mechanism of As(V)/As(III) on the chitosan−GO and regeneration of adsorbent (chitosan−GO) for recyclability. *(A) Reproduced with permission from Mohseni Kafshgari M, Tahermansouri H. Development of a graphene oxide/chitosan nanocomposite for the removal of picric acid from aqueous solutions: study of sorption parameters. Colloids Surf B Biointerfaces 2017;160:671−81. Copyright 2017, Elsevier Science Ltd. (B) Reproduced with permission from Kumar ASK, Jiang S-J. Chitosan-functionalized graphene oxide: A novel adsorbent an efficient adsorption of arsenic from aqueous solution. J Environ Chem Eng. 2016;4:1698−713. Copyright 2016, Elsevier Science Ltd.*

NaOH (Fig. 12.1B) with >99% adsorption efficiency for three consecutive adsorption—desorption cycles.

Doping of porous chitosan onto graphene oxide (CSGO1) enhanced the affinity of the latter for anionic contaminants, in combination with the porosity of the aerogel, and the absorption capacity of CSGO1 for MO was the highest that had been observed, at the publication of the report [55]. Ray et al. reported a porous 3D nanoarchitecture of chitosan/gold/graphene oxide with a high active surface area ($780 \ m^2/g$) and pore volume ($0.730 \ cm^3/g$), and a 1300 nm average pore diameter, which contributed to its efficacy for absorption applications. The nanocomposite was employed for effective separation via adsorption and label-free surface-enhanced Raman spectroscopy (SERS) detection of pharmaceutical contaminants (tetracycline antibiotics and doxorubicin, a drug used in chemotherapy) and for suppression of methicillin-resistant *Staphylococcus aureus* (MRSA) superbugs from environmental samples [56]. The high adsorption efficacies of the chitosan/gold/graphene oxide nanocomposite for the antibiotic and chemotherapy drug were attributed to a combination of interactions including $\pi-\pi$ interactions between GO and the pharmaceutical contaminants, hydrophobic interactions, electrostatic interactions and hydrogen bonding between chitosan and the drugs, and the pore-filling mechanism.

Separation of adsorbent material for regeneration and recyclability after the adsorption process is another issue of concern from the economic point of view. The magnetization of the adsorbent material is the simple solution for such problems. On functionalization of the adsorbent with iron oxide, it can easily be separated from the solution by the means of an external magnetic field. Sarkar et al. developed a GO—chitosan nanocomposite using cross-linker N,N'-methylenebis(acrylamide) (MBA), which was functionalized with in situ generated Fe_3O_4 [57]. Magnetic chitosan-functionalized GO ($cl-CS-p(MA)/Fe_3O_4$) was further used for the selective adsorption of cationic methylene blue (MB) dye in a mixture of cationic and anionic dyes (Fig. 12.2A) and could easily be separated from the aqueous medium after adsorption using a magnet [57]. GO ($cl-CS-p(MA)/Fe_3O_4$) nanocomposite exhibited a 3D sponge-like structure with an interconnected microporous network that was examined with scanning electron microscopy (SEM) and transmission electron microscopy (TEM) techniques. The high porosity of GO ($cl-CS-p(MA)/Fe_3O_4$) resulted in the high adsorption capacity of cationic dye at pH 8.5. The pH_{pzc} value of GO ($cl-CS-p(MA)/Fe_3O_4$) was found to be 6.5, which supported the negatively charged behavior of GO ($cl-CS-p(MA)/Fe_3O_4$) at pH >6.5. Therefore the high electrostatic interaction between cationic dye and negative charged adsorbent, is majorly responsible for the MB dye removal. Additionally, the mapping of elemental sulfur on the GO ($cl-CS-p(MA)/Fe_3O_4$) surface after MB adsorption shows the distribution of sulfur across the adsorbent surface, demonstrating the adsorption of MB. On using GO ($cl-CS-p(MA)/Fe_3O_4$) as an adsorbent for the mixture of cationic and anionic

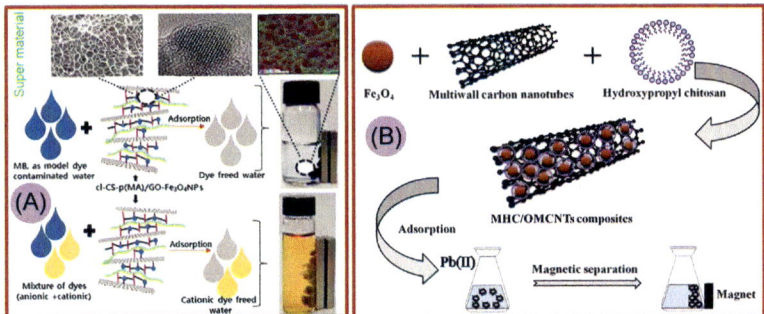

Figure 12.2 (A) SEM and TEM images of cl—CS—p(MA)/Fe₃O₄ before adsorbing dye, mapping of elemental sulfur of cl—CS—p(MA)/Fe₃O₄ after adsorbing dye (MB), selective adsorption of cationic dye (MB), and the separation of cl—CS—p(MA)/Fe₃O₄ using external magnet after adsorption. (B) Formation of the magnetic MHC/OMCNT composite for adsorption of Pb (II). *(A) Reproduced with permission from Sarkar AK, Bediako JK, Choi J-W, Yun Y-S. Functionalized magnetic biopolymeric graphene oxide with outstanding performance in water purification. NPG Asia Mater 2019;11:4. Copyright 2019 Springer Nature Publishers. (B) Reproduced with permission from Wang Y, Shi L, Gao L, Wei Q, Cui L, Hu L, et al. The removal of lead ions from aqueous solution by using magnetic hydroxypropyl chitosan/oxidized multiwalled carbon nanotubes composites. J Colloid Interface Sci. 2015;451:7—14. Copyright 2015, Elsevier Science Ltd.*

dye, there was significant uptake of the cationic dye, leaving behind the anionic dye, which was due to electrostatic attraction between the adsorbent and cationic dye and electrostatic repulsion between the adsorbent and anionic dye. The adsorbent was successfully removed using an external magnet after the adsorption process.

A magnetic hydroxypropyl chitosan (MHC) oxidized multiwalled carbon nanotube nanocomposite was designed for the adsorption of Pb(II) from aqueous solution (Fig. 12.2B) [58]. The effect of pH on the adsorption behavior of Pb(II) onto MHC/OMCNTs was studied (pH = 2.7—6), and the process was found to follow the Sips isotherm model. At low pH (pH = 2), the adsorption of Pb(II) was low. This was attributed to the competition between Pb(II) and H^+ ions for adsorption onto MHC/OMCNTs. With an increase in pH, the concentration of H^+ ions decreased, and subsequently the adsorption of Pb(II) increased. Optimal adsorption of the metal ion was observed at pH 5.

MWCNT—chitosan nanocomposites have also been investigated for the removal of water contaminants via adsorption [59—62]. Tahermansouri et al. removed the picric acid from wastewater via adsorption using chitosan-functionalized MWCNTs (MWCNT-Chi) [60]. The adsorption capacity of MWCNT-Chi (1250 mg/g) was observed to be much higher than that of MWCNT-COOH (222.2 mg/g). Minteer et al. modified MWCNTs with chitosan, for fast and efficient removal of the heavy metal ion Cr(VI) from wastewater [63]. They reported that adsorption of anionic Cr (VI) species onto the nanocomposite occurred via three mechanisms. First, at low pH the amine groups of chitosan and carboxylic acid groups of MWCNTs were protonated, which led to the adsorption of the anionic Cr(VI) groups via electrostatic attraction. Second, the reduction of the adsorbed Cr(VI) to Cr(III) was facilitated by the high concentration of protons in solution, which was confirmed by inductively coupled plasma mass spectroscopy (ICP-MS). Third, a chemical redox reaction directly reduced the Cr(VI) to Cr(III), which was adsorbed onto the chitosan−MWCNT nanocomposite via electrostatic interactions. The adsorption process followed pseudo-second-order kinetics and was best represented by the Langmuir isotherm model.

12.3 Lignin-functionalized carbon nanomaterials

Lignin is one of the main sources of naturally occurring phenolic macromolecules on Earth. It is the second most abundant biopolymer after cellulose and is easily available in the form of lignocellulosic biomass (amount varying in the range of 10% − 30% by weight), together with hemicellulose and cellulose [64]. It can also be obtained at paper pulp industries as a by-product of cellulose. The phenolic macromolecules of lignin are complex structures consisting of phenyl propane-like sections: coniferyl alcohol, p-coumaryl alcohol, and sinapyl alcohol [65]. These complex structures fluctuate considerably based on the plant species, growth duration, growing location, and isolation process. Due to its excessive active sites, unique physiochemical properties, low cost, biocompatibility, and high abundance, lignin has garnered significant attention as an adsorbent for the removal of various pollutants from water. Intrinsic lignin is not appropriate for practical applications due to its low adsorption capacity and low selectivity. Furthermore, the modification of lignin with various functional groups ($-OH$, $-NH_2$, $-COOH$/Na, and $-S$ carrying moieties) has shown enhanced affinity toward heavy metal ions and the adsorption of organic contaminants (Fig. 12.3) [66]. The modification of lignin can also tailor the specific physiochemical properties such as hydrophilicity, hydrophobicity, stability, and adsorption capability.

Moreover, lignin nanocomposites have been proved as a promising alternative for outdated adsorbents for effluents removal. Lignin-functionalized carbon nanomaterials are generally employed for the removal of heavy metal ions. Lignin-grafted CNT nanocomposite (L-CNTs) exhibited an excellent adsorption capacity (235.0 mg/L) for Pb(II), at a pH 5.8 with 120 minutes of contact time [67] (Fig. 12.4). It was attributed to the 3D structure, high surface area, enhanced pore size, large number of oxygen functional groups, and mechanical stability of the nanocomposite. The L-CNT lignin contains many oxygen functional groups that provide negative sites at higher pH and enhance the cationic contaminant adsorption via electrostatic interactions.

Similarly other lignin-functionalized CNMs were also reported for the removal of Pb(II) and other heavy metal ions [68,69].

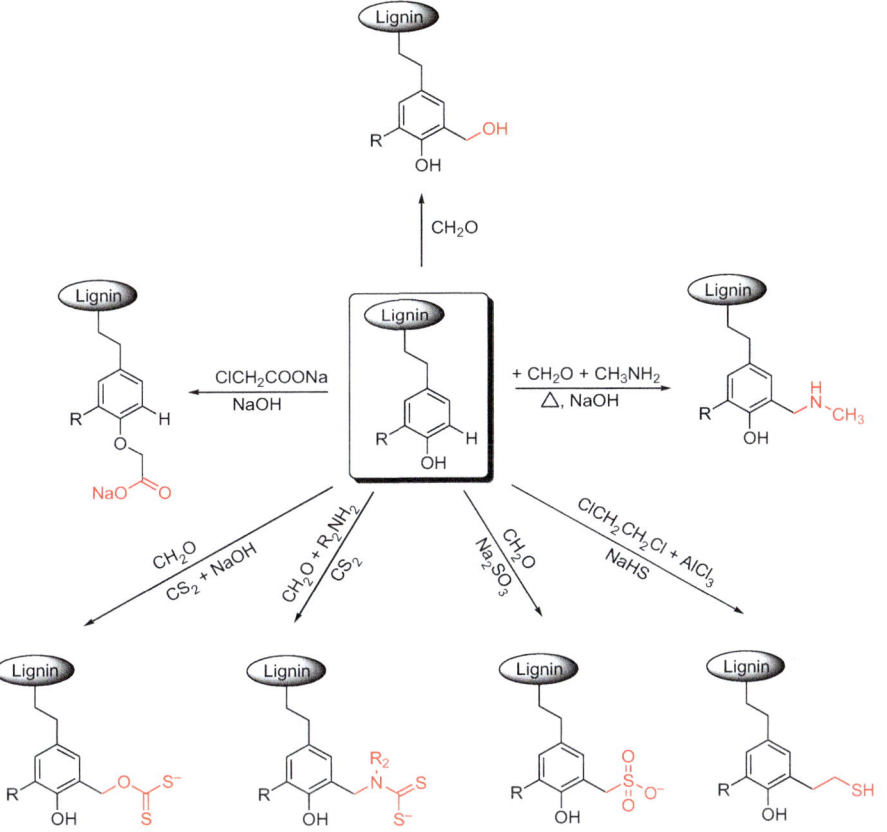

Figure 12.3 Schematic illustration of lignin synthetic modification with hydroxyl, carboxyl, amine, and sulfur functional groups.

12.4 Cellulose-functionalized carbon nanomaterials

Cellulose is an abundantly available, renewable, and linear chain biopolymer with a number of −OH groups. It is also the main waste product of many agroindustries, such as banana peel, corn cob, and sawdust, and can be used widely for adsorption purposes [70,71]. It is a linear polysaccharide biopolymer with two anhydroglucose rings which are connected by β-1,4 glycoside bonds in a repetitive mode. The cellulose structure exhibits a highly interactive surface area owing to many hydroxyl groups, which are highly chemically reactive and also provide hydrophilicity, degradability, and chirality, which promote its use as an adsorbent [72]. In the recent past, the isolation of cellulosic material into cellulose nanocrystals (NCC) and nanofibers (NFC) has been attractive for a wide range of applications such as catalysis,

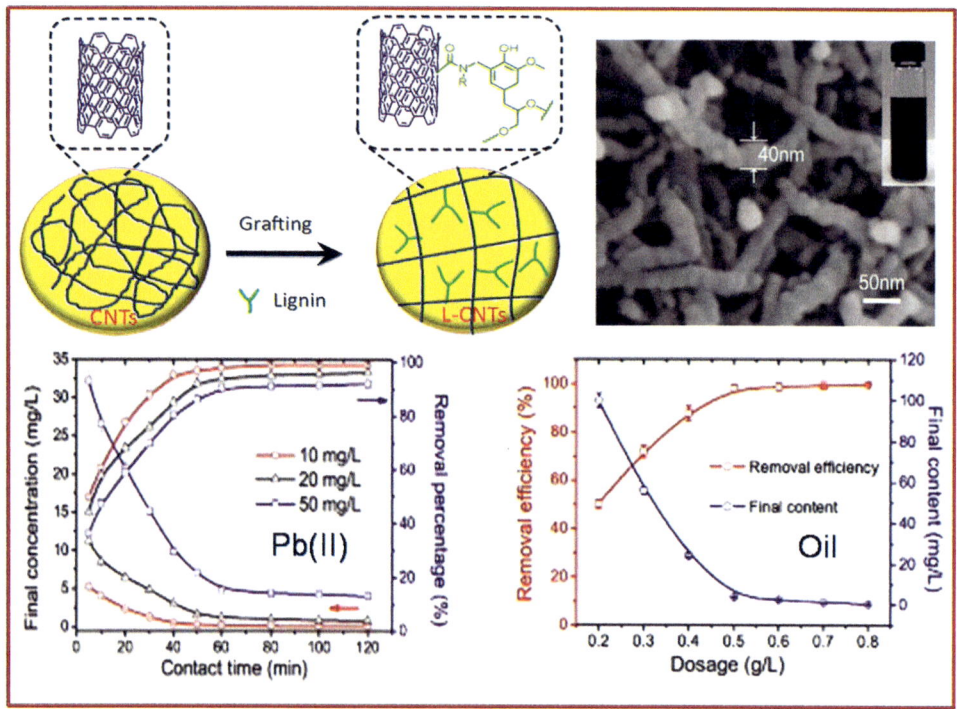

Figure 12.4 Lignin-grafted CNT nanocomposite (L-CNTs) for adsorption of Pb(II) and oil droplets from wastewater. *Reproduced with permission from Li Z, Chen J, Ge Y. Removal of lead ion and oil droplet from aqueous solution by lignin-grafted carbon nanotubes. Chem Eng J. 2017;308:809—17. Copyright 2017, Elsevier Science Ltd.*

adsorption, drug delivery, bioimaging and biosensors and so on [73]. High aspect ratio, large active surface area, excellent mechanical stability, ease of functionalization, and nontoxicity make it a preferred choice for targeted application including adsorption of contaminants from wastewater. Unlike other biopolymers, unmodified cellulose has also a low stability and low adsorption capacity for water pollutants [32]. Therefore chemical modification is a necessary step in order to enhance the adsorption capacity of cellulose effectively and selectively for various adsorbates.

Cellulose is rich in oxygen functionalities, such as carboxyl and hydroxyl groups, which facilitate its functionalization. Dai et al. prepared GO and bentonite-reinforced hydrogels of polyvinyl alcohol (PVA)/carboxymethyl cellulose obtained from pineapple peel (PCMC) using green freeze—thaw cycles for the efficient adsorption of MB from water [74] (Fig. 12.5A). On adding GO and bentonite into the hydrogels, the removal capacity was significantly changed from 83.33 to 172.14 mg/g at 30°C for MB adsorption. Adsorption of MB by PVP/PCMC/GO/bentonite hydrogels was well fitted by Langmuir isotherm and pseudo–second-order kinetic models. Tufenkji et al. prepared a partially reduced graphene oxide-cellulose sponge (rGO-CNC) using vitamin C as a reducing agent and employed it for the adsorption of dyes, heavy metal ions, and pharmaceuticals compounds [75] (Fig. 12.5B). The calculated surface area of wet RGO-CNC was found to be exceptionally high (~ 2080 m^2/g) which suggests its potential for adsorption applications. The adsorption capacity of the sponge was compared with those of GO and rGO sponges and observed to be the highest (850 mg/g). The adsorption capacities of the rGO-CNC sponge for pharmaceuticals, for example, SMX (107 mg/g), DCF (129 mg/g), tetracycline (TC) (149 mg/g), 17-α-ethynylestradiol (EE2, 117 mg/g), and cyanotoxin (22 mg/g), and heavy metal ions, for example, Cu (II) (65 mg/g) and Cd (II) (76 mg/g), were analyzed, and, except for TC and cyanotoxin, found to be higher than the reported adsorption capacity values of granular activated carbon. Pan et al. reported the adsorption characteristics of cellulose/graphene oxide (CGC) for the adsorption of six different triazine pesticides from wastewater and compared them with those of other sorbents such as primary secondary amine (PSA), graphene, cellulose, graphite carbon black, and graphitic carbon [76]. The adsorption of the six triazine pesticides on CGC was found to decrease in the following order: cyprazine > prometryn > simazine > ametryn > atrazine > simeton. All of the pesticides have a π-bonding network and an equal number of N atoms which contribute to adsorption onto the CGC surface. Therefore

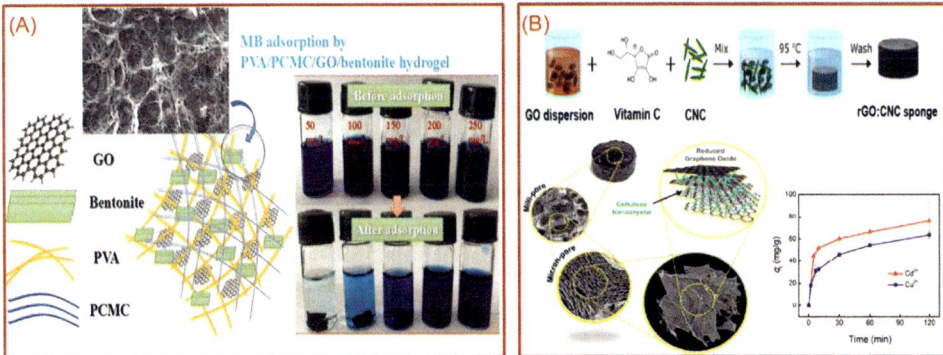

Figure 12.5 (A) PVP/PCMC/GO/bentonite hydrogels prepared using green freeze—thaw cycles for MB adsorption. (B) Hierarchical porous rGO—cellulose nanocrystals sponge (rGO-CNC) synthesized using vitamin C as a reducing agent for the removal of metal ions, pharmaceuticals, and dyes. *(A) Reproduced with permission from Dai H, Huang Y, Huang H. Eco-friendly polyvinyl alcohol/carboxymethyl cellulose hydrogels reinforced with graphene oxide and bentonite for enhanced adsorption of methylene blue. Carbohydr Polym 2018;185:1—11. Copyright 2018, Elsevier Science Ltd. (B) Reproduced with permission from Yousefi N, Wong KKW, Hosseinidoust Z, Sørensen HO, Bruns S, Zheng Y, et al. Hierarchically porous, ultra-strong reduced graphene oxide-cellulose nanocrystal sponges for exceptional adsorption of water contaminants. Nanoscale 2018;10:7171—84. Copyright 2018, Royal Society of Chemistry.*

the difference in the adsorption capacities for the pesticides was due to their differing S, O, and Cl functional groups and to van der Waals-type interactions [77].

12.5 Gelatin-functionalized carbon nanomaterials

Gelatin (Gel) is a water-soluble polypeptide derived from the hydrolysis of collagen. It has valuable characteristics, such as biodegradability, hydrophilicity, low cost, nontoxicity, and numerous active functionalities of amino, carboxyl, and hydroxyl groups. The amounts of amino acids vary in gelatin molecules but in most of the cases glycine, proline, and hydroxyproline dominate the composition [78,79]. The heating of collagen is responsible for a change in the conformation and formation of helix-like structures, while the original shape can be regained partly on cooling. Gelatin interacts with water molecules by inter- and intramolecular hydrogen bonding, resulting in the formation of a Gel-like structure (Fig. 12.6). On the basis of space arrangements and interactions of gelatin helixes, gelatin structures can be divided into primary, secondary, and tertiary structures. The parameters, such as temperature, gelatin concentration, and energy, control the formation of these structures. Besides complex structures, gelatin has drawbacks of weak mechanical strength and fast degradability in a wet atmosphere. The flaws can be improved by making a nanocomposite and grafting with polymers or cross-linker molecules.

Gelatin-functionalized carbon materials have also been investigated for water pollutant adsorption. Due to the presence of hydroxyl, amino, and carboxyl active

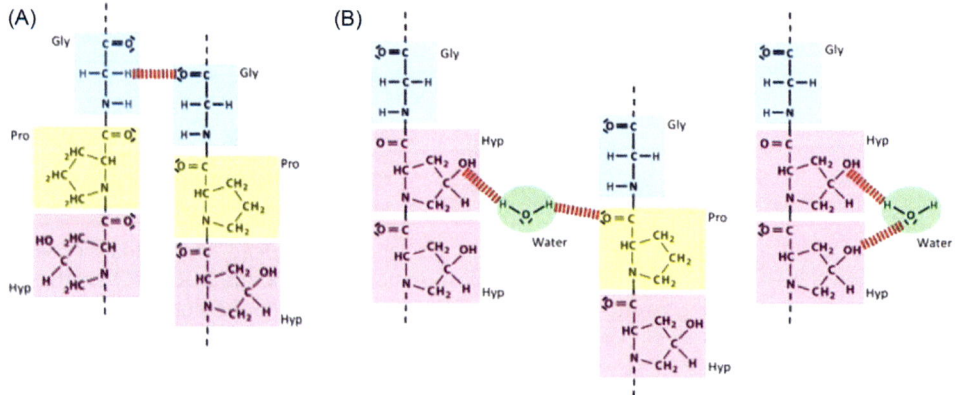

Figure 12.6 Illustrates the hydrogen bonding: (A) between Gel strands; (B) water molecules and Gel strands. *Reproduced with permission from Duconseille A, Astruc T, Quintana N, Meersman F, Sante-Lhoutellier V. Gelatin structure and composition linked to hard capsule dissolution: a review. Food Hydrocoll 2015;43:360—76 [80]. Copyright 2015, Elsevier Science Ltd.*

groups, gelatin acts as an efficient adsorbent for the purification of water by the removal of metal ions and organic pollutants. A composite of gelatin/activated carbon (GE-AC) with a pore size of 3.25 nm was prepared and its adsorption capacity for rhodamine dye in water was investigated [81]. The adsorption capacity of GE-AC was observed to be 256 mg/g for RhB at pH 4, which is significantly higher than that of AC. Reinforcement of gelatin with nanomaterials (such as CNTs, GO, etc.) can improve its mechanical properties and slow down its degradation rate due to the high hydrothermal stability of carbon-based materials.

Functionalization with magnetic nanoparticles assists in the separation of the adsorbent from the aqueous medium after the completion of the adsorption process. Magnetic beads composed of a gelatin—CNT—iron oxide nanocomposite (Gel-CNT-MNPs) were developed for the removal of cationic methylene blue (MB) dye and anionic Direct Red80 (DR) dye from wastewater [82]. Fig. 12.7 illustrates the morphological features of Gel beads, and the Gel-CNT-MNP nanocomposite. Strong magnetic properties can facilitate the retrieval of the adsorbent after the dye removal process, using an external magnet. The adsorption data followed pseudo-second-order

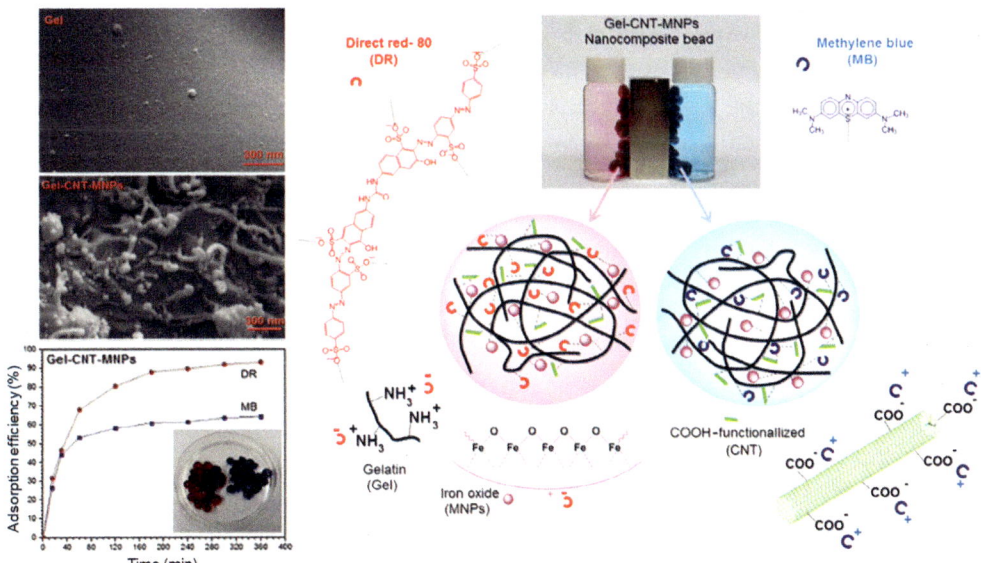

Figure 12.7 SEM images of the gelatin beads (Gel), and Gel-CNT-MNP nanocomposite; suggested mechanisms for adsorption of Direct Red80 (DR) and methylene blue (MB) on Gel-CNT-MNP nanocomposite. *Reproduced with permission from Saber-Samandari S, Saber-Samandari S, Joneidi-Yekta H, Mohseni M. Adsorption of anionic and cationic dyes from aqueous solution using gelatin-based magnetic nanocomposite beads comprising carboxylic acid functionalized carbon nanotube. Chem Eng J. 2017;308:1133—44. Copyright 2017, Elsevier Science Ltd.*

kinetics. Adsorption of dyes onto the Gel–CNT–MNP nanocomposite was found to be thermodynamically favorable and spontaneous. The intrinsic positive charge of gelatin is responsible for the adsorption of DR to the nanocomposite surface. Moreover, the embedded −COOH-functionalized CNTs in the gelatin matrix interact with MB dye via electrostatic attraction (Fig. 12.7).

12.6 Starch-functionalized carbon nanomaterials

Starch is a low-cost, environment-friendly, biodegradable, abundantly available biopolymer with high chemical stability and reactivity, which is sourced from renewable material and all of these qualities make it an interesting alternate adsorbent for wastewater treatment [21,83]. Corn, maize, rice, wheat, potato, peas, and so on are good renewable sources of starches. The starch extracted from potato contains phosphate ester and imparts a negative charge on the starch, which does not make it a good choice for anionic water contaminants [84]. Maywald et al. also noted that potato and carboxylated starches can selectively and successfully adsorb the cationic dyes but failed to adsorb anionic dyes due to a partial negative charge on these starches [85]. Therefore potato starch and carboxylated starch are particularly used for cationic dyes. Starch has many active surface sites with hydrophilic functional groups and interacts very well with the other nanomaterials through physical interactions (hydrogen bonding and electrostatic interactions) and chemical bondings. Ramin et al. synthesized a series of poly(acrylamide) grafted, starch-based graphene oxide/hydroxyapatite nanocomposite hydrogels (NHA) to remove the cationic dye (MG) from contaminated water [86]. An NHA composite synthesized without hydroxyapatite (nHAP) nanoparticles (NHA0) had a higher porosity than those which were synthesized with nHAP (NHA1, NHA3, and NHA5). The incorporation of nHAP bound the polymer chains, and reduced the free space and thus the porosity of the nanocomposite. However, the adsorption capacities of the NHA nanocomposites increased with an increase in hydroxyapatite concentration, and MG adsorptions were found to decrease in the order: NHA5 > NHA3> NHA1 > NHA0. The high adsorption capacity of NHA5 was attributed to its higher phosphate ion concentration. Ma et al. compared the adsorption capacity of MWCNT−starch−iron oxide and MWCNT−iron oxide for the adsorption of MB (cationic dye) and MO (anionic dye) from water [87]. MWCNT−iron oxide was insoluble in water and did not interact with the dyes. However, the starch in MWCNT−starch−iron oxide improved the hydrophilicity of the nanocomposite, allowing good dispersion in water, which enhanced the interactions between adsorbate and adsorbent. The iron oxide assisted in the magnetic separation of the adsorbent after the completion of adsorption.

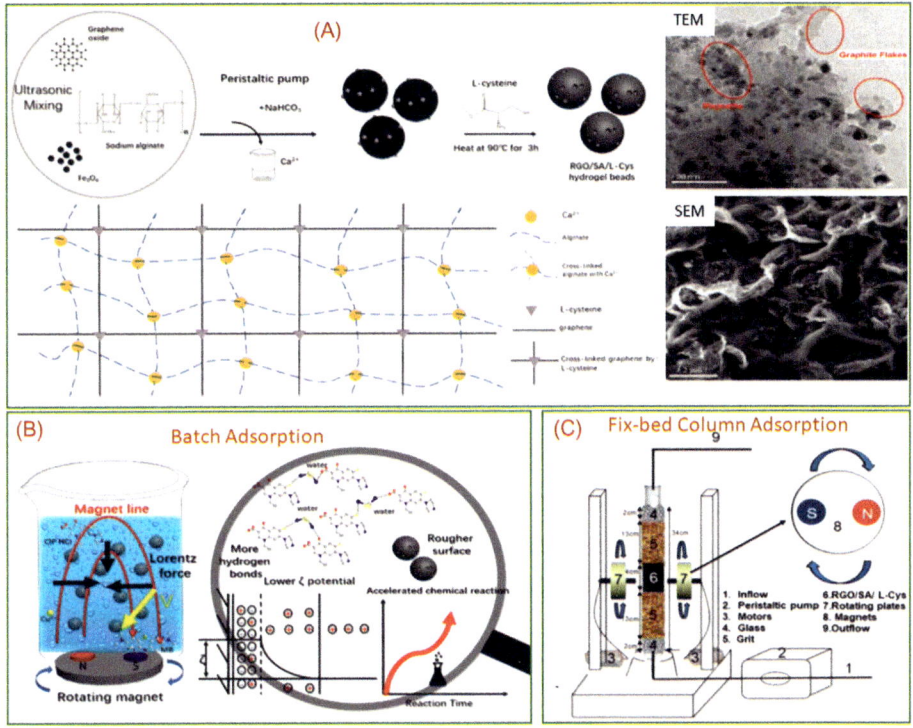

Figure 12.8 (A) Schematic synthesis steps, SEM and TEM image for bioadsorbent magnetic beads of sodium alginate/graphene/ʟ-cysteine (SA/GR/ʟ-Cys). It was used for the removal of CIP, MB, and Cu(II) ions from wastewater using a static magnetic field (SMF) and rotating magnetic field (RMF). (B) Batch adsorption study. (C) Fix-bed column study under RMF conditions. *Reproduced with permission from Ma J, Ma Y, Yu F, Dai X. Rotating magnetic field-assisted adsorption mechanism of pollutants on mechanically strong sodium alginate/graphene/L-cysteine beads in batch and fixed-bed column systems. Environ Sci Technol 2018;52:13925–34. Copyright 2018, American Chemical Society.*

12.7 Alginate-functionalized carbon nanomaterials

Alginate is a biodegradable, biocompatible, and nontoxic naturally occurring anionic polysaccharide. It is extracted from brown seaweed cell walls and algae. The backbone of the structure comprises 1–4 linked β-ᴅ-mannuronic and α-ʟ-guluronic acids [88,89]. Alginates can easily react with divalent cations (e.g., Ca^{2+}, Mg^{2+} ion) to form cross-link gel matrixes, which can be converted into sodium alginate by neutralization using HCl and again precipitated with $Na_2CO_3/NaOH$ [89]. The cross-linked matrixes of alginate have been used for adsorbents in the gel form due to their easy handling [90]. Alginate and sodium alginate based formulations are considered as efficient adsorbents due to the presence of hydroxyl and carboxyl groups, which enable them to have chelating activity for metal ions and organic effluents in aqueous solution [91].

Recently, magnetic beads of sodium alginate-graphene-L-cysteine (SA/GR/L-Cys) was synthesized for magnetic field-assisted removal of pharmaceutical (CIP), dye (MB), and metal ions (Cu(II)) (Fig. 12.8A). The adsorption performance in a rotating magnetic field (RMF) was much better than a static magnetic field (SMF). For instance, adsorption capacity of CIP reached up to 357.48 mg/g in RMF and 273.60 mg/g in SMF conditions. In RMF, adsorption was intensely influenced by ζ potential, formation of new H-bonds, and acceleration in reactions. In the case of fixed-bed column adsorption, adsorption performance enhanced by 30.66%, and the increment in breakthrough time of column adsorption was also noted (Fig. 12.8B and C). The cost and practical utility of bioadsorbent beads in sewage plants associated with CIP removal under RMF conditions were evaluated [92].

Gan et al. prepared GO-incorporated alginate hydrogel beads (SA/GO) for the removal of organic pollutants (MO, MB, rhodamine B (RhB), and vat green 1 (VG1) organic dyes) from wastewater [93]. In SA/GO hydrogel, GO exhibits carboxylic and hydroxyl functional groups on the surface, which behave as negatively charged in an aqueous medium. Therefore it can easily interact with the cationic and neutral pollutants rather than the anionic pollutants. SA/GO displayed a higher adsorption affinity toward MB, RhB, and VG1 dye than toward anionic MO dye. It was also employed to adsorb Bisphenol A (BPA) from wastewater. The interaction between the organic pollutants and SA/GO was due to electrostatic interaction, n−π interaction, π−π interaction, and hydrogen bonding.

12.8 Gum-functionalized carbon nanomaterials

Gum is a polysaccharide biopolymer and can be obtained from natural sources. On the basis of their origin gums can be classified into four classes: (1) obtained from seaweeds, for example, agar, alginic acid, and carrageenans; (2) obtained from botanical resources, for example, gum ghatti, gum arabic, gum karaya, and tragacanth; (3) obtained from animal resources, for example, chondroitin sulfate, chitin, and chitosan; and (4) obtained from microbial resources, for example, xanthan, curdian, dextran, and pullulan. Owing to their low cost, natural and abundant availability, nontoxicity, biodegradability, and different functional groups, the use of gums is supported in diverse industries such as in food industries, adsorption, drug delivery, and so on [36,94−96]. Functional groups, such as hydroxyl, amino, and carboxylic, present on the gum interact with the contaminants present in the water and aid their adsorption onto the gum surface and thus their removal from water [28,95]. However, natural gum has poor mechanical strength and physical properties, and therefore to improve such characteristics natural gums are modified through grafting with other materials.

Natural gum exhibits strong water wettability and after the application of gum in water treatment it is difficult to separate it. To improve the chemical strength and easy separation of gum after adsorption process in aqueous medium, Ma et al. prepared magnetic guar gum covalently grafted carbon nanotubes (GG$-$MWCNT$-$Fe$_3$O$_4$) and investigated the adsorption of organic dyes (MB and neutral red (NR)) [97]. The adsorption experiments reveal that GG$-$MWCNT$-$Fe$_3$O$_4$ follows Langmuir isotherm data, which support the monolayer adsorption of MB and NR on GG$-$MWCNT$-$Fe$_3$O$_4$. The adsorption capacity of GG$-$MWCNT$-$Fe$_3$O$_4$ toward MB and NR was found to be 61.92 and 89.85 mg/g, respectively. Strong hydrophilicity of guar gum promotes the diffusion of organic dyes from aqueous medium to the surface of MWCNT and enhanced the adsorption. Later with the use of an external magnetic field, the adsorbent could be easily recollected, and thus it can be further recycled and reused.

Magnetic gum tragacanth (GT) nanocomposite beads with GO were found to simultaneously adsorb inorganic heavy metal ions and organic dyes [98]. 1-Vinylimidazole and 2-acryamido-2-methyl-1-propanesulfonic acid were copolymerized onto the surface of GT in the presence of a polymerization initiator potassium persulfate (KPS), and further magnetized with Fe$_3$O$_4$ in GO dispersion to prepare magnetic poly(AMPS-co-VI)-g-GT/GO beads. The adsorption of cationic dyes (Congo red (CR) and crystal violet (CV)) and heavy metal ions (Pb(II) and Cu(II)) was found to be pH dependent. The removal efficiency of water pollutants was increased by increasing the pH of the aqueous medium. This might be due to the deprotonation of the functional groups, such as sulfonic and carboxylic groups, of the adsorbent material in an alkaline medium and the specific interactions between the negatively charged adsorbent and cationic adsorbates. However, in an acidic medium the groups get protonated and the competition between the protons and cationic pollutants reduced the adsorption efficiency. Furthermore, the electrostatic repulsion between protonated groups and cationic pollutants restricts the adsorption. After adsorption the adsorbent could easily be regenerated in a nitric acid solution for heavy metal ions, sodium hydroxide solution for CR, and in ethanol for CV for 8 hours. The regenerated adsorbent could be reused several times.

12.9 Other biopolymer-functionalized carbon nanomaterials

Biopolymer/carbon-based nanocomposites have been compared with several other commercial and noncommercial adsorbents to evaluate their potential in

Table 12.1 Adsorption capacities of PDA-GH and HT-GA for Pd(II), Cd(II), RhB, and
p-nitrophenol [104].

Adsorbent	Contaminant	Adsorption capacity (mg/g)
PDA-GH	Pd (II)	336
	Cd (II)	145
	RhB	207
	p-Nitrophenol	260
HT-GA	Pd (II)	119
	Cd (II)	53
	RhB	145
	p-Nitrophenol	324

Reaction conditions: pH = 6, contact time = 12 h, and temperature = 298K.

adsorption applications [99]. Carbon-based nanomaterials, such as GO, rGO, CNTs, and g-C_3N_4, can be also functionalized with other biopolymers, including polyacrylamide, gum, and L-cysteine, and applied to water purification [48,100−103]. Duan et al. reported the one-step synthesis of a polydopamine (PDA)-functionalized graphene hydrogel (PDA-GH),for the adsorption of organic dyes, heavy metal ions, and aromatic contaminants from water [104]. The active surface area of PDA-GH was measured to be 310.6 m^2/g. The adsorption properties of PDA-GH were compared with those of hydrothermally treated graphene (HT-GH). The isoelectric points (IEPs) for PDA-GH and HT-GH were measured using the zeta potential and occurred at pH 3.5 and 4.2, respectively. The difference is due to the surface charge of the hydrogel, which influenced the adsorption of ionic adsorbents. At a lower pH (pH < pH$_{iep}$), the adsorption of heavy metal ions was very low, due to a positive surface charge on the adsorbent via protonation of surface functional species. However, the adsorption capacity of PDA-GH for Pd(II) and Cd(II) dramatically increased with an increase in pH (pH = 6). The adsorption of RhB was efficient at pH values higher than the IEPs (pH = 6), as RhB is a cationic dye. Therefore the main interaction involved in the adsorption of the cationic contaminants Pd(II), Cd(II), and RhB (Table 12.1) was electrostatic interaction between PDA-GH, and the contaminant. RhB was more efficiently adsorbed on PDA-GH than HT-GH as the former has a large number of oxygen functionalities, which enhance the electrostatic interactions with positive species. Further, the adsorption of RhB on these adsorbents was not as high as those of the metal ions, due to the large size of the RhB molecule relative to the pores of the adsorbent. However, p-nitrophenol showed a good affinity for HT-GH at high pH, as π−π interactions and hydrogen bonding are the main interactions involved in the adsorption process. Thus the functional groups, pore size, and surface area of the adsorbent influence the response of the adsorbent to different adsorbates.

Tables 12.2 and 12.3 present the maximum adsorption capacities of various biopolymer−carbon–based nanocomposite adsorbents for a wide range of pollutants from

Table 12.2 Various biopolymer−carbon-based nanoadsorbent and their adsorption capacities for organic pollutants.

Biopolymer−carbon nanohybrid-based material	Organic pollutant	Adsorption conditions	Adsorption isotherm model	Adsorption kinetics	Maximum adsorption capacity (mg/g)	References
Cellulose/GO aerogel	MB	pH = 6 t = 30 min	Langmuir	Pseudo-second-order	68	[105]
Xanthan gum-cl-poly (acrylic acid)/rGO hydrogel	MV MB	pH = 5 T = 298K t = 90 min	Langmuir	Pseudo-second-order	1052 793	[106]
Biochar-supported rGO	Atrazine	pH = 6 T = 298K t = 24 h	Langmuir	Pseudo-second-order	67	[107]
Agricultural waste/GO	MB	pH = 12 T = 298K t = 3 h	Temkin	Pseudo-second-order	414	[108]
Polypyrrole/chitosan/GO	Ponceau 4R dye (P4R)	pH = 6 T = 303K t = 150 min	Langmuir	Pseudo-second-order	5.4	[109]
Chitosan/graphene nanoplates	MO AR1	pH = 3 and 4 T = 298K t = 60 min	Freundlich	Pseudo-second-order	230 133	[53]
Pentaerythritol-modified ox-MWCNT	ARS AYR	pH = 4−10 T = 298K t = 30 min	Langmuir	Pseudo-second-order	257 45	[110]
PVA/CMC/ GO/ bentonite hydrogels	MB	pH = 8 T = 303K t = 150 min	Langmuir	Pseudo-second-order	172	[74]
β-Cyclodextrin/chitosan/ GO hydrogel	MB	pH = 12 T = 298K	Freundlich	Pseudo-second-order	1134	[111]

(continued)

Table 12.2 (Continued)

Biopolymer–carbon nanohybrid-based material	Organic pollutant	Adsorption conditions	Adsorption isotherm model	Adsorption kinetics	Maximum adsorption capacity (mg/g)	References
Starch/poly(acrylamide)/GO/n-HAP hydrogel	MG	$t = 2$ h pH = 10 $T = 298$K $t = 60$ min	Langmuir	Pseudo-second-order	297	[86]
Graphene–chitosan hydrogel	CR	pH = 7 $T = $ RT $t = 10$ min	Langmuir	Pseudo-second-order	384	[112]
Graphene/poly(vinyl alcohol) (PVA) aerogels	Neutral red (NR) Indigo carmine (IC)	pH = 2 and 7 $T = 303$K $t = 24$ h	Langmuir	Pseudo-second-order	306 250	[113]
GO/chitosan	Picric acid	pH = 7 $T = 298$K $t = 5$ h	Freundlich	Pseudo-second-order	263	[50]
PMMA-rGO	MB	pH = 7 $t = 60$ min	Langmuir	Pseudo-second-order	698	[114]
Graphene/β-cyclodextrin	Phenolphthalein (Php) MB MO BF	pH = 2 and 7 $T = 298$K $t = 20$ min	Langmuir	Pseudo-second-order	468 580 328 425	[115]
GO–β-cyclodextrin (GO-CD)	BPA	pH = neutral $T = 298$K $t = 10$ min	Langmuir	Pseudo-second-order	373	[99]
β-Cyclodextrin/poly(l-glutamic acid)-supported magnetic GO	17β-estradiol (E2)	pH = 7 $T = 298$K $t = 12$ h	Langmuir	Pseudo-second-order	85.8	[116]
Magnetic chitosan/GO	Fluoxetine	pH = 4.5 $T = 298$K $t = 10$ min	Langmuir	Pseudo-second-order	66	[117]

Adsorbent	Dye	Conditions	Isotherm	Kinetics	q_{max}	Ref.
ZnFe$_2$O$_4$@CS/GO	BF	pH = 7 T = 313K t = 12 h	Langmuir	Pseudo-second-order	335	[118]
GO-PDA-PSPSH	MB	pH = 7 T = 298K t = 12 h	Langmuir	Pseudo-second-order	279	[119]
ILs-modified magnetic chitosan/GO	MB	pH = 12 T = 303K t = 60 min	Langmuir	Pseudo-second-order	243	[120]
Graphene/Fe$_3$O$_4$/chitosan	MB	pH = 9 t = 5 min	Langmuir	Pseudo-second-order	249	[121]
Konjac glucomannan/ GO hydrogel	MO	pH = 7 T = 298K t = 24 h	Freundlich	Pseudo-second-order	51	[122]
MWCNTs/ Fe$_3$O$_4$/PANI	MO CR	pH = 4.5 T = 298K t = 24 h	Langmuir	Pseudo-second-order	446 417	[123]
Magnetic β-cyclodextrin-GO	p-Phenylenediamine	pH = 8 T = 318K t = 120 min	Langmuir	Pseudo-second-order	1120	[124]
Magnetic β-cyclodextrin-GO	MG	pH = 7 T = 318K t = 120 min	Langmuir	Pseudo-second-order	990	[125]
Poly(acrylic acid)-functionalized magnetic GO	MB	pH = 7 T = RT t = 1500 min	Langmuir	Pseudo-second-order	291	[126]
GO/chitosan nanofibers	Fuchsin acid	T = RT t = 750 min	Langmuir	Pseudo-second-order	175	[127]

(continued)

Table 12.2 (Continued)

Biopolymer–carbon nanohybrid-based material	Organic pollutant	Adsorption conditions	Adsorption isotherm model	Adsorption kinetics	Maximum adsorption capacity (mg/g)	References
GO/calcium alginate	Ciprofloxacin	pH = 6 $T = 293K$ $t = 24$ h	Langmuir	Pseudo-second-order	39	[128]
Magnetic β-cyclodextrin–chitosan/GO	MB	pH = 2–11 $T = 298K$ $t = 80$ min	Langmuir	Pseudo-second-order	84	[129]
GO/magnetic cyclodextrin–chitosan	Hydroquinone	pH = 6 $T = 303K$ $t = 40$ min	Freundlich	Pseudo-second-order	458	[46]
GO/chitosan	RB5	pH = 2 $T = 298K$ $t = 24$ h	Langmuir–Freundlich	Brouers–Sotolongo	277	[130]
MCGO	MB	pH = 10 $T = 303K$	Langmuir–Freundlich	Pseudo-second-order	180	[131]

Table 12.3 Different types of biopolymer–carbon-based nanoadsorbent and their adsorption capacities for inorganic pollutants.

Biopolymer–carbon nanohybrid based material	Inorganic pollutant	Adsorption conditions	Adsorption isotherm model	Adsorption kinetics	Maximum adsorption capacity (mg/g)	References
Poly(tannin-tetrathylenepentamine)–polyacrylamide-rGO (P(TA–TEPA)–PAM-RGO)	Cr(VI)	pH = 2 T = 308K t = 48 h	Langmuir	Pseudo-second-order	394	[132]
Chitosan/GO	Cu(II)	pH = 7 T = 298K t = 70 min	Langmuir	Pseudo-second-order	217	[133]
Phytic acid-induced graphene	Hg(II)	pH = 7.2 T = 298K t = 5 h	Langmuir	Pseudo-second-order	361	[134]
Biochar-supported rGO	Pb(II)	pH = 6 T = 298K t = 24 h	Langmuir	Pseudo-second-order	26	[107]
GO/chitosan	U(VI)	pH = 5 t = 70 min	Freundlich	Pseudo-second-order	50	[51]
EDTA-functionalized magnetic chitosan/GO	Pb (II) Cu(II) As(II)	pH = 5, 5.5, 8 T = 298K t = 12 h	Langmuir and Freundlich	Pseudo-second-order	206 207 42	[135]
Methyl-β-cyclodextrin modified GO	Pb(II)	pH = 6 T = RT t = 2 h	Langmuir	Pseudo-second-order	312	[136]
Magnetic GO grafted polymaleicamide dendrimers	Pb(II)	pH = 5.3 T = 298K t = 10 h	Langmuir	Pseudo-second-order	181	[137]

(continued)

Table 12.3 (Continued)

Biopolymer–carbon nanohybrid based material	Inorganic pollutant	Adsorption conditions	Adsorption isotherm model	Adsorption kinetics	Maximum adsorption capacity (mg/g)	References
Gum tragacanth/GO	Pb(II) Cd(II) Ag(I)	pH = 6 T = RT t = 3 h	Langmuir	Pseudo-first order	142 112 132	[44]
GO/chitosan aerogel	Pb (II) Cu(II) Cr(VI)	pH = 5–7 T = 298K t = 6 h	Langmuir	Pseudo-second-order	747 457 292	[45]
Magnetic hydroxypropyl chitosan/GO	Pb(II)	pH = 5.5 T = RT t = 300 min	Freundlich	Pseudo-second-order	79	[138]
GO/chitosan/poly(acrylic acid)	Pb(II)	pH = 5 T = RT t = 1400 min	Langmuir	Pseudo-second-order	138	[139]
Polyvinyl alcohol/chitosan/GO hydrogel	Cu(II)	pH = 5.5 T = 303K t = 24 h	Langmuir	Pseudo-second-order	162	[140]
Chitosan/sulfydryl-functionalized GO	Cd(II) Pb (II) Cu(II)	pH = 5 T = 293K t = 90 min	Freundlich	Pseudo-second-order	177 447 425	[141]
GO/magnetic cyclodextrin–chitosan	Cr(VI)	pH = 3 T = 303K	Langmuir	Pseudo-second-order	67	[142]

aqueous solution. Biopolymers have comparatively low adsorption efficiencies but their modification with carbon-based nanomaterials leads to enhanced adsorption capacities due to increases in surface functionalities and surface areas, and thus to improved physicochemical properties.

12.10 Conclusion

Wastewater treatment is a global issue and various techniques have been employed to solve this, including adsorption. The use of biopolymers in wastewater treatment is an effective and affordable step due to their abundant availability, non-toxicity, and renewable sources. Their high water affinity and various functional groups enhance their popularity in wastewater treatment, whereas their low thermal and mechanical stability and difficulty in separation after adsorption process are drawbacks. These issues can be solved by the functionalization of the biopolymer with other nanomaterials. Carbon-based nanomaterials are widely suitable and studied nanoadsorbents due to their ease of functionalization, high chemical and thermal stability, high mechanical strength, and abundant availability. The high mechanical strength of carbon nanomaterials and the functional groups of biopolymers make biopolymer-functionalized carbon nanomaterials (biopolymer—CNMs) an acceptable and affordable choice as adsorbents in wastewater treatment.

Grafting of biopolymers with CNMs improves the chemical, thermal, and mechanical stability and avoids the aggregation of biopolymers, which enhances its adsorption efficiency as an adsorbent in wastewater treatment. Chitosan, lignin, cellulose, gelatin, alginate, starch and so on biopolymers-functionalized CNMs are the common examples of biopolymer—CNM materials and have been studied for the removal of various organic and inorganic toxic compounds from wastewater. Furthermore, the separation of biopolymer—CNM adsorbents can be enhanced via the magnetization of the adsorbent on functionalization with Fe_3O_4. After successful separation, the adsorbent can be regenerated and reused several times and thus this contributes to the economical efficiency of the water pollutants' removal method. However, there is still major work to be contributed to the use of graphene and CNT-based biopolymer—CNMs as adsorbents for wastewater treatment. These nanocomposite materials are economically feasible, environment-friendly, and nontoxic, which encourage their application in water treatment.

References

[1] Aravamudhan A, Ramos DM, Nada AA, Kumbar SG. Chapter 4 - natural polymers: polysaccharides and their derivatives for biomedical applications. Natural and Synthetic Biomedical Polymers. Oxford: Elsevier; 2014. p. 67—89.

[2] Hernandez N, Williams RC, Cochran EW. The battle for the "green" polymer. Different approaches for biopolymer synthesis: bioadvantaged vs. bioreplacement. Org Biomol Chem 2014;12:2834–49.

[3] Preeti Yadav HY, Shah VG, Shah G, Dhaka G. Biomedical biopolymers, their origin and evolution in biomedical sciences: a systematic review. J Clin Diag Res [serial online] 2015;9 ZE21–ZE5.

[4] Reddy N, Reddy R, Jiang Q. Crosslinking biopolymers for biomedical applications. Trends Biotechnol 2015;33:362–9.

[5] Houghton JI, Quarmby J. Biopolymers in wastewater treatment. Curr OpBiotechnol 1999;10:259–62.

[6] Castro GR, Panilaitis B, Bora E, Kaplan DL. Controlled release biopolymers for enhancing the immune response. Mol Pharm 2007;4:33–46.

[7] Iurian S, Dinte E, Iuga C, Bogdan C, Spiridon I, Barbu-Tudoran L, et al. The pharmaceutical applications of a biopolymer isolated from *Trigonella foenum-graecum* seeds: focus on the freeze-dried matrix forming capacity. Saudi Pharm J SPJ 2017;25:1217–25.

[8] Radoslav G, Vukic M, Gojkovic V. Application of biopolymers in the food industry; *Advances in Applications of Industrial Biomaterials*, 2017, Springer International Publishing.

[9] Cao L, Yang M, Wu D, Lyu F, Sun Z, Zhong X, et al. Biopolymer-chitosan based supramolecular hydrogels as solid state electrolytes for electrochemical energy storage. Chem Commun 2017;53:1615–18.

[10] Sebati W, Ray SS. Advances in nanostructured metal-encapsulated porous organic-polymer composites for catalyzed organic chemical synthesis. Catalysts. 2018;8:492.

[11] Mittal H, Ray SS, Kaith BS, Bhatia JK, Sukriti, Sharma J, et al. Recent progress in the structural modification of chitosan for applications in diversified biomedical fields. Eur Polym J 2018;109:402–34.

[12] Motshekga SC, Ray SS. Highly efficient inactivation of bacteria found in drinking water using chitosan-bentonite composites: modelling and breakthrough curve analysis. Water Res 2017;111:213–23.

[13] Seki H, Suzuki A. Adsorption of lead ions on composite biopolymer adsorbent. Ind & Eng Chem Res 1996;35:1378–82.

[14] Wan Ngah WS, Teong LC, Hanafiah MAKM. Adsorption of dyes and heavy metal ions by chitosan composites: a review. Carbohydr Polym 2011;83:1446–56.

[15] Torres JD, Faria EA, SouzaDe JR, Prado AGS. Preparation of photoactive chitosan–niobium (V) oxide composites for dye degradation. J Photochem Photobiol A Chem 2006;182:202–6.

[16] Mittal H, Maity A, Sinha Ray S. The adsorption of Pb^{2+} and Cu^{2+} onto gum ghatti-grafted poly (acrylamide-*co*-acrylonitrile) biodegradable hydrogel: isotherms and kinetic models. J Phys Chem B 2015;119:2026–39.

[17] Mittal H, Maity A, Ray SS. Gum ghatti and poly(acrylamide-*co*-acrylic acid) based biodegradable hydrogel-evaluation of the flocculation and adsorption properties. Polym Degrad Stab 2015;120:42–52.

[18] Mittal H, Ray SS. A study on the adsorption of methylene blue onto gum ghatti/TiO_2 nanoparticles-based hydrogel nanocomposite. Int J Biol Macromol 2016;88:66–80.

[19] Mittal H, Kumar V, Saruchi, Ray SS. Adsorption of methyl violet from aqueous solution using gum xanthan/Fe_3O_4 based nanocomposite hydrogel. Int J Biol Macromol 2016;89:1–11.

[20] Kera NH, Bhaumik M, Pillay K, Ray SS, Maity A. m-Phenylenediamine-modified polypyrrole as an efficient adsorbent for removal of highly toxic hexavalent chromium in water. Mater Today Commun 2018;15:153–64.

[21] Yang Y, Wei X, Sun P, Wan J. Preparation, characterization and adsorption performance of a novel anionic starch microsphere. Molecules. 2010;15:2872.

[22] Kera NH, Bhaumik M, Ballav N, Pillay K, Ray SS, Maity A. Selective removal of Cr(VI) from aqueous solution by polypyrrole/2,5-diaminobenzene sulfonic acid composite. J Colloid Interface Sci 2016;476:144–57.

[23] Crini G. Recent developments in polysaccharide-based materials used as adsorbents in wastewater treatment. Prog Polym Sci 2005;30:38–70.

[24] Mittal H, Maity A, Ray SS. Synthesis of co-polymer-grafted gum karaya and silica hybrid organic—inorganic hydrogel nanocomposite for the highly effective removal of methylene blue. Chem Eng J 2015;279:166—79.

[25] Mittal H, Maity A, Ray SS. Gum karaya based hydrogel nanocomposites for the effective removal of cationic dyes from aqueous solutions. Appl Surf Sci 2016;364:917—30.

[26] Mittal H, Kumar V, Alhassan SM, Ray SS. Modification of gum ghatti via grafting with acrylamide and analysis of its flocculation, adsorption, and biodegradation properties. Int J Biol Macromol 2018;114:283—94.

[27] Kera NH, Bhaumik M, Pillay K, Ray SS, Maity A. Selective removal of toxic Cr(VI) from aqueous solution by adsorption combined with reduction at a magnetic nanocomposite surface. J Colloid Interface Sci 2017;503:214—28.

[28] Kumar N, Mittal H, Alhassan SM, Ray SS. Bionanocomposite hydrogel for the adsorption of dye and reusability of generated waste for the photodegradation of ciprofloxacin: a demonstration of the circularity concept for water purification. ACS Sustain Chem Eng 2018;6:17011—25.

[29] Janaki V, Vijayaraghavan K, Ramasamy A, Lee K-J, Oh B-T, Kamala-Kannan S. Competitive adsorption of reactive orange 16 and reactive brilliant blue R on polyaniline/bacterial extracellular polysaccharides composite—a novel eco-friendly polymer. J Hazard Mater 2012;241:110—17.

[30] Pal S, Ghorai S, Das C, Samrat S, Ghosh A, Panda AB. Carboxymethyl tamarind-g-poly(acrylamide)/silica: a high performance hybrid nanocomposite for adsorption of methylene blue dye. Ind Eng Chem Res 2012;51:15546—56.

[31] Huang L, Xiao C, Chen B. A novel starch-based adsorbent for removing toxic Hg(II) and Pb(II) ions from aqueous solution. J Hazard Mater 2011;192:832—6.

[32] O'Connell DW, Birkinshaw C, O'Dwyer TF. Heavy metal adsorbents prepared from the modification of cellulose: a review. Bioresour Technol 2008;99:6709—24.

[33] Banerjee SS, Chen D-H. Fast removal of copper ions by gum arabic modified magnetic nano-adsorbent. J Hazard Mater 2007;147:792—9.

[34] Bhatnagar A, Sillanpää M. Applications of chitin- and chitosan-derivatives for the detoxification of water and wastewater—a short review. Adv Colloid Interface Sci 2009;152:26—38.

[35] Suhas Carrott PJM, Ribeiro Carrott MML. Lignin — from natural adsorbent to activated carbon: a review. Bioresour Technol 2007;98:2301—12.

[36] Gusain R, Kumar N, Ray SS. Recent advances in carbon nanomaterial-based adsorbents for water purification. Coord Chem Rev 2020;405:213111.

[37] Qi X, Wei W, Su T, Zhang J, Dong W. Fabrication of a new polysaccharide-based adsorbent for water purification. Carbohydr Polym 2018;195:368—77.

[38] Dragan ES, Apopei DF. Synthesis and swelling behavior of pH-sensitive semi-interpenetrating polymer network composite hydrogels based on native and modified potatoes starch as potential sorbent for cationic dyes. Chem Eng J 2011;178:252—63.

[39] Zhu HY, Jiang R, Xiao L, Zeng GM. Preparation, characterization, adsorption kinetics and thermodynamics of novel magnetic chitosan enwrapping nanosized γ-Fe$_2$O$_3$ and multi-walled carbon nanotubes with enhanced adsorption properties for methyl orange. Bioresour Technol 2010;101:5063—9.

[40] Rodrigo AM, Martínez ME, Escudero ML, Ruíz J, Martınez P, Saldaña L, et al. Influence of particle size in the effect of polyethylene on human osteoblastic cells. Biomaterials. 2001;22:755—62.

[41] Botlhoko OJ, Ramontja J, Ray SS. Morphological development and enhancement of thermal, mechanical, and electronic properties of thermally exfoliated graphene oxide-filled biodegradable polylactide/poly(ϵ-caprolactone) blend composites. Polymer. 2018;139:188—200.

[42] Botlhoko OJ, Makwakwa D, Ray SS, Ramontja J. Enzymatic degradation, electronic, and thermal properties of graphite- and graphene oxide-filled biodegradable polylactide/poly(ϵ-caprolactone) blend composites. J Appl Polym Sci 2019;136:47387.

[43] Huang Q, Li G, Chen M, Dong S. Graphene oxide functionalized O-(carboxymethyl)-chitosan membranes: fabrication using dialysis and applications in water purification. Colloids Surf A Physicochem Eng Asp 2018;554:27—33.

[44] Sahraei R, Ghaemy M. Synthesis of modified gum tragacanth/graphene oxide composite hydrogel for heavy metal ions removal and preparation of silver nanocomposite for antibacterial activity. Carbohydr Polym 2017;157:823—33.

[45] Yu R, Shi Y, Yang D, Liu Y, Qu J, Yu Z-Z. Graphene Oxide/chitosan aerogel microspheres with honeycomb-cobweb and radially oriented microchannel structures for broad-spectrum and rapid adsorption of water contaminants. ACS Appl Mater Interfaces 2017;9:21809—19.

[46] Li L, Fan L, Sun M, Qiu H, Li X, Duan H, et al. Adsorbent for hydroquinone removal based on graphene oxide functionalized with magnetic cyclodextrin—chitosan. Int J Biol Macromol 2013;58:169—75.

[47] Ravi Kumar MNV. A review of chitin and chitosan applications. Reactive Funct Polym 2000;46:1—27.

[48] Zhou G, Wang KP, Liu HW, Wang L, Xiao XF, Dou DD, et al. Three-dimensional polylactic acid@graphene oxide/chitosan sponge bionic filter: highly efficient adsorption of crystal violet dye. Int J Biol Macromol 2018;113:792—803.

[49] Qi C, Zhao L, Lin Y, Wu D. Graphene oxide/chitosan sponge as a novel filtering material for the removal of dye from water. J Colloid Interface Sci 2018;517:18—27.

[50] Mohseni Kafshgari M, Tahermansouri H. Development of a graphene oxide/chitosan nanocomposite for the removal of picric acid from aqueous solutions: study of sorption parameters. Colloids Surf B Biointerfaces 2017;160:671—81.

[51] Yang A, Yang P, Huang CP. Preparation of graphene oxide—chitosan composite and adsorption performance for uranium. J Radioanal Nucl Chem 2017;313:371—8.

[52] Banerjee P, Barman SR, Mukhopadhayay A, Das P. Ultrasound assisted mixed azo dye adsorption by chitosan—graphene oxide nanocomposite. Chem Eng Res Des 2017;117:43—56.

[53] Zhang C, Chen Z, Guo W, Zhu C, Zou Y. Simple fabrication of chitosan/graphene nanoplates composite spheres for efficient adsorption of acid dyes from aqueous solution. Int J Biol Macromol 2018;112:1048—54.

[54] Kumar ASK, Jiang S-J. Chitosan-functionalized graphene oxide: a novel adsorbent an efficient adsorption of arsenic from aqueous solution. J Environ Chem Eng 2016;4:1698—713.

[55] Wang Y, Xia G, Wu C, Sun J, Song R, Huang W. Porous chitosan doped with graphene oxide as highly effective adsorbent for methyl orange and amido black 10B. Carbohydr Polym 2015;115:686—93.

[56] Jones S, Pramanik A, Kanchanapally R, Viraka Nellore BP, Begum S, Sweet C, et al. Multifunctional three-dimensional chitosan/gold nanoparticle/graphene oxide architecture for separation, label-free sers identification of pharmaceutical contaminants, and effective killing of superbugs. ACS Sustain Chem Eng 2017;5:7175—87.

[57] Sarkar AK, Bediako JK, Choi J-W, Yun Y-S. Functionalized magnetic biopolymeric graphene oxide with outstanding performance in water purification. NPG Asia Mater 2019;11:4.

[58] Wang Y, Shi L, Gao L, Wei Q, Cui L, Hu L, et al. The removal of lead ions from aqueous solution by using magnetic hydroxypropyl chitosan/oxidized multiwalled carbon nanotubes composites. J Colloid Interface Sci 2015;451:7—14.

[59] Huang Y, Lee X, Grattieri M, Macazo FC, Cai R, Minteer SD. A sustainable adsorbent for phosphate removal: modifying multi-walled carbon nanotubes with chitosan. J Mater Sci 2018;53:12641—9.

[60] Khakpour R, Tahermansouri H. Synthesis, characterization and study of sorption parameters of multi-walled carbon nanotubes/chitosan nanocomposite for the removal of picric acid from aqueous solutions. Int J Biol Macromol 2018;109:598—610.

[61] Wasim M, Sagar S, Sabir A, Shafiq M, Jamil T. Decoration of open pore network in Polyvinylidene fluoride/MWCNTs with chitosan for the removal of reactive orange 16 dye. Carbohydr Polym 2017;174:474—83.

[62] Mallakpour S, Nezamzadeh Ezhieh A. Effect of Starch-MWCNT@Valine nanocomposite on the optical, morphological, thermal, and adsorption properties of chitosan. J Polym Environ 2017;25:875—83.

[63] Huang Y, Lee X, Macazo FC, Grattieri M, Cai R, Minteer SD. Fast and efficient removal of chromium (VI) anionic species by a reusable chitosan-modified multi-walled carbon nanotube composite. Chem Eng J 2018;339:259—67.

[64] Thakur VK, Thakur MK, Raghavan P, Kessler MR. Progress in green polymer composites from lignin for multifunctional applications: a review. ACS Sustain Chem Eng 2014;2:1072—92.

[65] Upton BM, Kasko AM. Strategies for the conversion of lignin to high-value polymeric materials: review and perspective. Chem Rev 2016;116:2275—306.

[66] Ge Y, Li Z. Application of lignin and its derivatives in adsorption of heavy metal ions in water: a review. ACS Sustain Chem Eng 2018;6:7181—92.

[67] Li Z, Chen J, Ge Y. Removal of lead ion and oil droplet from aqueous solution by lignin-grafted carbon nanotubes. Chem Eng J 2017;308:809—17.

[68] Li F, Wang X, Yuan T, Sun R. A lignosulfonate-modified graphene hydrogel with ultrahigh adsorption capacity for Pb(ii) removal. J Mater Chem A 2016;4:11888—96.

[69] Zhou F, Feng X, Yu J, Jiang X. High performance of 3D porous graphene/lignin/ sodium alginate composite for adsorption of Cd(II) and Pb(II). Environ Sci Pollut Res 2018;25:15651—61.

[70] Ma X, Liu X, Anderson DP, Chang PR. Modification of porous starch for the adsorption of heavy metal ions from aqueous solution. Food Chem 2015;181:133—9.

[71] Wan Ngah WS, Hanafiah MAKM. Removal of heavy metal ions from wastewater by chemically modified plant wastes as adsorbents: a review. Bioresour Technol 2008;99:3935—48.

[72] Dieter K, Brigitte H, Hans-Peter F, Andreas B. Cellulose: fascinating biopolymer and sustainable raw material. Angew Chem Int Ed 2005;44:3358—93.

[73] Lam E, Male KB, Chong JH, Leung ACW, Luong JHT. Applications of functionalized and nanoparticle-modified nanocrystalline cellulose. Trends Biotechnol 2012;30:283—90.

[74] Dai H, Huang Y, Huang H. Eco-friendly polyvinyl alcohol/carboxymethyl cellulose hydrogels reinforced with graphene oxide and bentonite for enhanced adsorption of methylene blue. Carbohydr Polym 2018;185:1—11.

[75] Yousefi N, Wong KKW, Hosseinidoust Z, Sørensen HO, Bruns S, Zheng Y, et al. Hierarchically porous, ultra-strong reduced graphene oxide-cellulose nanocrystal sponges for exceptional adsorption of water contaminants. Nanoscale. 2018;10:7171—84.

[76] Zhang C, Zhang RZ, Ma YQ, Guan WB, Wu XL, Liu X, et al. Preparation of cellulose/graphene composite and its applications for triazine pesticides adsorption from water. ACS Sustain Chem Eng 2015;3:396—405.

[77] Bermudez VM, Robinson JT. Effects of molecular adsorption on the electronic structure of single-layer graphene. Langmuir. 2011;27:11026—36.

[78] Muyonga JH, Cole CGB, Duodu KG. Extraction and physico-chemical characterisation of Nile perch (Lates niloticus) skin and bone gelatin. Food Hydrocoll 2004;18:581—92.

[79] Farris S, Song J, Huang Q. Alternative reaction mechanism for the cross-linking of gelatin with glutaraldehyde. J Agric Food Chem 2010;58:998—1003.

[80] Duconseille A, Astruc T, Quintana N, Meersman F, Sante-Lhoutellier V. Gelatin structure and composition linked to hard capsule dissolution: a review. Food Hydrocoll 2015;43:360—76.

[81] Hayeeye F, Sattar M, Chinpa W, Sirichote O. Kinetics and thermodynamics of Rhodamine B adsorption by gelatin/activated carbon composite beads. Colloids Surf A: Physicochem Eng Asp 2017;513:259—66.

[82] Saber-Samandari S, Saber-Samandari S, Joneidi-Yekta H, Mohseni M. Adsorption of anionic and cationic dyes from aqueous solution using gelatin-based magnetic nanocomposite beads comprising carboxylic acid functionalized carbon nanotube. Chem Eng J 2017;308:1133—44.

[83] Mittal H, Alhassan SM, Ray SS. Efficient organic dye removal from wastewater by magnetic carbonaceous adsorbent prepared from corn starch. J Environ Chem Eng 2018;6:7119—31.

[84] Nanninga HJ, Gottschal JC. Anaerobic purification of waste water from a potato-starch producing factory. Water Res 1986;20:97—103.

[85] Bates RG, Bower VE. Alkaline solutions for pH control. Anal Chem 1956;28:1322—4.

[86] Hosseinzadeh H, Ramin S. Fabrication of starch-graft-poly(acrylamide)/graphene oxide/hydroxyapatite nanocomposite hydrogel adsorbent for removal of malachite green dye from aqueous solution. Int J Biol Macromol 2018;106:101—15.

[87] Chang PR, Zheng P, Liu B, Anderson DP, Yu J, Ma X. Characterization of magnetic soluble starch-functionalized carbon nanotubes and its application for the adsorption of the dyes. J Hazard Mater 2011;186:2144—50.

[88] Esmat M, Farghali AA, Khedr MH, El-Sherbiny IM. Alginate-based nanocomposites for efficient removal of heavy metal ions. Int J Biol Macromol 2017;102:272—83.

[89] Pawar SN, Edgar KJ. Alginate derivatization: a review of chemistry, properties and applications. Biomaterials. 2012;33:3279—305.

[90] Cataldo S, Gianguzza A, Milea D, Muratore N, Pettignano A. Pb(II) adsorption by a novel activated carbon - alginate composite material. A kinetic and equilibrium study. Int J Biol Macromol 2016;92:769—78.

[91] Cataldo S, Muratore N, Orecchio S, Pettignano A. Enhancement of adsorption ability of calcium alginate gel beads towards Pd(II) ion. A kinetic and equilibrium study on hybrid laponite and montmorillonite—alginate gel beads. Appl Clay Sci 2015;118:162—70.

[92] Ma J, Ma Y, Yu F, Dai X. Rotating magnetic field-assisted adsorption mechanism of pollutants on mechanically strong sodium alginate/graphene/L-cysteine beads in batch and fixed-bed column systems. Environ Sci Technol 2018;52:13925—34.

[93] Gan L, Li H, Chen L, Xu L, Liu J, Geng A, et al. Graphene oxide incorporated alginate hydrogel beads for the removal of various organic dyes and bisphenol A in water. Colloid Polym Sci 2018;296:607—15.

[94] Patel S, Goyal A. Applications of natural polymer gum arabic: a review. Int J Food Prop 2015;18:986—98.

[95] Kumar N, Mittal H, Parashar V, Ray SS, Ngila JC. Efficient removal of rhodamine 6G dye from aqueous solution using nickel sulphide incorporated polyacrylamide grafted gum karaya bionanocomposite hydrogel. RSC Adv 2016;6:21929—39.

[96] Prajapati VD, Jani GK, Moradiya NG, Randeria NP. Pharmaceutical applications of various natural gums, mucilages and their modified forms. Carbohydr Polym 2013;92:1685—99.

[97] Yan L, Chang PR, Zheng P, Ma X. Characterization of magnetic guar gum-grafted carbon nanotubes and the adsorption of the dyes. Carbohydr Polym 2012;87:1919—24.

[98] Sahraei R, Pour ZS, Ghaemy M. Novel magnetic bio-sorbent hydrogel beads based on modified gum tragacanth/graphene oxide: removal of heavy metals and dyes from water. J Clean Prod 2017;142:2973—84.

[99] Gupta VK, Agarwal S, Sadegh H, Ali GAM, Bharti AK, Hamdy Makhlouf AS. Facile route synthesis of novel graphene oxide-β-cyclodextrin nanocomposite and its application as adsorbent for removal of toxic bisphenol A from the aqueous phase. J Mol Liq 2017;237:466—72.

[100] Hu H, Chang M, Zhang M, Wang X, Chen D. A new insight into PAM/graphene-based adsorption of water-soluble aromatic pollutants. J Mater Sci 2017;52:8650—64.

[101] Sitko R, Musielak M, Zawisza B, Talik E, Gagor A. Graphene oxide/cellulose membranes in adsorption of divalent metal ions. RSC Adv 2016;6:96595—605.

[102] Dong Z, Zhang F, Wang D, Liu X, Jin J. Polydopamine-mediated surface-functionalization of graphene oxide for heavy metal ions removal. J Solid State Chem 2015;224:88—93.

[103] Xiao J, Lv W, Xie Z, Song Y, Zheng Q. L-cysteine-reduced graphene oxide/poly(vinyl alcohol) ultralight aerogel as a broad-spectrum adsorbent for anionic and cationic dyes. J Mater Sci 2017;52:5807—21.

[104] Gao H, Sun Y, Zhou J, Xu R, Duan H. Mussel-inspired synthesis of polydopamine-functionalized graphene hydrogel as reusable adsorbents for water purification. ACS Appl Mater Interfaces 2013;5:425—32.

[105] Ren F, Li Z, Tan W-Z, Liu X-H, Sun Z-F, Ren P-G, et al. Facile preparation of 3D regenerated cellulose/graphene oxide composite aerogel with high-efficiency adsorption towards methylene blue. J Colloid Interface Sci 2018;532:58—67.

[106] Makhado E, Pandey S, Ramontja J. Microwave assisted synthesis of xanthan gum-cl-poly (acrylic acid) based-reduced graphene oxide hydrogel composite for adsorption of methylene blue and methyl violet from aqueous solution. Int J Biol Macromol 2018;119:255—69.

[107] Zhang Y, Cao B, Zhao L, Sun L, Gao Y, Li J, et al. Biochar-supported reduced graphene oxide composite for adsorption and coadsorption of atrazine and lead ions. Appl Surf Sci 2018;427:147—55.

[108] Liu S, Ge H, Wang C, Zou Y, Liu J. Agricultural waste/graphene oxide 3D bio-adsorbent for highly efficient removal of methylene blue from water pollution. Sci Total Environ 2018;628—629:959—68.

[109] Salahuddin N, El-Daly H, El Sharkawy RG, Nasr BT. Synthesis and efficacy of PPy/CS/GO nanocomposites for adsorption of ponceau 4R dye. Polymer. 2018;146:291—303.

[110] Yang J-Y, Jiang X-Y, Jiao F-P, Yu J-G. The oxygen-rich pentaerythritol modified multi-walled carbon nanotube as an efficient adsorbent for aqueous removal of alizarin yellow R and alizarin red S. Appl Surf Sci 2018;436:198—206.

[111] Liu Y, Huang S, Zhao X, Zhang Y. Fabrication of three-dimensional porous β-cyclodextrin/chitosan functionalized graphene oxide hydrogel for methylene blue removal from aqueous solution. Colloids Surf A Physicochem Eng Asps 2018;539:1—10.

[112] Omidi S, Kakanejadifard A. Eco-friendly synthesis of graphene—chitosan composite hydrogel as efficient adsorbent for Congo red. RSC Adv 2018;8:12179—89.

[113] Xiao J, Zhang J, Lv W, Song Y, Zheng Q. Multifunctional graphene/poly(vinyl alcohol) aerogels: in situ hydrothermal preparation and applications in broad-spectrum adsorption for dyes and oils. Carbon. 2017;123:354—63.

[114] Mercante LA, Facure MHM, Locilento DA, Sanfelice RC, Migliorini FL, Mattoso LHC, et al. Solution blow spun PMMA nanofibers wrapped with reduced graphene oxide as an efficient dye adsorbent. N J Chem 2017;41:9087—94.

[115] Tan P, Hu Y. Improved synthesis of graphene/β-cyclodextrin composite for highly efficient dye adsorption and removal. J Mol Liq 2017;242:181—9.

[116] Jiang L, Liu Y, Liu S, Hu X, Zeng G, Hu X, et al. Fabrication of β-cyclodextrin/poly (L-glutamic acid) supported magnetic graphene oxide and its adsorption behavior for 17β-estradiol. Chem Eng J 2017;308:597—605.

[117] Barati A, Kazemi E, Dadfarnia S, Haji Shabani AM. Synthesis/characterization of molecular imprinted polymer based on magnetic chitosan/graphene oxide for selective separation/preconcentration of fluoxetine from environmental and biological samples. J Ind Eng Chem 2017;46:212—21.

[118] Wu X-L, Xiao P, Zhong S, Fang K, Lin H, Chen J. Magnetic ZnFe2O4@chitosan encapsulated in graphene oxide for adsorptive removal of organic dye. RSC Adv 2017;7:28145—51.

[119] Wan Q, Liu M, Xie Y, Tian J, Huang Q, Deng F, et al. Facile and highly efficient fabrication of graphene oxide-based polymer nanocomposites through mussel-inspired chemistry and their environmental pollutant removal application. J Mater Sci 2017;52:504—18.

[120] Li L, Liu F, Duan H, Wang X, Li J, Wang Y, et al. The preparation of novel adsorbent materials with efficient adsorption performance for both chromium and methylene blue. Colloids Surf B Biointerfaces 2016;141:253—9.

[121] Van Hoa N, Khong TT, Thi Hoang Quyen T, Si Trung T. One-step facile synthesis of mesoporous graphene/Fe3O4/chitosan nanocomposite and its adsorption capacity for a textile dye. J Water Process Eng 2016;9:170—8.

[122] Gan L, Shang S, Hu E, Yuen CWM, Jiang S-x. Konjac glucomannan/graphene oxide hydrogel with enhanced dyes adsorption capability for methyl blue and methyl orange. Appl Surf Sci 2015;357:866—72.

[123] Zhao Y, Chen H, Li J, Chen C. Hierarchical MWCNTs/Fe3O4/PANI magnetic composite as adsorbent for methyl orange removal. J Colloid Interface Sci 2015;450:189—95.

[124] Wang D, Liu L, Jiang X, Yu J, Chen X, Chen X. Adsorbent for p-phenylenediamine adsorption and removal based on graphene oxide functionalized with magnetic cyclodextrin. Appl Surf Sci 2015;329:197—205.

[125] Wang D, Liu L, Jiang X, Yu J, Chen X. Adsorption and removal of malachite green from aqueous solution using magnetic β-cyclodextrin-graphene oxide nanocomposites as adsorbents. Colloids Surf A Physicochem Eng Asp 2015;466:166—73.

[126] Zhang J, Azam MS, Shi C, Huang J, Yan B, Liu Q, et al. Poly(acrylic acid) functionalized magnetic graphene oxide nanocomposite for removal of methylene blue. RSC Adv 2015;5:32272—82.

[127] Li Y, Sun J, Du Q, Zhang L, Yang X, Wu S, et al. Mechanical and dye adsorption properties of graphene oxide/chitosan composite fibers prepared by wet spinning. Carbohydr Polym 2014;102:755—61.

[128] Wu S, Zhao X, Li Y, Zhao C, Du Q, Sun J, et al. Adsorption of ciprofloxacin onto biocomposite fibers of graphene oxide/calcium alginate. Chem Eng J 2013;230:389—95.

[129] Fan L, Luo C, Sun M, Qiu H, Li X. Synthesis of magnetic β-cyclodextrin—chitosan/graphene oxide as nanoadsorbent and its application in dye adsorption and removal. Colloids Surf B Biointerfaces 2013;103:601—7.

[130] Travlou NA, Kyzas GZ, Lazaridis NK, Deliyanni EA. Graphite oxide/chitosan composite for reactive dye removal. Chem Eng J 2013;217:256—65.

[131] Fan L, Luo C, Sun M, Li X, Lu F, Qiu H. Preparation of novel magnetic chitosan/graphene oxide composite as effective adsorbents toward methylene blue. Bioresour Technol 2012;114:703—6.

[132] Zhang Z, Gao T, Si S, Liu Q, Wu Y, Zhou G. One-pot preparation of P(TA-TEPA)-PAM-RGO ternary composite for high efficient Cr(VI) removal from aqueous solution. Chem Eng J 2018;343:207—16.

[133] Hosseinzadeh H, Ramin S. Effective removal of copper from aqueous solutions by modified magnetic chitosan/graphene oxide nanocomposites. Int J Biol Macromol 2018;113:859—68.

[134] Tan B, Zhao H, Zhang Y, Quan X, He Z, Zheng W, et al. Amphiphilic PA-induced three-dimensional graphene macrostructure with enhanced removal of heavy metal ions. J Colloid Interface Sci 2018;512:853—61.

[135] Shahzad A, Miran W, Rasool K, Nawaz M, Jang J, Lim S-R, et al. Heavy metals removal by EDTA-functionalized chitosan graphene oxide nanocomposites. RSC Adv 2017;7:9764—71.

[136] Nyairo WN, Eker YR, Kowenje C, Zor E, Bingol H, Tor A, et al. Efficient removal of lead(II) ions from aqueous solutions using methyl-β-cyclodextrin modified graphene oxide. Water Air Soil Pollut 2017;228:406.

[137] Ma Y-X, Kou Y-L, Xing D, Jin P-S, Shao W-J, Li X, et al. Synthesis of magnetic graphene oxide grafted polymaleicamide dendrimer nanohybrids for adsorption of Pb(II) in aqueous solution. J Hazard Mater 2017;340:407—16.

[138] Wang Y, Yan T, Gao L, Cui L, Hu L, Yan L, et al. Magnetic hydroxypropyl chitosan functionalized graphene oxide as adsorbent for the removal of lead ions from aqueous solution. Desalination Water Treat 2016;57:3975—84.

[139] Medina RP, Nadres ET, Ballesteros FC, Rodrigues DF. Incorporation of graphene oxide into a chitosan—poly(acrylic acid) porous polymer nanocomposite for enhanced lead adsorption. Environ Sci Nano 2016;3:638—46.

[140] Li L, Wang Z, Ma P, Bai H, Dong W, Chen M. Preparation of polyvinyl alcohol/chitosan hydrogel compounded with graphene oxide to enhance the adsorption properties for Cu(II) in aqueous solution. J Polym Res 2015;22:150.

[141] Li X, Zhou H, Wu W, Wei S, Xu Y, Kuang Y. Studies of heavy metal ion adsorption on chitosan/sulfydryl-functionalized graphene oxide composites. J Colloid Interface Sci 2015;448:389—97.

[142] Li L, Fan L, Sun M, Qiu H, Li X, Duan H, et al. Adsorbent for chromium removal based on graphene oxide functionalized with magnetic cyclodextrin—chitosan. Colloids Surf B Biointerfaces 2013;107:76—83.

CHAPTER THIRTEEN

Conducting polymer-functionalized carbon nanomaterials-based adsorbents

13.1 Introduction

Polymers are extremely popular in various applications due to their controllable mechanical, viscoelastic, and surface properties [1−6]. Organic polymers are generally a suitable choice as insulators in multiple industries. Conductive polymers are a special class of organic polymers, which are conductive in nature due to their unusual structure. Organic polymers with a conjugated π-electron system exhibit extraordinary characteristics, such as low-energy optical transition, high mechanical strength, low ionization potentials, and high electron attraction. These polymers can be modified and restored more easily than conventional polymers. Polyaniline (PANI), polypyrrole (PPy), polyacetylene (PAc), and polythiophene (PTh) are typical examples of the conducting polymers. Due to the combination of the properties of polymers and the semiconducting nature of their materials, conducting polymers have been used in a large range of applications, such as sensors, biomedicals, catalysis, energy storage, electric nanodevices, and water purification [6−10]. Conducting polymers are garnering great attention in both academia and industry sectors due to their low cost, ease of preparation and modification, and excellent conducting properties compared with conventional polymers.

Conducting polymers can be generally divided into three different classes: (1) ion conductive polymers, (2) electron conductive polymers, and (3) proton conductive polymers [9]. Ion conductive polymers have been applied in electrochemical devices as cell separators. Electron conductive polymers can easily be modified and switched into insulating and conducting stages by changes in the potential by electrochemical redox reactions [11]. Therefore electron conductive polymers are gaining interest in various practical applications, such as optics, sensors, and batteries [11]. Proton or cation conductive polymers perform conductivity on the backbone of the polymer due to the presence of carboxylate or sulfonated groups along with a cationic counterion. Fig. 13.1 presents the molecular structure of a few conducting polymers. The presence of heteroatoms, lone pairs of electrons, and π-electrons enhance the potential

Carbon Nanomaterial-Based Adsorbents for Water Purification.
DOI: https://doi.org/10.1016/B978-0-12-821959-1.00013-1

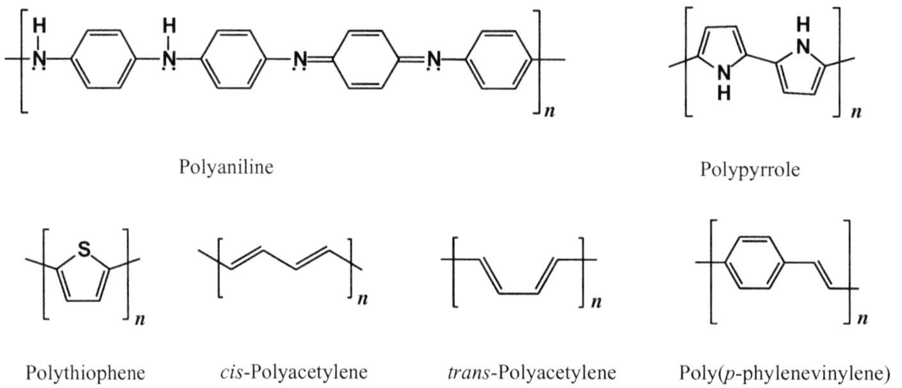

Figure 13.1 Examples of a few conductive polymers.

of conductive polymers as absorbent in wastewater treatment application. The low cost, environmental stability, presence of functional groups, and electrical conductivity further promote the implementation of conductive polymers in water purification.

Among all of them, the leading candidates for water purification are PANI, PPy, and PTh. Although conductive polymers are cost-effective and biocompatible, the adsorption capacity of conductive polymers is not significantly high. Furthermore, the separation and regeneration of standalone conductive polymers after the adsorption process are the prime issues in their use as adsorbents. However, on combining conductive polymers with large surface area organic/inorganic nanomaterials, a high adsorption capacity toward the various water pollutants can be demonstrated and the separation after adsorption can also be improved. Conductive polymer-functionalized carbon nanomaterials (CNMs)-based nanoadsorbents are attracting the attention of researchers. Primarily, conductive polymers are functionalized with graphene, graphene oxide (GO), and reduced graphene oxide (rGO) and studied for water purification applications as adsorbents. Except for PANI and PPy, conductive polymers are not well explored in the wastewater treatment application as adsorbents. This chapter is focused on the application of low-cost conductive polymer (PANI, PPy and PTh)-functionalized CNMs-based adsorbent material for water purification.

13.2 Polyaniline-functionalized carbon nanomaterials

PANI, also known as aniline black, was first acknowledged in the mid-19th century and produced via the oxidation reaction of aniline in an acidic medium [9]. PANI can be synthesized via the chemical or electrochemical oxidation of aniline. Several oxidizing agents, such as H_2O_2 [12], $FeCl_3$ [13], $(NH_3)_2S_2O_8H_2$ [14], $HAuCl_4$

[15], and $K_2Cr_2O_7$ [16,17], have been used for the synthesis of PANI following chemical oxidation. Among all the conductive polymers, the extraordinary characteristics of PANI, such as excellent environmental stability, presence of functional groups, economic feasibility of monomer aniline, relatively high conductivity, and easy and facile route of preparation, have allowed the wide spectrum of PANI applications such as catalysis [18], sensors [19], batteries [20], capacitors [21], adsorption [22], composite fabrication [23], and light-emitting diodes [24]. Therefore PANI is attracting interest in both academia and industry in many research fields.

PANI has been enormously studied as an adsorbent for the removal of water contaminants due to its unique structure and extraordinary properties [9]. The presence of nitrogen atoms and plenty of π electrons in PANI facilitate the interaction with the pollutant molecules for adsorption [25,26]. However, poor mechanical stability and low solubility in aqueous medium are significant disadvantages of PANI for its use in wastewater treatment, reducing its performance. This issue can be overcome via nanocomposite formation with other organic/inorganic materials. PANI nanocomposites with carbon-based nanomaterials such as graphene oxide (GO), reduced graphene oxide (rGO), and carbon nanotubes (CNTs) have been widely studied in wastewater treatment via adsorption. CNMs improve the mechanical stability of the polymer, and PANI enhances the adsorption capacity of the nanocomposite due to presence of nitrogen atoms and π electrons. Numerous studies have been done on PANI–CNMs nanocomposite toward the adsorption of several water pollutants.

PANI-modified GO-SrTiO$_3$ nanocomposite (GOPSr-2) successfully adsorbed cationic and anionic dyes simultaneously (Fig. 13.2A). Oxygen functionalities on the GO nanosheets promoted the adsorption of cationic dyes, and nitrogen functionalities in PANI, including the imine ($-N=$) and amine ($-N<$) groups, contributed an overall positive charge to the nanocomposite backbone, promoting adsorption of anionic dyes [27]. Cationic dye [methylene blue (MB)] and anionic dye [methylene orange (MO)] are adsorbed successfully on the GOPSr-2, via π–π interactions, electrostatic interactions, and hydrogen bonding. SrTiO$_3$ also assisted in the segregation of the GO nanosheets and hence in increasing the active surface area of the adsorbate. Wang et al. grafted PANI chains onto the edges of GO nanosheets and reported the adsorption capacity of the GO–PANI nanocomposite (1416 mg/g) for Pb(II) adsorption compared with those of GO (555 mg/g) and PANI (625 mg/g) at 298K [28]. Carboxyl groups are generated on the GO sheets due to harsh oxidation, which react with N-hydroxysuccinimide and p-phenylenediamine and are converted into amines. PANI is introduced at the GO sheets' edges because of the in situ oxidation reaction. The nanocomposite successfully removed both dyes and heavy metal ions from a mixed aqueous solution (Fig. 13.2B). The adsorption of heavy metal ions and dyes by GO–PANI followed the Langmuir isotherm, which suggests the formation of a monolayer of adsorbate on the adsorbent surface.

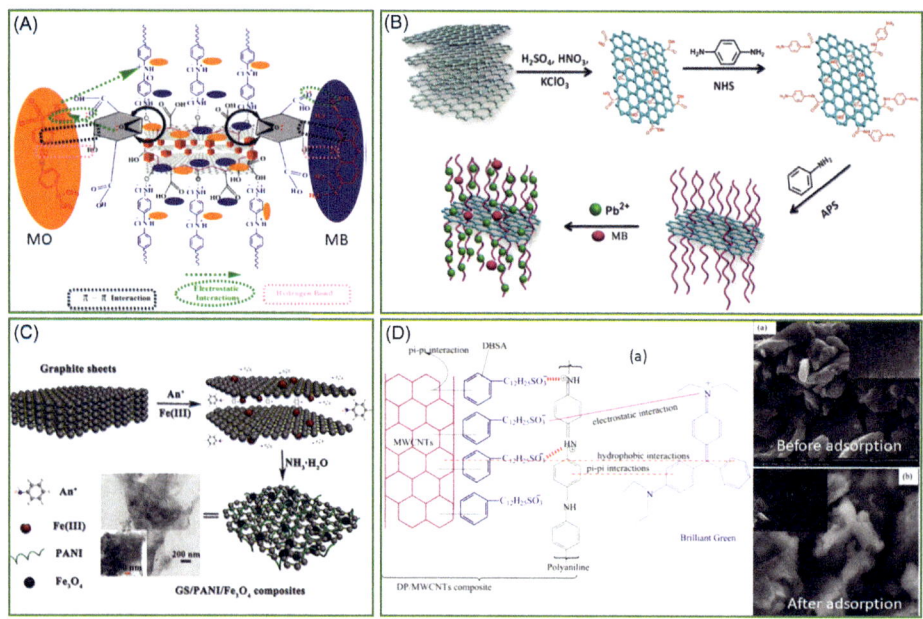

Figure 13.2 (A) Proposed adsorption mechanism of MO and MB adsorption on GOPSr-2. (B) Illustration of the synthesis of PANI chains onto the edges of GO nanosheets and adsorption of heavy metal ion (Pb(II)) and organic dye (MB). (C) Synthesis of magnetic graphene sheet/polyaniline/Fe$_3$O$_4$ (GS/PANI/Fe$_3$O$_4$) nanocomposites for adsorption of CR and BG dye. (D) Proposed mechanism of BG adsorption on DBSA-doped PANI/MWCNT (DP-MWCNT) nanocomposite and the FESEM images of DBSA-doped PANI/MWCNT before and after adsorption of BG dye. *(A) Reproduced with permission from Shahabuddin S, Sarih NM, Afzal Kamboh M, Rashidi Nodeh H, Mohamad S. Synthesis of polyaniline-coated graphene oxide@SrTiO$_3$ nanocube nanocomposites for enhanced removal of carcinogenic dyes from aqueous solution. Polymers 2016;8:305. Copyright 2016, MDPI. (B) Reproduced with permission from Yang Y, Wang W, Li M, Wang H, Zhao M, Wang C. Preparation of PANI grafted at the edge of graphene oxide sheets and its adsorption of Pb(II) and methylene blue. Polym Compos 2018;39:1663−73. Copyright 2016, John Wiley and Sons. (C) Reproduced with permission from Mu B, Tang J, Zhang L, Wang A. Facile fabrication of superparamagnetic graphene/polyaniline/Fe$_3$O$_4$ nanocomposites for fast magnetic separation and efficient removal of dye. Sci Rep 2017;7:5347. Copyright 2017, Nature. (D) Reproduced with permission from Kumar R, Ansari MO, Barakat MA. Adsorption of brilliant green by surfactant doped polyaniline/MWCNTs composite: evaluation of the kinetic, thermodynamic, and isotherm. Ind Eng Chem Res 2014;53:7167−75. Copyright 2014, American Chemical Society Publications.*

To improve the separation of the adsorbent after the adsorption process, magnetic graphene sheet/polyaniline/Fe$_3$O$_4$ (GS/PANI/Fe$_3$O$_4$) nanocomposites were prepared by a green industrial method [29]. The intercalation polymerization of PANI prepared superparamagnetic GS/PANI/Fe$_3$O$_4$ into the graphite sheets using Fe(III) as oxidant (Fig. 13.2C). It showed 92.4% adsorption toward Congo Red (CR) (100 ppm) dye within 2 hours. In the mixture of cationic (CR) and anionic [Brilliant Green (BG)] dyes, it could preferentially remove the anionic dye from a mixture of, and later, cation dye from aqueous solution and the adsorbent could also be removed with the help of a magnet. The easy separation of the adsorbent enhanced the probability of regeneration and reusability, which makes it cost-effective. GS/PANI/Fe$_3$O$_4$ can easily be regenerated in 0.5 M NaOH solution and be further employed for the adsorption of water pollution. No significant reduction in adsorption ratio was noticed for five consecutive adsorption—desorption cycles. The adsorption of BG dye was efficiently observed with the use of dodecyl-benzene-sulfonic-acid (DBSA)-doped PANI/multiwalled carbon nanotubes (MWCNTs) nanocomposite [30]. The adsorption was found to be highly selective at pH 3 and increased with an increase in the temperature. The study of the thermodynamic parameters concluded that the adsorption of BG on DBSA-doped PANI/MWCNT is spontaneous and endothermic in nature. The maximum adsorption capacity of DBSA-doped PANI/MWCNT for BD dye was noted to be 434.78, 454.54, and 476.19 mg/g at a temperature of 30°C, 40°C, and 50°C, respectively. Electrostatic interactions, $\pi-\pi$ interactions, and hydrogen bonding are mainly involved in the adsorption of BG on DBSA-doped PANI/MWCNT (Fig. 13.2D). In an aqueous medium, BG acts as a cationic molecule and interacts with the adsorbent electrostatically through a sulfonic group of a DBSA molecule. The MWCNT has a high density of π-electrons, which are involved in the $\pi-\pi$ bonding with the benzene ring of BG and hence contribute to the adsorption. The N atoms in the PANI as amine and imine groups interact with the BG molecule via hydrogen bonding. The excellent adsorption capacity and magnetic characteristics of the nanocomposite beads indicate that they have the potential as adsorbents for the treatment of industrial wastewater [31].

Table 13.1 presents the various examples of PANI—CNM based adsorbents for the removal of several water pollutants from aqueous medium.

13.3 Polypyrrole functionalized carbon nanomaterials

PPy is one of the most extensively studied conducting polymers in various applications due to its high environmental stability, low cost, simple preparation strategies, ease of modification, high conductivity, high capacitance, excellent mechanical stability, and redox properties [42—45]. Pyrrole is generally polymerized in the presence of oxidants such as FeCl$_3$, Fe(ClO$_4$)$_3$, and (NH$_4$)$_2$S$_2$O$_8$ to prepare PPy [46,47]. Due to the presence of nitrogen atoms and π-electrons in the backbone of the PPy polymer, it is used in the adsorption of various water contaminants. The N atom in the PPy backbone causes a positive charge on the adsorbent molecule, which enhances the adsorption via electrostatic interaction and ion exchange [48,49]. Furthermore, it is also affected by the pH of the solution and undergoes protonation and deprotonation

Table 13.1 Examples of PANI–CNMs-based adsorbents for the removal of various water pollutants.

PANI–CNM based adsorbent	Water pollutant	Adsorption conditions	Adsorption isotherm models	Adsorption kinetics	Maximum adsorption capacity (mg/g)	References
PANI/RGO	Hg(II)	pH = 4; T = 304K; t = 24 h	Langmuir and Freundlich	Pseudo-second-order	1000	[32]
PANI/GO	Cr(VI)	pH = 3; T = 298K; t = 160 min	Langmuir	Pseudo-second-order	1149.4	[33]
pTSA–PANI@ GO-CNT	Cr(VI) CR	pH = 2 and 5; T = 303K; t = 640 min	Langmuir	Pseudo-second-order	142.85 66.66	[34]
pTSA–PANI@ GN-PVC	CR	pH = 4.5; T = 303K; t = 6 h	Langmuir	—	45.454	[35]
CNT–PANI	Malachite Green (MG)	pH = 4; T = 303K; t = 120 min	Langmuir	Pseudo-second-order	15.23	[36]
Fe$_3$O$_4$/G/PANI	Cr(VI)	pH = 6.5; T = 303K; t = 2 h	Langmuir and Freundlich	Pseudo-second-order	153.54	[37]
GO–PANI	Zn(II)	pH = 7; T = 298K; t = 20 min	Langmuir and Freundlich	—	588	[38]
Magnetic PANI/GO	Cu(II)	pH = 5.8; T = 293K; t = 15 min	Langmuir	Pseudo-second-order	101.94	[39]
PANI/MWCNT	Cr(VI)	pH = 4.5; T = 298K; t = 24 h	Freundlich	Pseudo-second-order	31.75	[40]
PANI/ MWCNTs	Pb(II)	pH = 5; T = 293K; t = 24 h	Langmuir	—	22.2	[41]

exercises, which change the surface charge of the adsorbent and hence affect the adsorption efficiency. However, the individual PPy is not a perfect choice for adsorption. The reduced dispersion of PPy in an aqueous medium and the high affinity to self-agglomerate via $\pi-\pi$ interaction downgrade the application of PPy in water purification applications [50,51]. Furthermore, the separation and regeneration of individual PPy after the adsorption process is one of the significant challenges. Therefore many efforts have been made to use various PPy nanocomposites with other nanomaterials, including carbon-based nanomaterials (CNMs) to adsorb water pollutants with enhanced adsorption capacity and separation efficiency [52,53]. The functionalization of PPy matrix on the CNMs improves the adsorption efficiency in comparison to individual CNMs and PPy.

Several reports are justifying the role of the enhanced adsorption capacity of GO nanosheets on functionalization with PPy. Recently, Wang et al. synthesized the PPy/GO nanocomposite following a sacrificial template polymerization method to remove Cr(VI) from the wastewater [54]. MnO_2 nanoslices were employed as an oxidant and sacrificial template for the preparation and deposition of PPy on the GO nanosheets as nanoslices. The adsorption capacity of PPy/GO was observed to be two times greater than the pristine commercial PPy. Another nanocomposite of GO-α cyclodextrin$-$PPy (GO-αCD$-$PPy) has also shown a high affinity towards the adsorption of Cr(VI) [55]. In this nanocomposite cyclodextrin (CD) exhibits a hydrophobic inner cavity and a hydrophilic exterior surface, and can easily adsorb organic and inorganic pollutants. Therefore attaching α-CD to GO$-$PPy enhanced the removal efficiency of the adsorbent. GO-α-CD$-$PPy nanocomposite possesses a high surface area, with a high adsorption capacity (666.67 mg/g) for Cr(VI). Several other PPy/GO nanocomposites were also studied for the removal of Cr(VI) from the wastewater, owing to the ion-exchange mechanism [56,57]. A novel PPy-decorated magnetic rGO nanocomposite was fabricated by in situ polymerization (Fig. 13.3A) [51]. The polymerization of pyrrole was done in the presence of ammonium persulfate (APS) as the oxidant. Hexadecyltrimethylammonium bromide (CTAB) solution was used as a medium for the polymerization and decoration of PPy on Fe_3O_4/rGO to prepare PPy$-Fe_3O_4$/rGO. As shown in the TEM image (Fig. 13.3A), PPy is either decorated on the rGO surface or circled the Fe_3O_4 nanoparticles to obtain a three-dimensional structure. This nanocomposite has also been employed for the adsorption of Cr(VI) ions. The adsorption capacity for PPy$-Fe_3O_4$/rGO was found to be greater than Fe_3O_4/rGO and was highly influenced by the pH of the solution. This indicates that the adsorption behavior was dominated by the electrostatic interaction along with ion exchange and chemical reduction. The Cr(VI) removal mechanism can be explained by the synergistic effects of all three components in the ternary hybrid nanocomposite PPy$-Fe_3O_4$/rGO as follows: (1) electrostatic interaction between the negatively charged Cr(VI) and protonated N of PPy in PPy$-Fe_3O_4$/rGO;

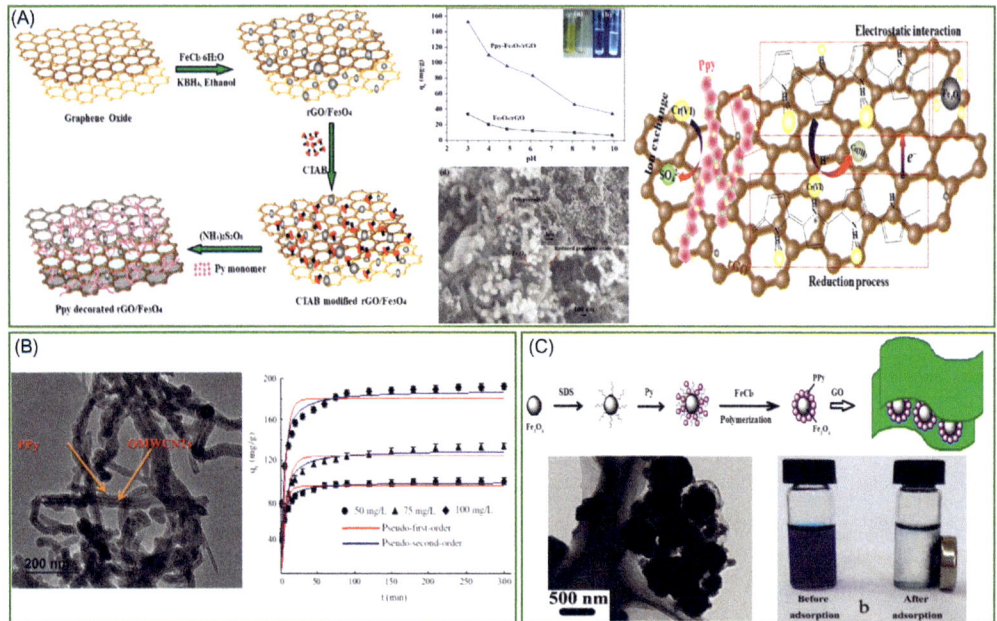

Figure 13.3 (A) Synthesis route of ternary nanocomposite PPy—Fe$_3$O$_4$/rGO, TEM image of PPy—Fe$_3$O$_4$/rGO, effect of pH on the PPy—Fe$_3$O$_4$/rGO and Fe$_3$O$_4$/rGO to remove Cr(VI) from wastewater and the mechanism illustration of Cr(VI) removal using PPy—Fe$_3$O$_4$/rGO. (B) HRTEM image of PPy/OMWCNTs nanocomposite and the effect of time on Cr(VI) adsorption using PPy/OMWCNTs nanocomposite and fit the data on the nonlinear kinetic models pseudo-first-order and pseudo-second-order. (C) Schematic representation of Fe$_3$O$_4$@PPy/RGO synthesis, TEM image of Fe$_3$O$_4$@PPy/RGO and magnetic separation of adsorbent after the adsorption process. *(A) Reproduced and reprinted with permission from Wang H, Yuan X, Wu Y, Chen X, Leng L, Wang H, et al. Facile synthesis of polypyrrole decorated reduced graphene oxide—Fe$_3$O$_4$ magnetic composites and its application for the Cr (VI) removal. Chem Eng J 2015;262:597—606. Copyright 2015, Elsevier Science Ltd. (B) Reproduced and reprinted with permission from Bhaumik M, Agarwal S, Gupta VK, Maity A. Enhanced removal of Cr (VI) from aqueous solutions using polypyrrole wrapped oxidized MWCNTs nanocomposites adsorbent. J Colloid Interface Sci 2016;470:257—67. Copyright 2016, Elsevier Science Ltd. (C) Reproduced and reprinted with permission from Bai L, Li Z, Zhang Y, Wang T, Lu R, Zhou W, et al. Synthesis of water-dispersible graphene-modified magnetic polypyrrole nanocomposite and its ability to efficiently adsorb methylene blue from aqueous solution. Chem Eng J 2015;279:757—66 [53]. Copyright 2015, Elsevier Science Ltd.*

(2) ion—exchange route by replacing the SO$_4^{2-}$ ions by Cr(VI) anions; and (3) Cr(VI) is chemically reduced to Cr(III).

Cr(VI) removal was also studied using the PPy wrapped oxidized MWCNTs nanocomposite (PPy—OMWCNTs) [58]. TEM images of the PPy—OMWCNTs show that the OMWCNTs are nicely and uniformly wrapped (Fig. 13.3B) with the PPy and the surface of the nanocomposite material being comparatively rough after PPy wrapping. The diameters of the PPy—OMWCNTs were found to be

60−80 nm, which was calculated to be larger than the individual OMWCNT. The adsorption of Cr(VI) on PPy−OMWCNTs nicely follows the pseudo-second-order kinetic model and Langmuir adsorption isotherm model. The maximum adsorption of Cr(VI) was observed at low pH (pH = 2, 294.18 mg/g). Bai et al. also prepared the magnetic PPy/rGO nanocomposite (Fe$_3$O$_4$@PPy/RGO) for the MB dye removal [53]. Fig. 13.3C illustrates the synthesis of Fe$_3$O$_4$@PPy/RGO, in which first Fe$_3$O$_4$@PPy was prepared by the solvothermal method using sodium dodecyl sulfate (SDS) to enhance the dispersion of material in water. Fe$_3$O$_4$@PPy nanocomposite was grafted onto the GO nanosheets under an acidic medium and then GO was reduced to RGO using hydrazine to produce Fe$_3$O$_4$@PPy/RGO. The TEM image of Fe$_3$O$_4$@PPy/RGO shows that the RGO exhibits one to several layers thickness and the magnetic beads (Fe$_3$O$_4$@PPy nanoparticles) are decorated on the RGO sheets. The adsorption capacity of Fe$_3$O$_4$@PPy/RGO was found to be 270.3 mg/g for MB and after removal, the adsorbent can easily be separated using the external magnet, which can further be used after the regeneration.

PPy−RGO nanocomposite has also been utilized for the selective removal of Hg(II) ions [59]. The polymerization of pyrrole on the GO nanosheets increased the surface area of the nanocomposite. The large surface area of PPy−RGO increases the binding sites for Hg(II) removal with higher adsorption capacity (980 mg/g) and desorption capacity (92.3%), which promotes the regeneration and reusability of the adsorbent.

13.4 Polythiophene functionalized carbon nanomaterials

Polythiophene (PTh) is similar to PPy in structure, the only difference is the presence of the heteroatom "S" in the place of "N" atom in the aromatic ring. Oxidative polymerization of thiophene resulted in the formation of positively charged PTh [60]. The positive charge from the PTh polymer can be modified by subsequent treatments, for example, after alkaline pH (pH >8.5) treatment, PTh exhibits negative zeta potential, whereas at lower pH (pH <8.5) it exhibits a positive zeta potential [61]. Similar to PPy, it also exhibits high chemical and environmental stability, high capacitance, low cost, high conductivity, excellent mechanical stability, and redox properties [60,62,63]. Due to these properties and ease in scaling up the synthesis process, PTh has been employed in various applications [64−67]. Recently, PTh-based adsorbents have emerged with promising attributes in wastewater purification application.

The PTh structure contains a number of S atoms, bonded to sp^2 hybridized carbons, which exhibit two lone pairs of electrons that are available to interact with other pollutant molecules. Thus it can easily donate the lone pair of electrons to the electron-deficient species, such as heavy metal ions. Ochieng et al. employed the

PTh/GO nanocomposite for the removal of Hg^{2+} ions from wastewater [68]. Hg^{2+} ions can behave as the electron acceptor and S as the electron donor due to the availability of the lone pair of electrons and thus can easily form a complex [69]. The main mechanism for Hg^{2+} removal using PTh/GO was attributed to the complexation formation, ion exchange, and neutralization. The adsorption was found to be endothermic in nature and followed Langmuir adsorption isotherm and pseudo-second-order kinetics.

Magnetic graphene polythiophene ($G/Fe_3O_4@PTh$) nanocomposite was prepared for the effective adsorption of polycyclic aromatic hydrocarbons (PAHs) [70]. The nanocomposite $G/Fe_3O_4@PTh$ exhibited a higher adsorption capacity for PAHs (e.g., naphthalene, acenaphthene, fluorene, phenanthrene, and anthracene) compared with the respective binary G/Fe_3O_4 and $Fe_3O_4@PTh$ nanocomposites. After the adsorption process, the magnetic separation of adsorbent provides the good regeneration and reusability of adsorbent (at least 17 times) with high adsorption capacity. The huge delocalized π-electron system of the adsorbent $G/Fe_3O_4@PTh$ molecule interacts with the aromatic rings of PAHs for the adsorption.

There is not much literature available on the adsorption potential of other conductive polymer nanocomposites with carbon-based nanomaterials.

13.5 Conclusion

Recycling of wastewater for drinking or other purposes is the need of the hour. Adsorption is considered as an economic, time-efficient, and easy operational method. Conductive polymers are conjugative electroactive polymers, which have been immensely studied in various applications including wastewater treatment as adsorbent. Conductive polymers are gaining the interest due to low cost, biocompatibility, environmental stability, and ease of synthesis. They can be easily synthesized and scaled up for industrial applications. The presence of functional groups and heteroatoms indicate its credibility for wastewater treatment. However, individual conductive polymers have not experienced successful progress in practical applications. Therefore the conductive polymer nanocomposites materials have been extensively investigated as an adsorbent for the removal of water pollutants. This review briefly explained the adsorption potential of conductive polymer—CNMs nanocomposites to remove the water pollutants. The high active surface area on the CNMs and the presence of conjugated π-electrons and functional groups makes conductive polymer-grafted CNMs an efficient and affordable adsorbent for water contaminant removal from wastewater. PANI, PPy, and PTh-grafted CNMs-based nanoadsorbents have gained the extensive attention of researchers in water purification to remove toxic contaminants.

The conductive polymer—CNMs nanocomposites can be easily prepared and the properties of nanocomposites are much improved with enhanced potential in adsorption. The $\pi-\pi$ interactions, hydrogen bonding, and electrostatic interactions are the driving forces for the adsorption of water pollutants on the conductive polymer—CNM nanocomposites. However, major work is still required to contribute to the use of PANI and PPy-grafted graphene as nanoadsorbent for wastewater treatment. These nanocomposite materials can be further explored for the environmental remediation applications as they are economically feasible, environmentally stable, and exhibit high mechanical strength, which encourages their use in water treatment.

References

[1] Kumar N, Mittal H, Alhassan SM, Ray SS. Bionanocomposite hydrogel for the adsorption of dye and reusability of generated waste for the photodegradation of ciprofloxacin: a demonstration of the circularity concept for water purification. ACS Sustain Chem Eng 2018;6:17011—25.

[2] Kumar N, Mittal H, Parashar V, Ray SS, Ngila JC. Efficient removal of rhodamine 6G dye from aqueous solution using nickel sulphide incorporated polyacrylamide grafted gum karaya bionanocomposite hydrogel. RSC Adv 2016;6:21929—39.

[3] Malkappa K, Ray SS, Kumar N. Enhanced thermo-mechanical stiffness, thermal stability, and fire retardant performance of surface-modified 2D MoS_2 nanosheet-reinforced polyurethane composites. Macromol Mater Eng 2019;304:1800562.

[4] Sinha Ray S, Bousmina M. Biodegradable polymers and their layered silicate nanocomposites: in greening the 21st century materials world. Prog Mater Sci 2005;50:962—1079.

[5] Mukwevho N, Kumar N, Fosso-Kankeu E, Waanders F, Bunt JR, Sinha Ray S. Visible light-excitable ZnO/2D graphitic-C_3N_4 heterostructure for the photodegradation of naphthalene. Desalination Water Treat 2019;163:286—96.

[6] Naveen MH, Gurudatt NG, Shim Y-B. Applications of conducting polymer composites to electrochemical sensors: a review. Appl Mater Today 2017;9:419—33.

[7] Nezakati T, Seifalian A, Tan A, Seifalian AM. Conductive polymers: opportunities and challenges in biomedical applications. Chem Rev 2018;118:6766—843.

[8] Das TK, Prusty S. Review on conducting polymers and their applications. Polym Technol Eng 2012;51:1487—500.

[9] Eskandari E, Kosari M, Davood Abadi Farahani MH, Khiavi ND, Saeedikhani M, Katal R, et al. A review on polyaniline-based materials applications in heavy metals removal and catalytic processes. Sep Purif Technol 2019;231:115901.

[10] Gusain R, Kumar N, Ray SS. Recent advances in carbon nanomaterial-based adsorbents for water purification. Coord Chem Rev 2020;405:213111.

[11] Inzelt G, Pineri M, Schultze JW, Vorotyntsev MA. Electron and proton conducting polymers: recent developments and prospects. Electrochim Acta 2000;45:2403—21.

[12] Surwade SP, Agnihotra SR, Dua V, Manohar N, Jain S, Ammu S, et al. Catalyst-free synthesis of oligoanilines and polyaniline nanofibers using H_2O_2. J Am Chem Soc 2009;131:12528—9.

[13] Zhang L, Wan M, Wei Y. Nanoscaled polyaniline fibers prepared by ferric chloride as an oxidant. Macromol Rapid Commun 2006;27:366—71.

[14] Rahy A, Sakrout M, Manohar S, Cho SJ, Ferraris J, Yang D. Polyaniline nanofiber synthesis by co-use of ammonium peroxydisulfate and sodium hypochlorite. Chem Mater 2008;20:4808—14.

[15] Wang Y, Liu Z, Han B, Sun Z, Huang Y, Yang G. Facile synthesis of polyaniline nanofibers using chloroaurate acid as the oxidant. Langmuir. 2005;21:833—6.

[16] Zoromba MS, Alghool S, Abdel-Hamid S, Bassyouni M, Abdel-Aziz M. Polymerization of aniline derivatives by $K_2Cr_2O_7$ and production of Cr_2O_3 nanoparticles. Polym Adv Technol 2017;28:842—8.

[17] Toptaş N, Karakışla M, Saçak M. Conductive polyaniline/polyacrylonitrile composite fibers: effect of synthesis parameters on polyaniline content and electrical surface resistivity. Polym Compos 2009;30:1618−24.

[18] He D, Zeng C, Xu C, Cheng N, Li H, Mu S, et al. Polyaniline-functionalized carbon nanotube supported platinum catalysts. Langmuir. 2011;27:5582−8.

[19] Kaempgen M, Roth S. Transparent and flexible carbon nanotube/polyaniline pH sensors. J Electroanal Chem 2006;586:72−6.

[20] Karami H, Mousavi MF, Shamsipur M. A new design for dry polyaniline rechargeable batteries. J Power Sour 2003;117:255−9.

[21] Belanger D, Ren X, Davey J, Uribe F, Gottesfeld S. Characterization and long-term performance of polyaniline-based electrochemical capacitors. J Electrochem Soc 2000;147:2923−9.

[22] Ayad MM, El-Nasr AA. Adsorption of cationic dye (methylene blue) from water using polyaniline nanotubes base. J Phys Chem C 2010;114:14377−83.

[23] Zhao Y-P, Cai Z-S, Zhou Z-Y, Fu X-L. Fabrication of conductive network formed by polyaniline−ZnO composite on fabric surfaces. Thin Solid Films 2011;519:5887−91.

[24] Zhu W, Chen X-L, Chang J, Yu R-M, Li H, Liang D, et al. Doped polyaniline-hybridized tungsten oxide nanocrystals as hole injection layers for efficient organic light-emitting diodes. J Mater Chem C 2018;6:7242−8.

[25] Wei J, Zhang X, Sheng Y, Shen J, Huang P, Guo S, et al. Simple one-step synthesis of water-soluble fluorescent carbon dots from waste paper. N J Chem 2014;38:906−9.

[26] Zhou Q, Wang Y, Xiao J, Fan H. Adsorption and removal of bisphenol A, α-naphthol and β-naphthol from aqueous solution by $Fe_3O_4@$ polyaniline core−shell nanomaterials. Synth Met 2016;212:113−22.

[27] Shahabuddin S, Sarih NM, Afzal Kamboh M, Rashidi Nodeh H, Mohamad S. Synthesis of polyaniline-coated graphene oxide@$SrTiO_3$ nanocube nanocomposites for enhanced removal of carcinogenic dyes from aqueous solution. Polymers. 2016;8:305.

[28] Yang Y, Wang W, Li M, Wang H, Zhao M, Wang C. Preparation of PANI grafted at the edge of graphene oxide sheets and its adsorption of Pb(II) and methylene blue. Polym Compos 2018;39:1663−73.

[29] Mu B, Tang J, Zhang L, Wang A. Facile fabrication of superparamagnetic graphene/polyaniline/Fe_3O_4 nanocomposites for fast magnetic separation and efficient removal of dye. Sci Rep 2017;7:5347.

[30] Kumar R, Ansari MO, Barakat MA. Adsorption of brilliant green by surfactant doped polyaniline/ MWCNTs composite: evaluation of the kinetic, thermodynamic, and isotherm. Ind Eng Chem Res 2014;53:7167−75.

[31] Saber-Samandari S, Saber-Samandari S, Joneidi-Yekta H, Mohseni M. Adsorption of anionic and cationic dyes from aqueous solution using gelatin-based magnetic nanocomposite beads comprising carboxylic acid functionalized carbon nanotube. Chem Eng J 2017;308:1133−44.

[32] Li R, Liu L, Yang F. Preparation of polyaniline/reduced graphene oxide nanocomposite and its application in adsorption of aqueous Hg(II). Chem Eng J 2013;229:460−8.

[33] Zhang S, Zeng M, Xu W, Li J, Li J, Xu J, et al. Polyaniline nanorods dotted on graphene oxide nanosheets as a novel super adsorbent for Cr(vi). Dalton Trans 2013;42:7854−8.

[34] Ansari MO, Kumar R, Ansari SA, Ansari SP, Barakat MA, Alshahrie A, et al. Anion selective pTSA doped polyaniline@graphene oxide-multiwalled carbon nanotube composite for Cr(VI) and Congo red adsorption. J Colloid Interface Sci 2017;496:407−15.

[35] Ansari MO, Kumar R, Parveen N, Barakat MA, Cho MH. Facile strategy for the synthesis of non-covalently bonded and para-toluene sulfonic acid-functionalized fibrous polyaniline@graphene−PVC nanocomposite for the removal of Congo red. N J Chem 2015;39:7004−11.

[36] Zeng Y, Zhao L, Wu W, Lu G, Xu F, Tong Y, et al. Enhanced adsorption of malachite green onto carbon nanotube/polyaniline composites. J Appl Polym Sci 2013;127:2475−82.

[37] Harijan DKL, Chandra V. Magnetite/graphene/polyaniline composite for removal of aqueous hexavalent chromium. J Appl Polym Sci 2016;133.

[38] Ramezanzadeh M, Asghari M, Ramezanzadeh B, Bahlakeh G. Fabrication of an efficient system for Zn ions removal from industrial wastewater based on graphene oxide nanosheets decorated with

highly crystalline polyaniline nanofibers (GO-PANI): experimental and ab initio quantum mechanics approaches. Chem Eng J 2018;337:385−97.

[39] Liu Y, Chen L, Li Y, Wang P, Dong Y. Synthesis of magnetic polyaniline/graphene oxide composites and their application in the efficient removal of Cu(II) from aqueous solutions. J Environ Chem Eng 2016;4:825−34.

[40] Wang J, Yin X, Tang W, Ma H. Combined adsorption and reduction of Cr(VI) from aqueous solution on polyaniline/multiwalled carbon nanotubes composite. Korean J Chem Eng 2015;32:1889−95.

[41] Shao D, Chen C, Wang X. Application of polyaniline and multiwalled carbon nanotube magnetic composites for removal of Pb (II). Chem Eng J 2012;185:144−50.

[42] Wang L-X, Li X-G, Yang Y-L. Preparation, properties and applications of polypyrroles. Reactive Funct Polym 2001;47:125−39.

[43] Shi Y, Pan L, Liu B, Wang Y, Cui Y, Bao Z, et al. Nanostructured conductive polypyrrole hydrogels as high-performance, flexible supercapacitor electrodes. J Mater Chem A 2014;2:6086−91.

[44] Huang Y, Tao J, Meng W, Zhu M, Huang Y, Fu Y, et al. Super-high rate stretchable polypyrrole-based supercapacitors with excellent cycling stability. Nano Energy 2015;11:518−25.

[45] Zhao Ce Wu J, Kjelleberg S, Loo JSC, Zhang Q. Employing a flexible and low-cost polypyrrole nanotube membrane as an anode to enhance current generation in microbial fuel cells. Small. 2015;11:3440−3.

[46] Nishio K, Fujimoto M, Ando O, Ono H, Murayama T. Characteristics of polypyrrole chemically synthesized by various oxidizing reagents. J Appl Electrochem 1996;26:425−9.

[47] Zhang X, Zhang J, Song W, Liu Z. Controllable synthesis of conducting polypyrrole nanostructures. J Phys Chem B 2006;110:1158−65.

[48] Ballav N, Maity A, Mishra SB. High efficient removal of chromium (VI) using glycine doped polypyrrole adsorbent from aqueous solution. Chem Eng J 2012;198:536−46.

[49] Li J, Feng J, Yan W. Excellent adsorption and desorption characteristics of polypyrrole/TiO$_2$ composite for Methylene Blue. Appl Surf Sci 2013;279:400−8.

[50] Nezhad AA, Alimoradi M, Ramezani M. One-step preparation of graphene oxide/polypyrrole magnetic nanocomposite and its application in the removal of methylene blue dye from aqueous solution. Mater Res Express 2018;5:025508.

[51] Wang H, Yuan X, Wu Y, Chen X, Leng L, Wang H, et al. Facile synthesis of polypyrrole decorated reduced graphene oxide−Fe$_3$O$_4$ magnetic composites and its application for the Cr (VI) removal. Chem Eng J 2015;262:597−606.

[52] Muhammad Ekramul Mahmud HN, Huq AKO, Yahya Rb. The removal of heavy metal ions from wastewater/aqueous solution using polypyrrole-based adsorbents: a review. RSC Adv 2016;6:14778−91.

[53] Bai L, Li Z, Zhang Y, Wang T, Lu R, Zhou W, et al. Synthesis of water-dispersible graphene-modified magnetic polypyrrole nanocomposite and its ability to efficiently adsorb methylene blue from aqueous solution. Chem Eng J 2015;279:757−66.

[54] Li S, Lu X, Xue Y, Lei J, Zheng T, Wang C. Fabrication of polypyrrole/graphene oxide composite nanosheets and their applications for Cr(VI) removal in aqueous solution. PLoS One 2012;7:e43328.

[55] Chauke VP, Maity A, Chetty A. High-performance towards removal of toxic hexavalent chromium from aqueous solution using graphene oxide-alpha cyclodextrin-polypyrrole nanocomposites. J Mol Liq 2015;211:71−7.

[56] Yao W, Ni T, Chen S, Li H, Lu Y. Graphene/Fe$_3$O$_4$@polypyrrole nanocomposites as a synergistic adsorbent for Cr(VI) ion removal. Compos Sci Technol 2014;99:15−22.

[57] Setshedi KZ, Bhaumik M, Onyango MS, Maity A. High-performance towards Cr(VI) removal using multi-active sites of polypyrrole−graphene oxide nanocomposites: batch and column studies. Chem Eng J 2015;262:921−31.

[58] Bhaumik M, Agarwal S, Gupta VK, Maity A. Enhanced removal of Cr (VI) from aqueous solutions using polypyrrole wrapped oxidized MWCNTs nanocomposites adsorbent. J Colloid Interface Sci 2016;470:257−67.

[59] Chandra V, Kim KS. Highly selective adsorption of Hg^{2+} by a polypyrrole−reduced graphene oxide composite. Chem Commun 2011;47:3942−4.

[60] Roncali J. Conjugated poly(thiophenes): synthesis, functionalization, and applications. Chem Rev 1992;92:711—38.

[61] Murugavel S, Malathi M. Structural, photoconductivity, and dielectric studies of polythiophene-tin oxide nanocomposites. Mater Res Bull 2016;81:93—100.

[62] Mehmood U, Al-Ahmed A, Hussein IA. Review on recent advances in polythiophene based photovoltaic devices. Renew Sustain Energy Rev 2016;57:550—61.

[63] Kaloni TP, Giesbrecht PK, Schreckenbach G, Freund MS. Polythiophene: from fundamental perspectives to applications. Chem Mater 2017;29:10248—83.

[64] Ansari MO, Khan MM, Ansari SA, Cho MH. Polythiophene nanocomposites for photodegradation applications: past, present and future. J Saudi Chem Soc 2015;19:494—504.

[65] Zhang M, Guo X, Ma W, Ade H, Hou J. A polythiophene derivative with superior properties for practical application in polymer solar cells. Adv Mater 2014;26:5880—5.

[66] Wang F, Gu H, Swager TM. Carbon nanotube/polythiophene chemiresistive sensors for chemical warfare agents. J Am Chem Soc 2008;130:5392—3.

[67] Pernites RB, Ponnapati RR, Advincula RC. Superhydrophobic—superoleophilic polythiophene films with tunable wetting and electrochromism. Adv Mater 2011;23:3207—13.

[68] Muliwa AM, Onyango MS, Maity A, Ochieng A. Batch equilibrium and kinetics of mercury removal from aqueous solutions using polythiophene/graphene oxide nanocomposite. Water Sci Technol 2017;75:2841—51.

[69] Wajima T, Sugawara K. Adsorption behaviors of mercury from aqueous solution using sulfur-impregnated adsorbent developed from coal. Fuel Process Technol 2011;92:1322—7.

[70] Mehdinia A, Khodaee N, Jabbari A. Fabrication of graphene/Fe$_3$O$_4$@polythiophene nanocomposite and its application in the magnetic solid-phase extraction of polycyclic aromatic hydrocarbons from environmental water samples. Anal Chim Acta 2015;868:1—9.

CHAPTER FOURTEEN

Carbon-based nano/micromotors for adsorption

14.1 Introduction

Over the past two decades, the emerging applications of nanotechnology have brought remarkable improvements in various applications including environmental remediation. Carbon nanomaterials (CNMs) have been extensively studied in almost all dimensional forms in several environmental remediation applications [1,2]. Graphene is a two-dimensional (2D) form of CNM and the most widely studied material of all CNMs with many applications in sensing, energy storage, catalysis, medical, environmental, photonics, and electrochemical devices [3−10]. A promising application of nanotechnology has now added a new structural form of CNMs for alternative sustainable environmental applications, that is, carbon-based nano/micromotors. A synthetic nano/micromotor is an advanced nanoscale device that can self-propel in an aqueous medium by transforming the physical and biochemical energy into movement and mechanical force [11,12]. These nano/micromotors are powered either by fuel (H_2O_2, acid, water, $NaBH_4$) [13,14] or by fuel-free sources, such as light, acoustics, electric fields, and magnetic fields [15−17]. Precise movement control, ease of fabrication, high speed, and self-mixing ability have allowed nanomotors to be adapted in various applications, including wastewater treatment [18−21]. Recently, carbon-based self-propelled nano/micromotors have gained extensive attention as alternatives to remove different pollutants from wastewater [18,22]. Fig. 14.1 represents the detection of water pollutants and removal by means of the self-propulsion mechanism of micro/nanomotors [12].

Diffusion is one of the limiting factors for enhanced removal efficiency using conventional wastewater treatment methods, which requires external stirring to accelerate the treatment method. However, in carbon nano/micromotors, the constant self-propelled movements of carbon nano/micromotors target continuous mixing in the solution without the help of external stirring, leading to the higher removal efficiency of water pollutants in a shorter time [23]. Several methods have been considered for the fabrication of nano/micromotors, such as 3D printing, sputtering, electroplating, rolled-up technology, etc. [11,23−26]. The concept of carbon-based nano/micromotors is quite young and only a few publications have been reported regarding the

Carbon Nanomaterial-Based Adsorbents for Water Purification.
DOI: https://doi.org/10.1016/B978-0-12-821959-1.00014-3

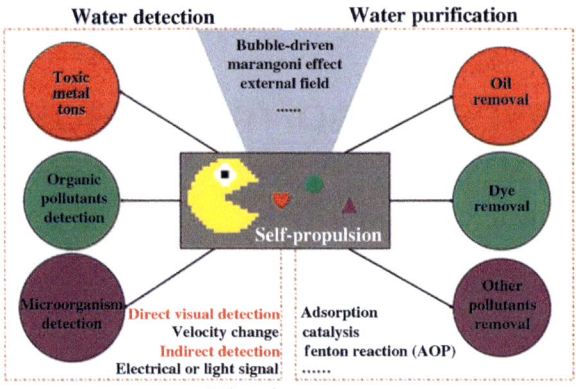

Figure 14.1 Proposed representation of detection and removal of water pollutants using self-propulsion nanomotors for water purification. *Reproduced and reprinted with permission from Ying Y, Pumera M. Micro/nanomotors for water purification. Chem A Eur J 2019;25:106—21. Copyright 2019, John Wiley and Sons Publications.*

design and synthesis of carbon–based nano/micromotors with environmental applications in wastewater treatment.

14.2 Carbon-based nano/micromotors as adsorbent

Carbon–based nano/micromotors show a higher removal efficiency of water pollutants compared with their static counterparts, as they not only exhibit the outstanding properties of nanomaterials but also display the autonomous motion capacity which enhances performance. Nanomotors can also move in a predetermined direction under the influence of an external magnetic field. Wang et al. fabricated self-propelling activated carbon–based Janus micromotors with a 60-nm-thick Pt layer on the half sphere of activated carbon particles, which can be used to efficiently remove pollutants from wastewater [23]. The activity of the bubble-propelled Janus micromotor depends on the symmetric deposition of a catalytic Pt layer on the activated carbons. The continuous motion of micromotors produced due to fast propulsion results in increased fluid dynamics, which significantly enhances the water purification efficiency. Activated carbon also exhibits a high adsorption capacity, which results in the high removal efficiency of micromotors compared with the static adsorbents. Similarly, Merkoçi et al. prepared rGO–coated silica (SiO$_2$)-Pt Janus magnetic micromotors for the efficient removal of polybrominated diphenyl ethers (PBDEs) and 5-chloro-2-(2,4-dichlorophenoxy)phenol (triclosan) from seawater samples [27]. SiO$_2$@rGO-Pt Janus micromotors consist of a SiO$_2$-rGO core—shell with a hemisphere of a catalytic Pt layer (Fig. 14.2). On introducing Janus micromotors into contaminated water, the Pt face of the motor starts self-propelling and generates O_2 bubbles via decomposition of H_2O_2, and this accelerates the adsorption of pollutants on the rGO surface in a short period. The adsorption capacity of these micromotors was found to be superior to their static and dynamic counterparts. The synergistic effects of improved interactions between adsorbate and adsorbent stimulated due to the self-propelled micromotors' movement, and superior adsorbent characteristics of rGO, led to the enhanced removal of pollutants ($\sim 90\%$) within 10 minutes.

Nanographene-based micromotors (n–rGO/Ni/Pt) were also effectively engaged in the removal of nitroaromatic explosives, such as 2,4,6-trinitrotoluene (TNT), 2,4–dinitrotoluene (DNT), and 2,4,6-trinitrophenol (TNP), and heavy metal ions via an adsorption process [28,29]. Fig. 14.3A represents the SEM images of the hollow tubular morphology of GOx microbots with an average outer and inner diameter of 4.6 ± 0.1 and $2.5 \pm 0.1 \ \mu m$, respectively [29]. These microbots were synthesized following the template–directed electrodeposition method. The surface of GOx microbots was not found to be homogeneous due to a high number of defects developed during the fabrication. In these microbots, the outer GOx layer adsorbs the pollutant, the inner Pt layer propels the motor by decomposing H_2O_2, and the ferromagnetic Ni controls the micromotors by an external magnetic field. The advantage of self-propulsion by microbots

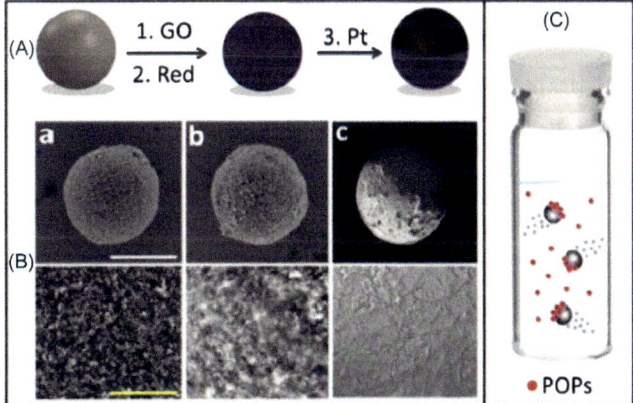

Figure 14.2 Janus micromotors fabrication and POPs removal process. (A) Sketch of Janus micromotors fabrication, which includes a coating of silica particles with GO (1) followed by the reduction (2) and 60-nm-thick coating of catalytic Pt on the hemisphere of SiO₂@GO (3). (B) SEM images of silica particles (a), SiO₂@rGO (b), and SiO₂@rGO-Pt Janus micromotors (c). Upper line-scale bars represent 4 μm, and bottom lines represent 500 nm size. (C) Sketch of the pollutant removal process using Janus micromotors, which are self-propelled in a POPs solution. *Reproduced with permission from Orozco J, Mercante LA, Pol R, Merkoçi A. Graphene-based Janus micromotors for the dynamic removal of pollutants. J Mater Chem A 2016;4:3371−8. Copyright 2016, Royal Society of Chemistry.*

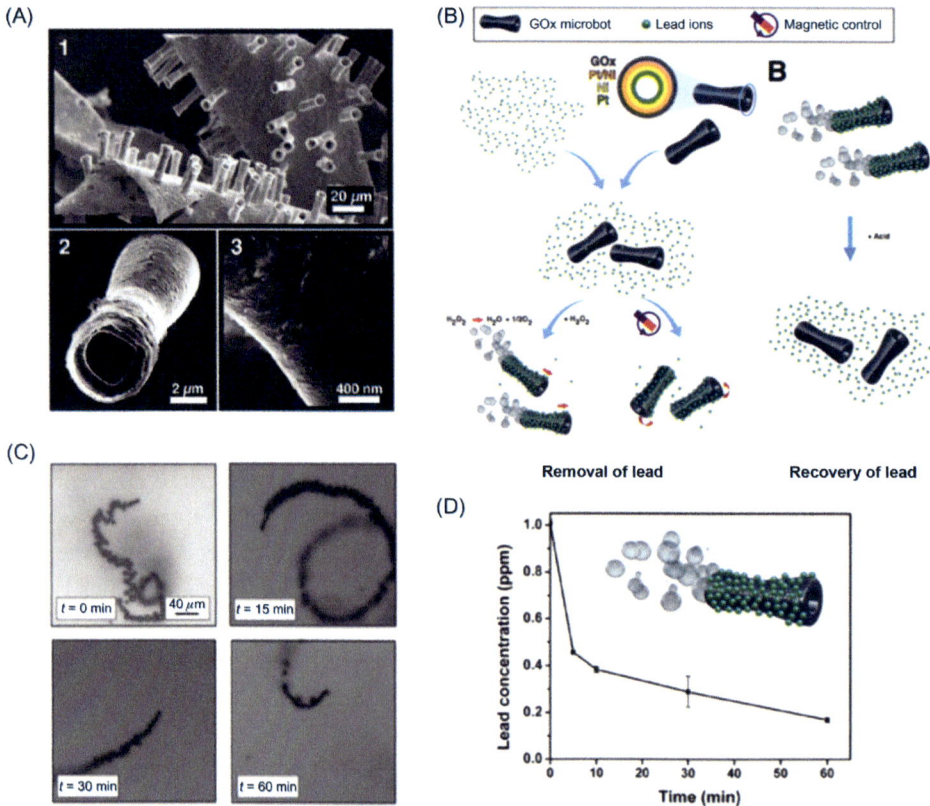

Figure 14.3 (A) Different SEM images of GOx microbots: (1) tubular GOx microbots are attached to the Au layer, (2) single tubular GOx-microbot, and (3) GOx-microbot surface. (B) Adsorption of Pb(II) ions on the GOx microbots followed by the desorption for the regeneration of GOx microbots. (C) Optical images of GOx microbots movements during Pb(II) removal at different times. (D) Removal of Pb(II) using GOx microbots as a function of time. *Reproduced and reprint with permission from Vilela D, Parmar J, Zeng Y, Zhao Y, Sánchez S. Graphene-based microbots for toxic heavy metal removal and recovery from water. Nano Lett 2016;16:2860—6. Copyright 2016, American Chemical Society Publications.*

and the magnetic characteristics of the microbots provide the two approaches for the decontamination of water by adsorbing Pb(II) (Fig. 14.3B). Further, the Pb(II)–adsorbed GOx microbots can easily be recovered in an acidic medium and reused for adsorption purposes. Fig. 14.3C represents the optical snapshots of the microbots at different time intervals, swimming in contaminated water. The Pb(II) was adsorbed efficiently (>80%) in the presence of mobile GOx microbots in 60 minutes (Fig. 14.3D). These micromotors were found to be 10 times more effective than static GOx.

Escarpa et al. introduced the multifunctional carbon nanotube ferrite—manganese dioxide tubular micromotors for the removal of organic pollutants (Remazol Brilliant

Blue R (RBB) dye and 4-chlorophenol (4-CP)) from industrial wastewater [20]. The inner MnO_2 layer in the micromotors catalytically decomposes the H_2O_2 to produce O_2 bubbles and hydroxyl radicals, which further efficiently mineralizes the water pollutants into CO_2 and H_2O. The backbone of the micromotors is incorporated with carbon and Fe_2O_3 nanoparticles, which introduces the defects and enhances the roughness of the outer surface for an increased speed and high rate of radical production for degradation. The outer layer with Fe_2O_3 also improves the recovery of the micromotor from an aqueous medium for reusability. Another multifunctional high-speed self-propulsion carbon/MnO_2 micromotor was hydrothermally synthesized and also found to be extremely effective for the removal of water pollutants [30]. Methylene blue (MB) dyes and Ag ions were successfully removed via adsorption followed by catalytic degradation using carbon/MnO_2 micromotors. These motors require a shorter time and a low concentration of H_2O_2. However, carbon-based micromotors are found to be much more efficient in the wastewater treatment process, but have still not been much explored for adsorption applications, probably due to their high cost and specific requirements for fabrication. However, self-propelled motion enhanced the activity of micromotors via improving the interactions of pollutants and the adsorbent surface, making them more effective than static adsorbents.

14.3 Conclusion

Self-propulsion of nano/micromotors represents the opportunities and challenges in many applications including wastewater treatment. Immense efforts have been put to fabricate the nanodevices that can be propelled by different mechanisms, such as bubble propulsion [31], self-electrophoresis [32], light [33], magnet [34], diffusiophoresis [35], or ultrasound [36]. The extraordinary characteristics of the synthetic nano/micromotors opened the new dimensions in environmental remediation and have shown considerable potential in wastewater treatment. This chapter illustrates the recent progress in the application of various carbon-based nano/micromotors with the accelerated speed in water purification through adsorption. The high surface area of CNMs, such as graphene and CNTs, promote the adsorption characteristics in carbon-based nano/micromotors. These micromotors enhance the diffusion rate, which improves the adsorption capacity. A few micromotors also generate hydroxyl radicals, which on adsorption of water pollutants degrade them completely, and clean the water without generating secondary pollutants. However, there are a few challenges to commercializing this product for industrial application. The main challenges are to scale up the synthesis process and reduce the cost of the micromotors. The coverage and uptake of a wide range of contaminants with high efficiency in large volumes of industrial wastewaters is another challenge. Furthermore, the attention on the evaluation of the toxicity of micromotors also requires additional efforts. The further study of the regeneration and reusability of micromotors should also be a requirement for future practical applications. Carbon-based nano/micromotors are new materials and require more attention from researchers for their thorough investigation, that is, toxicity, scaling up of the process, reusability, and economic feasibility, with the aim of practical application.

References

[1] Mauter MS, Elimelech M. Environmental applications of carbon-based nanomaterials. Environ Sci Technol 2008;42:5843–59.

[2] Gusain R, Kumar N, Fosso-Kankeu E, Ray SS. Efficient removal of Pb(II) and Cd(II) from industrial mine water by a hierarchical MoS$_2$/SH-MWCNT nanocomposite. ACS Omega 2019;4:13922–35.

[3] Chabot V, Higgins D, Yu A, Xiao X, Chen Z, Zhang J. A review of graphene and graphene oxide sponge: material synthesis and applications to energy and the environment. Energy Environ Sci 2014;7:1564–96.

[4] Choi W, Lahiri I, Seelaboyina R, Kang YS. Synthesis of graphene and its applications: a review. Crit Rev Solid State Mater Sci 2010;35:52–71.

[5] Wang T, Huang D, Yang Z, Xu S, He G, Li X, et al. A review on graphene-based gas/vapor sensors with unique properties and potential applications. Nano-Micro Lett 2016;8:95–119.

[6] Gusain R, Kumar P, Sharma OP, Jain SL, Khatri OP. Reduced graphene oxide–CuO nanocomposites for photocatalytic conversion of CO$_2$ into methanol under visible light irradiation. Appl Catal B Environ 2016;181:352–62.

[7] Bonaccorso F, Sun Z, Hasan T, Ferrari A. Graphene photonics and optoelectronics. Nat Photonics 2010;4:611.

[8] Verma S, Mungse HP, Kumar N, Choudhary S, Jain SL, Sain B, et al. Graphene oxide: an efficient and reusable carbocatalyst for aza-Michael addition of amines to activated alkenes. Chem Commun 2011;47:12673–5.

[9] Mukwevho N, Gusain R, Fosso-Kankeu E, Kumar N, Waanders F, Ray SS. Removal of naphthalene from simulated wastewater through adsorption-photodegradation by ZnO/Ag/GO nanocomposite. J Ind Eng Chem 2020;81:393–404.

[10] Ama OM, Kumar N, Adams FV, Ray SS. Efficient and cost-effective photoelectrochemical degradation of dyes in wastewater over an exfoliated graphite-MoO$_3$ nanocomposite electrode. Electrocatalysis. 2018;9:623–31.

[11] Wang H, Pumera M. Fabrication of micro/nanoscale motors. Chem Rev 2015;115:8704–35.

[12] Ying Y, Pumera M. Micro/nanomotors for water purification. Chem A Eur J 2019;25:106–21.

[13] Gao W, Uygun A, Wang J. Hydrogen-bubble-propelled zinc-based microrockets in strongly acidic media. J Am Chem Soc 2011;134:897–900.

[14] Wang Y-S, Xia H, Lv C, Wang L, Dong W-F, Feng J, et al. Self-propelled micromotors based on Au–mesoporous silica nanorods. Nanoscale. 2015;7:11951–5.

[15] Dong R, Zhang Q, Gao W, Pei A, Ren B. Highly efficient light-driven TiO$_2$–Au Janus micromotors. ACS Nano 2015;10:839–44.

[16] Ahmed S, Wang W, Bai L, Gentekos DT, Hoyos M, Mallouk TE. Density and shape effects in the acoustic propulsion of bimetallic nanorod motors. ACS Nano 2016;10:4763–9.

[17] Solovev AA, Sanchez S, Pumera M, Mei YF, Schmidt OG. Magnetic control of tubular catalytic microbots for the transport, assembly, and delivery of micro-objects. Adv Funct Mater 2010;20:2430–5.

[18] Eskandarloo H, Kierulf A, Abbaspourrad A. Nano- and micromotors for cleaning polluted waters: focused review on pollutant removal mechanisms. Nanoscale. 2017;9:13850–63.

[19] Gao W, Wang J. Synthetic micro/nanomotors in drug delivery. Nanoscale. 2014;6:10486−94.

[20] Maria-Hormigos R, Pacheco M, Jurado-Sánchez B, Escarpa A. Carbon nanotubes-ferrite-manganese dioxide micromotors for advanced oxidation processes in water treatment. Environ Sci Nano 2018;5:2993−3003.

[21] Singh VV, Wang J. Nano/micromotors for security/defense applications. A review. Nanoscale. 2015;7:19377−89.

[22] Gusain R, Kumar N, Ray SS. Recent advances in carbon nanomaterial-based adsorbents for water purification. Coord Chem Rev 2020;405:213111.

[23] Jurado-Sánchez B, Sattayasamitsathit S, Gao W, Santos L, Fedorak Y, Singh VV, et al. Self-propelled activated carbon Janus micromotors for efficient water purification. Small. 2015;11:499−506.

[24] Gao W, Sattayasamitsathit S, Orozco J, Wang J. Highly efficient catalytic microengines: template electrosynthesis of polyaniline/platinum microtubes. J Am Chem Soc 2011;133:11862−4.

[25] Mei Y, Solovev AA, Sanchez S, Schmidt OG. Rolled-up nanotech on polymers: from basic perception to self-propelled catalytic microengines. Chem Soc Rev 2011;40:2109−19.

[26] Soler L, Magdanz V, Fomin VM, Sanchez S, Schmidt OG. Self-propelled micromotors for cleaning polluted water. Acs Nano 2013;7:9611−20.

[27] Orozco J, Mercante LA, Pol R, Merkoçi A. Graphene-based Janus micromotors for the dynamic removal of pollutants. J Mater Chem A 2016;4:3371−8.

[28] Khezri B, Mousavi SMB, Sofer Z, Pumera M. Recyclable nanographene-based micromachines for the on-the-fly capture of nitroaromatic explosives. Nanoscale. 2019;11:8825−34.

[29] Vilela D, Parmar J, Zeng Y, Zhao Y, Sánchez S. Graphene-based microbots for toxic heavy metal removal and recovery from water. Nano Lett 2016;16:2860−6.

[30] He X, Büchel R, Figi R, Zhang Y, Bahk Y, Ma J, et al. High-performance carbon/MnO_2 micromotors and their applications for pollutant removal. Chemosphere. 2019;219:427−35.

[31] Sanchez S, Solovev AA, Mei Y, Schmidt OG. Dynamics of biocatalytic microengines mediated by variable friction control. J Am Chem Soc 2010;132:13144−5.

[32] Paxton WF, Kistler KC, Olmeda CC, Sen St. A, Angelo SK, Cao Y, et al. Catalytic nanomotors: autonomous movement of striped nanorods. J Am Chem Soc 2004;126:13424−31.

[33] Ibele M, Mallouk TE, Sen A. Schooling behavior of light-powered autonomous micromotors in water. Angew Chem Int Ed 2009;48:3308−12.

[34] Ghosh A, Fischer P. Controlled propulsion of artificial magnetic nanostructured propellers. Nano Lett 2009;9:2243−5.

[35] Gao W, Pei A, Feng X, Hennessy C, Wang J. Organized self-assembly of Janus micromotors with hydrophobic hemispheres. J Am Chem Soc 2013;135:998−1001.

[36] Wang W, Castro LA, Hoyos M, Mallouk TE. Autonomous motion of metallic microrods propelled by ultrasound. ACS Nano 2012;6:6122−32.

CHAPTER FIFTEEN

Regeneration and recyclability of carbon nanomaterials after adsorption

15.1 Introduction

The contamination of water resources, due to natural and anthropogenic activities, is a significant problem and an alarming issue, which is deteriorating water quality. The contaminants can be organic (dyes, pharmaceutical ingredients, pesticides, polycyclic aromatic hydrocarbons) and/or inorganic (heavy metal ions, radioactive substances). The ingestion of these toxic substances by consuming polluted water is causing severe damage to human body organs, such as the central nervous system, reproductive system, respiratory system, kidney, liver, lungs, and so on [1–5]. For example, mercury (Hg) is a highly toxic metal, which on consumption through Hg-contaminated water can be a neurotoxin, permanently damage lungs and kidney function, and cause chest pain and dyspnea [1]. Another heavy metal ion, Pb(II) is most commonly used in various industries and enters into water sources through industrial effluents. It is one of the essential trace element in the human body, but an excessive amount of Pb(II) can cause acute and chronic poisoning. It severely affects the nervous system, reproductive system, neurological system, lungs, liver, kidney, and other vital organs and also it can be carcinogenic [5–7]. Similarly, organic dyes are commonly used in textile, paper, and pulp industries to color the material, but the discharge from these industries containing those dyes is released into water sources which is fatal for the ecosystem [8–10].

Therefore to remove such contaminants from the polluted water or industrial discharges, several physical, chemical, and biological technologies, such as flocculation, coagulation, reverse osmosis, filtration, photocatalysis, photo-fenton, and adsorption, have been utilized [11–18]. Adsorption is one of the most popular, safe, and easy to handle technologies [13]. This process requires an active high surface area adsorbent, which on being introduced into the polluted water can adsorb the contaminants leaving behind clean water. However, the efficiency of the adsorption process in wastewater treatment depends on the high adsorption capacity and regeneration and reusability of the adsorbent material. Regeneration and recyclability of the adsorbent material are important parts of promoting adsorption on the industrial scale and enhance the efficacy of the adsorption process from an economic point of view. The main aim of the

Carbon Nanomaterial-Based Adsorbents for Water Purification.
DOI: https://doi.org/10.1016/B978-0-12-821959-1.00015-5

regeneration process is the removal of adsorbed material from the adsorbent without any cost or change to the initial physical, chemical, and morphological characteristics of the adsorbent material. The adsorbent material after the adsorption process can be collected via cross-flow filtration, field-flow fractionation, centrifugation, filtration, magnetic and electric field methods [19−21]. After that, desorption, collection, and decomposition of the contaminant and then the regeneration and the degree of reusability of the adsorbent material are required in order to make the adsorption process more environment-friendly and economical [22]. This chapter is primarily dedicated to the several techniques for the regeneration of the adsorbent material and its reusability to remove contaminants from the wastewater over several cycles.

15.2 Chemical regeneration method

The chemical regeneration technique of the adsorbent material is most commonly used and is a cost-effective technology to regenerate the adsorbent [23]. In this method after the adsorption process, the adsorbent material with adsorbed species is introduced into the solvent or chemicals and regenerated via either by the decomposition of the adsorbed species or desorption of all the adsorbed material from the surface of the adsorbent. Chemical regeneration of adsorbent can be achieved via the degradation of adsorbed material by oxidation, by the use of acid/alkali, and complex formation. During the chemical regeneration method, it should be kept in the mind that the adsorption capacity of the adsorbent should not be compromised via the attack of chemicals during regeneration.

15.2.1 Chemical regeneration via degradation

Chemical regeneration of spent adsorbent material is limited to the specific adsorbed species which can be oxidized. This process involves several advanced oxidation methods such as Fenton oxidation, ozonation, and permanganate oxidation. In the advanced oxidation process (AOP), a necessary step is the formation of the reactive intermediate radical (e.g., hydroxyl radical), which is a strong oxidizing species and can easily degrade the organic adsorbed species. Hydroxyl radical is one of the popular intermediate oxidizing agents, with a high oxidation potential (2.8 eV). It attacks the organic molecules either by extracting the hydrogen atom or introducing a double bond. Codony et al. employed AOP for the regeneration of the activated carbon adsorbent using H_2O_2 and O_3 [24]. Activated carbon was used for the removal of octamethylcyclotetrasiloxane and could recover successfully up to 40% of the original adsorption capacity. Inse et al. also regenerated the granular activated carbon (GAC) using a UV/H_2O_2 AOP method after the adsorption of phenol [25]. Phenol removal from wastewater was observed by total organic carbon (TOC) removal in the solution. Several parameters were also observed during the regeneration of activated carbon. After the adsorption of phenol, the activated carbon was collected and the organic matter on the surface of activated carbon surface was removed by destructive oxidation on exposure of UV/H_2O_2 under specific conditions, and the regenerated activated carbon was again for adsorption. Fig. 15.1A shows the TOC removal capacity of activated carbon for several cycles of regeneration using UV/H_2O_2 and without the regeneration process. The regenerated GAC exhibited higher TOC removal capability.

Ozone can also be used to degrade the organic contaminants adsorbed on the adsorbent and help in the regeneration of adsorbent material. Ozone is also one of the strongest oxidants with a high oxidation potential (2.058 V). It is widely used for the oxidation of organic contaminants. Therefore the adsorption of organic matter from wastewater on the adsorbent surface is the first step and the oxidation of those organic contaminants through ozonation is the second step. It can be performed either by direct oxidation of organic matter in the presence of molecular ozone or by using hydroxyl radicals generated due to the decomposition of ozone in alkaline conditions. An FAU type zeolite was used for the efficient removal of 2,4,6-trichlorophenol (TCP) from wastewater and was again regenerated via ozonation to reuse for the adsorption of TCP [26]. After the adsorption of TCP on FAU zeolite, the O_3 was bubbled in the solution for 30 minutes so that the TCP adsorbed on the surface of the adsorbent and in the

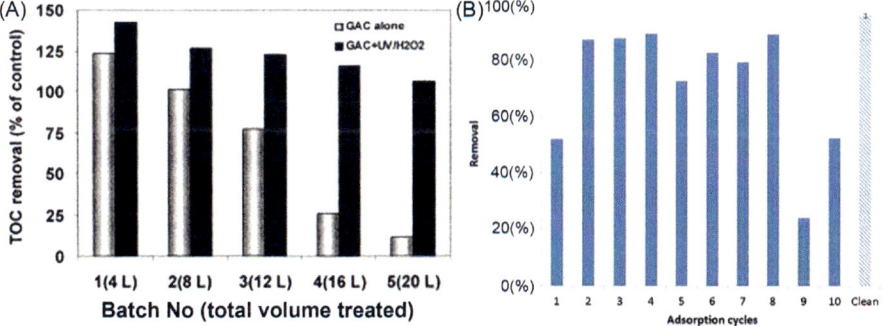

Figure 15.1 (A) Percentage of TOC removal using GAC and GAC + UV/H_2O_2 upon use for five successive cycles. (B) Adsorption/desorption cycles for TCP removal from the aqueous cycle for 10 cycles using Faujasite (FAU) zeolite. *(A) Reproduced with permission from Ince NH, Apikyan IG. Combination of activated carbon adsorption with light-enhanced chemical oxidation via hydrogen peroxide. Water Res 2000;34:4169—76 [25]. Copyright 2000, Elsevier Science Ltd. (B) Reproduced with permission from Zhang Y, Mancke RG, Sabelfeld M, Geißen S-U. Adsorption of trichlorophenol on zeolite and adsorbent regeneration with ozone. J Hazard Mater 2014;271:178—84 [26]. Copyright 2014 Elsevier Science Ltd.*

water was oxidized. After the first regeneration process, the adsorption of TCP was increased remarkably (Fig. 15.1B) and could be reused for several cycles.

15.2.2 Chemical degradation via acid/alkali treatment

After the adsorption process, desorption of contaminants can also be done by using acid or alkali solvents by changing the pH of the solution. The change in the pH also alters the charge on the adsorbate species or adsorbent matter. The alteration of the pH of the solution replaced the adsorbed pollutants via the exchange of cations or anions present in the acid/alkali solution. Therefore this regeneration process is also termed as ion-exchange regeneration of adsorbent material. Optimization and controlled pH of the solution is most important for the efficient adsorption and desorption of the contaminants. Huang et al. used dithiocarbamate (DTC)-functionalized magnetic reduced graphene oxide (rGO-DTC/Fe_3O_4) as an adsorbent material to remove heavy metal ions [Pb(II), Cu(II), Cd(II), and Hg(II)] from wastewater [27]. The adsorption process was followed by the pseudo-second-order kinetic model, indicating the chemisorption through ion-exchange of ions on the rGO-DTC/Fe_3O_4 surface. The adsorption capacity of rGO-DTC/Fe_3O_4 was also observed to be high for the heavy metal ion adsorption. For the regeneration of rGO-DTC/Fe_3O_4 after the adsorption process, the adsorbent was treated in 0.1 M HCl solution (pH = 1) to remove all the adsorbed heavy metal ions. During this exercise, the DTC group on the adsorbent material decomposed to the polyethyleneimine (PEI) group and provided rGO-PEI/Fe_3O_4 along with CS_2 molecules in the acidic solution. To regenerate rGO-DTC/Fe_3O_4 from rGO-PEI/Fe_3O_4 was successfully achieved by a one-step synthesis procedure using a 5 mol/L NaOH in the same solution for 20 hours at 40°C and this was further used for the removal of heavy metal ions from wastewater. Fig. 15.2A shows the reusability of rGO-DTC/Fe_3O_4 for five consecutive cycles. The little reduction in adsorption capacity might be due to the loss of a few active binding sites on the adsorbent surface due to the desorption and regeneration process. Alternatively, 1.0 mol/L NaOH solution was used as the eluent to remove the As(III/V) from the surface of chitosan−GO adsorbent to regenerate the adsorbent [28]. Arsenic desorption from the adsorbent surface is mostly performed in an alkaline medium [30−32]. The adsorption efficiency of chitosan−GO was intact up to >99% for three consecutive adsorption−desorption cycles (Fig. 15.2B) [28]. The gradual reduction in the adsorption efficiency of adsorbent after three cycles might be due to the constant use of the acidic and alkaline medium for the adsorption, and the desorption of As(III/V) might have reduced the performance of the adsorbent. The high pH of the solution weakens the electrostatic interaction between the chitosan−GO and As(III/V) and it undergoes deprotonation to regenerate the chitosan−GO. Also, the alkaline medium disturbs the equilibrium

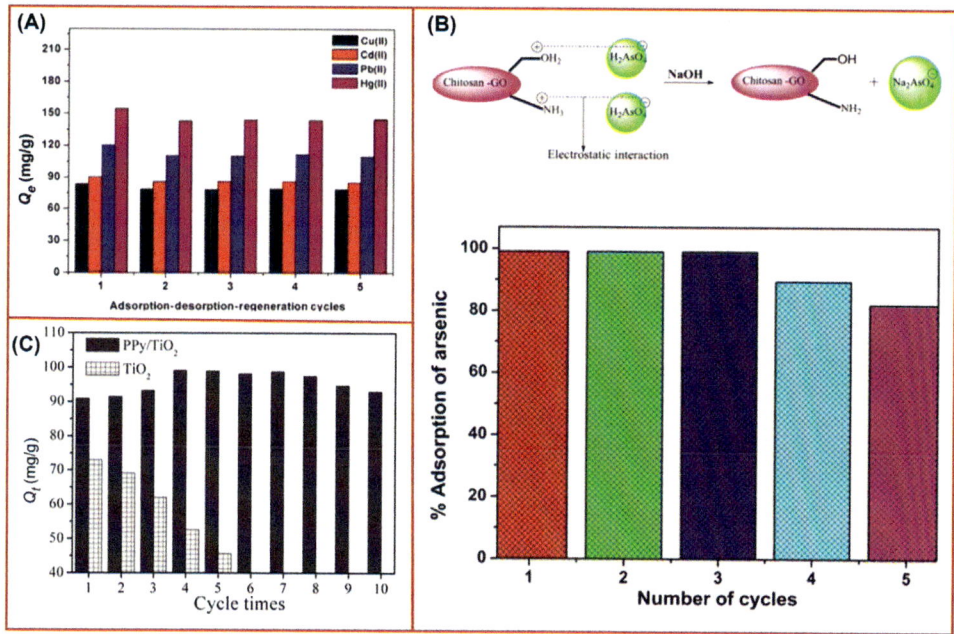

Figure 15.2 (A) Regeneration and recyclability studies of rGO-DTC/Fe$_3$O$_4$ adsorbent for five consecutive cycles to remove heavy metal ions (Pb(II), Cu(II), Cd(II) and Hg(II)) from wastewater. (B) Schematic mechanism of desorption of As(III/V) and regeneration of adsorbent (chitosan−GO) in alkaline medium and the recyclability of chitosan−GO in terms of As removal for five cycles. (C) Recyclability studies of polypyrrole-modified TiO$_2$ (PPy/TiO$_2$) and TiO$_2$ for 10 cycles to remove acid Red G dye from the aqueous solution. *(A) Reproduced with permission from Fu W, Huang Z. Magnetic dithiocarbamate functionalized reduced graphene oxide for the removal of Cu(II), Cd(II), Pb(II), and Hg(II) ions from aqueous solution: synthesis, adsorption, and regeneration. Chemosphere 2018;209:449−56 [27]. Copyright 2018, Elsevier Science Ltd. (B) Reproduced with permission from Kumar ASK, Jiang S-J. Chitosan-functionalized graphene oxide: a novel adsorbent an efficient adsorption of arsenic from aqueous solution. J Environ Chem Eng 2016;4:1698−713 [28]. Copyright 2016, Elsevier Science Ltd. (C) Reproduced with permission from Li J, Feng J, Yan W. Synthesis of polypyrrole-modified TiO$_2$ composite adsorbent and its adsorption performance on acid Red G. J Appl Polym Sci 2013;128:3231−9 [29]. Copyright 2012, John Wiley and Sons Publications.*

phase of adsorption and the dominant monovalent ($H_2AsO_4^-$) and divalent ($HAsO_4^{2-}$) anions could be removed quantitatively as sodium salts in the alkaline medium (Fig. 15.2B).

Yan et al. compared the adsorption potential of PPy/TiO$_2$ and TiO$_2$ to remove the acid Red G dye from an aqueous medium and the PPy/TiO$_2$ performed better than the TiO$_2$ [29]. To regenerate the adsorbent after the adsorption process, it was treated with 0.1 M NaOH solution, which acted as a desorption agent and 0.1 M HNO$_3$ solution, which acted as an activation agent. The regenerated adsorbent was reused for 10 cycles following adsorption−desorption activity. Fig. 15.2C shows that the adsorption capacity of PPy/TiO$_2$ was found to be more than 90% for 10 cycles, which indicates the excellent stability and adsorption efficiency of the adsorbent. However, TiO$_2$ followed a regular decreasing trend for dye removal indicating the poor stability and adsorption capacity of TiO$_2$. These results also revealed that the functionalization of polypyrrole) on the TiO$_2$ significantly improves the adsorption capacity due to the presence of functional groups and also enhances the stability of TiO$_2$. The rising trend of dye adsorption for PPy/TiO$_2$ for another cycle might be due to the activation in an acidic medium, which lifts the positive charge on the regenerated adsorbent and improves the dye adsorption qualities.

15.3 Microbial regeneration method

Microbial regeneration of the adsorbent material involves the treatment of the contaminant-loaded adsorbent material by microbial activities. It can be achieved by the biodegradation of the adsorbed material using microbe colonies [33]. Generally, two pathways are established for the microbial regeneration of adsorbents: (1) inoculation of microorganisms on the adsorbent before adsorption and (2) treatment of pollutant-adsorbed adsorbent in a microorganism environment after adsorption. Several reports have been published on the bioregeneration or microbial regeneration of adsorbent material [34−36]. The degree of microbial regeneration of the adsorbent depends on various factors, such as the biodegradability of contaminant, adsorbate−adsorbent interaction strength, physicochemical characteristics of the adsorbent, and the process configuration [35]. Andel et al. used powdered activated carbon (PAC) for the removal of aromatic compounds from aqueous media and after the adsorption process again regenerated the PAC by bioregeneration [37]. The microbial regeneration of nonphenolic (3-chlorobenzoic acid)-loaded PAC was readily achieved using *Pseudomonas* B13 bacteria, whereas it did not work efficiently for the phenolic o-cresol-loaded PAC. This might be because of the effect of the activity of a particular microbe depends on the structure of the degradable material. The nature of chemical substitution and the presence of functional groups also significantly affects the rate of bioregeneration. For example, on comparing the phenol-loaded PAC, *p*-methylphenol-loaded PAC, *p*-ethylphenol-loaded PAC, and *p*-isopropyl phenol-loaded PAC, the bioregeneration of phenol-loaded PAC was achieved efficiently in a laboratory-scale sequencing batch reactor [38]. Increasing the side alkyl chain on the adsorbate (phenol) molecule reduces the bioregeneration efficiency of the adsorbent [38−40]. This might be due to the fact that alkylated phenolic molecule is adsorbed on the activated carbon more strongly than the phenol, which restricts the reversible desorption for the regeneration of adsorbent, and the rate of regeneration of the adsorbent depends on the degree of desorption [39].

Alternatively, aromatic adsorbates with electron-attracting groups allowed reversible desorption from the adsorbent surface and readily biodegrade in the presence of microbes to regenerate the adsorbents compared with the aromatic adsorbates with electron-donating substitutes [35,41]. Therefore microbial regeneration is preferred for the regeneration of aromatic contaminants with electron-attracting groups-loaded adsorbent.

15.4 Ultrasound regeneration method

The ultrasonic regeneration technique involves the desorption of adsorbed material from the surface with the aid of ultrasonic waves. This method has garnered much attention due to its advantages such as the removal of contaminants from the surface, decomposition of organic contaminants, and energy efficiency [42]. However, during this process, it should always be kept in mind that the intensity of ultrasonic waves should not be too high and cross the threshold value to damage the adsorbent as well [43]. The diffusion–convection system in ultrasound regeneration helps in the desorption of adsorbate and then decomposition. The cavitation involves the nucleation and growth of tiny bubbles which collapse driven by the ultrasonic waves and this process helps in the desorption of adsorbed material loaded on the adsorbent. The key to the desorption of adsorbate from the adsorbent under ultrasonic treatment is always a topic for debate. A few researchers suggest that the desorption during ultrasonic treatment is only because of thermal effects [44], whereas according to others it might be due to mass transfer by cavitation and acoustic vortex microstreaming and spot energy effects [45–48].

Derakhshani et al. explored the effect of pH and ultrasonic irradiation time on the ultrasonic regeneration of graphene nanoparticles saturated with humic acid along with the effect of a number of regeneration cycles on the activity of graphene nanoparticles [42]. A regeneration study was performed at two frequencies, that is, 37 and 60 kHz, respectively. Regeneration of graphene was performed more efficiently at the higher (60 kHz) frequency. Similar observations were also reported in other literature [45,46]. The enhanced regeneration activity at high frequency might be due to the enhancement of surface diffusivity [46]. Also, increases in pH value enhance the regeneration of graphene [42]. On increasing the pH, the humic acid is easily desorbed due to the increase in charge repulsion between the graphene and humic acid. Results also show that the regeneration of humic acid-saturated graphene increased with the increase in ultrasonication time. Fachinger et al. successfully desorbed *p*-chlorophenol from the GAC under ultrasonic treatment at different frequencies and temperature [49]. The higher the intensity of ultrasonic waves, the higher were the rates of desorption of p-chlorophenol from the surface of GAC. Additionally, the increase in temperature also improved the desorption rates.

15.5 Thermal regeneration method

The thermal regeneration technique is energy driven and requires high-temperature treatment to regenerate the toxic contaminant-loaded adsorbent. This technique is quite popular for the regeneration of the activated carbon at a large scale in industries [50−52]. Before the thermal regeneration treatment, the study of the thermal stability of the adsorbent is necessary. It cannot be used for the adsorbents with low thermal stability. It is widely accepted for the regeneration of carbonaceous adsorbents. Thermal regeneration of adsorbent involves a three-step reaction: (1) drying of adsorbent and desorption only of adsorbed matter in its original form ($\sim 105°C$); (2) desorption and subsequently decomposition of a part of the adsorbed material under inert atmosphere, leaving behind a charred residue ($500°C−900°C$); and (3) a chemical reaction to remove the charred residue to regenerate the adsorbent with enhanced activity [53]. The chemical reaction generally happens in the presence of mild gaseous oxidants such as steam and CO_2 at high temperatures [54]. High-temperature events in thermal regeneration provide sufficient energy to break the adsorbate−adsorbent interactions, which helps in desorption. However, the main disadvantages of this process are the loss of 5%−10% of carbon during thermal treatment via oxidation and also the consumption of a high amount of energy.

Wei et al. studied the regeneration of diclofenac sodium (DS) and carbamazepine (CBZ)-adsorbed granular carbon nanotubes/alumina ($CNTs/Al_2O_3$) hybrid adsorbents at $400°C$ under air atmosphere [55]. The DS and CBZ adsorbed on the surface of $CNTs/Al_2O_3$ were decomposed during the regeneration process of the adsorbent. High thermal stability of the CNTs allowed the thermal regeneration of the adsorbent. Fig. 15.3 depicts the removal percentage of DS and CBZ using $CNTs/Al_2O_3$ for the 10 cycles. The trend of removal percentage of DS was noted to have slightly increased after the first cycle and then was almost constant for the other consecutive cycles (Fig. 15.3A), whereas an opposite trend was observed for the removal of CBZ (Fig. 15.3B). After the regeneration processes, the specific surface area of the $CNTs/Al_2O_3$ was reduced, whereas the average pore size and volume had increased, which is due to the loss of some unstable CNTs and Al_2O_3. Therefore the larger molecular size of DS (1.1 nm) prefers adsorption on regenerated $CNTs/Al_2O_3$ in comparison to

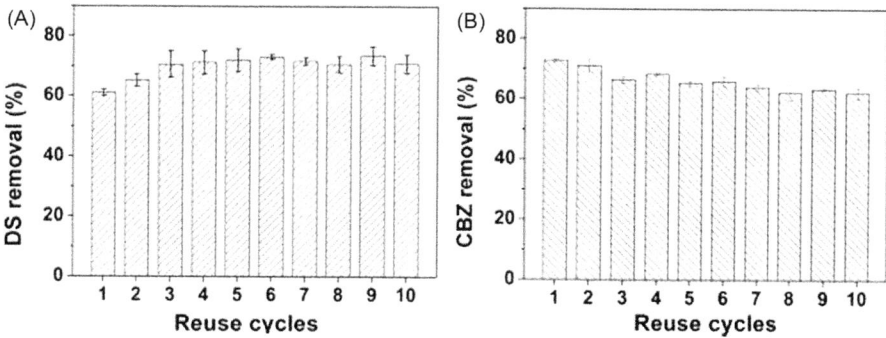

Figure 15.3 Adsorption of DS (A) and CBZ (B) on the thermally regenerated CNTs/Al$_2$O$_3$ for consecutive 10 cycles. *Reproduced with permission from Wei H, Deng S, Huang Q, Nie Y, Wang B, Huang J, et al. Regenerable granular carbon nanotubes/alumina hybrid adsorbents for diclofenac sodium and carbamazepine removal from aqueous solution. Water Res 2013;47:4139—47 [55]. Copyright 2013, Elsevier Science Ltd.*

CBZ (0.9 nm) adsorption. A little reduction in the weight of the adsorbent material was also measured after the regeneration process, which illustrates that some of the unstable CNTs and Al$_2$O$_3$ were washed out.

Several other reports have been published on the potential thermal regeneration of pollutant-loaded adsorbents [56—58]. However, thermal regeneration is an energy-demanding process, increasing the global carbon footprint and also is limited to the high-thermally stable adsorbents.

15.6 Electrochemical regeneration method

Electrochemical regeneration methods involve desorption and decomposition of adsorbed organic material loaded on the surface of an adsorbent in an electrochemical cell. Owan and Barry were the first to report the electrochemical regeneration of the activated carbon for the reuse purposes [59]. Since then, several articles have been published on the electrochemical regeneration of the adsorbents, especially carbonaceous adsorbents with regeneration efficiencies up to 95% [60—63]. Regeneration of carbonaceous adsorbents can be performed at the cathode and anode [64,65]. However, the oxidation of contaminant adsorbed carbonaceous material is efficient and attractive at the anode due to the strong oxidizing behavior of the anode [66].

Sharif et al. prepared the reduced graphene oxide/magnetite (rGO-IO) nanocomposite for the adsorption of methylene blue (MB) dye and further regenerated the rGO-IO electrochemically using anodic oxidation for recycling purposes (Fig. 15.4) [66]. The adsorbent was recovered after the adsorption process using an external

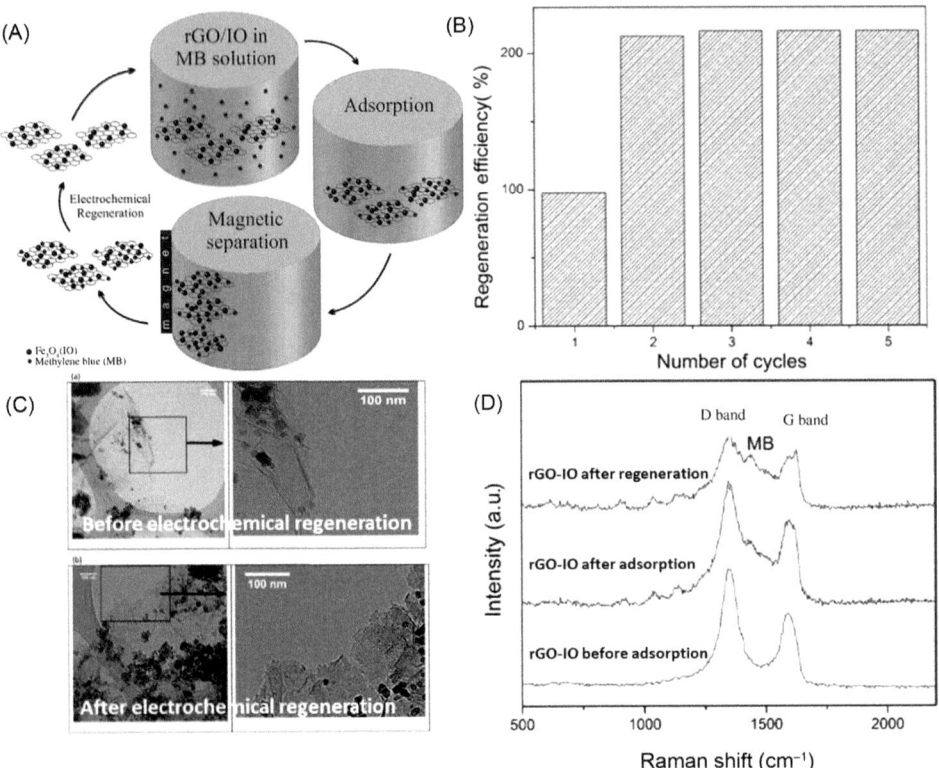

Figure 15.4 (A) Schematic representation of adsorption of MB from the aqueous solution using rGO-IO, magnetic separation of rGO-IO, electrochemical regeneration, and recycling of rGO-IO; (B) Regeneration efficiency of rGO-IO after electrochemical regeneration for five cycles. (C) TEM images of the rGO-IO before and after regeneration and (D) Raman spectra of rGO-IO before adsorption, after adsorption, and after regeneration. *Reproduced and reprint from Sharif F, Gagnon LR, Mulmi S, Roberts EPL. Electrochemical regeneration of a reduced graphene oxide/magnetite composite adsorbent loaded with methylene blue. Water Res 2017;114:237−45 [66]. Copyright 2017, Elsevier Science Ltd.*

magnetic field (Fig. 15.4A). The effect of regeneration time was also studied following the regeneration efficiency at constant current density. With the increase in the regeneration time, the regeneration efficiency was also found to be enhanced. After the regeneration of the rGO-IO, the adsorption efficiency was found to be significantly high in the second and other consecutive cycles (Fig. 15.4B). During anodic oxidation of MB in the regeneration process, the oxidation of rGO also occurred, which resulted in the generation of more adsorptive sites. TEM images of the adsorbent before and after the electrochemical regeneration are shown in Fig. 15.4C. rGO-IO exhibits soft edges before regeneration and after electrochemical regeneration the edges become ragged, which indicates the oxidation/corrosion of the rGO.

Similarly, the Raman spectra of rGO-IO before adsorption, after adsorption, and after regeneration are represented in Fig. 15.4D. The additional peak in rGO-IO after MB adsorption, except for the D and G band, at $1440\,cm^{-1}$ is due to the adsorbed MB. This peak could also be seen in the Raman spectrum of rGO-IO after regeneration, indicating that some of MB or its derivatives are still present on the surface of the adsorbent.

15.7 Photo-assisted regeneration method

The photo-assisted regeneration method involves the photocatalytic and photosensitized oxidation of adsorbed species on the adsorbent. In this method, the degradation of organic contaminants occurs due to the generation of free radicals without the use of chemicals [67,68]. In this regeneration technique, a photocatalyst or photosensitizer is introduced into the suspension containing the organic contaminant-loaded adsorbent. The organic contaminants are first desorbed from the surface and start degrading under light illumination. Zhao et al. illustrated that the photosensitized degradation of 2,4,6-trichlorophenol (TCP) could be efficiently performed over palladium(II) phthalocyaninesulfonate (PdPcS)-modified organic clay under UV light irradiation [69].

Zhang et al. have modified the ethylenediaminetetraacetic dianhydride (EDTAD) with sugarcane bagasse (SB) and used it for the adsorption of MB [70]. Due to the presence of a large number of carboxyl groups, the adsorption efficiency of EDTAD-SB was found to be higher than SB for MB adsorption. The photo-assisted regeneration of spent EDTAD-SB was performed using TiO_2 under UV light. During regeneration the blue color of EDTAD-SB becomes slowly colorless due to the degradation of adsorbed MB. The regenerated EDTAD-SB exhibits 85% adsorption efficiency for MB adsorption from aqueous solution for subsequent adsorption cycles. The reduced efficiency of EDTAD-SB was due to the blocking of a few active sites of the EDTAD-SB with TiO_2 during the regeneration process [71]. Therefore the photo-assisted regeneration process involves the degradation of the adsorbed organic contaminant in the presence of light.

15.8 Conclusion

Regeneration of the spent adsorbent is an important aspect of the economy as well as being important from an environmental point of view. After the adsorption exercise, the disposal of the adsorbent is one of the rising problems. Adsorbate-loaded adsorbents give rise to secondary pollutants and the disposal of such pollutants is creating other kinds of pollution, such as the dumping of secondary pollutants inland causing land pollution. Therefore the environment-friendly technique to get rid of the secondary pollutant in the budget is the need of the hour. The regeneration of adsorbents without the cost of losing active sites can reduce the requirements of preparing new adsorbents and give a solution to reduce the stress of the disposal of used adsorbents. Various regeneration techniques, such as ultrasound, biological (microbial), chemical, electrochemical, and photochemical, have been employed and have achieved a degree of success in regenerating the adsorbent materials. This process involves desorption or complete degradation of the toxic adsorbent molecules from the adsorbent, which can further be used to clean the wastewater through adsorption. This leads to the recycling of the adsorbent material for several cycles with slightly enhanced or lowered adsorption efficiency. Regeneration and recycling of the adsorbent are economical approaches that make the adsorption process more viable, efficient, and cost-effective.

References

[1] Carocci A, Rovito N, Sinicropi MS, Genchi G. Mercury toxicity and neurodegenerative effects. Reviews of environmental contamination and toxicology. Springer; 2014. p. 1–18.
[2] Cvetnic M, Perisic DJ, Kovacic M, Ukic S, Bolanca T, Rasulev B, et al. Toxicity of aromatic pollutants and photooxidative intermediates in water: a QSAR study. Ecotoxicol Environ Saf 2019;169:918–27.
[3] Saxena G, Chandra R, Bharagava RN. Environmental pollution, toxicity profile and treatment approaches for tannery wastewater and its chemical pollutants. Reviews of environmental contamination and toxicology, vol. 240. Springer; 2016. p. 31–69.
[4] Markandeya S, Shukla P, Mohan D. Toxicity of disperse dyes and its removal from wastewater using various adsorbents: a review. Res J Environ Toxicol 2017;11:72–89.
[5] Gusain R, Kumar N, Ray SS. Recent advances in carbon nanomaterial-based adsorbents for water purification. Coord Chem Rev 2020;405:213111.
[6] Gusain R, Kumar N, Fosso-Kankeu E, Ray SS. Efficient removal of Pb(II) and Cd(II) from industrial mine water by a hierarchical MoS2/SH-MWCNT nanocomposite. ACS Omega 2019;4:13922–35.
[7] Kumar N, Fosso-Kankeu E, Ray SS. Achieving controllable MoS2 nanostructures with increased interlayer spacing for efficient removal of Pb(II) from aquatic systems. ACS Appl Mater Interfaces 2019;11:19141–55.
[8] Padhi B. Pollution due to synthetic dyes toxicity & carcinogenicity studies and remediation. Int J Environ Sci 2012;3:940–55.
[9] Ama OM, Kumar N, Adams FV, Ray SS. Efficient and cost-effective photoelectrochemical degradation of dyes in wastewater over an exfoliated graphite-MoO3 nanocomposite electrode. Electrocatalysis 2018;9:623–31.
[10] Kumar N, Reddy L, Parashar V, Ngila JC. Controlled synthesis of microsheets of ZnAl layered double hydroxides hexagonal nanoplates for efficient removal of Cr(VI) ions and anionic dye from water. J Environ Chem Eng 2017;5:1718–31.
[11] Kumar N, Mittal H, Parashar V, Ray SS, Ngila JC. Efficient removal of rhodamine 6G dye from aqueous solution using nickel sulphide incorporated polyacrylamide grafted gum karaya bionanocomposite hydrogel. RSC Adv 2016;6:21929–39.

[12] Kumar N, Mittal H, Reddy L, Nair P, Ngila JC, Parashar V. Morphogenesis of ZnO nanostructures: role of acetate (COOH−) and nitrate (NO3−) ligand donors from zinc salt precursors in synthesis and morphology dependent photocatalytic properties. RSC Adv 2015;5:38801−9.

[13] Kumar N, Mittal H, Alhassan SM, Ray SS. Bionanocomposite hydrogel for the adsorption of dye and reusability of generated waste for the photodegradation of ciprofloxacin: a demonstration of the circularity concept for water purification. ACS Sustain Chem Eng 2018;6:17011−25.

[14] Kumar N, Sinha Ray S, Ngila JC. Ionic liquid-assisted synthesis of Ag/Ag2Te nanocrystals via a hydrothermal route for enhanced photocatalytic performance. N J Chem 2017;41:14618−26.

[15] Umukoro EH, Kumar N, Ngila JC, Arotiba OA. Expanded graphite supported p-n MoS2-SnO2 heterojunction nanocomposite electrode for enhanced photo-electrocatalytic degradation of a pharmaceutical pollutant. J Electroanal Chem 2018;827:193−203.

[16] Mukwevho N, Kumar N, Fosso-Kankeu E, Waanders F, Bunt JR, Sinha Ray S. Visible light-excitable ZnO/2D graphitic-C3N4 heterostructure for the photodegradation of naphthalene. Desalin water Treat 2019;163:286−96.

[17] Mukwevho N, Gusain R, Fosso-Kankeu E, Kumar N, Waanders F, Ray SS. Removal of naphthalene from simulated wastewater through adsorption-photodegradation by ZnO/Ag/GO nanocomposite. J Ind Eng Chem 2020;81:393−404.

[18] Ntakadzeni M, Anku WW, Kumar N, Govender PP, Reddy L. PEGylated MoS$_2$ nanosheets: a dual functional photocatalyst for photodegradation of organic dyes and photoreduction of chromium from aqueous solution. Bull Chem React Eng Catal 2019;14:142−152.

[19] Ali I, Alharbi OM, Tkachev A, Galunin E, Burakov A, Grachev VA. Water treatment by new-generation graphene materials: hope for bright future. Environ Sci Pollut Res 2018;25:7315−29.

[20] Kim S, Marion M, Jeong B-H, Hoek EM. Crossflow membrane filtration of interacting nanoparticle suspensions. J Membr Sci 2006;284:361−72.

[21] Moeser GD, Roach KA, Green WH, Alan Hatton T, Laibinis PE. High-gradient magnetic separation of coated magnetic nanoparticles. AIChE J 2004;50:2835−48.

[22] Omorogie DM, Babalola J, Francis T, Unuabonah E, Babalola J. Regeneration strategies for spent solid matrices used in adsorption of organic pollutants from surface water: a critical review. Desalin Water Treat 2014;57:518−44.

[23] Martin RJ, Ng WJ. Chemical regeneration of exhausted activated carbon—I. Water Res 1984;18:59−73.

[24] Cabrera-Codony A, Gonzalez-Olmos R, Martín MJ. Regeneration of siloxane-exhausted activated carbon by advanced oxidation processes. J Hazard Mater 2015;285:501−8.

[25] Ince NH, Apikyan IG. Combination of activated carbon adsorption with light-enhanced chemical oxidation via hydrogen peroxide. Water Res 2000;34:4169−76.

[26] Zhang Y, Mancke RG, Sabelfeld M, Geißen S-U. Adsorption of trichlorophenol on zeolite and adsorbent regeneration with ozone. J Hazard Mater 2014;271:178−84.

[27] Fu W, Huang Z. Magnetic dithiocarbamate functionalized reduced graphene oxide for the removal of Cu(II), Cd(II), Pb(II), and Hg(II) ions from aqueous solution: synthesis, adsorption, and regeneration. Chemosphere 2018;209:449−56.

[28] Kumar ASK, Jiang S-J. Chitosan-functionalized graphene oxide: a novel adsorbent an efficient adsorption of arsenic from aqueous solution. J Environ Chem Eng 2016;4:1698−713.

[29] Li J, Feng J, Yan W. Synthesis of polypyrrole-modified TiO$_2$ composite adsorbent and its adsorption performance on acid Red G. J Appl Polym Sci 2013;128:3231−9.

[30] Thirunavukkarasu O, Viraraghavan T, Subramanian K. Arsenic removal from drinking water using iron oxide-coated sand. Water Air Soil Pollut 2003;142:95−111.

[31] Xu Y-h, Nakajima T, Ohki A. Adsorption and removal of arsenic (V) from drinking water by aluminum-loaded Shirasu-zeolite. J Hazard Mater 2002;92:275−87.

[32] Kamala C, Chu K, Chary N, Pandey P, Ramesh S, Sastry A, et al. Removal of arsenic (III) from aqueous solutions using fresh and immobilized plant biomass. Water Res 2005;39:2815−26.

[33] Salvador F, Martin-Sanchez N, Sanchez-Hernandez R, Sanchez-Montero MJ, Izquierdo C. Regeneration of carbonaceous adsorbents. Part II: Chemical, microbiological and vacuum regeneration. Microporous Mesoporous Mater 2015;202:277−96.

[34] Ha S, Vinitnantharat S. Competitive removal of phenol and 2, 4-dichlorophenol in biological activated carbon system. Environ Technol 2000;21:387—96.

[35] Aktaş Ö, Çeçen F. Adsorption, desorption and bioregeneration in the treatment of 2-chlorophenol with activated carbon. J Hazard Mater 2007;141:769—77.

[36] El Gamal M, Mousa HA, El-Naas MH, Zacharia R, Judd S. Bio-regeneration of activated carbon: a comprehensive review. Sep Purif Technol 2018;197:345—59.

[37] de Jonge RJ, Breure AM, van Andel JG. Bioregeneration of powdered activated carbon (PAC) loaded with aromatic compounds. Water Res 1996;30:875—82.

[38] Lee KM, Lim PE. Bioregeneration of powdered activated carbon in the treatment of alkyl-substituted phenolic compounds in simultaneous adsorption and biodegradation processes. Chemosphere 2005;58:407—16.

[39] Vinitnantharat S, Baral A, Ishibashi Y, Ha S. Quantitative bioregeneration of granular activated carbon loaded with phenol and 2, 4-dichlorophenol. Environ Technol 2001;22:339—44.

[40] Ha S-R, Vinitnantharat S, Ishibashi Y. A modeling approach to bioregeneration of granular activated carbon loaded with phenol and 2, 4-dichlorophenol. J Environ Sci Health, Part A 2001;36:275—92.

[41] Tanthapanichakoon W, Ariyadejwanich P, Japthong P, Nakagawa K, Mukai S, Tamon H. Adsorption—desorption characteristics of phenol and reactive dyes from aqueous solution on mesoporous activated carbon prepared from waste tires. Water Res 2005;39:1347—53.

[42] Naghizadeh A, Momeni F, Derakhshani E. Efficiency of ultrasonic process in regeneration of graphene nanoparticles saturated with humic acid. Desalin Water Treat 2017;70:290—3.

[43] Breitbach M, Bathen D. Influence of ultrasound on adsorption processes. Ultrason Sonochem 2001;8:277—83.

[44] Breitbach M, Bathen D, Schmidt-Traub H. Effect of ultrasound on adsorption and desorption processes. Ind Eng Chem Res 2003;42:5635—46.

[45] Rege SU, Yang RT, Cain CA. Desorption by ultrasound: phenol on activated carbon and polymeric resin. AIChE J 1998;44:1519—28.

[46] Schueller BS, Yang RT. Ultrasound enhanced adsorption and desorption of phenol on activated carbon and polymeric resin. Ind Eng Chem Res 2001;40:4912—18.

[47] Qin W, Yuan Y, Dai Y. Effect of ultrasound on desorption equilibrium. Chin J Chem Eng 2001;9:427—30.

[48] Hamdaoui O, Naffrechoux E, Tifouti L, Pétrier C. Effects of ultrasound on adsorption—desorption of p-chlorophenol on granular activated carbon. Ultrason Sonochem 2003;10:109—14.

[49] Hamdaoui O, Naffrechoux E, Suptil J, Fachinger C. Ultrasonic desorption of p-chlorophenol from granular activated carbon. Chem Eng J 2005;106:153—61.

[50] Sabio E, González E, González J, González-García C, Ramiro A, Ganan J. Thermal regeneration of activated carbon saturated with p-nitrophenol. Carbon 2004;42:2285—93.

[51] Bagreev A, Rahman H, Bandosz TJ. Thermal regeneration of a spent activated carbon previously used as hydrogen sulfide adsorbent. Carbon 2001;39:1319—26.

[52] Alvarez P, Beltran F, Gomez-Serrano V, Jaramillo J, Rodrıguez E. Comparison between thermal and ozone regenerations of spent activated carbon exhausted with phenol. Water Res 2004;38:2155—65.

[53] Salvador F, Martin-Sanchez N, Sanchez-Hernandez R, Sanchez-Montero MJ, Izquierdo C. Regeneration of carbonaceous adsorbents. Part I: Thermal regeneration. Microporous Mesoporous Mater 2015;202:259—76.

[54] Chihara K, Matsui I, Smith J. Regeneration of powdered activated carbon. Part II: Steam-carbon reaction kinetics. AIChE J 1981;27:220—5.

[55] Wei H, Deng S, Huang Q, Nie Y, Wang B, Huang J, et al. Regenerable granular carbon nanotubes/alumina hybrid adsorbents for diclofenac sodium and carbamazepine removal from aqueous solution. Water Res 2013;47:4139—47.

[56] Ledesma B, Román S, Álvarez-Murillo A, Sabio E, González JF. Cyclic adsorption/thermal regeneration of activated carbons. J Anal Appl Pyrolysis 2014;106:112—17.

[57] Chestnutt TE, Bach MT, Mazyck DW. Improvement of thermal reactivation of activated carbon for the removal of 2-methylisoborneol. Water Res 2007;41:79—86.

[58] Pełech R, Milchert E, Wróblewska A. Desorption of chloroorganic compounds from a bed of activated carbon. J Colloid Interface Sci 2005;285:518—24.

[59] Owen P, Barry J. Electrochemical carbon regeneration. Rep number PB 1972;239156.

[60] Doniat D, Corajoud J-M, Mosetti J, Porta A. Process for regenerating contaminated activated carbon. Google Patents; 1980.

[61] Narbaitz RM, Cen J. Electrochemical regeneration of granular activated carbon. Water Res 1994;28:1771—8.

[62] Zhang H, Ye L, Zhong H. Regeneration of phenol-saturated activated carbon in an electrochemical reactor. J Chem Technol Biotechnol: Int Res Process Environ Clean Technol 2002;77:1246—50.

[63] Brown NW, Roberts EPL, Chasiotis A, Cherdron T, Sanghrajka N. Atrazine removal using adsorption and electrochemical regeneration. Water Res 2004;38:3067—74.

[64] Zhang H. Regeneration of exhausted activated carbon by electrochemical method. Chem Eng J 2002;85:81—5.

[65] Berenguer R, Marco-Lozar J, Quijada C, Cazorla-Amorós D, Morallón E. Electrochemical regeneration and porosity recovery of phenol-saturated granular activated carbon in an alkaline medium. Carbon 2010;48:2734—45.

[66] Sharif F, Gagnon LR, Mulmi S, Roberts EPL. Electrochemical regeneration of a reduced graphene oxide/magnetite composite adsorbent loaded with methylene blue. Water Res 2017;114:237—45.

[67] Hoffmann MR, Martin ST, Choi W, Bahnemann DW. Environmental applications of semiconductor photocatalysis. Chem Rev 1995;95:69—96.

[68] Xiong Z, Xu Y, Zhu L, Zhao J. Photosensitized oxidation of substituted phenols on aluminum phthalocyanine-intercalated organoclay. Environ Sci Technol 2005;39:651—7.

[69] Xiong Z, Xu Y, Zhu L, Zhao J. Enhanced photodegradation of 2,4,6-trichlorophenol over palladium phthalocyaninesulfonate modified organobentonite. Langmuir 2005;21:10602—7.

[70] Xing Y, Liu D, Zhang L-P. Enhanced adsorption of Methylene Blue by EDTAD-modified sugarcane bagasse and photocatalytic regeneration of the adsorbent. Desalination 2010;259:187—91.

[71] Dhodapkar R, Rao N, Pande S, Nandy T, Devotta S. Adsorption of cationic dyes on Jalshakti®, super absorbent polymer and photocatalytic regeneration of the adsorbent. Reactive Funct Polym 2007;67:540—8.

CHAPTER SIXTEEN

Toxicity of carbon nanomaterials

16.1 Introduction

Since the remarkable debut of fullerenes in 1985, various carbon nanomaterials (CNMs) such as carbon nanotubes (CNTs), carbon nanohorns (CNHs), graphene, graphene oxide (GO), carbon nanofibers (CNFs), nanodiamonds (NDs), and carbon quantum dots (CDs) have been discovered subsequently in the fast-growing field of nanotechnology. They have garnered huge interest in a wide array of socioeconomic applications [1–7]. The unique characteristics of CNMs are attributed to the sp^2-hybridized C atoms that are mostly organized in a hexagonal arrangement. The combination of the distinct features of C atoms' arrangement in CNMs and their extraordinary physical and chemical properties have resulted in various applications in a variety of areas, such as drug delivery, therapeutics, biosensing, bioimaging, energy storage and conversion, electronics, catalysis, and water purification. Additionally, they are considered to be performance enhancers in concrete, catalysis, composites, and flame-retardants [8–14].

Overall, the discovery of CNMs has opened up a new horizon in material science. However, various findings have demonstrated that CNMs can have adverse effects on humans and the environment [15,16]. The toxicity of CNMs originates from their leakage into the environment, the consumption of high doses by costumers, not being standardized by manufactures, and industrial exposure to workers. CNMs possess distinctive toxicities, which depend on various factors such as particle size and morphology, type of CNM, structural defects, and amount of functionalization. The development of critical CNMs regulations for production, interaction with, or use for different purposes or final products are most urgently needed to mitigate their adverse effects. To achieve this, the possible toxicological nature of each CNMs should be determined via conducting many experiments considering the effects on the environment and living organisms. These studies have been conducted in vivo and in vitro on the tissue or cells and organisms [17]. In addition to laboratory experiments, numerous simulation or modeling approaches using computer technology have been developed to determine the nanomaterial–biomolecules interactions [18]. Moreover, the increasing large-scale production and a wide array of applications of CNMs give rise to the probability that they might be introduced into the carbon cycle. CNMs can enter the carbon cycle via food chains or physicochemical processes, which increase the

Carbon Nanomaterial-Based Adsorbents for Water Purification.
DOI: https://doi.org/10.1016/B978-0-12-821959-1.00016-7

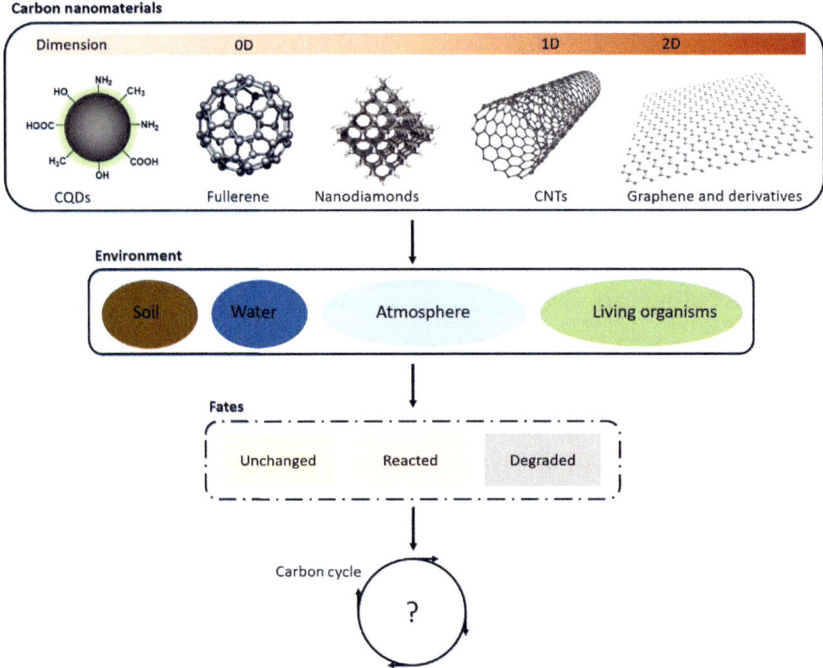

Figure 16.1 Schematic illustration of CNMs based on dimensionality and their environmental fates.

complexity of determining carbon-transfer mechanisms [19]. CNMs in the environment face three fates: decompose, react, or remain unchanged (Fig. 16.1). A few microorganisms can use CNMs as a carbon source, producing CO_2 that can ultimately enter the atmosphere [20]. In some cases, CNMs can directly interact with environmental substances, which can pose difficulties in analyzing them correctly [19]. It is difficult to track carbon transfer associated with CNMs in the environment. Therefore carbon transfer paths for CNMs, and how these assimilate with the entire carbon cycle, must be explored in detail. The environmental release of CNMs has negative effects and might affect the carbon cycle as well. This chapter aims to discuss the toxicity of CNMs toward the environment, including microbes, animals, and human cells. It will also focus on the mechanisms of the nan−bio interactions in the environment and the ecological safety of CNMs.

16.2 Nanotoxicology and toxicity mechanism

Nanotoxicology is described as a division of bionanoscience that determines the toxic characteristics of nanomaterials/particles on various biological systems and living organisms [21,22]. The toxic behavior of nanoscale materials depends on physical and chemical features such as structure, surface charge, particle size, shape, catalytic activity, solubility, active functionalities on the surface, surface area, and surface coatings [23,24]. Due to their small size, these nanoparticles (NPs) can easily penetrate through cell membranes and other biointerfaces into living organisms resulting in cell death/damage. Nanomaterials such as CNMs offer high specific surface area, which allows the attachment of more chemical or biological substances on the surface; consequently their toxicity and reactivity are increased. In 2015 Srikanth et al. studied the cytotoxicity of four types of CNMs (carbon nanowires, fullerene, CNTs, and graphene) on L929 mouse fibroblast cancerous cells via evaluation by MTT Assay [25]. Based on concentration, contact duration, and morphology, graphene showed the highest average toxicity (52.24%), followed by CNTs, fullerene, and carbon nanowires. The toxicity differences of various CNMs were attributed to their different aspect ratios and structural arrangements.

One of the most critical toxicity mechanism of nanomaterials is the generation of reactive oxygen species (ROS) that instigate oxidative stress, lipid peroxidation, inflammation, and lead to damage of cell membranes, proteins, and DNA [26−28]. Oxidative stress can be described as an imbalance between the living system's ability to detoxify the produced superoxide radical anions and hydroxyl radicals and the ROS generation. It may occur because of increased ROS production or through weakening of the biological system's defense mechanism, or a combination of both [29]. For example, Shvedova et al. used electron spin resonance spin trapping to explore the negative effects of single-walled carbon nanotubes (SWCNTs) on human epidermal keratinocytes (HaCaT). They observed that the exposure of HaCaT cells to SWCNT increased the oxidative stress in the system. The morphological and ultrastructural changes in cultured skin were also noted on CNTs exposure [30,31].

Furthermore, most research on CNMs-mediated immune cell responses has been performed on macrophages, including the effects on induction of inflammation, cellular uptake, and cellular viability. The important role of macrophages is to assist in the ingestion of microorganisms and tissue homeostasis. CNMs have demonstrated adverse effects on macrophages. The mechanisms of CNMs-induced toxicity on macrophages are explained in Fig. 16.2. CNMs can trigger cell death through pyroptosis, necrosis, and apoptosis. CNMs induce the mitochondrial reliant apoptotic cascades via the ROS-activated MAPKs pathway. Also, the transcription factor, for example, nuclear factor kappa-light-chain-enhancer of activated B cells (NF-κB), protein complex that controls the inflammatory response is activated by ROS. CNMs trigger the lysosomal membrane permeabilization (LMP), which results in the transfer of cathepsins to the cytoplasm. LMP might induce autophagy dysfunction. LMP and ROS show reciprocal causation creating an amplification circle. The breaking of caspase 1 and the release of downstream IL-1β induce the inflammasome-dependent pyroptosis [32]. In a

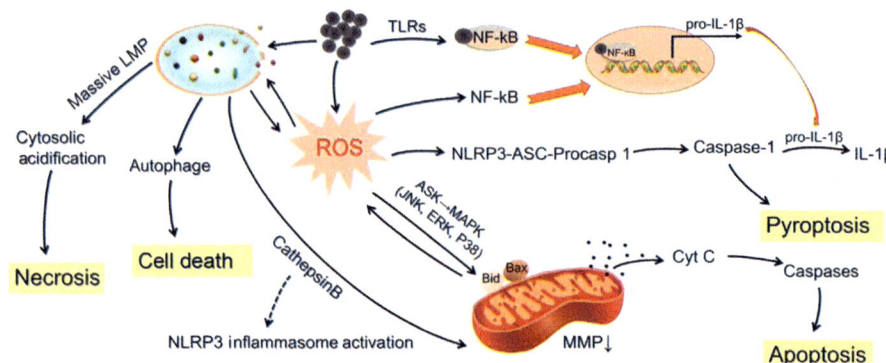

Figure 16.2 The mechanisms of CNMs-induced cytotoxicity of macrophages. Contact of CNMs to macrophages induces a series of cellular and molecular events, including lysosome damage and ROS generation, which is responsible for CNMs-triggered cell death through pyroptosis, necrosis, and apoptosis. *Reproduced with permission from Yuan X, Zhang X, Sun L, Wei Y, Wei X. Cellular toxicity and immunological effects of carbon-based nanomaterials. Part Fibre Toxicol 2019;16:18 [32]. Copyright 2019, Springer Nature.*

particular study, it was noticed that low concentrations of SWCNTs inhibited the function of human monocyte-derived macrophages due to suppressive effects [33]. This process also reduced the engulfment of apoptotic target cells by macrophages, resulting in preincubation with SWCNTs. Moreover, pristine multiwall carbon nanotubes (MWCNTs) and functionalized MWCNTs (with COOH, and polyethylene glycol) exhibited significant generation of inflammatory cytokines, including tumour necrosis factor–alpha (TNF–α), interleukin 1 beta (IL–1β), and IL–6, when varying the concentration of MWCNTs [34]. Palomäki et al. checked the potency of various materials (carbon black, long tangled CNTs, short CNTs, long needle-like CNTs, and crocidolite asbestos) to activate the secretion of IL-1 family cytokines based on shape and size of materials [35]. Asbestos and asbestos-like CNTs (long needle-like) induced the secretion of IL-1β from lipopolysaccharide (LPS)-primed macrophages, while only long needle-like CNTs activated the IL-1α secretion. They concluded that CNT-triggered nod-like receptor family pyrin domain containing protein 3 (NLRP3) inflammasome activation that depended on cathepsin B activity, P2X7 receptor, ROS production, Src and Syk tyrosine kinases.

16.3 Methods for assessing the toxicity of carbon nanomaterials

Due to its multidisciplinary nature, nanotechnology has provided a wide range of innovations, leading to crucial challenges in determining cell or tissue toxicity. Thus advances and the validation of toxicity assessment approaches for CNMs will be of indispensable concern in the future. The exposure of CNMs to the environment, whether intended or unintended, leads to perplexing and unanticipated results. Generally, both in vivo and in vitro, assays are used to assess the toxicity of CNMs. The in vivo studies deliver more relevant and accurate results with a wide range of parameters such as metabolism, distribution, elimination, and also offer the opportunity for the evaluation of long-term chronic effects [36]. In this approach, the toxicity of CNMs can be evaluated based on the route of administration, dose, exposure time, and exposure conditions. In total there are six types of nanotoxicity assessment used in in vivo experiments: analysis of blood serum chemistry, median lethal dose (LD50), histology, tissue morphology, cell population, and overall NPs biodistribution and clearance [37]. Overall, in vivo toxicity is related to different concerns such as the selection of NPs and characterization, exposure route, choice of model system, dose amount, selection of target organ, cell or tissue hematological and histological analysis [38]. The most commonly studied in vivo assays and their examples to assess the toxicity of CNMs are presented in Table 16.1.

Table 16.1 Common in vivo assay approaches for the evaluation of nanomaterials toxicity.

Category	Principle	Techniques	Examples
Biodistribution and clearance	Following the NPs localization, metabolism, and passage through the animal body	ICP-AES, ICP-MS, TEM, Raman spectroscopy, Radiolabeling, Fluorescence imaging, Hematoxylin-Eosin staining, Helium-3/proton MRI,	[39−46]
Lethality	The median lethal dose (LD50)		
Hematology serum chemistry	Examining blood composition, cell population (RBC, WBC, T cell, macrophages)	Hematology Analyzers, Flow Cytometry	[40,45]
Histology/ histopathology	Changes in proteins and enzymes levels: albumin, alanine aminotransferase (ALT), aspartate transaminase (AST), alkaline phosphatase, hemoglobin, and total protein Changes in the tissue or cell morphology	Turbidity or nephelometric measurements, protein-specific labeling methods capillary electrophoresis Light microscopy	[46,47]

ICP-AES, Inductively coupled plasma atomic emission spectroscopy; *ICP-MS*, inductively coupled plasma mass spectrometry; *MRI*, magnetic resonance imaging; *RBC*, red blood cells; *TEM*, transmission electron microscopy; *WBC*, white blood cells.

In recent studies, in vitro assay methods have been more popular due to their minimal ethical concerns, low-cost, and quick procedure. These assays can be categorized into three common subdivisions: viability assays, uptake analysis, and functional assays. The viability assays exclusively evaluate the induction of cell death, while functional assays assess the effects of NPs on various cellular processes. The most commonly studied in vitro assays and their examples to assess the toxicity of CNMs are presented in Table 16.2. In vivo toxicity studies may lead to a number of conflicting observations because of the complexity associated with the proper optimization of NPs dispersion, dose determination, and their interaction mechanism with biological components after administration [37]. Whereas in vitro assessment of toxicity of CNMs is done in laboratory conditions considering particular cell lines, which offers easy control of dose, NPs dispersion, and prediction of NPs activated toxicity. The concentration of introduced NPs must mimic the definite dose amount that living organisms are being exposed.

Moreover, increasing dose amount from a specific dose limit increases the possibility of agglomeration. Due to the small size and high surface area of CNMs, there is the risk of agglomeration at physiological salt and pH concentration. Thus the selection of suitable nontoxic and isotonic dispersants and sorting appropriate dispersion conditions based on the method of administration are still challenging [37]. The exposed NPs in in vitro conditions encounter different cell types, inflammatory conditions, matrices, and numerous surface-active biomolecules [107]. The conformational changes in protein folding and disturbances in biological functions and different signaling pathways may be noted due to the interaction of CNMs with reactive biostructures [37,108].

16.4 Toxicity evaluation of carbon nanomaterials

16.4.1 Graphene

Graphene is the two-dimensional material, which can be prepared with different aspect ratios [109,110]. The graphene-based nanomaterials (GNMs) can be either toxic or biocompatible to biological cells. The toxicity or biocompatibility of GNMs depend significantly on their lateral size, dose, purity, number of layer, hydrophilicity, and surface characteristics. The surface characteristics of GNMs vary hugely due to the availability of different synthesis approaches and different functionalization strategies using polymer and other organic molecules. The proper knowledge of the interaction of GNMs with normal/cancer cells is necessary to understand their medical utility as well as environmental implications. Due to the large surface area and excellent

Table 16.2 Common in vitro assay approaches for the evaluation of nanomaterials toxicity.

Classification	Category	Principle	Techniques	Examples
Viability	Metabolic activity	Assessment of metabolically active cells	MTT, XTT, WST-1, MTS, Alamar blue	[48–50]
	Hemolysis	Membrane disruption and necrosis	Spectrophotometric detection of hemoglobin	[51,52]
	Necrosis	Measurement of membrane integrity	Uptake of dyes: Neutral Red and Trypan Blue, LDH assay	[34,53–56]
	Apoptosis	Membrane alterations	Annexin-V, Propidium iodide	[57–62]
		DNA fragmentation	Comet assay (SCGE), TUNEL, DNA laddering	
Uptake	–	Localization of particles	TEM	[63–65]
		Quantitative measurement of acceptance and localization within the cell	Fluorescence imaging, ICP-AES, ICP-MS	[40,66,67]
Functional	DNA damage	Single-stranded break	Comet assay	[68–70]
	DNA synthesis	Cell proliferation/cell cycle arrest	BrdU incorporation, Thymidine incorporation	[71–74]
	Proliferation	Double-stranded break	TUNEL	[59,75,76]
	Gene expression/ immunogenicity	Increased expression and activation of DNA repair-related proteins	Immunohistochemistry, DASL assay, qRT-PCR, Western blotting	[77–81]
	Exocytosis	Number of cell colonies	CFE assay, Clonogenic assay	[73,82,83]
		DNA synthesis	Thymidine incorporation	[84]
		Altered expression of functional genes in cellular processes	DNA microarray, PCR, q-PCR ELISA, q-PCR	[85–92]
		Cell counting	Flow cytometry, high content image analyzers	[93–95]
		Secretion of electro-active small molecules (e.g., serotonin, epinephrine)	DCFDA assay, DHE assay, NBT assay, Dihydrorhodamine 123 assay	[96–99]
		Cytokines levels	Carbon-fiber microelectrode amperometry	[96,100–102]
	Oxidative stress	Directly: ROS	C11-BODIPY, TBA assay	[39,92,103–106]
		Indirectly: secondary effects of increased cellular ROS	DTNB, Amplex Red	

BODIPY, Boron dipyrromethene; *CFE*, colony forming efficiency; *DASL*, DNA-mediated annealing, extension, selection and ligation; *DCFDA*, 2,7′-dichlorodihydrofluorescein diacetate; *DHE*, dihydroethidium; *DTNB*, 5,5′-dithiobis-2-nitrobenzoic acid; *ELISA*, Enzyme Linked Immunosorbent assay; *LDH*, lactate dehydrogenase; *MTS*, 3-(4,5-dimethylthiazol-2-yl)-5-(3-carboxymethoxyphenyl)-2-(4-sulfophenyl)-2H-tetrazolium inner salt; *MTT*, 3-(4,5-Dimethyl-2-thiazolyl)-2,5-diphenyl-2H-tetrazolium bromide; *NBT*, nitroblue tetrazolium; *ROS*, reactive oxygen species; *SCGE*, single-cell gel electrophoresis; *TBA*, thiobarbituric acid; *TUNEL*, terminal deoxynucleotidyl transferase dUTP nick end labeling; *WST-1*, water soluble tetrazolium-1; *XTT*, 2,3-bis-(2-ethoxy-4-nitro-5-sulfophenyl)-2H-tetrazolium-5-carboxanilide.

antibacterial properties, graphene has crucial effects on microbes in the soil environment compared to CNTs and fullerenes [111]. Various studies have highlighted that graphene has a substantial influence on the microbial community structure and number of microorganisms in soil and it depends on contact time between microorganisms and graphene [112]. A small concentration of graphene (<100 mg/kg) could enhance the bacterial biomass and soil microbial enzyme activity in a short period, which results in an increased uptake rate of contaminants in the soil. When the concentration of graphene was extremely high, the bacteria participating in soil ammoxidation (viz. black pine fungus and digestive spirochete) were considerably reduced, which might cause adverse effects on the nitrogen cycle in the soil. In another study, a low dose of GO showed an increase in proliferation and cell attachment, which might stimulate the growth of cells [113]. Furthermore, based on investigating the effects of graphene on microorganisms in water environment and in soil, it was noted that the adverse effects of graphene on microorganisms in soil are greater than on microorganisms in water bodies due to its extremely low solubility in water. The toxicity of GO is much higher than that of graphene. GO is also difficult to degrade in the environment, compared to graphene. Additionally, organic matter can influence the movement of graphene in the soil and it can be used as an adsorption carrier and transport carrier for graphene [114].

The toxicity of graphene and derivatives has been assessed in vitro using various human cell lines such as epithelial cells [53], alveolar basal epithelial cells [115], fibroblasts [116], cervical cells [117], pheochromocytoma cells, oligodendroglia cells, fetal osteoblasts [118], epithelial breast cancer cells [119], skin fibroblasts [120], red blood cells [120], and neuronal cells [121]. Previous studies show that lung diseases and severe cytotoxicity might be induced by exposure to graphene and derivatives. Wang et al. exhibited that upon exposure GO could enter human lung fibroblasts cytoplasm and nuclei, resulting in cell floating, a decrease in cell adhesion, and apoptosis at doses higher than $20\,\mu g/mL$ after 24 hours [116]. The GO less than $20\,\mu g/mL$ did not show any toxicity to human fibroblast cells. The results also suggested that the introduced graphene and derivatives could enter the lung tissues and stay there and activate lung inflammation and subsequent granulomas, which were highly dependent on injected dose. Duch et al. studied in vivo toxicity of Pluronic (block copolymer) dispersed graphene, pristine graphene, and GO in the lungs of six C57BL mice [122]. The directly introduced GO in lung resulted in severe and consistent lung injury. The results showed that GO in lungs increased the rate of ROS generation and mitochondrial respiration, inducing the inflammatory and apoptotic pathways. The toxicity of pristine graphene was considerably reduced after liquid-phase exfoliation and was further reduced when it was well dispersed with block copolymer Pluronic. Thus oxidation of graphene is mainly responsible for pulmonary toxicity and good dispersion in Pluronic suggests a potential pathway for safe handling and biomedical applications.

In another study, the toxicity of graphite nanoplatelets was evaluated in the model living organism such as *Caenorhabditis elegans* (a nematode) [123]. They observed no acute or chronic toxicity of graphite nanoplatelets in this system.

Recently, Mendes et al. examined the consequence of the size of GO on the viability or toxicity of macrophages and cervical cancer HeLa cells [124]. They observed that GO with large flakes (277 nm) induced higher levels of toxicity than small GO flakes (89 nm), at a more prolonged exposure of 48 hours (Fig. 16.3A). A TEM investigation showed that different sizes of GO are internalized by macrophage (Fig. 16.3B1−B4) and HeLa cells at 12 and 48 hours exposure time. Irrespective of the GO sizes, the macrophage observed the morphological change for phagocytosis after incubation for 48 hours. This morphological change resulted in the formation of large vacuoles carrying GO.

16.4.2 Fullerene

Fullerene (C60), a hollow spherical carbon structure was the first known carbon nanomaterial, discovered by Kroto et al. in 1985 [1]. Since then, various types of fullerenes have been developed and used in a wide array of applications, including biomedical applications. Thus toxicity assessments and the environmental impact of fullerenes were studied from the beginning. A fullerene molecule is not soluble in water; it forms colloidal suspensions in water. Fullerene with hydroxyl groups as well as other larger functional groups was also synthesized. Fullerene soot may enter an atmosphere by various combustion processes [125]. A series of transformations were noted when fullerene is introduced to water [126]. It can be added to the environment as an organic solvent, and induced effects may be different from those of its aqueous dispersions. The aqueous suspension of C60 aggregates had a size-dependent impact on soil bacterial composition in the low–organic matter contents. It triggered a negligible change in the microbial biomass and metabolic activity in soils in both low and high organic matter systems [127].

Cytotoxicity assessment of fullerene was conducted with different types of cells, and the different cell lines provided various toxicity observations. Fiorito et al. described that the cellular uptake of fullerene from macrophage cells was deficient, and its toxicity was lower than that of graphite [128]. Jia et al. tested the toxicity of pristine C60 to macrophages and demonstrated that the toxicity of C60 was lower than that of MWCNTs and SWCNTs [129] (Fig. 16.4). In vitro SWCNTs showed profound cytotoxicity in alveolar macrophages after a 6-hour incubation period. No considerable toxicity was noted for C60 up to a concentration of $226.0\,\mu g/cm^2$. Furthermore, the toxicity of C60 was assessed by conducting animal experiments using fish and rats. The oral administration of water-soluble polyalkylsulfonated C60 in rats did not show any lethal damage, but intraperitoneal administration resulted in a

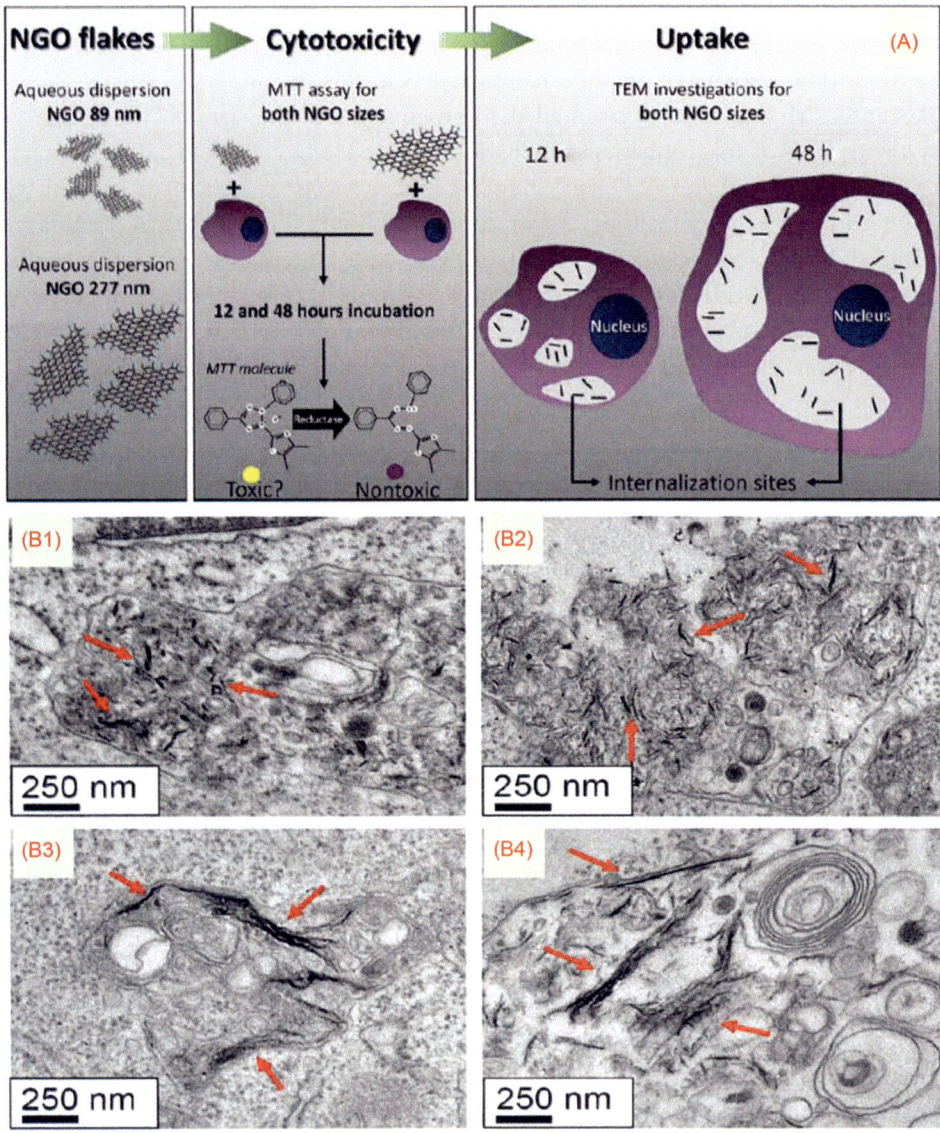

Figure 16.3 (A) Schematic presenting the approaches used to evaluate size-dependent cytotoxicity and internalization of GO. TEM images of a macrophage incubated with small GOs (B1 and B2) and large GO (B3 and B4) for 12 and 48 h, respectively. *Reproduced with permission from Mendes RG, Koch B, Bachmatiuk A, Ma X, Sanchez S, Damm C, et al. A size dependent evaluation of the cytotoxicity and uptake of nanographene oxide. J Mater Chem B 2015;3:2522—9 [124]. Copyright 2015, the Royal Society of Chemistry.*

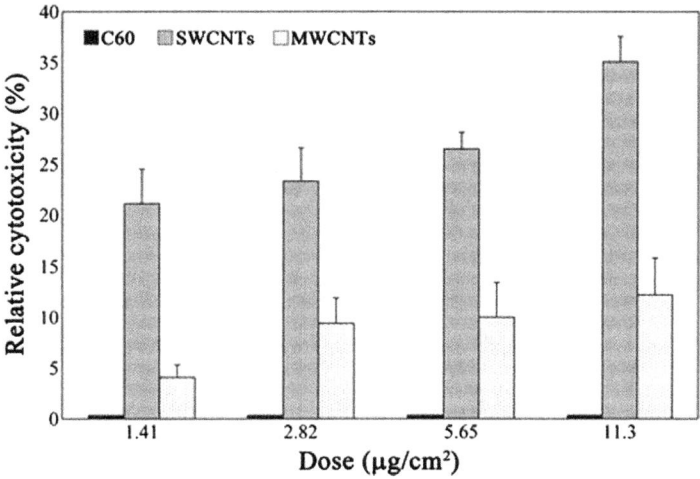

Figure 16.4 Comparison of cytotoxicity to alveolar macrophage cells using C60, SWCNTs, and MWCNTs at different dosages. *Reproduced with permission from Jia G, Wang H, Yan L, Wang X, Pei R, Yan T, et al. Cytotoxicity of carbon nanomaterials: single-wall nanotube, multi-wall nanotube, and fullerene. Environ Sci Technol 2005;39:1378–83 [129]. Copyright 2005, American Chemical Society.*

median lethal dose of 600 mg/kg [130]. The intraperitoneally or intravenously injected C60 amassed in the kidney and induced nephropathy. In vivo, the toxicological nature of C60 on fishes was studied and significant adverse effects were observed. It demonstrated a decrease in the hatching rate of the zebrafish embryo, lipid peroxidation in the brain, and fin malformation [131–133].

16.4.3 Carbon nanotubes

CNTs have emerged as a popular adsorbent for water treatment. Thus the risks and toxicological evaluation of CNTs are very crucial in order to employ them in a safe manner. Different types of CNTs can have different physicochemical properties that must be evaluated separately. Various kinds of CNTs exist in the market with various parameters such as different shapes, length, layers (single, bi-, or multilayers), surface charges and surface functionalities, which illustrate the intricacy of the CNTs in the environment. Sometimes, pristine CNTs can be risky because of their generic impurities, such as residue of carbonaceous materials and metal catalysts that may induce nanosafety concerns [134]. To avoid this, researchers have purified CNTs using acid treatment and functionalized them using various methodologies, but a recent report showed that functionalized CNTs may enhance the metal acceptance and toxicity levels on the living cells in a dose-dependent manner [135].

Cytotoxicity of SWCNTs was evaluated on different human cells, such as human keratinocyte cells (HaCaT), alveolar epithelial cells (A549), and macrophages, due to their relationship with dermatological, respiratory, and immunological toxicity. Herzog et al. demonstrated that EC50 (50% reduction concentration in cell viability) of SWCNTs for A549 cells was higher than $400 \, \mu g/mL$, which is 10 times higher than for HaCaT and bronchial epithelial cells (BEAS-2B) [136]. The solubility of SWCNTs is less than for C60, and it can be improved by surface functionalization. Dumortier et al. discussed the cytotoxicity of various functionalized SWCNTs (using oxidation, amidation, and 1,3-dipolar cycloaddition reaction) to B/T lymphocytes and macrophages. They concluded that highly water-soluble functionalized SWCNTs showed low toxicity as they were taken up into cells without affecting the cell viability [137]. In another study, surfactant (polyoxyethylene sorbitan monooleate) was used to enhance the water solubility of SWCNTs. It was noted that the surfactant-modified SWCNTs had a suppressed cytotoxicity [138]. Thus, agglomeration and surface functionalities on CNTs can strongly control the cytotoxicity.

The previous studies indicate that the toxicity of MWCNTs was strongly regulated by their surface conditions, size, and purity. Bottini et al. [54]. and Vittorio et al. [139] demonstrated that surface oxidized and purified MWCNTs could suppress cell viability. Wang et al. showed that MWCNTs with smaller diameters revealed less cytotoxicity than large diameter MWCNTs at an equal dosage [139]. During the toxicity assessment of MWCNTs, many researchers compared CNTs with asbestos due to their similar shape [140−142]. Muller et al. noted that intratracheally administrated MWCNTs in rats were effective even after 60 days, and induced fibrotic reactions, inflammation, and granulomas. They claimed that the cytotoxicity of MWCNTs was approximately similar to that of chrysotile [140]. Poland et al. reported that exposure to long MWCNTs by the mesothelial lining of the body cavity of mice resulted in length-dependent, pathogenic behavior similar to that seen from exposure to asbestos [142]. They also induced the inflammation and generation of lesions known as granulomas (Fig. 16.5). MWCNTs with different lengths induced different degrees of granuloma generation. Long MWCNTs ($>20 \, \mu m$) were considerably more risky than tangled and low-aspect-ratio MWCNTs because macrophages could not entirely engulf longer tubes [143]. Previous reports indicate that MWCNTs have a cytotoxicity comparable to or higher than asbestos. However, the morphology of CNTs is more interconnected than asbestos, which allows easy generation after instillation/injection into animals. Additionally, the CNTs toxicity assessment did not always demonstrate pulmonary toxicity, which is commonly seen with asbestos [144].

16.4.4 Other carbon nanomaterials

Other carbon nanomaterials such as carbon nanofibers (CNFs), carbon nanohorns (CNHs), nanodiamonds (NDs), and carbon dots (CDs) also have been assessed for

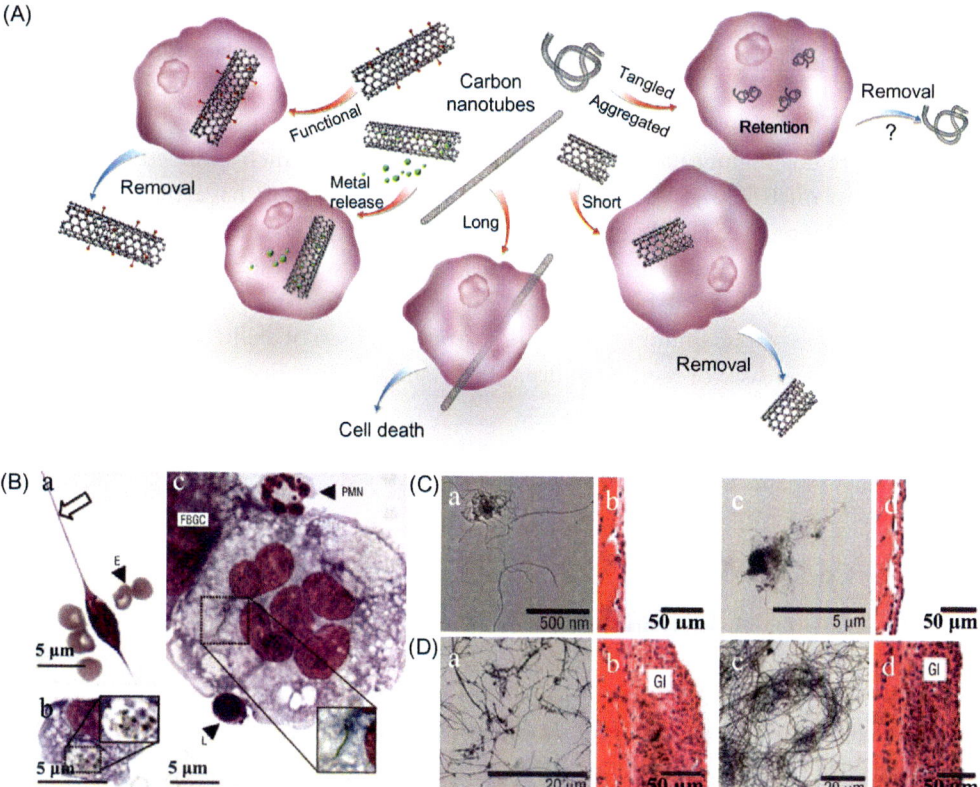

Figure 16.5 Schematic illustration of the physiochemical characteristics of CNTs, which can affect cell uptake and responses. (A) The different types of CNTs can influence phagocytosis and cytotoxicity; (B) (a,b) long MWCNTs result in frustrated phagocytosis, whereas short MWCNTs are easily phagocytosed. (c) Foreign body giant cells (FBGCs) after administration of long MWCNTs (L, lymphocyte; PMN, polymorphonuclear leukocyte); (C) images display the presence of small granuloma response in mice exposed to tangled MWCNTs, whereas (D) shows granulomatous inflammation (GI) in mice exposed to long MWCNTs. *Reproduced with permission from Poland CA, Duffin R, Kinloch I, Maynard A, Wallace WAH, Seaton A, et al. Carbon nanotubes introduced into the abdominal cavity of mice show asbestos-like pathogenicity in a pilot study. Nat Nanotechnol 2008;3:423−8 [142]. Copyright 2008 Nature Publishing Group.*

toxicity. CNHs are horn-shaped graphene tubules, which exhibited low cytotoxicity. They did not show any mutation and stimulation of the skin, eyes, and respiration tract [145,146]. Yokoyama et al. evaluated the subcutaneous tissue (rats) response to hat-stacked carbon nanofibers (h-CNFs) [147]. No severe inflammatory reaction such as necrosis was detected during exposure to h-CNFs, although some h-CNFs were noted in lysosomal vacuoles of phagocytes. In another study, water-soluble h-CNFs showed no specific toxicity in the cytotoxicity experiment [148]. NDs have been demonstrated to induce cardiotoxic effects in animals [149]. As compared to other

CNMs, NDs have exhibited lower cytotoxicity and higher biocompatibility in both in vivo and in vitro studies [149]. The exposure to NDs via inhalation in mice triggered a huge increase of ROS generation, disruption of mitochondrial membrane potential (MMP), lipid peroxidation, high levels of glutathione disulfide (GSSG), and low levels of reduced glutathione (GSH) in the heart tissue, which induced the cardiotoxic effects in mice [150]. Various in vivo and in vitro studies have been published on the cytotoxicity of CDs [151]. In a particular report, CDs with a particle size of ~ 7 nm and hydrodynamic diameter of ~ 60.3 nm were synthesized and used to evaluate their toxicity to human bronchial epithelial cells (16HBE) [152]. CDs exposure induced oxidative stress, draining the antioxidant defenses of cells, and subsequently resulting in a decrease of cell viability. Thus surface functionalization of CDs is required to reduce their toxicity.

16.5 Conclusion

In this chapter, the toxicity of CNMs has been discussed using different cell lines and animal models. It was noted that the toxicity of CNMs originates from numerous factors including dispersion, size, agglomeration, morphology, surface functionalization, and cell permeability. Among graphene and its derivatives, GO demonstrated high toxicity due to the presence of highly reactive oxygen functionalities. The dispersion of graphene in surfactants or functionalization using organic molecules can significantly reduce its toxicity. Among different types of CNTs, long CNTs and pristine CNTs exhibited high toxicity over different cell lines. Low agglomeration of SWCNTs leads to low toxicity. Therefore surface functionalization is the best approach to reduce the toxicity of CNMs and offer a pathway for safe handling, and crucial medical and environmental applications. Although various studies have exclusively been published on the toxicity of CNMs in both in vitro and in vivo, there are still a number of issues to be solved: (1) a comparative detail study is needed for different animal species and cell lines for the same kind of CNMs; (2) a systematic study is required to substantiate the toxicity results with the surface charge (negative or positive) of the CNMs with subtle differences; (3) the toxicity investigations of the same types of CNMs obtained from different preparation routes should be explored; (4) it is necessary to determine the toxicity limitations and prove the effectiveness of CNMs; and (5) an organized study to confirm that the CNMs show toxicity to cancerous cells only, but not normal cells at the standardized concentration range. Importantly, the understanding of the toxicity of CNMs will motivate the interdisciplinary researchers to discover more potential biocompatible materials with excellent functionalities for advanced applications.

References

[1] Kroto HW, Heath JR, O'Brien SC, Curl RF, Smalley RE. C60: Buckminsterfullerene. Nature 1985;318:162—3.

[2] Yang Z, Tian J, Yin Z, Cui C, Qian W, Wei F. Carbon nanotube- and graphene-based nanomaterials and applications in high-voltage supercapacitor: a review. Carbon 2019;141:467—80.

[3] 15 years of graphene electronics. Nat Electron 2019;2:369.

[4] Gusain R, Kumar N, Ray SS. Recent advances in carbon nanomaterial-based adsorbents for water purification. Coord Chem Rev 2020;405:213111.

[5] Gusain R, Kumar N, Fosso-Kankeu E, Ray SS. Efficient removal of Pb(II) and Cd(II) from industrial mine water by a hierarchical MoS2/SH-MWCNT nanocomposite. ACS Omega 2019;4:13922—35.

[6] Mukwevho N, Gusain R, Fosso-Kankeu E, Kumar N, Waanders F, Ray SS. Removal of naphthalene from simulated wastewater through adsorption-photodegradation by ZnO/Ag/GO nanocomposite. J Ind Eng Chem 2020;81:393—404.

[7] Mukwevho N, Kumar N, Fosso-Kankeu E, Waanders F, Bunt JR, Sinha Ray S. Visible light-excitable ZnO/2D graphitic-C3N4 heterostructure for the photodegradation of naphthalene. Desalin Water Treat 2019;163:286—96.

[8] Norhasri MSM, Hamidah MS, Fadzil AM. Applications of using nano material in concrete: a review. Constr Build Mater 2017;133:91—7.

[9] Morsy MS, Alsayed SH, Aqel M. Hybrid effect of carbon nanotube and nano-clay on physico-mechanical properties of cement mortar. Constr Build Mater 2011;25:145—9.

[10] Huang C, Li C, Shi G. Graphene based catalysts. Energy Environ Sci 2012;5:8848—68.

[11] Ama OM, Kumar N, Adams FV, Ray SS. Efficient and cost-effective photoelectrochemical degradation of dyes in wastewater over an exfoliated graphite-MoO3 nanocomposite electrode. Electrocatalysis 2018;9:623—31.

[12] Kumar N, Sinha Ray S. Synthesis and functionalization of nanomaterials. In: Sinha Ray S, editor. Processing of polymer-based nanocomposites: introduction. Cham: Springer International Publishing; 2018. p. 15—55.

[13] Mungse HP, Verma S, Kumar N, Sain B, Khatri OP. Grafting of oxo-vanadium Schiff base on graphene nanosheets and its catalytic activity for the oxidation of alcohols. J Mater Chem 2012;22:5427—33.

[14] Yuan B, Fan A, Yang M, Chen X, Hu Y, Bao C, et al. The effects of graphene on the flammability and fire behavior of intumescent flame retardant polypropylene composites at different flame scenarios. Polym Degrad Stab 2017;143:42—56.

[15] Chen M, Zeng G, Xu P, Zhang Y, Jiang D, Zhou S. Understanding enzymatic degradation of single-walled carbon nanotubes triggered by functionalization using molecular dynamics simulation. Environ Sci: Nano 2017;4:720—7.

[16] Chen M, Zhou S, Zeng G, Zhang C, Xu P. Putting carbon nanomaterials on the carbon cycle map. Nano Today 2018;20:7—9.

[17] Hischier R, Walser T. Life cycle assessment of engineered nanomaterials: state of the art and strategies to overcome existing gaps. Sci Total Environ 2012;425:271—82.

[18] Ding H-m, Ma Y-q. Computational approaches to cell—nanomaterial interactions: keeping balance between therapeutic efficiency and cytotoxicity. Nanoscale Horiz 2018;3:6—27.

[19] Larue C, Pinault M, Czarny B, Georgin D, Jaillard D, Bendiab N, et al. Quantitative evaluation of multi-walled carbon nanotube uptake in wheat and rapeseed. J Hazard Mater 2012;227:155—63.

[20] Zhang L, Petersen EJ, Habteselassie MY, Mao L, Huang Q. Degradation of multiwall carbon nanotubes by bacteria. Environ Pollut 2013;181:335—9.

[21] Zhao Y, Wang B, Feng W, Bai C. Nanotoxicology: toxicological and biological activities of nanomaterials. Nanosci Nanotechnol 2012;1—53.

[22] Ai J, Biazar E, Jafarpour M, Montazeri M, Majdi A, Aminifard S, et al. Nanotoxicology and nanoparticle safety in biomedical designs. Int J Nanomed 2011;6:1117.

[23] Sukhanova A, Bozrova S, Sokolov P, Berestovoy M, Karaulov A, Nabiev I. Dependence of nanoparticle toxicity on their physical and chemical properties. Nanoscale Res Lett 2018;13:44.

[24] Krug HF, Wick P. Nanotoxicology: an interdisciplinary challenge. Angew Chem Int Ed 2011;50:1260−78.

[25] Srikanth M, Misak H, Yang SY, Asmatulu R. Effects of morphology, concentration and contact duration of carbon-based nanoparticles on cytotoxicity of L929 cells. ASME. 2015. International Mechanical Engineering Congress and Exposition 2015.

[26] Khalili Fard J, Jafari S, Eghbal MA. A review of molecular mechanisms involved in toxicity of nanoparticles. Adv Pharm Bull 2015;5:447−54.

[27] Plackal Adimuriyil George B, Kumar N, Abrahamse H, Ray SS. Apoptotic efficacy of multifaceted biosynthesized silver nanoparticles on human adenocarcinoma cells. Sci Rep 2018;8:14368.

[28] Kumar N, George BPA, Abrahamse H, Parashar V, Ngila JC. Sustainable one-step synthesis of hierarchical microspheres of PEGylated MoS2 nanosheets and MoO3 nanorods: their cytotoxicity towards lung and breast cancer cells. Appl Surf Sci 2017;396:8−18.

[29] Walters C, Pool E, Somerset V. Nanotoxicology: a review. Toxicology—new aspects to this scientific conundrum. InTech; 2016. p. 45−63.

[30] Tsuruoka S, Takeuchi K, Koyama K, Noguchi T, Endo M, Tristan F, et al. ROS evaluation for a series of CNTs and their derivatives using an ESR method with DMPO. J Phys: Conf Ser 2013;429:012029.

[31] Shvedova A, Castranova V, Kisin E, Schwegler-Berry D, Murray A, Gandelsman V, et al. Exposure to carbon nanotube material: assessment of nanotube cytotoxicity using human keratinocyte cells. J Toxicol Environ Health, Part A 2003;66:1909−26.

[32] Yuan X, Zhang X, Sun L, Wei Y, Wei X. Cellular toxicity and immunological effects of carbon-based nanomaterials. Part Fibre Toxicol 2019;16:18.

[33] Witasp E, Shvedova AA, Kagan VE, Fadeel B. Single-walled carbon nanotubes impair human macrophage engulfment of apoptotic cell corpses. Inhalation Toxicol 2009;21:131−6.

[34] Zhang T, Tang M, Kong L, Li H, Zhang T, Zhang S, et al. Comparison of cytotoxic and inflammatory responses of pristine and functionalized multi-walled carbon nanotubes in RAW 264.7 mouse macrophages. J Hazard Mater 2012;219:203−12.

[35] Palomäki J, Välimäki E, Sund J, Vippola M, Clausen PA, Jensen KA, et al. Long, needle-like carbon nanotubes and asbestos activate the NLRP3 inflammasome through a similar mechanism. ACS Nano 2011;5:6861−70.

[36] Dhawan A, Sharma V. Toxicity assessment of nanomaterials: methods and challenges. Anal Bioanal Chem 2010;398:589−605.

[37] Kumar V, Sharma N, Maitra SS. In vitro and in vivo toxicity assessment of nanoparticles. Int Nano Lett 2017;7:243−56.

[38] Marquis BJ, Love SA, Braun KL, Haynes CL. Analytical methods to assess nanoparticle toxicity. Analyst 2009;134:425−39.

[39] Han SG, Andrews R, Gairola CG. Acute pulmonary response of mice to multi-wall carbon nanotubes. Inhalation Toxicol 2010;22:340−7.

[40] Nurunnabi M, Khatun Z, Huh KM, Park SY, Lee DY, Cho KJ, et al. In vivo biodistribution and toxicology of carboxylated graphene quantum dots. ACS Nano 2013;7:6858−67.

[41] Elgrabli D, Floriani M, Abella-Gallart S, Meunier L, Gamez C, Delalain P, et al. Biodistribution and clearance of instilled carbon nanotubes in rat lung. Part Fibre Toxicol 2008;5:20.

[42] Jia F, Wu L, Meng J, Yang M, Kong H, Liu T, et al. Preparation, characterization and fluorescent imaging of multi-walled carbon nanotube−porphyrin conjugate. J Mater Chem 2009;19:8950−7.

[43] Al Faraj A, Cieslar K, Lacroix G, Gaillard S, Canet-Soulas E, Crémillieux Y. In vivo imaging of carbon nanotube biodistribution using magnetic resonance imaging. Nano Lett 2009;9:1023−7.

[44] Al Faraj A, Bessaad A, Cieslar K, Lacroix G, Canet-Soulas E, Crémillieux Y. Long-term follow-up of lung biodistribution and effect of instilled SWCNTs using multiscale imaging techniques. Nanotechnology 2010;21:175103.

[45] Yang S-T, Wang X, Jia G, Gu Y, Wang T, Nie H, et al. Long-term accumulation and low toxicity of single-walled carbon nanotubes in intravenously exposed mice. Toxicol Lett 2008;181:182−9.

[46] Das B, Chattopadhyay P, Upadhyay A, Gupta K, Mandal M, Karak N. Biophysico-chemical interfacial attributes of Fe3O4 decorated MWCNT nanohybrid/bio-based hyperbranched polyurethane

nanocomposite: an antibacterial wound healing material with controlled drug release potential. N J Chem 2014;38.

[47] Régnier J-F, Pothmann-Krings D, Simar S, Dony E, Net J-LL, Beausoleil J. Graphistrength© C100 MultiWalled Carbon Nanotubes (MWCNT): thirteen-week inhalation toxicity study in rats with 13- and 52-week recovery periods combined with comet and micronucleus assays. J Phys: Conf Ser 2017;838:012030.

[48] Vallabani N, Mittal S, Shukla RK, Pandey AK, Dhakate SR, Pasricha R, et al. Toxicity of graphene in normal human lung cells (BEAS-2B). J Biomed Nanotechnol 2011;7:106−7.

[49] Chng ELK, Pumera M. The toxicity of graphene oxides: dependence on the oxidative methods used. Chem−A Eur J 2013;19:8227−35.

[50] van Berlo D, Wilhelmi V, Boots AW, Hullmann M, Kuhlbusch TA, Bast A, et al. Apoptotic, inflammatory, and fibrogenic effects of two different types of multi-walled carbon nanotubes in mouse lung. Arch Toxicol 2014;88:1725−37.

[51] Lodhi N, Mehra NK, Jain NK. Development and characterization of dexamethasone mesylate anchored on multi walled carbon nanotubes. J drug Target 2013;21:67−76.

[52] Mehra NK, Jain N. Cancer targeting propensity of folate conjugated surface engineered multi-walled carbon nanotubes. Colloids Surf B: Biointerfaces 2015;132:17−26.

[53] Chang Y, Yang S-T, Liu J-H, Dong E, Wang Y, Cao A, et al. In vitro toxicity evaluation of graphene oxide on A549 cells. Toxicol Lett 2011;200:201−10.

[54] Bottini M, Bruckner S, Nika K, Bottini N, Bellucci S, Magrini A, et al. Multi-walled carbon nanotubes induce T lymphocyte apoptosis. Toxicol Lett 2006;160:121−6.

[55] Ali-Boucetta H, Al-Jamal KT, Kostarelos K. Cytotoxic assessment of carbon nanotube interaction with cell cultures. Biomedical nanotechnology. Springer; 2011. p. 299−312.

[56] Diabaté S, Bergfeldt B, Plaumann D, Übel C, Weiss C. Anti-oxidative and inflammatory responses induced by fly ash particles and carbon black in lung epithelial cells. Anal Bioanal Chem 2011;401:3197−212.

[57] Pulskamp K, Wörle-Knirsch JM, Hennrich F, Kern K, Krug HF. Human lung epithelial cells show biphasic oxidative burst after single-walled carbon nanotube contact. Carbon 2007;45:2241−9.

[58] Liu Z, Dong X, Song L, Zhang H, Liu L, Zhu D, et al. Carboxylation of multiwalled carbon nanotube enhanced its biocompatibility with L02 cells through decreased activation of mitochondrial apoptotic pathway. J Biomed Mater Res Part A 2014;102:665−73.

[59] Jiang Y, Zhang H, Wang Y, Chen M, Ye S, Hou Z, et al. Modulation of apoptotic pathways of macrophages by surface-functionalized multi-walled carbon nanotubes. PLoS One 2013;8.

[60] Cancino J, Paino I, Micocci K, Selistre-de-Araujo H, Zucolotto V. In vitro nanotoxicity of single-walled carbon nanotube−dendrimer nanocomplexes against murine myoblast cells. Toxicol Lett 2013;219:18−25.

[61] Han Y-g, Xu J, Li Z-g, Ren G-g, Yang Z. In vitro toxicity of multi-walled carbon nanotubes in C6 rat glioma cells. Neurotoxicology 2012;33:1128−34.

[62] Qin Y, Zhou Z-W, Pan S-T, He Z-X, Zhang X, Qiu J-X, et al. Graphene quantum dots induce apoptosis, autophagy, and inflammatory response via p38 mitogen-activated protein kinase and nuclear factor-κB mediated signaling pathways in activated THP-1 macrophages. Toxicology 2015;327:62−76.

[63] Zhang X, Hu W, Li J, Tao L, Wei Y. A comparative study of cellular uptake and cytotoxicity of multi-walled carbon nanotubes, graphene oxide, and nanodiamond. Toxicol Res 2012;1:62−8.

[64] Sweeney S, Hu S, Ruenraroengsak P, Chen S, Gow A, Schwander S, et al. Carboxylation of multi-walled carbon nanotubes reduces their toxicity in primary human alveolar macrophages. Environ Sci: Nano 2016;3:1340−50.

[65] Zhu S, Zhu B, Huang A, Hu Y, Wang G, Ling F. Toxicological effects of multi-walled carbon nanotubes on Saccharomyces cerevisiae: the uptake kinetics and mechanisms and the toxic responses. J Hazard Mater 2016;318:650−62.

[66] Kostarelos K, Lacerda L, Pastorin G, Wu W, Wieckowski S, Luangsivilay J, et al. Cellular uptake of functionalized carbon nanotubes is independent of functional group and cell type. Nat Nanotechnol 2007;2:108−13.

[67] Gao L, Nie L, Wang T, Qin Y, Guo Z, Yang D, et al. Carbon nanotube delivery of the GFP gene into mammalian cells. ChemBioChem 2006;7:239—42.

[68] De LM, Ottaviano L, Perrozzi F, Nardone M, Santucci S, De JL, et al. Flake size-dependent cyto and genotoxic evaluation of graphene oxide on in vitro A549, CaCo2 and vero cell lines. J Biol Regulators Homeost Agents 2014;28:281—9.

[69] Wang A, Pu K, Dong B, Liu Y, Zhang L, Zhang Z, et al. Role of surface charge and oxidative stress in cytotoxicity and genotoxicity of graphene oxide towards human lung fibroblast cells. J Appl Toxicol 2013;33:1156—64.

[70] Sasidharan A, Swaroop S, Chandran P, Nair S, Koyakutty M. Cellular and molecular mechanistic insight into the DNA-damaging potential of few-layer graphene in human primary endothelial cells. Nanomedicine: Nanotechnol Biol Med 2016;12:1347—55.

[71] Hackbarth A, Schaudien D, Bellmann B, Ernst H, Ziemann C, Leonhardt A, et al. Toxic effects of multiwall carbon nanotubes (MWCNT) in vivo and in vitro. Pneumologie 2013;67:A2.

[72] Porwal H, Estili M, Grünewald A, Grasso S, Detsch R, Hu C, et al. 45S5 Bioglass®—MWCNT composite: processing and bioactivity. J Mater Sci: Mater Med 2015;26:199.

[73] Vankoningsloo S, Piret J-P, Saout C, Noel F, Mejia J, Zouboulis CC, et al. Cytotoxicity of multi-walled carbon nanotubes in three skin cellular models: effects of sonication, dispersive agents and corneous layer of reconstructed epidermis. Nanotoxicology 2010;4:84—97.

[74] Piret J-P, Detriche S, Vigneron R, Vankoningsloo S, Rolin S, Mendoza JM, et al. Dispersion of multi-walled carbon nanotubes in biocompatible dispersants. J Nanopart Res 2010;12:75—82.

[75] Zhu X, Zhang H, Huang H, Zhang Y, Hou L, Zhang Z. Functionalized graphene oxide-based thermosensitive hydrogel for magnetic hyperthermia therapy on tumors. Nanotechnology 2015;26:365103.

[76] Ravichandran P, Baluchamy S, Sadanandan B, Gopikrishnan R, Biradar S, Ramesh V, et al. Multiwalled carbon nanotubes activate NF-κB and AP-1 signaling pathways to induce apoptosis in rat lung epithelial cells. Apoptosis 2010;15:1507—16.

[77] Liu Y, Luo Y, Wu J, Wang Y, Yang X, Yang R, et al. Graphene oxide can induce in vitro and in vivo mutagenesis. Sci Rep 2013;3:3469.

[78] Brown TA, Lee JW, Holian A, Porter V, Fredriksen H, Kim M, et al. Alterations in DNA methylation corresponding with lung inflammation and as a biomarker for disease development after MWCNT exposure. Nanotoxicology 2016;10:453—61.

[79] Xing Y, Xiong W, Zhu L, Osawa E, Hussin S, Dai L. DNA damage in embryonic stem cells caused by nanodiamonds. ACS Nano 2011;5:2376—84.

[80] Chowdhury SM, Lalwani G, Zhang K, Yang JY, Neville K, Sitharaman B. Cell specific cytotoxicity and uptake of graphene nanoribbons. Biomaterials 2013;34:283—93.

[81] Ponti J, Broggi F, Mariani V, De Marzi L, Colognato R, Marmorato P, et al. Morphological transformation induced by multiwall carbon nanotubes on Balb/3T3 cell model as an in vitro end point of carcinogenic potential. Nanotoxicology 2013;7:221—33.

[82] Sun Z, Liu Z, Meng J, Duan J, Xie S, Lu X, et al. Carbon nanotubes enhance cytotoxicity mediated by human lymphocytes in vitro. PLoS One 2011;6:e21073.

[83] Bianco A, Hoebeke J, Godefroy S, Chaloin O, Pantarotto D, Briand J-P, et al. Cationic carbon nanotubes bind to CpG oligodeoxynucleotides and enhance their immunostimulatory properties. J Am Chem Soc 2005;127:58—9.

[84] Mu Q, Du G, Chen T, Zhang B, Yan B. Suppression of human bone morphogenetic protein signaling by carboxylated single-walled carbon nanotubes. ACS Nano 2009;3:1139—44.

[85] Carvalho S, Ferrini M, Herritt L, Holian A, Jaffar Z, Roberts K. Multi-walled carbon nanotubes augment allergic airway eosinophilic inflammation by promoting cysteinyl leukotriene production. Front Pharmacol 2018;9:585.

[86] Visalli G, Currò M, Iannazzo D, Pistone A, Ciarello MP, Acri G, et al. In vitro assessment of neurotoxicity and neuroinflammation of homemade MWCNTs. Environ Toxicol Pharmacol 2017;56:121—8.

[87] Tzeng S-F, Lee J-L, Kuo J-S, Yang C-S, Murugan P, Tai LA, et al. Effects of malonate C60 derivatives on activated microglia. Brain Res 2002;940:61—8.

[88] Pacurari M, Yin XJ, Zhao J, Ding M, Leonard SS, Schwegler-Berry D, et al. Raw single-wall carbon nanotubes induce oxidative stress and activate MAPKs, AP-1, NF-κB, and Akt in normal and malignant human mesothelial cells. Environ Health Perspect 2008;116:1211−17.

[89] Li Y, Liu Y, Fu Y, Wei T, Le Guyader L, Gao G, et al. The triggering of apoptosis in macrophages by pristine graphene through the MAPK and TGF-beta signaling pathways. Biomaterials 2012;33:402−11.

[90] Kagan V, Tyurina Y, Tyurin V, Konduru N, Potapovich A, Osipov A, et al. Direct and indirect effects of single walled carbon nanotubes on RAW 264.7 macrophages: role of iron. Toxicol Lett 2006;165:88−100.

[91] Tahara Y, Nakamura M, Yang M, Zhang M, Iijima S, Yudasaka M. Lysosomal membrane destabilization induced by high accumulation of single-walled carbon nanohorns in murine macrophage RAW 264.7. Biomaterials 2012;33:2762−9.

[92] Lin C, Fugetsu B, Su Y, Watari F. Studies on toxicity of multi-walled carbon nanotubes on Arabidopsis T87 suspension cells. J Hazard Mater 2009;170:578−83.

[93] Ding L, Stilwell J, Zhang T, Elboudwarej O, Jiang H, Selegue JP, et al. Molecular characterization of the cytotoxic mechanism of multiwall carbon nanotubes and nano-onions on human skin fibroblast. Nano Lett 2005;5:2448−64.

[94] Poulsen SS, Saber AT, Williams A, Andersen O, Købler C, Atluri R, et al. MWCNTs of different physicochemical properties cause similar inflammatory responses, but differences in transcriptional and histological markers of fibrosis in mouse lungs. Toxicol Appl Pharmacol 2015;284:16−32.

[95] Pacurari M, Qian Y, Porter D, Wolfarth M, Wan Y, Luo D, et al. Multi-walled carbon nanotube-induced gene expression in the mouse lung: association with lung pathology. Toxicol Appl Pharmacol 2011;255:18−31.

[96] Lin C, Su Y, Takahiro M, Fugetsu B. Multi-walled carbon nanotubes induce oxidative stress and vacuolar structure changes to Arabidopsis T87 suspension cells. Nano Biomed 2010;2:170−81.

[97] Sharma CS, Sarkar S, Periyakaruppan A, Barr J, Wise K, Thomas R, et al. Single-walled carbon nanotubes induces oxidative stress in rat lung epithelial cells. J Nanosci Nanotechnol 2007;7:2466−72.

[98] Kagan VE, Konduru NV, Feng W, Allen BL, Conroy J, Volkov Y, et al. Carbon nanotubes degraded by neutrophil myeloperoxidase induce less pulmonary inflammation. Nat Nanotechnol 2010;5:354−9.

[99] Andón FT, Kapralov AA, Yanamala N, Feng W, Baygan A, Chambers BJ, et al. Biodegradation of single-walled carbon nanotubes by eosinophil peroxidase. Small 2013;9:2721−9.

[100] Sayes CM, Gobin AM, Ausman KD, Mendez J, West JL, Colvin VL. Nano-C60 cytotoxicity is due to lipid peroxidation. Biomaterials 2005;26:7587−95.

[101] Wang Y, Okazaki Y, Shi L, Kohda H, Tanaka M, Taki K, et al. Role of hemoglobin and transferrin in multi-wall carbon nanotube-induced mesothelial injury and carcinogenesis. Cancer Sci 2016;107:250−7.

[102] Reddy ARN, Reddy YN, Krishna DR, Himabindu V. Multi wall carbon nanotubes induce oxidative stress and cytotoxicity in human embryonic kidney (HEK293) cells. Toxicology 2010;272:11−16.

[103] Awasthi KK, John P, Awasthi A, Awasthi K. Multi walled carbon nano tubes induced hepatotoxicity in Swiss albino mice. Micron 2013;44:359−64.

[104] Guo Y-Y, Zhang J, Zheng Y-F, Yang J, Zhu X-Q. Cytotoxic and genotoxic effects of multi-wall carbon nanotubes on human umbilical vein endothelial cells in vitro. Mutat Res/Genetic Toxicol Environ Mutagenesis 2011;721:184−91.

[105] Imai K, Suese K, Takashima H, Senuma M, Watari F. Effects of in vitro new capillary formation by C60 fullerene. Nano Biomed 2010;2:123−9.

[106] Xu J, Alexander DB, Futakuchi M, Numano T, Fukamachi K, Suzui M, et al. Size- and shape-dependent pleural translocation, deposition, fibrogenesis, and mesothelial proliferation by multi-walled carbon nanotubes. Cancer Sci 2014;105:763−9.

[107] Jones CF, Grainger DW. In vitro assessments of nanomaterial toxicity. Adv Drug Deliv Rev 2009;61:438−56.

[108] Madannejad R, Shoaie N, Jahanpeyma F, Darvishi MH, Azimzadeh M, Javadi H. Toxicity of carbon-based nanomaterials: reviewing recent reports in medical and biological systems. Chemico-Biol Interact 2019;307:206−22.

[109] Verma S, Mungse HP, Kumar N, Choudhary S, Jain SL, Sain B, et al. Graphene oxide: an efficient and reusable carbocatalyst for aza-Michael addition of amines to activated alkenes. Chem Commun 2011;47:12673−5.

[110] Wang H, Maiyalagan T, Wang X. Review on recent progress in nitrogen-doped graphene: synthesis, characterization, and its potential applications. ACS Catal 2012;2:781−94.

[111] Qu Y, Ma Q, Deng J, Shen W, Zhang X, He Z, et al. Responses of microbial communities to single-walled carbon nanotubes in phenol wastewater treatment systems. Environ Sci Technol 2015;49:4627−35.

[112] Ren W, Ren G, Teng Y, Li Z, Li L. Time-dependent effect of graphene on the structure, abundance, and function of the soil bacterial community. J Hazard Mater 2015;297:286−94.

[113] Ruiz ON, Fernando KS, Wang B, Brown NA, Luo PG, McNamara ND, et al. Graphene oxide: a nonspecific enhancer of cellular growth. ACS Nano 2011;5:8100−7.

[114] Chen M, Sun Y, Liang J, Zeng G, Li Z, Tang L, et al. Understanding the influence of carbon nanomaterials on microbial communities. Environ Int 2019;126:690−8.

[115] Hu W, Peng C, Lv M, Li X, Zhang Y, Chen N, et al. Protein corona-mediated mitigation of cytotoxicity of graphene oxide. ACS Nano 2011;5:3693−700.

[116] Wang K, Ruan J, Song H, Zhang J, Wo Y, Guo S, et al. Biocompatibility of Graphene Oxide. Nanoscale Res Lett 2010;6:8.

[117] Gollavelli G, Ling Y-C. Multi-functional graphene as an in vitro and in vivo imaging probe. Biomaterials 2012;33:2532−45.

[118] Agarwal S, Zhou X, Ye F, He Q, Chen GC, Soo J, et al. Interfacing live cells with nanocarbon substrates. Langmuir 2010;26:2244−7.

[119] Robinson JT, Tabakman SM, Liang Y, Wang H, Sanchez Casalongue H, Vinh D, et al. Ultrasmall reduced graphene oxide with high near-infrared absorbance for photothermal therapy. J Am Chem Soc 2011;133:6825−31.

[120] Liao K-H, Lin Y-S, Macosko CW, Haynes CL. Cytotoxicity of graphene oxide and graphene in human erythrocytes and skin fibroblasts. ACS Appl Mater Interfaces 2011;3:2607−15.

[121] Zhang Y, Ali SF, Dervishi E, Xu Y, Li Z, Casciano D, et al. Cytotoxicity effects of graphene and single-wall carbon nanotubes in neural phaeochromocytoma-derived PC12 cells. ACS Nano 2010;4:3181−6.

[122] Duch MC, Budinger GS, Liang YT, Soberanes S, Urich D, Chiarella SE, et al. Minimizing oxidation and stable nanoscale dispersion improves the biocompatibility of graphene in the lung. Nano Lett 2011;11:5201−7.

[123] Zanni E, De Bellis G, Bracciale MP, Broggi A, Santarelli ML, Sarto MS, et al. Graphite nanoplatelets and Caenorhabditis elegans: insights from an in vivo model. Nano Lett 2012;12:2740−4.

[124] Mendes RG, Koch B, Bachmatiuk A, Ma X, Sanchez S, Damm C, et al. A size dependent evaluation of the cytotoxicity and uptake of nanographene oxide. J Mater Chem B 2015;3:2522−9.

[125] Wang J, Onasch TB, Ge X, Collier S, Zhang Q, Sun Y, et al. Observation of Fullerene Soot in Eastern China. Environ Sci Technol Lett 2016;3:121−6.

[126] Snow SD, Kim KC, Moor KJ, Jang SS, Kim J-H. Functionalized fullerenes in water: a closer look. Environ Sci Technol 2015;49:2147−55.

[127] Tong Z-H, Bischoff M, Nies LF, Carroll NJ, Applegate B, Turco RF. Influence of fullerene (C60) on soil bacterial communities: aqueous aggregate size and solvent co-introduction effects. Sci Rep 2016;6:28069.

[128] Fiorito S, Serafino A, Andreola F, Bernier P. Effects of fullerenes and single-wall carbon nanotubes on murine and human macrophages. Carbon 2006;44:1100−5.

[129] Jia G, Wang H, Yan L, Wang X, Pei R, Yan T, et al. Cytotoxicity of carbon nanomaterials: single-wall nanotube, multi-wall nanotube, and fullerene. Environ Sci Technol 2005;39:1378−83.

[130] Chen HHC, Yu C, Ueng TH, Chen S, Chen BJ, Huang KJ, et al. Acute and subacute toxicity study of water-soluble polyalkylsulfonated C60 in rats. Toxicol Pathol 1998;26:143−51.

[131] Oberdörster E. Manufactured nanomaterials (fullerenes, C60) induce oxidative stress in the brain of juvenile largemouth bass. Environ Health Perspect 2004;112:1058−62.

[132] Zhu X, Zhu L, Li Y, Duan Z, Chen W, Alvarez PJ. Developmental toxicity in zebrafish (Danio rerio) embryos after exposure to manufactured nanomaterials: buckminsterfullerene aggregates (nC60) and fullerol. Environ Toxicol Chem Int J 2007;26:976−9.

[133] Usenko CY, Harper SL, Tanguay RL. Fullerene C60 exposure elicits an oxidative stress response in embryonic zebrafish. Toxicol Appl Pharmacol 2008;229:44−55.

[134] Das R, Ali M, Bee Abd Hamid S, Annuar M, Ramakrishna S. Common wet chemical agents for purifying multiwalled carbon nanotubes. J Nanomater 2014;2014.

[135] Wang C, Wei Z, Feng M, Wang L, Wang Z. The effects of hydroxylated multiwalled carbon nanotubes on the toxicity of nickel to Daphnia magna under different pH levels. Environ Toxicol Chem 2014;33:2522−8.

[136] Herzog E, Casey A, Lyng F, Chambers G, Byrne H, Davoren M. A new approach to the toxicity testing of carbon-based nanomaterials-the clonogenic assay. Toxicol Lett 2007;174:49−60.

[137] Dumortier H, Lacotte S, Pastorin G, Marega R, Wu W, Bonifazi D, et al. Functionalized carbon nanotubes are non-cytotoxic and preserve the functionality of primary immune cells. Nano Lett 2006;6:1522−8.

[138] Wick P, Manser P, Limbach LK, Dettlaff-Weglikowska U, Krumeich F, Roth S, et al. The degree and kind of agglomeration affect carbon nanotube cytotoxicity. Toxicol Lett 2007;168:121−31.

[139] Vittorio O, Raffa V, Cuschieri A. Influence of purity and surface oxidation on cytotoxicity of multiwalled carbon nanotubes with human neuroblastoma cells. Nanomedicine: Nanotechnol Biol Med 2009;5:424−31.

[140] Muller J, Huaux F, Moreau N, Misson P, Heilier J-F, Delos M, et al. Respiratory toxicity of multi-wall carbon nanotubes. Toxicol Appl Pharmacol 2005;207:221−31.

[141] Murr L, Garza K, Soto K, Carrasco A, Powell T, Ramirez D, et al. Cytotoxicity assessment of some carbon nanotubes and related carbon nanoparticle aggregates and the implications for anthropogenic carbon nanotube aggregates in the environment. Int J Environ Res Public Health 2005;2:31−42.

[142] Poland CA, Duffin R, Kinloch I, Maynard A, Wallace WAH, Seaton A, et al. Carbon nanotubes introduced into the abdominal cavity of mice show asbestos-like pathogenicity in a pilot study. Nat Nanotechnol 2008;3:423−8.

[143] Liu Y, Zhao Y, Sun B, Chen C. Understanding the toxicity of carbon nanotubes. Acc Chem Res 2013;46:702−13.

[144] Uo M, Akasaka T, Watari F, Sato Y, Tohji K. Toxicity evaluations of various carbon nanomaterials. Dental Mater J 2011; 1105140126-.

[145] Miyawaki J, Yudasaka M, Azami T, Kubo Y, Iijima S. Toxicity of single-walled carbon nanohorns. ACS Nano 2008;2:213−26.

[146] Matsuoka A, Önfelt A, Matsuda Y, Nakaoka R, Haishima Y, Yudasaka M, et al. Development of an in vitro screening method for safety evaluation of nanomaterials. Bio-Medical Mater Eng 2009;19:19−27.

[147] Yokoyama A, Sato Y, Nodasaka Y, Yamamoto S, Kawasaki T, Shindoh M, et al. Biological behavior of hat-stacked carbon nanofibers in the subcutaneous tissue in rats. Nano Lett 2005;5:157−61.

[148] Sato Y, Shibata K-i, Kataoka H, Ogino S-i, Bunshi F, Yokoyama A, et al. Strict preparation and evaluation of water-soluble hat-stacked carbon nanofibers for biomedical application and their high biocompatibility: influence of nanofiber-surface functional groups on cytotoxicity. Mol Biosyst 2005;1:142−5.

[149] Wierzbicki M, Sawosz E, Grodzik M, Hotowy A, Prasek Kutwin M, Jaworski S, et al. Carbon nanoparticles downregulate expression of basic fibroblast growth factor in the heart during embryogenesis. Int J Nanomed 2013;8:3427−35.

[150] Khosravi Y, Salimi A, Pourahmad J, Naserzadeh P, Seydi E. Inhalation exposure of nano diamond induced oxidative stress in lung, heart and brain. Xenobiotica 2018;48:860−6.

[151] Raja IS, Song S-J, Kang MS, Lee YB, Kim B, Hong SW, et al. Toxicity of zero- and one-dimensional carbon nanomaterials. Nanomaterials. 2019;9:1214.

[152] Zhang X, He X, Li Y, Ma Y, Li F, Liu J. A cytotoxicity study of fluorescent carbon nanodots using human bronchial epithelial cells. J Nanosci Nanotechnol 2013;13:5254−9.

CHAPTER SEVENTEEN

Outlook and future research, development, and innovation directions

17.1 General conclusion

The problem of water pollution has become a significant concern at each wplatform for everybody—researchers, industry, and the public, as well as to policy-makers at national and international levels. The treatment of industrial wastewater before discharge and reutilization is a top priority. The provision of a universal method for the removal of all types of pollutants from wastewater is a challenge. Adsorption is now well-established in the industry, as it achieves satisfactory results for the elimination of different kinds of pollutant from different effluents. The adsorption process is affected by many factors, including the nature of the adsorbate (i.e., ionic, hydrophilic, and hydrophobic), the type of adsorbent, and the variables of the process. It can be measured using different kinetic models, isotherms, and thermodynamic parameters, and principally occurs via physisorption, chemisorption, and ion exchange mechanisms.

Recently, several adsorbents have been investigated for the elimination of toxic inorganic and organic pollutants. Carbon nanomaterials (CNMs), including graphene (G), graphene oxide (GO), reduced graphene oxide (rGO), carbon nanotubes (CNTs), fullerenes, nanodiamonds (NDs), carbon-dots (CDs), carbon nanofibers (CNFs), graphitic carbon nitride (g-C_3N_4), and nanoporous carbons (NPCs), have received extensive attention for the adsorptive removal of pollutants due to their extraordinary physicochemical properties and eco-friendly behavior. This book has summarized the advancements in synthesis, functionalization, and adsorption performance of CNMs for dyes, polycyclic aromatic hydrocarbons (PAHs), heavy metal ions, radioactive substances, pharmaceuticals, antibiotics, and pesticides. Initially, it thoroughly discusses the classification of water contaminants, water purification using various technologies and their advantages and disadvantages, adsorption in the context of water purification, adsorption equilibrium isotherms, kinetics and thermodynamics, and the effect of reaction parameters on the adsorption. The later part of the book is dedicated to the synthesis, functionalization, and properties of CNMs, zero/one/two/three-dimensional (0D/1D/2D/3D)-based CNMs adsorbents, biopolymer and

Carbon Nanomaterial-Based Adsorbents for Water Purification.
DOI: https://doi.org/10.1016/B978-0-12-821959-1.00017-9

387

conducting polymer-functionalized carbon nanomaterials—based adsorbents, carbon-based nano/micromotors for adsorption, the regeneration and recyclability of carbon nanomaterials after adsorption, and the toxicity of carbon nanomaterials.

Adsorption process parameters, mechanisms, and surface modification strategies, with past and recent examples, have been discussed in detail. Pristine CNMs have high adsorption capacities for aromatic and hydrophobic pollutants. Functionalization and modification of CNMs can improve adsorption performance for a broad spectrum of pollutants, inorganic ions, and ionic organic molecules, at the expense of adsorption capacity for aromatic and hydrophobic molecules. Nanocomposite CNMs, including both inorganic and organic materials, further enhance the ability for pollutant removal. Polymers are not very efficient adsorbents; however, their composites/hydrogels with CNMs are useful for the removal of a variety of pollutants from wastewater. Magnetic nanocomposites of CNMs have an extra advantage in magnetic separation and reutilization of the adsorbents. High adsorption capacities rely on high adsorbent surface areas and accessible porosities. Oxidized forms of CNMs (such as GO, rGO, and o-CNTs) possess low specific surface areas, which can be increased by chemical and physical activation.

Activation approaches lead to the formation of exceedingly porous carbons from diverse carbon sources. Chemical activation treatments involve the reaction of CNMs with KOH, NaOH, K_2CO_3, $KHCO_3$, $Mg(OH)_2$, O_3, and H_2SO_4/HNO_3. Overall, the activation of carbon nanostructures (commonly using KOH) is effective, as it results in the generation of novel micro-/mesopores by burning carbon atoms at corners, by coalescence of nearby pores, and by opening closed pores present in the materials. The choice of activation process or agent can control the adsorption mechanism of functionalized CNMs; for example, the use of KOH might produce materials that adsorb via physisorption, while the use of $H_2SO_4/HNO_3/O_3$ might promote chemisorption by the resulting materials. Despite many recent reports, research into activation strategies is still at an early stage, and more investigation is required. The number and type of oxygen functionalities on adsorbent surfaces is governed by the oxidation or reduction process used in the synthetic procedure. Adsorption performance is considerably affected by the quantity of oxygen on the CNM surface. Reports concerning the adsorption performances of oxidized/reduced CNMs for a number of pollutants are contradictory. Thus the connection between oxygen species of CNMs and their adsorption behavior should be reinvestigated to achieve an in-depth understanding.

The use of pristine CNMs, similar to many adsorbents, has several disadvantages, including the difficulty in the segregation of nanoadsorbents from water, low adsorption capacities, and fouling effects. These problems can be solved by the development of multifunctional macroscopic 3D architectures, using CNMs as building blocks. For instance, graphene-based 3D network superstructures have demonstrated improved adsorption performances for a wide range of contaminants, due to extra functionalities contributed by additives, and the porous network itself. These 3D/2D networks integrate the macroscale and nanoscale structures of the building blocks, maintaining their intrinsic properties while introducing unique features such as easy recycling, additional binding sites, and more straightforward operation, to CNMs. Many CNM superstructures/macrostructures, including rGO-CNT, GO-CNT, G-CNT/Fe_3O_4, rGO-CNT-p-phenylenediamine, CNF/GO, and 3D CNMs (3D sponges or foams, hydrogels, aerogels) have been applied to the removal of water contaminations (i.e., oils, dyes, solvents, metals, and organics). The problems associated with these superstructures are fragility and poor stability. The mechanical properties of superstructures could be improved by surface modification with nanofillers or fibrous materials, developing better cross-linking conditions, and assimilation of the polymer matrix. Moreover, research efforts are required to improve our understanding of self-assembly and growth mechanisms to produce stable, size-controllable, processable, and cheap CNM superstructures. Additionally, it would be interesting to investigate the adsorption performance of CNM superstructures in multicomponent pollutant systems, continuous adsorption via packed-bed columns, and pilot-plants.

The extraordinary characteristic (self-propulsion/bubble propulsion) of the synthetic nano/micromotors has opened new horizons in environmental remediation and these have shown considerable potential in wastewater treatment. The high surface area of CNMs,such as graphene and CNTs, promote the adsorption characteristics in the carbon-based nano/micromotors. Importantly, these micromotors enhance the diffusion rate, which improves the adsorption capacity. A few micromotors also generate hydroxyl radicals, which on adsorption of water pollutants degrade them completely, and thus clean the water without generating secondary pollutants. Nevertheless, there are many challenges to commercializing these motors for practical industrial water treatments. The foremost challenges are to reduce the cost of the micromotors and to scale up the synthesis process. The uptake of a broad array of pollutants with high efficiency in large volumes of industrial wastewaters is another challenge. Furthermore, the consideration and evaluation of the toxicity of micromotors also necessitates additional efforts.

CNMs are highly biocompatible with biological systems and environment [1]. Several findings recommend that the CNMs are safe for water purification as they are just utilized for the elimination and degradation of contaminants, and are not used by living beings directly.

17.2 Future research, development, and innovation directions

A reduction in the costs of production and a thorough review of the environmental impacts of CNMs are critical to the development of CNM-based water purification technologies. The large-scale production of CNMs, especially of G, GO, CNTs, NDs, fullerene, and CNFs is very costly. Additional research efforts and developments in CNM synthesis are imperative for the commercialization of nanomaterials. However, fabrication costs will decrease with improved synthetic methods, for example, utilizing eco-friendly, low-cost precursors, scaling up production, and optimizing the supply chain. Many industries have already begun large-scale CNT production using CVD and arc-discharge methods, but the cost is still high. Large-scale production methods for high-quality G and GO have not yet been developed. Further investigation into the low-cost production of NDs, CNMs, and fullerene for environmental applications is undoubtedly warranted.

In line with the development of CNM-based nanoadsorbents and similar methods, research is required to understand the fate and transport behavior of CNMs in the environment. Under severe conditions or with long exposure, CNMs can undergo physical and chemical alterations (i.e., dissolution, oxidation, aggregation, sulfidation, and deposition) in natural ecosystems. The commonly occurring dissolution process could produce soluble and highly bioavailable toxic species that may be a threat to human beings and aquatic life. The risk is higher if the materials are nondegradable and persist in the environment. The sharp edges, lateral sizes, defects, tubes, and fibrous morphologies of CNMs pose possible risks to living beings, and to the ecosystem. Before practical applications can be considered, a thorough assessment of risk to health and the environment should be performed, with proper management and controls. Extensive research is required to determine the toxicity of each CNM and understand its effects and mechanisms, for safer use in environmental applications.

There are limited studies related to the use of nano/micromotors in water purification. The research on nano/micromotors is still in its early stages and more dedicated research is needed to unfurl its complete potential of providing a fast and smart solution for water purification. Study into the regeneration and reusability of micromotors should also be concerned with future practical applications. Carbon-based nano/ micromotors are new materials and require more attention from the researchers for thorough investigations, that is, toxicity, scale-up of the process, reusability, and economic feasibility, which can be aimed toward their practical application.

Although adsorption is a promising method of water purification, it generates secondary environmental toxic waste [2,3]. It is critical to mitigate this problem sustainably. The solution lies in the reusability of spent materials in various applications, including catalysis, sensors, and antimicrobial and energy applications [4−6]. Furthermore, a circular approach may be envisaged in water purification, based on the circular economy concept, which rests on no-waste or minimal waste production. In the circular path, the residue or waste material from any process is treated as a resource for a new process, unless the material has lost all of its functionality or activity [4]. Dedicated future research is needed to add value to secondary toxic waste via reusability and the proposed circular approach.

Nanotechnology is contributing extensively to the provision of affordable solutions for wastewater purification with ultrafast, low-cost detection and treatment approaches. The nanotechnologies principally involve adsorption and membrane separation techniques. Recently, a few nanoadsorbents have been commercialized for the removal of arsenic from water. In pilot-scale tests, their costs have been comparable with those of available adsorbents. ArsenXnp is a commercial nanoproduct (hybrid ion exchange resin), which consists of hydrous iron oxide nanoparticles and polymers, that is used for the elimination of arsenic (arsenate and arsenite) from water (https://www.systematixusa. com/products/media/active_media/arsenx.htm). It is perfect for municipal wastewater treatment plants, as well as point-of-use and point-of-entry devices.

ADSORBSIATM is a nanoproduct in the form of beads (0.25−1.2 mm diameter) composed of nanocrystalline TiO_2 [7]. The first nanotechnology-based water purifier for the removal of arsenic, iron, and turbidity was successfully developed by the Indian Institute of Technology, Madras, and is called AMRIT—Arsenic and Metal Removal through Indian Technology (http://www.dstuns.iitm.ac.in/filesdec2015/1.%20Amrit% 20Brochure.pdf). The purifier operates using gravity, without electricity. This technology supplies water with an arsenic level of less than 2 ppb at a cost of US$0.0007 per liter. The purification process works in two stages: first, microbial pollutants are removed using an Ag−polymer nanocomposite [8]. Next, metals ions (e.g., As and Fe), and colored impurities are removed by supported iron oxyhydroxide nanoparticles (size <3 nm). The system is effectively employed for the purification of arsenic-rich wastewater in various parts of India. A Taiwan-based company, Green-Tak, has introduced Nano X-Plus purification technology, which provides clean, mineral-rich water from municipal water using a single-stage process. The purifier is nonelectric and environmentally friendly. It readily removes bacteria (99%), heavy metals (99%), chlorine, pesticides, and bad odors, while retaining the minerals in the water. The purifier operates using advanced technology, in which an antibacterial cloth is attached to a high-quality carbon block filter for the purification of water. The antibacterial fabric consists of Ag NPs, Cu NPs, and Ti alloy films. The Tata Swach is an efficient, nonelectric water purifier launched by Tata Chemicals, India. This purifier uses processed rice-husk ash impregnated with Ag NPs to kill bacteria and other microorganisms in polluted water. Superwetting nanofiber membranes composed of $K_{2-x}Mn_8O_{16}$ were developed to remove toxic substances from water and clean up oil spills [9].

CNMs such as CNTs and graphene/GO have proven to be favorable materials for desalination and membrane technologies. Many new functional CNMs have concurrently demonstrated excellent adsorption performance for wastewater treatments due to high surface area, porosity, stability in extreme conditions, mechanical strength, and excessive functionalities. Overall, CNMs have already established their potential in water purification at the laboratory scale. While the research is still in progress, it is expected that CNMs will lead to more exciting discoveries, and possibly low-cost industrial adsorbents for water remediation. They might be converted into new products and nanotechnologies in the near future.

References

[1] Baby R, Saifullah B, Hussein MZ. Carbon nanomaterials for the treatment of heavy metal-contaminated water and environmental remediation. Nanoscale Res Lett 2019;14:341.

[2] Gusain R, Kumar N, Ray SS. Recent advances in carbon nanomaterial-based adsorbents for water purification. Coord Chem Rev 2020;405:213111.

[3] Kumar N, Mittal H, Parashar V, Ray SS, Ngila JC. Efficient removal of rhodamine 6G dye from aqueous solution using nickel sulphide incorporated polyacrylamide grafted gum karaya bionanocomposite hydrogel. RSC Adv 2016;6:21929−39.

[4] Kumar N, Mittal H, Alhassan SM, Ray SS. Bionanocomposite hydrogel for the adsorption of dye and reusability of generated waste for the photodegradation of ciprofloxacin: a demonstration of the circularity concept for water purification. ACS Sustain Chem Eng 2018;6:17011−25.

[5] Das R, Giri S, King Abia AL, Dhonge B, Maity A. Removal of noble metal ions (Ag +) by mercapto group-containing polypyrrole matrix and reusability of its waste material in environmental applications. ACS Sustain Chem Eng 2017;5:2711−24.

[6] Kumar N, Fosso-Kankeu E, Ray SS. Achieving controllable MoS2 nanostructures with increased interlayer spacing for efficient removal of Pb(II) from aquatic systems. ACS Appl Mater Interfaces 2019;11:19141−55.

[7] Aragon MJ, Everett RL, Siegel MD, Aragon AR, Kottenstette RJ, Holub Jr WE, et al. Arsenic pilot plant operation and results: Anthony, New Mexico. Sandia Natl Laboratories; 2007.

[8] Swathy JR, Sankar MU, Chaudhary A, Aigal S, Anshup, Pradeep T. Antimicrobial silver: an unprecedented anion effect. Sci Rep 2014;4:7161.

[9] Yuan J, Liu X, Akbulut O, Hu J, Suib SL, Kong J, et al. Superwetting nanowire membranes for selective absorption. Nat Nanotechnol 2008;3:332−6.

Index

Note: Page numbers followed by "*f*" and "*t*" refer to figures and tables, respectively.

Printed and bound by CPI Group (UK) Ltd, Croydon, CR0 4YY

17/05/2025

01874710-0001